Roger Arthur Sheldon,
Isabel Arends,
and Ulf Hanefeld

**Green Chemistry
and Catalysis**

1807–2007 Knowledge for Generations

Each generation has its unique needs and aspirations. When Charles Wiley first opened his small printing shop in lower Manhattan in 1807, it was a generation of boundless potential searching for an identity. And we were there, helping to define a new American literary tradition. Over half a century later, in the midst of the Second Industrial Revolution, it was a generation focused on building the future. Once again, we were there, supplying the critical scientific, technical, and engineering knowledge that helped frame the world. Throughout the 20th Century, and into the new millennium, nations began to reach out beyond their own borders and a new international community was born. Wiley was there, expanding its operations around the world to enable a global exchange of ideas, opinions, and know-how.

For 200 years, Wiley has been an integral part of each generation's journey, enabling the flow of information and understanding necessary to meet their needs and fulfill their aspirations. Today, bold new technologies are changing the way we live and learn. Wiley will be there, providing you the must-have knowledge you need to imagine new worlds, new possibilities, and new opportunities.

Generations come and go, but you can always count on Wiley to provide you the knowledge you need, when and where you need it!

William J. Pesce
President and Chief Executive Officer

Peter Booth Wiley
Chairman of the Board

Roger Arthur Sheldon, Isabel Arends, and Ulf Hanefeld

Green Chemistry and Catalysis

BICENTENNIAL
BICENTENNIAL
1807
⊛WILEY
2007
BICENTENNIAL
BICENTENNIAL

WILEY-VCH Verlag GmbH & Co. KGaA

The Authors
Prof. Dr. Roger Sheldon
Dr. Isabel W. C. E. Arends
Dr. Ulf Hanefeld
Biocatalysis and Organic Chemistry
Delft University of Technology
Julianalaan 136
2628 BL Delft
The Netherlands

Library of Congress Card No.: applied for

British Library Cataloguing-in-Publication Data
A catalogue record for this book is available from the British Library.

Bibliographic information published by the Deutsche Nationalbibliothek
The Deutsche Nationalbibliothek lists this publication in the Deutsche Nationalbibliografie; detailed bibliographic data are available in the Internet at http://dnb.d-nb.de.

Composition K+V Fotosatz GmbH, Beerfelden
Printing betz-druck GmbH, Darmstadt
Bookbinding Litges & Dopf GmbH, Heppenheim
Cover Design Schulz Grafik-Design, Fußgönheim
Wiley Bicentennial Logo Richard J. Pacifico

Printed in the Federal Republic of Germany
Printed on acid-free paper

ISBN 978-3-527-30715-9

Contents

Green Chemistry and Catalysis. I. Arends, R. Sheldon, U. Hanefeld
Copyright © 2007 WILEY-VCH Verlag GmbH & Co. KGaA, Weinheim
ISBN: 978-3-527-30715-9

Preface

The publication of books such as *Silent Spring* by Rachel Carson and *The Closing Circle* by Barry Commoner, in the late 1960s and early 1970s, focused attention on the negative side effects of industrial and economic growth on our natural environment. However, the resulting environmental movement was not fully endorsed by the chemical industry. It was the publication, two decades later in 1987, of the report *Our Common Future* by the World Commission on Environment and Development which marked a turning point in industrial and societal impact. The report emphasized that industrial development must be sustainable over time, that is it must meet the needs of the present generation without compromising the ability of future generations to meet their own needs. Sustainable development has subsequently become a catch phrase of industry in the 21st century and the annual reports and mission statements of many corporations endorse the concept of sustainability.

At roughly the same time, in the early 1990s, the concept of green chemistry emerged. A workable definition is *Green chemistry efficiently utilizes (preferably renewable) raw materials, eliminates waste and avoids the use of toxic and/or hazardous reagents and solvents in the manufacture and application of chemical products*. On reflection it becomes abundantly clear that if sustainability is the goal of society in the 21st century then green chemistry is the means to achieve it, in the chemical industry at least. A key feature is that greenness and sustainability need to be included in designing the product and/or process from the outset (benign by design). The basic premise of this book is that a major driver towards green chemistry is the widespread application of clean, catalytic methodologies in the manufacture of chemicals, hence, its title: Green Chemistry and Catalysis.

The book is primarily aimed at researchers, in industry or academia, who are involved in developing processes for chemicals manufacture. It could also form the basis for an advanced undergraduate or post-graduate level course on green chemistry. Indeed, the material is used by the authors for this very purpose in Delft.

The main underlying theme is the application of catalytic methodologies – homogeneous, heterogeneous and enzymatic – to industrial organic synthesis. The material is divided based on the type of transformation rather than the type of catalyst. This provides for a comparison of the different methodologies, *e.g.* chemo- *vs* biocatalytic for achieving a particular conversion.

Green Chemistry and Catalysis. I. Arends, R. Sheldon, U. Hanefeld
Copyright © 2007 WILEY-VCH Verlag GmbH & Co. KGaA, Weinheim
ISBN: 978-3-527-30715-9

Chapter 1 gives an introduction to the overall theme of green chemistry and catalysis, emphasizing the importance of concepts such as atom efficiency and E factors (kg waste/kg product) for assessing the environmental impact or 'greenness' of a process. In this context, it is emphasized that it is not really a question of green *vs* not green but rather one of one process being greener than another, that is there are 'many shades of green'. Chapter 2 is concerned with replacing traditional acids and bases with solid recyclable ones, thereby dramatically reducing the amount of waste formed. Chapters 3 and 4 cover the application of catalytic methodologies to reductions and oxidations, two pivotal processes in organic synthesis.

Chapter 5 is concerned with C-C bond forming reactions and Chapter 6 with solvolytic processes, which mainly consist of enzyme-catalyzed transformations.

Chapter 7 addresses another key topic in the context of green chemistry: the replacement of traditional, environmentally unattractive organic solvents by greener alternative reaction media such as water, supercritical carbon dioxide, ionic liquids and perfluorous solvents. The use of liquid/liquid biphasic systems provides the additional benefit of facile catalyst recovery and recycling.

Another key topic – renewable resources – is addressed in Chapter 8. The utilization of fossil resources as sources of energy and chemical feedstocks is clearly unsustainable and needs to be replaced by the use of renewable biomass. Such a switch is also desirable for other reasons such as biocompatibility, biodegradability and lower toxicity of the products compared to their oil-derived counterparts. Here again, the application of chemo- and biocatalysis will play an important underlying role.

Chapter 9 is concerned with the integration of catalytic methodologies in organic synthesis and with downstream processing. The ultimate green process involves the integration of several catalytic steps into catalytic cascade processes, thus emulating the multi-enzyme processes which are occurring in the living cell. Chapter 10 consists of the Conclusions and Future Prospects.

The principal literature has been covered up to 2006 and an extensive list of references is included at the end of each chapter. The index has been thoroughly cross-referenced, and the reader is encouraged to seek the various topics under more than one entry.

Finally, we wish to acknowledge the important contribution of our colleague, Fred van Rantwijk, to writing Chapter 8.

Roger Sheldon
Isabel Arends
Ulf Hanefeld

Delft, November 2006

Foreword

Green Chemistry was born around 1990. However it was in 1992, with the invention of the E-factor by Roger Sheldon, that Green Chemistry really cut its teeth. For the first time Chemists had a simple metric for judging the environmental impact of their processes.

As the subject has developed, it has become increasingly clear that catalysis is one of the central tools for the greening of chemistry. Catalysis not only can improve the selectivity of a reaction but also, by doing so, it can decrease or even remove the need for downstream processing, thereby reducing the material and energy waste associated with purification. This is particularly important because the solvent requirements for recrystallization, chromatography and other purification processes are often much greater than the usage of solvent in the original reaction!

Thus, it gives me great pleasure to write this Foreword, because this book is one of the first to bring together catalysis and Green Chemistry in a single volume. Furthermore, it is a real book not a mere collection of separate chapters thrown together by a whole football team of authors. So as you read the book, you can see the links, the synergies and even the occasional contradictions between the different topics.

Roger Sheldon and his coauthors, Isabel Arends and Ulf Hanefeld, have a very broad knowledge of current industrial processes and this knowledge comes through clearly over and over again in this book. This means that readers are getting two books for the price of one, firstly a detailed summary of the various types of catalysis which can be applied to Green Chemistry and, secondly, a concise snapshot of new catalytic processes which have been commercialised in recent years. Indeed it is quite possible that this snapshot will become a benchmark by which the rate of future industrial innovation can be judged.

Green Chemistry is now in its teens and, like any teenager, it shows promise but has not yet reached its full potential. No longer are 10 kilograms of solvent required to make each 100 mg tablet of a leading antidepressant; now we need only 4 kilograms! So there is still a long way to go till we reach truly sustainable chemical production. Thus, I believe that this book has a potentially much greater value than just describing the *status quo*. By showing what has been achieved, it hints at what could be achieved if only chemical intuition and imagination were properly focussed in an attack on the dirtier aspects of current

Green Chemistry and Catalysis. I. Arends, R. Sheldon, U. Hanefeld
Copyright © 2007 WILEY-VCH Verlag GmbH & Co. KGaA, Weinheim
ISBN: 978-3-527-30715-9

chemical manufacture. I sincerely hope that this book will inspire a whole new generation of Green Chemists and enable them to realise their own vision of a sustainable chemical future.

Martyn Poliakoff
School of Chemistry
The University of Nottingham, UK
November 2006

1
Introduction: Green Chemistry and Catalysis

1.1
Introduction

It is widely acknowledged that there is a growing need for more environmentally acceptable processes in the chemical industry. This trend towards what has become known as 'Green Chemistry' [1–9] or 'Sustainable Technology' necessitates a paradigm shift from traditional concepts of process efficiency, that focus largely on chemical yield, to one that assigns economic value to eliminating waste at source and avoiding the use of toxic and/or hazardous substances.

The term 'Green Chemistry' was coined by Anastas [3] of the US Environmental Protection Agency (EPA). In 1993 the EPA officially adopted the name 'US Green Chemistry Program' which has served as a focal point for activities within the United States, such as the Presidential Green Chemistry Challenge Awards and the annual Green Chemistry and Engineering Conference. This does not mean that research on green chemistry did not exist before the early 1990s, merely that it did not have the name. Since the early 1990s both Italy and the United Kingdom have launched major initiatives in green chemistry and, more recently, the Green and Sustainable Chemistry Network was initiated in Japan. The inaugural edition of the journal Green Chemistry, sponsored by the Royal Society of Chemistry, appeared in 1999. Hence, we may conclude that Green Chemistry is here to stay.

A reasonable working definition of green chemistry can be formulated as follows [10]: *Green chemistry efficiently utilizes (preferably renewable) raw materials, eliminates waste and avoids the use of toxic and/or hazardous reagents and solvents in the manufacture and application of chemical products.*

As Anastas has pointed out, the guiding principle is the *design* of environmentally benign products and processes (benign by design) [4]. This concept is embodied in the 12 Principles of Green Chemistry [1, 4] which can be paraphrased as:

1. Waste prevention instead of remediation
2. Atom efficiency
3. Less hazardous/toxic chemicals
4. Safer products by design
5. Innocuous solvents and auxiliaries

Green Chemistry and Catalysis. I. Arends, R. Sheldon, U. Hanefeld
Copyright © 2007 WILEY-VCH Verlag GmbH & Co. KGaA, Weinheim
ISBN: 978-3-527-30715-9

6. Energy efficient by design
7. Preferably renewable raw materials
8. Shorter syntheses (avoid derivatization)
9. Catalytic rather than stoichiometric reagents
10. Design products for degradation
11. Analytical methodologies for pollution prevention
12. Inherently safer processes

Green chemistry addresses the environmental impact of both chemical products and the processes by which they are produced. In this book we shall be concerned only with the latter, i.e. the product is a given and the goal is to design a green process for its production. Green chemistry eliminates waste at source, i.e. it is primary pollution prevention rather than waste remediation (end-of-pipe solutions). Prevention is better than cure (the first principle of green chemistry, outlined above).

An alternative term, that is currently favored by the chemical industry, is Sustainable Technologies. Sustainable development has been defined as [11]: *Meeting the needs of the present generation without compromising the ability of future generations to meet their own needs.*

One could say that Sustainability is the goal and Green Chemistry is the means to achieve it.

1.2.
E Factors and Atom Efficiency

Two useful measures of the potential environmental acceptability of chemical processes are the E factor [12–18], defined as the mass ratio of waste to desired product and the atom efficiency, calculated by dividing the molecular weight of the desired product by the sum of the molecular weights of all substances produced in the stoichiometric equation. The sheer magnitude of the waste problem in chemicals manufacture is readily apparent from a consideration of typical E factors in various segments of the chemical industry (Table 1.1).

The E factor is the actual amount of waste produced in the process, defined as everything but the desired product. It takes the chemical yield into account and includes reagents, solvents losses, all process aids and, in principle, even fuel (although this is often difficult to quantify). There is one exception: water is generally not included in the E factor. For example, when considering an aqueous waste stream only the inorganic salts and organic compounds contained in the water are counted; the water is excluded. Otherwise, this would lead to exceptionally high E factors which are not useful for comparing processes [8].

A higher E factor means more waste and, consequently, greater negative environmental impact. The ideal E factor is zero. Put quite simply, it is kilograms (of raw materials) in, minus kilograms of desired product, divided by kilograms

Table 1.1 The E factor.

Industry segment	Product tonnage[a]	kg waste[b]/kg product
Oil refining	10^6–10^8	<0.1
Bulk chemicals	10^4–10^6	<1–5
Fine chemicals	10^2–10^4	5->50
Pharmaceuticals	10–10^3	25->100

a) Typically represents annual production volume of a product at one site (lower end of range) or world-wide (upper end of range).
b) Defined as everything produced except the desired product (including all inorganic salts, solvent losses, etc.).

of product out. It can be easily calculated from a knowledge of the number of tons of raw materials purchased and the number of tons of product sold, for a particular product or a production site or even a whole company. It is perhaps surprising, therefore, that many companies are not aware of the E factors of their processes. We hasten to point out, however, that this situation is rapidly changing and the E factor, or an equivalent thereof (see later), is being widely adopted in the fine chemicals and pharmaceutical industries (where the need is greater). We also note that this method of calculation will automatically exclude water used in the process but not water formed.

Other metrics have also been proposed for measuring the environmental acceptability of processes. Hudlicky and coworkers [19], for example, proposed the effective mass yield (EMY), which is defined as the percentage of product of all the materials used in its preparation. As proposed, it does not include so-called environmentally benign compounds, such as NaCl, acetic acid, etc. As we shall see later, this is questionable as the environmental impact of such substances is very volume-dependent. Constable and coworkers of GlaxoSmithKline [20] proposed the use of mass intensity (MI), defined as the total mass used in a process divided by the mass of product, i.e. MI = E factor + 1 and the ideal MI is 1 compared with zero for the E factor. These authors also suggest the use of so-called mass productivity which is the reciprocal of the MI and, hence, is effectively the same as EMY.

In our opinion none of these alternative metrics appears to offer any particular advantage over the E factor for giving a mental picture of how wasteful a process is. Hence, we will use the E factor in further discussions.

As is clear from Table 1.1, enormous amounts of waste, comprising primarily inorganic salts, such as sodium chloride, sodium sulfate and ammonium sulfate, are formed in the reaction or in subsequent neutralization steps. The E factor increases dramatically on going downstream from bulk to fine chemicals and pharmaceuticals, partly because production of the latter involves multi-step syntheses but also owing to the use of stoichiometric reagents rather than catalysts (see later).

The atom utilization [13–18], atom efficiency or atom economy concept, first introduced by Trost [21, 22], is an extremely useful tool for rapid evaluation of the amounts of waste that will be generated by alternative processes. It is calculated by dividing the molecular weight of the product by the sum total of the molecular weights of all substances formed in the stoichiometric equation for the reaction involved. For example, the atom efficiencies of stoichiometric (CrO_3) vs. catalytic (O_2) oxidation of a secondary alcohol to the corresponding ketone are compared in Fig. 1.1.

In contrast to the E factor, it is a theoretical number, i.e. it assumes a yield of 100% and exactly stoichiometric amounts and disregards substances which do not appear in the stoichiometric equation. A theoretical E factor can be derived from the atom efficiency, e.g. an atom efficiency of 40% corresponds to an E factor of 1.5 (60/40). In practice, however, the E factor will generally be much higher since the yield is not 100% and an excess of reagent(s) is used and solvent losses and salt generation during work-up have to be taken into account.

An interesting example, to further illustrate the concepts of E factors and atom efficiency is the manufacture of phloroglucinol [23]. Traditionally, it was produced from 2,4,6-trinitrotoluene (TNT) as shown in Fig. 1.2, a perfect example of nineteenth century organic chemistry.

This process has an atom efficiency of <5% and an E factor of 40, i.e. it generates 40 kg of solid waste, containing $Cr_2(SO_4)_3$, NH_4Cl, $FeCl_2$ and $KHSO_4$ per kg of phloroglucinol (note that water is not included), and obviously belongs in a museum of industrial archeology.

All of the metrics discussed above take only the mass of waste generated into account. However, what is important is the environmental impact of this waste, not just its amount, i.e. the nature of the waste must be considered. One kg of sodium chloride is obviously not equivalent to one kg of a chromium salt. Hence, the term 'environmental quotient', EQ, obtained by multiplying the E factor with an arbitrarily assigned unfriendliness quotient, Q, was introduced [15]. For example, one could arbitrarily assign a Q value of 1 to NaCl and, say, 100–1000 to a heavy metal salt, such as chromium, depending on its toxicity, ease of recycling, etc. The magnitude of Q is obviously debatable and difficult to quantify but, importantly, 'quantitative assessment' of the environmental im-

$$3\ PhCH(OH)CH_3 + 2\ CrO_3 + 3\ H_2SO_4 \longrightarrow 3\ PhCOCH_3 + Cr_2(SO_4)_3 + 6\ H_2O$$

atom efficiency = 360 / 860 = 42 %

$$Ph\ CH(OH)CH_3 + 1/2\ O_2 \xrightarrow{\text{catalyst}} Ph\ COCH_3 + H_2O$$

atom efficiency = 120 / 138 = 87 %

Fig. 1.1 Atom efficiency of stoichiometric vs. catalytic oxidation of an alcohol.

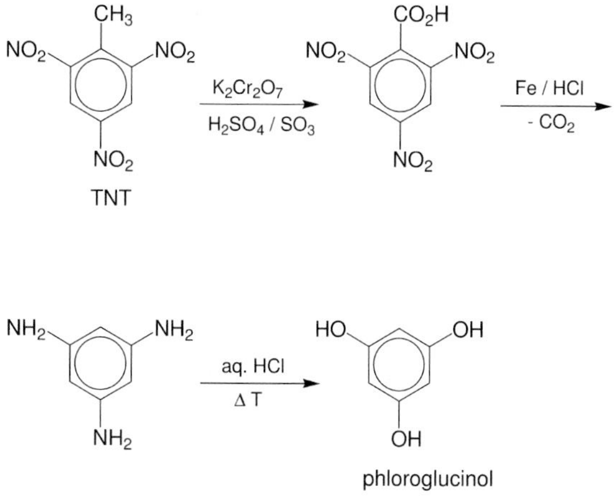

Fig. 1.2 Phloroglucinol from TNT.

pact of chemical processes is, in principle, possible. It is also worth noting that Q for a particular substance can be both volume-dependent and influenced by the location of the production facilities. For example, the generation of 100–1000 tons per annum of sodium chloride is unlikely to present a waste problem, and could be given a Q of zero. The generation of 10000 tons per annum, on the other hand, may already present a disposal problem and would warrant assignation of a Q value greater than zero. Ironically, when very large quantities of sodium chloride are generated the Q value could decrease again as recycling by electrolysis becomes a viable proposition, e.g. in propylene oxide manufacture via the chlorohydrin route. Thus, generally speaking the Q value of a particular waste will be determined by its ease of disposal or recycling. Hydrogen bromide, for example, could warrant a lower Q value than hydrogen chloride as recycling, via oxidation to bromine, is easier. In some cases, the waste product may even have economic value. For example, ammonium sulfate, produced as waste in the manufacture of caprolactam, can be sold as fertilizer. It is worth noting, however, that the market could change in the future, thus creating a waste problem for the manufacturer.

1.3
The Role of Catalysis

As noted above, the waste generated in the manufacture of organic compounds consists primarily of inorganic salts. This is a direct consequence of the use of stoichiometric inorganic reagents in organic synthesis. In particular, fine chemicals and pharmaceuticals manufacture is rampant with antiquated 'stoichiometric' technologies. Examples, which readily come to mind are stoichiometric reductions with metals (Na, Mg, Zn, Fe) and metal hydride reagents (LiAlH$_4$,

NaBH$_4$), oxidations with permanganate, manganese dioxide and chromium(VI) reagents and a wide variety of reactions, e.g. sulfonations, nitrations, halogenations, diazotizations and Friedel-Crafts acylations, employing stoichiometric amounts of mineral acids (H$_2$SO$_4$, HF, H$_3$PO$_4$) and Lewis acids (AlCl$_3$, ZnCl$_2$, BF$_3$). The solution is evident: substitution of classical stoichiometric methodologies with cleaner catalytic alternatives. Indeed, a major challenge in (fine) chemicals manufacture is to develop processes based on H$_2$, O$_2$, H$_2$O$_2$, CO, CO$_2$ and NH$_3$ as the direct source of H, O, C and N. Catalytic hydrogenation, oxidation and carbonylation (Fig. 1.3) are good examples of highly atom efficient, low-salt processes.

The generation of copious amounts of inorganic salts can similarly be largely circumvented by replacing stoichiometric mineral acids, such as H$_2$SO$_4$, and Lewis acids and stoichiometric bases, such as NaOH, KOH, with recyclable solid acids and bases, preferably in catalytic amounts (see later).

For example, the technologies used for the production of many substituted aromatic compounds (Fig. 1.4) have not changed in more than a century and are, therefore, ripe for substitution by catalytic, low-salt alternatives (Fig. 1.5).

An instructive example is provided by the manufacture of hydroquinone (Fig. 1.6) [24]. Traditionally it was produced by oxidation of aniline with stoichiometric amounts of manganese dioxide to give benzoquinone, followed by reduction with iron and hydrochloric acid (Béchamp reduction). The aniline was derived from benzene via nitration and Béchamp reduction. The overall process generated more than 10 kg of inorganic salts (MnSO$_4$, FeCl$_2$, NaCl, Na$_2$SO$_4$) per kg of hydroquinone. This antiquated process has now been replaced by a more modern route involving autoxidation of *p*-diisopropylbenzene (produced by Friedel-Crafts alkylation of benzene), followed by acid-catalysed rearrangement of the bis-hydroperoxide, producing <1 kg of inorganic salts per kg of hydroquinone. Alternatively, hydroquinone is produced (together with catechol) by tita-

$$PhCOCH_3 + H_2 \xrightarrow{\text{catalyst}} PhCH(OH)CH_3$$

heterogeneous   100%

$$PhCH(OH)CH_3 + 1/2\,O_2 \xrightarrow{\text{catalyst}} PhCOCH_3 + H_2O$$
$$(H_2O_2) \qquad\qquad (2\,H_2O)$$

homo-/heterogeneous

120*100/138 = 87%
(120*100/156 = 77%)

$$PhCH(OH)CH_3 + CO \xrightarrow{\text{catalyst}} PhCH(CH_3)CO_2H$$

homogeneous   100%

Fig. 1.3 Atom efficient catalytic processes.

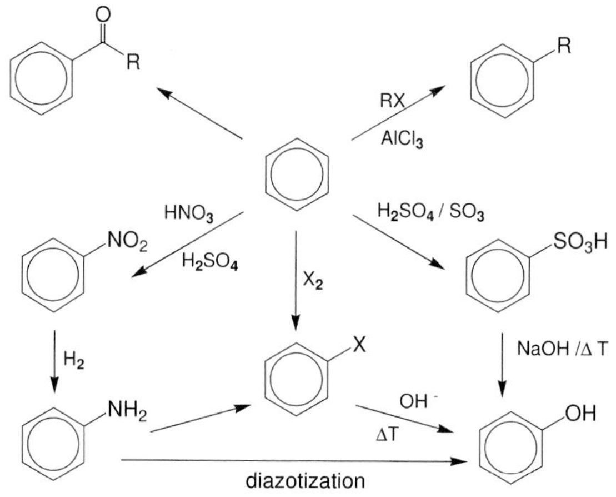

Fig. 1.4 Classical aromatic chemistry.

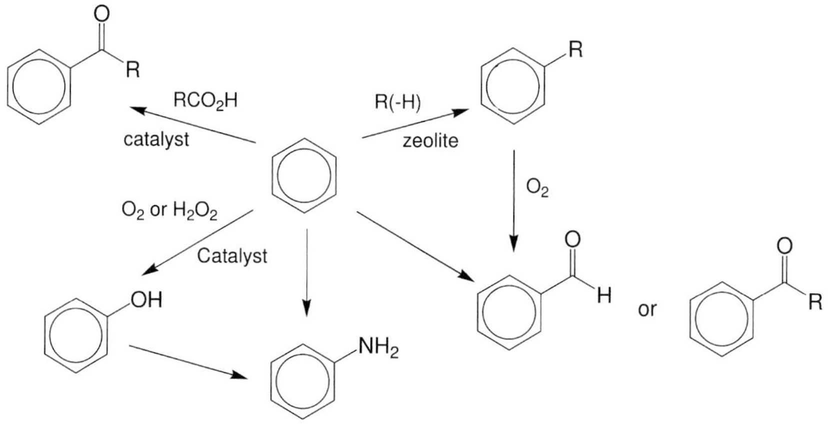

Fig. 1.5 Non-classical aromatic chemistry.

nium silicalite (TS-1)-catalysed hydroxylation of phenol with aqueous hydrogen peroxide (see later).

Biocatalysis has many advantages in the context of green chemistry, e.g. mild reaction conditions and often fewer steps than conventional chemical procedures because protection and deprotection of functional groups are often not required. Consequently, classical chemical procedures are increasingly being replaced by cleaner biocatalytic alternatives in the fine chemicals industry (see later).

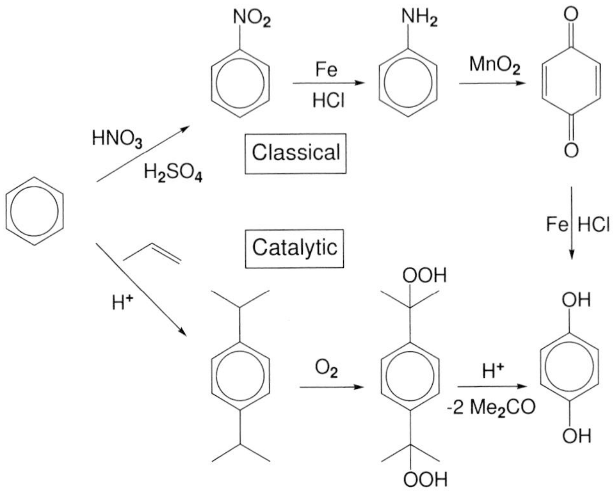

Fig. 1.6 Two routes to hydroquinone.

1.4
The Development of Organic Synthesis

If the solution to the waste problem in the fine chemicals industry is so obvious – replacement of classical stoichiometric reagents with cleaner, catalytic alternatives – why was it not applied in the past? We suggest that there are several reasons for this. First, because of the smaller quantities compared with bulk chemicals, the need for waste reduction in fine chemicals was not widely appreciated.

A second, underlying, reason is the more or less separate evolution of organic chemistry and catalysis (Fig. 1.7) since the time of Berzelius, who coined both terms, in 1807 and 1835, respectively [25]. Catalysis subsequently developed as a subdiscipline of physical chemistry, and is still often taught as such in university undergraduate courses. With the advent of the petrochemicals industry in the 1930s, catalysis was widely applied in oil refining and bulk chemicals manufacture. However, the scientists responsible for these developments, which largely involved heterogeneous catalysts in vapor phase reactions, were generally not organic chemists.

Organic synthesis followed a different line of evolution. A landmark was Perkin's serendipitous synthesis of mauveine (aniline purple) in 1856 [26] which marked the advent of the synthetic dyestuffs industry, based on coal tar as the raw material. The present day fine chemicals and pharmaceutical industries evolved largely as spin-offs of this activity. Coincidentally, Perkin was trying to synthesise the anti-malarial drug, quinine, by oxidation of a coal tar-based raw material, allyl toluidine, using stoichiometric amounts of potassium dichromate. Fine chemicals and pharmaceuticals have remained primarily the domain of

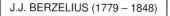

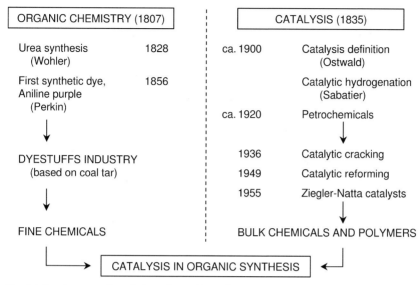

Fig. 1.7 Development of catalysis and organic synthesis.

synthetic organic chemists who, generally speaking, have clung to the use of classical "stoichiometric" methodologies and have been reluctant to apply catalytic alternatives.

A third reason, which partly explains the reluctance, is the pressure of time. Fine chemicals generally have a much shorter lifecycle than bulk chemicals and, especially in pharmaceuticals, 'time to market' is crucial. An advantage of many time-honored classical technologies is that they are well-tried and broadly applicable and, hence, can be implemented rather quickly. In contrast, the development of a cleaner, catalytic alternative could be more time consuming. Consequently, environmentally (and economically) inferior technologies are often used to meet market deadlines. Moreover, in pharmaceuticals, subsequent process changes are difficult to realise owing to problems associated with FDA approval.

There is no doubt that, in the twentieth century, organic synthesis has achieved a high level of sophistication with almost no molecule beyond its capabilities, with regard to chemo-, regio- and stereoselectivity, for example. However, little attention was focused on atom selectivity and catalysis was only sporadically applied. Hence, what we now see is a paradigm change: under the mounting pressure of environmental legislation, organic synthesis and catalysis, after 150 years in splendid isolation, have come together again. The key to waste minimisation is precision in organic synthesis, where every atom counts. In this chapter we shall briefly review the various categories of catalytic processes, with emphasis on fine chemicals but examples of bulk chemicals will also be discussed where relevant.

1.5
Catalysis by Solid Acids and Bases

As noted above, a major source of waste in the (fine) chemicals industry is derived from the widespread use of liquid mineral acids (HF, H_2SO_4) and a variety of Lewis acids. They cannot easily be recycled and generally end up, via a hydrolytic work-up, as waste streams containing large amounts of inorganic salts. Their widespread replacement by recyclable solid acids would afford a dramatic reduction in waste. Solid acids, such as zeolites, acidic clays and related materials, have many advantages in this respect [27–29]. They are often truly catalytic and can easily be separated from liquid reaction mixtures, obviating the need for hydrolytic work-up, and recycled. Moreover, solid acids are non-corrosive and easier (safer) to handle than mineral acids such as H_2SO_4 or HF.

Solid acid catalysts are, in principle, applicable to a plethora of acid-promoted processes in organic synthesis [27–29]. These include various electrophilic aromatic substitutions, e.g. nitrations, and Friedel-Crafts alkylations and acylations, and numerous rearrangement reactions such as the Beckmann and Fries rearrangements.

A prominent example is Friedel-Crafts acylation, a widely applied reaction in the fine chemicals industry. In contrast to the corresponding alkylations, which are truly catalytic processes, Friedel-Crafts acylations generally require more than one equivalent of, for example, $AlCl_3$ or BF_3. This is due to the strong complexation of the Lewis acid by the ketone product. The commercialisation of the first zeolite-catalysed Friedel-Crafts acylation by Rhône-Poulenc (now Rhodia) may be considered as a benchmark in this area [30, 31]. Zeolite beta is employed as a catalyst, in fixed-bed operation, for the acetylation of anisole with acetic anhydride, to give *p*-methoxyacetophenone (Fig. 1.8). The original process used acetyl chloride in combination with 1.1 equivalents of $AlCl_3$ in a chlorinated hydrocarbon solvent, and generated 4.5 kg of aqueous effluent, containing $AlCl_3$, HCl, solvent residues and acetic acid, per kg of product. The catalytic alternative, in stark contrast, avoids the production of HCl in both the acylation and in the synthesis of acetyl chloride. It generates 0.035 kg of aqueous effluent, i.e. more than 100 times less, consisting of 99% water, 0.8% acetic acid and <0.2% other organics, and requires no solvent. Furthermore, a product of higher purity is obtained, in higher yield (>95% vs. 85–95%), the catalyst is recyclable and the number of unit operations is reduced from twelve to two. Hence, the Rhodia process is not only environmentally superior to the traditional process, it has more favorable economics. This is an important conclusion; green, catalytic chemistry, in addition to having obvious environmental benefits, is also economically more attractive.

Another case in point pertains to the manufacture of the bulk chemical, caprolactam, the raw material for Nylon 6. The conventional process (Fig. 1.9) in-

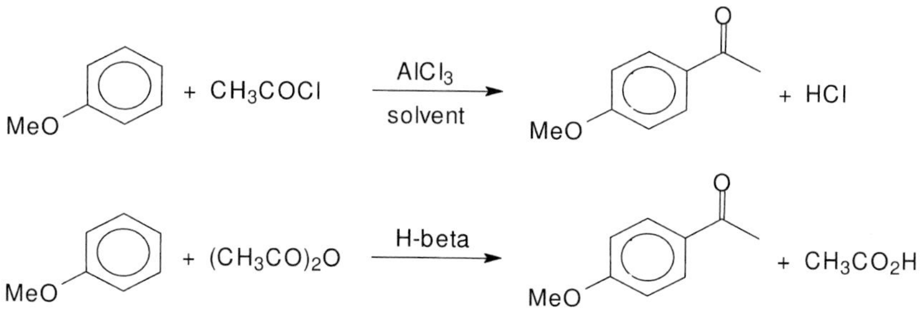

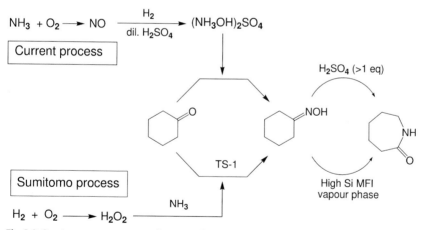

Homogeneous

AlCl$_3$ > 1 equivalent
Solvent
Hydrolysis of products
Phase separation
Distillation organic phase
Solvent recycle
85-95% yield
4.5 kg aqueous effluent per kg
12 unit operations

Heterogeneous

H-beta, catalytic & regenerable
No solvent
No water necessary
-
Distillation organic phase
-
> 95% yield, higher purity
0.035 kg aqueous effluent per kg
3 unit operations

Fig. 1.8 Zeolite-catalysed vs. classical Friedel-Crafts acylation.

Fig. 1.9 Sumitomo vs. conventional process for caprolactam manufacture.

volves the reaction of cyclohexanone with hydroxylamine sulfate (or another salt), producing cyclohexanone oxime which is subjected to the Beckmann rearrangement in the presence of stoichiometric amounts of sulfuric acid or oleum. The overall process generates ca. 4.5 kg of ammonium sulfate per kg of caprolactam, divided roughly equally over the two steps.

Ichihashi and coworkers at Sumitomo [32, 33] developed a catalytic vapor phase Beckmann rearrangement over a high-silica MFI zeolite. When this is combined with the technology, developed by Enichem [34], for the ammoximation of cyclohexanone with NH_3/H_2O_2 over the titanium silicalite catalyst (TS-1) described earlier, this affords caprolactam in >98% yield (based on cyclohexanone; 93% based on H_2O_2). The overall process generates caprolactam and two molecules of water from cyclohexanone, NH_3 and H_2O_2, and is essentially salt-free. This process is currently being commercialised by Sumitomo in Japan.

Another widely used reaction in fine chemicals manufacture is the acid-catalysed rearrangement of epoxides to carbonyl compounds. Lewis acids such as $ZnCl_2$ or $BF_3 \cdot OEt_2$ are generally used, often in stoichiometric amounts, to perform such reactions. Here again, zeolites can be used as solid, recyclable catalysts. Two commercially relevant examples are the rearrangements of α-pinene oxide [35, 36] and isophorone oxide [37] shown in Fig. 1.10. The products of these rearrangements are fragrance intermediates. The rearrangement of α-pinene oxide to campholenic aldehyde was catalysed by H-USY zeolite [35] and titanium-substituted zeolite beta [36]. With the latter, selectivities up to 89% in the liquid phase and 94% in the vapor phase were obtained, exceeding the best results obtained with homogeneous Lewis acids.

As any organic chemist will tell you, the conversion of an amino acid to the corresponding ester also requires more than one equivalent of a Brønsted acid. This is because an amino acid is a zwitterion and, in order to undergo acid catalysed esterification, the carboxylate anion needs to be protonated with one equivalent of acid. However, it was shown [38] that amino acids undergo esterification in the presence of a catalytic amount of zeolite H-USY, the very same catalyst that is used in naphtha cracking, thus affording a salt-free route to amino acid esters (Fig. 1.11). This is a truly remarkable reaction in that a basic compound (the amino ester) is formed in the presence of an acid catalyst. Esterification of optically active amino acids under these conditions (MeOH, 100 °C) un-

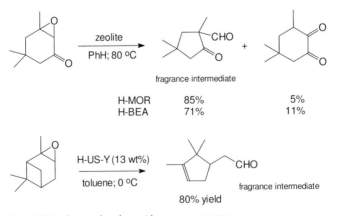

H-MOR	85%	5%
H-BEA	71%	11%

80% yield

Fig. 1.10 Zeolite-catalysed epoxide rearrangements.

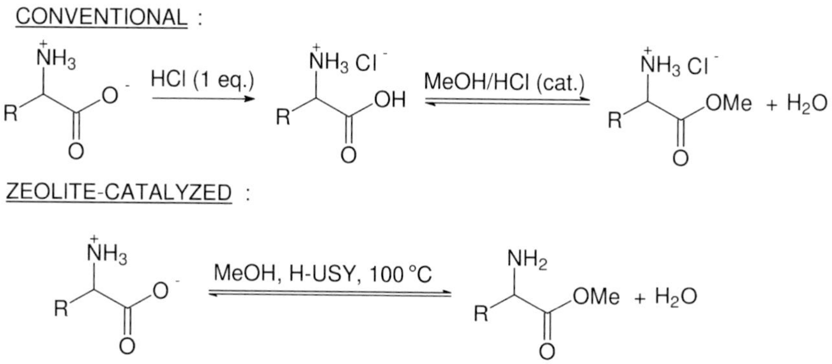

CONVENTIONAL :

ZEOLITE-CATALYZED :

Fig. 1.11 Salt-free esterification of amino acids.

fortunately led to (partial) racemisation. The reaction could be of interest for the synthesis of racemic phenylalanine methyl ester, the raw material in the DSM-Tosoh process for the artificial sweetener, aspartame.

In the context of replacing conventional Lewis acids in organic synthesis it is also worth pointing out that an alternative approach is to use lanthanide salts [39] that are both water soluble and stable towards hydrolysis and exhibit a variety of interesting activities as Lewis acids (see later).

The replacement of conventional bases, such as NaOH, KOH and NaOMe, by recyclable solid bases, in a variety of organic reactions, is also a focus of recent attention [27, 40]. For example, synthetic hydrotalcite clays, otherwise known as layered double hydroxides (LDHs) and having the general formula $Mg_{8-x}Al_x$ $(OH)_{16}(CO_3)_{x/2} \cdot nH_2O$, are hydrated aluminum-magnesium hydroxides possess-

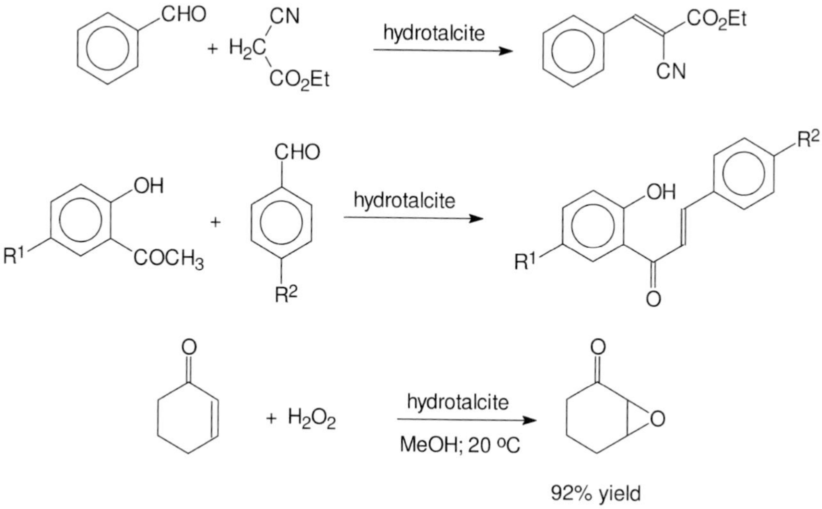

Fig. 1.12 Hydrotalcite-catalysed condensation reactions.

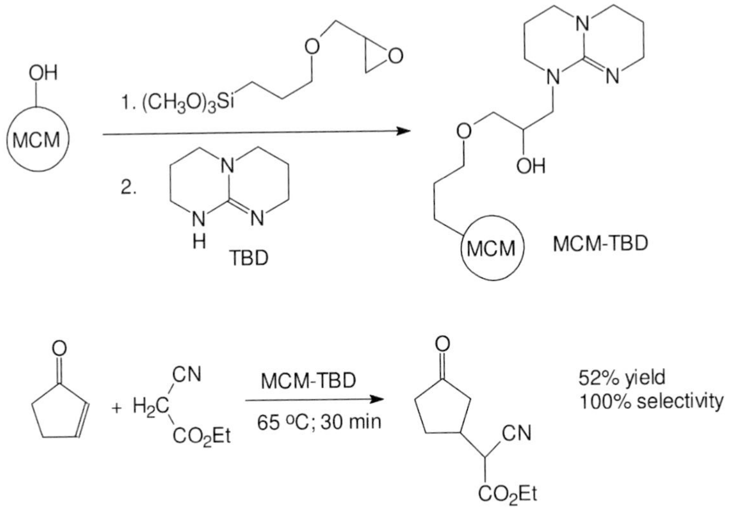

Fig. 1.13 Tethered organic bases as solid base catalysts.

ing a lamellar structure in which the excess positive charge is compensated by carbonate anions in the interlamellar space [41, 42]. Calcination transforms hydrotalcites, via dehydroxylation and decarbonation, into strongly basic mixed magnesium-aluminum oxides, that are useful recyclable catalysts for, inter alia, aldol [43], Knoevenagel [44, 45] and Claisen-Schmidt [45] condensations. Some examples are shown in Fig. 1.12.

Another approach to designing recyclable solid bases is to attach organic bases to the surface of, e.g. mesoporous silicas (Fig. 1.13) [46–48]. For example, aminopropyl-silica, resulting from reaction of 3-aminopropyl(trimethoxy)silane with pendant silanol groups, was an active catalyst for Knoevenagel condensations [49]. A stronger solid base was obtained by functionalisation of mesoporous MCM-41 with the guanidine base, 1,5,7-triazabicyclo-[4,4,0]dec-5-ene (TBD), using a surface glycidylation technique followed by reaction with TBD (Fig. 1.13). The resulting material was an active catalyst for Knoevenagel condensations, Michael additions and Robinson annulations [50].

1.6
Catalytic Reduction

Catalytic hydrogenation perfectly embodies the concept of precision in organic synthesis. Molecular hydrogen is a clean and abundant raw material and catalytic hydrogenations are generally 100% atom efficient, with the exception of a few examples, e.g. nitro group reduction, in which water is formed as a coproduct. They have a tremendously broad scope and exhibit high degrees of chemo-, regio-, diastereo and enantioselectivity [51, 52]. The synthetic prowess of catalytic hydrogenation is admirably rendered in the words of Rylander [51]:

"Catalytic hydrogenation is one of the most useful and versatile tools available to the organic chemist. The scope of the reaction is very broad; most functional groups can be made to undergo reduction, frequently in high yield, to any of several products. Multifunctional molecules can often be reduced selectively at any of several functions. A high degree of stereochemical control is possible with considerable predictability, and products free of contaminating reagents are obtained easily. Scale up of laboratory experiments to industrial processes presents little difficulty."
Paul Rylander (1979)

Catalytic hydrogenation is unquestionably the workhorse of catalytic organic synthesis, with a long tradition dating back to the days of Sabatier [53] who received the 1912 Nobel Prize in Chemistry for his pioneering work in this area. It is widely used in the manufacture of fine and specialty chemicals and a special issue of the journal Advanced Synthesis and Catalysis was recently devoted to this important topic [54]. According to Roessler [55], 10–20% of all the reaction steps in the synthesis of vitamins (even 30% for vitamin E) at Hoffmann-La Roche (in 1996) are catalytic hydrogenations.

Most of the above comments apply to heterogeneous catalytic hydrogenations over supported Group VIII metals (Ni, Pd, Pt, etc.). They are equally true, however, for homogeneous catalysts where spectacular progress has been made in the last three decades, culminating in the award of the 2001 Nobel Prize in Chemistry to W.S. Knowles and R. Noyori for their development of catalytic asymmetric hydrogenation (and to K.B. Sharpless for asymmetric oxidation catalysis) [56]. Recent trends in the application of catalytic hydrogenation in fine chemicals production, with emphasis on chemo-, regio- and stereoselectivity using both heterogeneous and homogeneous catalysts, is the subject of an excellent review by Blaser and coworkers [57].

A major trend in fine chemicals and pharmaceuticals is towards increasingly complex molecules, which translates to a need for high degrees of chemo-, regio- and stereoselectivity. An illustrative example is the synthesis of an intermediate for the Roche HIV protease inhibitor, Saquinavir (Fig. 1.14) [55]. It involves a chemo- and diastereoselective hydrogenation of an aromatic while avoiding racemisation at the stereogenic centre present in the substrate.

The chemoselective hydrogenation of one functional group in the presence of other reactive groups is a frequently encountered problem in fine chemicals manufacture. An elegant example of the degree of precision that can be achieved is the chemoselective hydrogenation of an aromatic nitro group in the presence of both an olefinic double bond and a chlorine substituent in the aromatic ring (Fig. 1.15) [58].

Although catalytic hydrogenation is a mature technology that is widely applied in industrial organic synthesis, new applications continue to appear, sometimes in unexpected places. For example, a time-honored reaction in organic

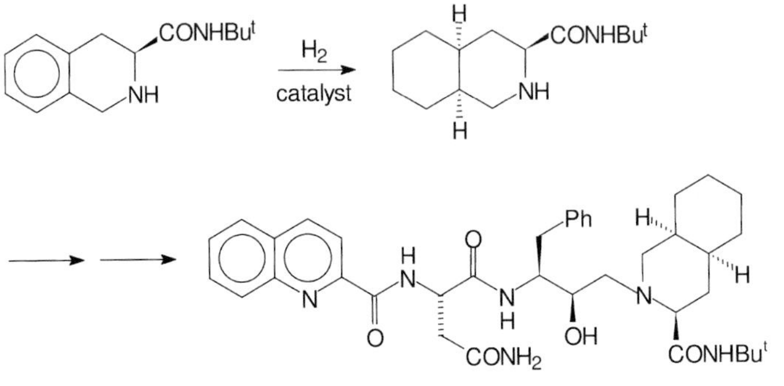

Fig. 1.14 Synthesis of a Saquinavir intermediate.

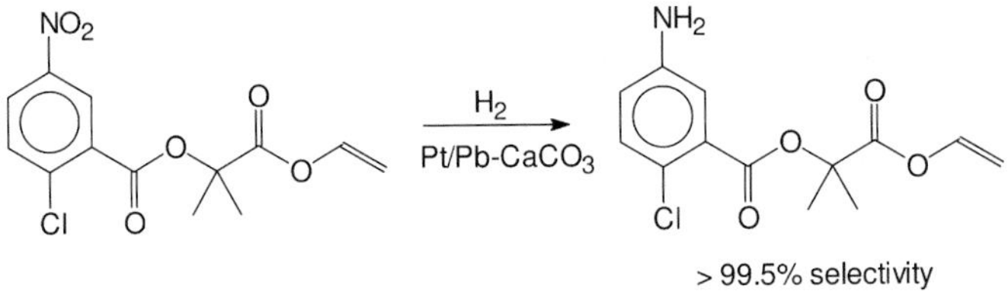

Fig. 1.15 Chemoselective hydrogenation of a nitro group.

synthesis is the Williamson synthesis of ethers, first described in 1852 [59]. A low-salt, catalytic alternative to the Williamson synthesis, involving reductive alkylation of an aldehyde (Fig. 1.16) has been reported [60]. This avoids the coproduction of NaCl, which may or may not be a problem, depending on the production volume (see earlier). Furthermore, the aldehyde may, in some cases, be more readily available than the corresponding alkyl chloride.

The Meerwein-Pondorff-Verley (MPV) reduction of aldehydes and ketones to the corresponding alcohols [61] is another example of a long-standing technology. The reaction mechanism involves coordination of the alcohol reagent, usually isopropanol, and the ketone substrate to the aluminum center, followed by hydride transfer from the alcohol to the carbonyl group. In principle, the re-

Williamson ether synthesis :

$$R^1CH_2Cl + R^2ONa \longrightarrow R^1CH_2OR^2 + Na\,Cl$$

Catalytic alternative :

$$R^1CHO + R^2OH + H_2 \xrightarrow{\text{catalyst}} R^1CH_2OR^2 + H_2O$$

Fig. 1.16 Williamson ether synthesis and a catalytic alternative.

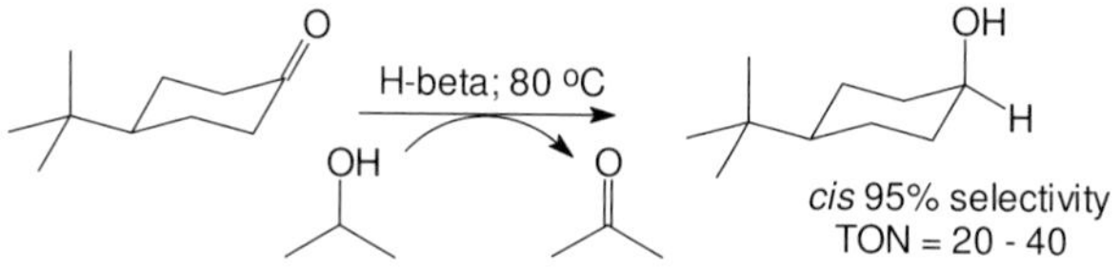

Fig. 1.17 Zeolite beta catalysed MPV reduction.

Fig. 1.18 Direct hydrogenation of carboxylic acids to aldehydes.

action is catalytic in aluminum alkoxide but, in practice, it generally requires stoichiometric amounts owing to the slow rate of exchange of the alkoxy group in aluminum alkoxides. Recently, van Bekkum and coworkers [62, 63] showed that Al- and Ti-Beta zeolites are able to catalyse MPV reductions. The reaction is truly catalytic and the solid catalyst can be readily separated, by simple filtration, and recycled. An additional benefit is that confinement of the substrate in the zeolite pores can afford interesting shape selectivities. For example, reduction of 4-*tert*-butylcyclohexanone led to the formation of the thermodynamically less stable *cis*-alcohol, an important fragrance intermediate, in high (>95%) selectivity (Fig. 1.17). In contrast, conventional MPV reduction gives the thermodynamically more stable, but less valuable, *trans*-isomer. Preferential formation of the *cis*-isomer was attributed to transition state selectivity imposed by confinement in the zeolite pores.

More recently, Corma and coworkers [64] have shown that Sn-substituted zeolite beta is a more active heterogeneous catalyst for MPV reductions, also showing high *cis*-selectivity (99–100%) in the reduction of 4-alkylcyclohexanones. The higher activity was attributed to the higher electronegativity of Sn compared to Ti.

The scope of catalytic hydrogenations continues to be extended to more difficult reductions. For example, a notoriously difficult reduction in organic synthesis is the direct conversion of carboxylic acids to the corresponding aldehydes. It is usually performed indirectly via conversion to the corresponding acid chloride and Rosenmund reduction of the latter over Pd/BaSO$_4$ [65]. Rhône-Poulenc [30] and Mitsubishi [66] have developed methods for the direct hydrogenation of aromatic, aliphatic and unsaturated carboxylic acids to the corresponding aldehydes, over a Ru/Sn alloy and zirconia or chromia catalysts, respectively, in the vapor phase (Fig. 1.18).

Finally, it is worth noting that significant advances have been made in the utilisation of biocatalytic methodologies for the (asymmetric) reduction of, for example, ketones to the corresponding alcohols (see later).

1.7
Catalytic Oxidation

It is probably true to say that nowhere is there a greater need for green catalytic alternatives in fine chemicals manufacture than in oxidation reactions. In contrast to reductions, oxidations are still largely carried out with stoichiometric inorganic (or organic) oxidants such as chromium(VI) reagents, permanganate, manganese dioxide and periodate. There is clearly a definite need for catalytic alternatives employing clean primary oxidants such as oxygen or hydrogen peroxide. Catalytic oxidation with O_2 is widely used in the manufacture of bulk petrochemicals [67]. Application to fine chemicals is generally more difficult, however, owing to the multifunctional nature of the molecules of interest. Nonetheless, in some cases such technologies have been successfully applied to the manufacture of fine chemicals. An elegant example is the BASF process [68] for the synthesis of citral (Fig. 1.19), a key intermediate for fragrances and vitamins A and E. The key step is a catalytic vapor phase oxidation over a supported silver catalyst, essentially the same as that used for the manufacture of formaldehyde from methanol.

This atom efficient, low-salt process has displaced the traditional route, starting from β-pinene, which involved, inter alia, a stoichiometric oxidation with MnO_2 (Fig. 1.19).

The selective oxidation of alcohols to the corresponding carbonyl compounds is a pivotal transformation in organic synthesis. As noted above, there is an urgent need for greener methodologies for these conversions, preferably employing O_2 or H_2O_2 as clean oxidants and effective with a broad range of substrates. One method which is finding increasing application in the fine chemicals industry employs the stable free radical, TEMPO 2,2′,6,6′-tetramethylpiperidine-N-oxyl) as a catalyst and NaOCl (household bleach) as the oxidant [69]. For example, this methodology was used, with 4-hydroxy TEMPO as the catalyst, as the key step in a new process for the production of progesterone from stigmasterol, a soy sterol (Fig. 1.20) [70].

This methodology still suffers from the shortcomings of salt formation and the use of bromide (10 mol%) as a cocatalyst and dichloromethane as solvent. Recently, a recyclable oligomeric TEMPO derivative, PIPO, derived from a commercially available polymer additive (Chimasorb 944) was shown to be an effective catalyst for the oxidation of alcohols with NaOCl in the absence of bromide ion using neat substrate or in e.g. methyl *tert*-butyl ether (MTBE) as solvent (Fig. 1.21) [71].

Another improvement is the use of a Ru/TEMPO catalyst combination for the selective aerobic oxidations of primary and secondary alcohols to the corresponding aldehydes and ketones, respectively (Fig. 1.22) [72]. The method is effective (>99% selectivity) with a broad range of primary and secondary aliphatic, allylic and benzylic alcohols. The overoxidation of aldehydes to the corresponding carboxylic acids is suppressed by the TEMPO which acts as a radical scavenger in preventing autoxidation.

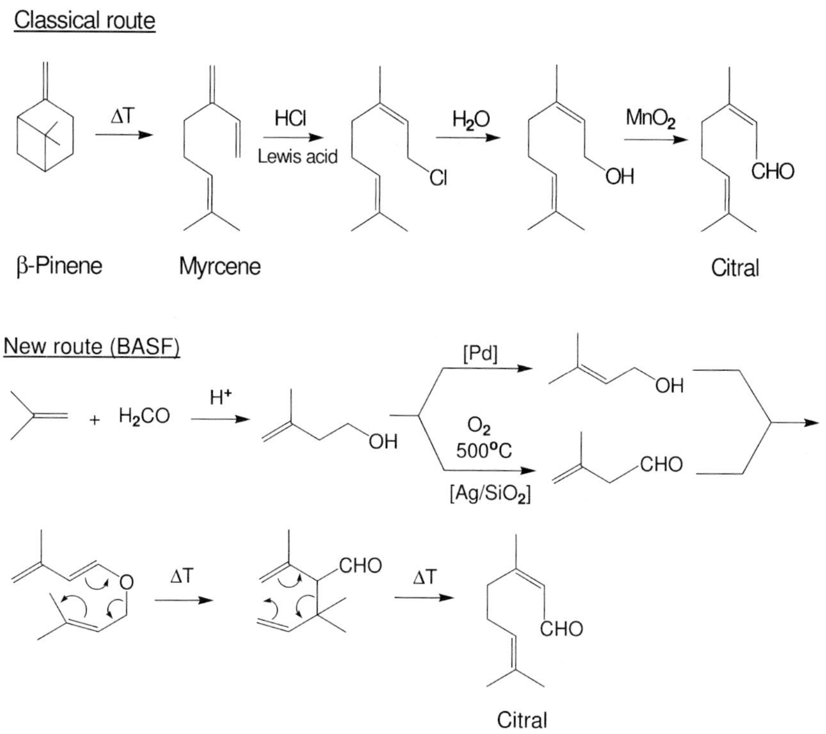
Classical route
ΔT
HCl
Lewis acid
Cl
H2O
OH
MnO2
CHO
β-Pinene
Myrcene
Citral
New route (BASF)
H+
+ H2CO
OH
[Pd]
OH
O2
500°C
[Ag/SiO2]
CHO
O
ΔT
CHO
ΔT
CHO
Citral

Fig. 1.19 Two routes to citral.

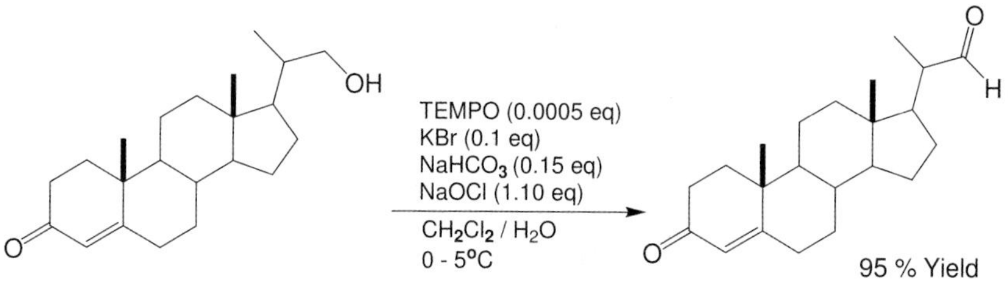
OH
TEMPO (0.0005 eq)
KBr (0.1 eq)
NaHCO3 (0.15 eq)
NaOCl (1.10 eq)
CH2Cl2 / H2O
0 - 5°C
O
H
O
O
95 % Yield

Fig. 1.20 Key step in the production of progesterone from stigmasterol.

Another recent development is the use of water soluble palladium complexes as recyclable catalysts for the aerobic oxidation of alcohols in aqueous/organic biphasic media (Fig. 1.22) [73].

In the fine chemicals industry, H_2O_2 is often the oxidant of choice because it is a liquid and processes can be readily implemented in standard batch equipment. To be really useful catalysts should be, for safety reasons, effective with 30% aqueous hydrogen peroxide and many systems described in the literature do not fulfill this requirement.

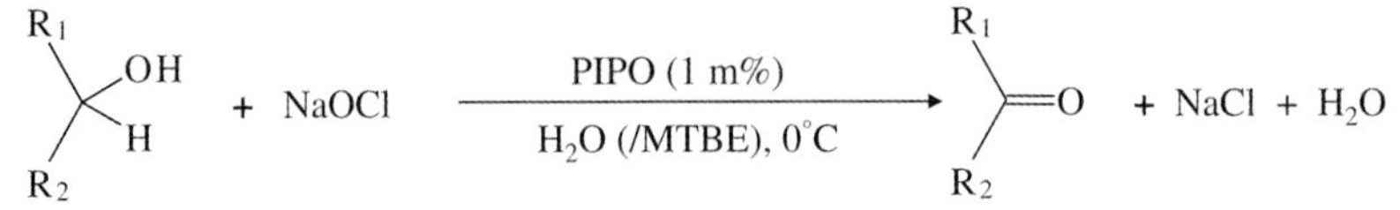

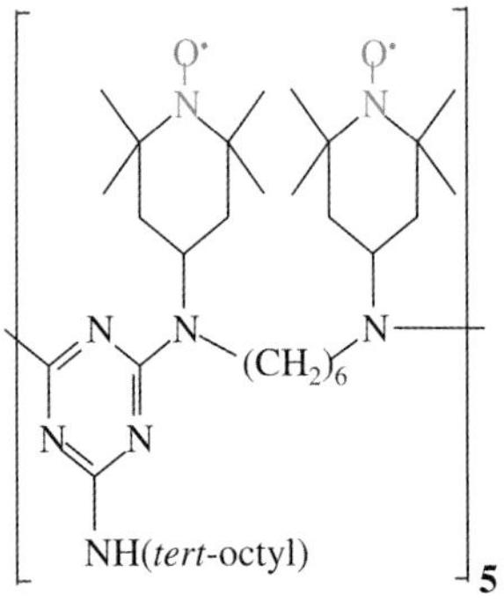

substrate	time (min)	conv. (%)	sel. (%)
1-octanol	45	90	50
	45	80	94[a]
1-hexanol	45	89	95[a]
2-octanol	45	99	>99
cyclooctanol	45	100	>99
benzyl alcohol	30	100	>99
1-phenylethanol	30	100	>99

a. in MTBE as solvent

Fig. 1.21 PIPO catalysed oxidation of alcohols with NaOCl.

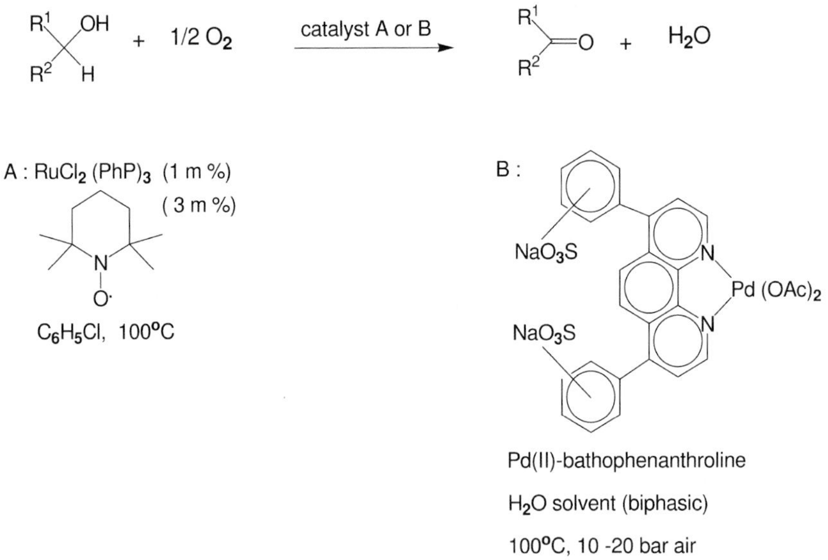

Fig. 1.22 Two methods for aerobic oxidation of alcohols.

In this context, the development of the heterogeneous titanium silicalite (TS-1) catalyst, by Enichem in the mid-1980s was an important milestone in oxidation catalysis. TS-1 is an extremely effective and versatile catalyst for a variety of synthe-

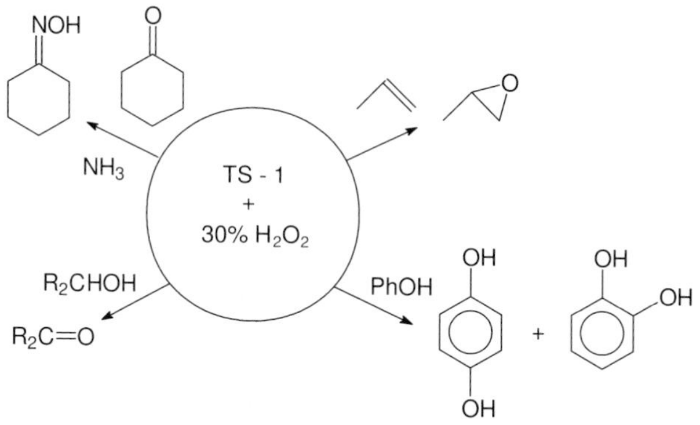

Fig. 1.23 Catalytic oxidations with TS-1/H$_2$O$_2$.

tically useful oxidations with 30% H$_2$O$_2$, e.g. olefin epoxidation, alcohol oxidation, phenol hydroxylation and ketone ammoximation (Fig. 1.23) [74].

A serious shortcoming of TS-1, in the context of fine chemicals manufacture, is the restriction to substrates that can be accommodated in the relatively small (5.1×5.5 Å^2) pores of this molecular sieve, e.g. cyclohexene is not epoxidised. This is not the case, however, with ketone ammoximation which involves *in situ* formation of hydroxylamine by titanium-catalysed oxidation of NH$_3$ with H$_2$O$_2$. The NH$_2$OH then reacts with the ketone in the bulk solution, which means that the reaction is, in principle, applicable to any ketone (or aldehyde). Indeed it was applied to the synthesis of the oxime of *p*-hydroxyacetophenone, which is converted, via Beckmann rearrangement, to the analgesic, paracetamol (Fig. 1.24) [75].

TS-1 was the prototype of a new generation of solid, recyclable catalysts for selective liquid phase oxidations, which we called "redox molecular sieves" [76]. A more recent example is the tin(IV)-substituted zeolite beta, developed by Corma and coworkers [77], which was shown to be an effective, recyclable catalyst

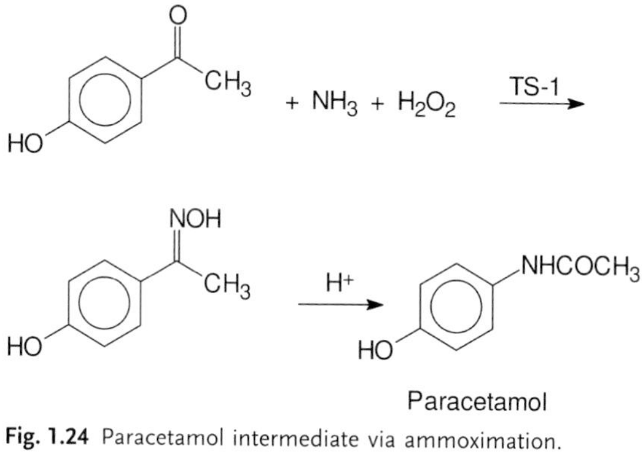

Paracetamol

Fig. 1.24 Paracetamol intermediate via ammoximation.

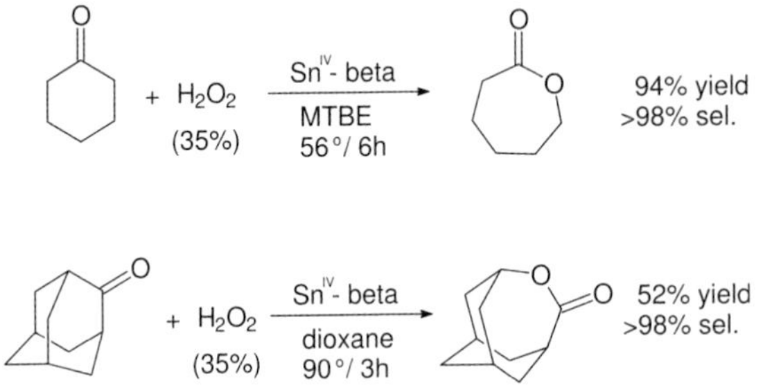

Fig. 1.25 Baeyer-Villiger oxidation with H_2O_2 catalysed by Sn-Beta.

for the Baeyer-Villiger oxidation of ketones and aldehydes [78] with aqueous H_2O_2 (Fig. 1.25).

At about the same time that TS-1 was developed by Enichem, Venturello and coworkers [79] developed another approach to catalysing oxidations with aqueous hydrogen peroxide: the use of tungsten-based catalysts under phase transfer conditions in biphasic aqueous/organic media. In the original method a tetra-alkylammonium chloride or bromide salt was used as the phase transfer agent and a chlorinated hydrocarbon as the solvent [79]. More recently, Noyori and coworkers [80] have optimised this methodology and obtained excellent results using tungstate in combination with a quaternary ammonium hydrogen sulfate as the phase transfer catalyst. This system is a very effective catalyst for the organic solvent- and halide-free oxidation of alcohols, olefins and sulfides with

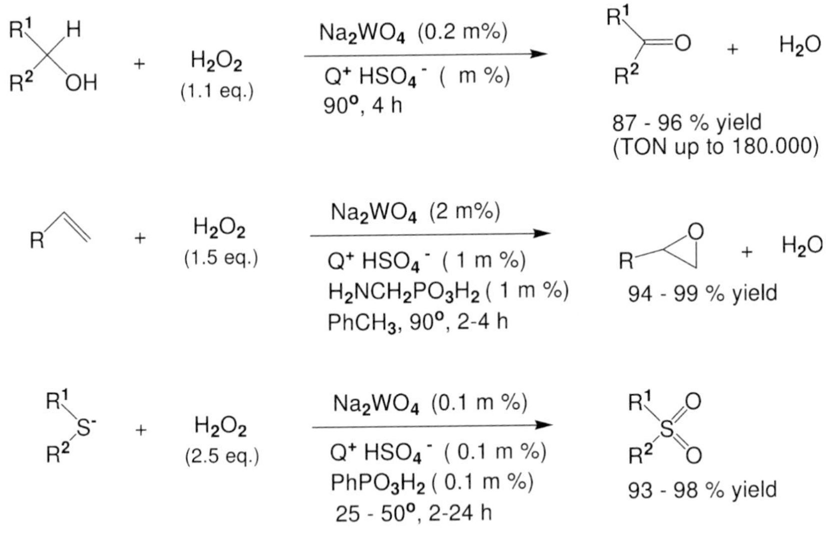

$Q = CH_3(C_8H_{17})_3N$

Fig. 1.26 Catalytic oxidations with hydrogen peroxide under phase transfer conditions.

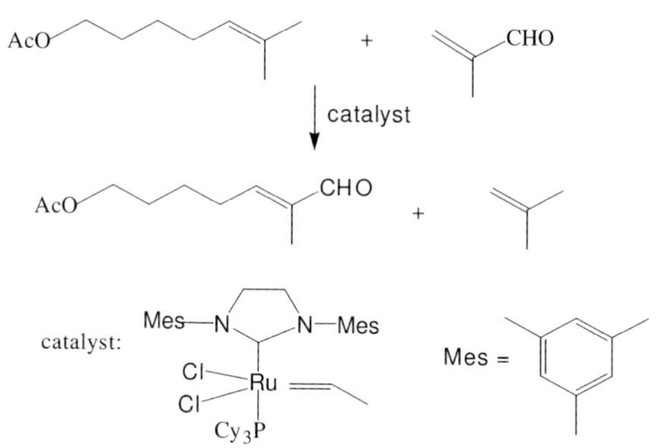

Fig. 1.27 The best oxidation is no oxidation.

aqueous H_2O_2, in an environmentally and economically attractive manner (Fig. 1.26).

Notwithstanding the significant advances in selective catalytic oxidations with O_2 or H_2O_2 that have been achieved in recent years, selective oxidation, especially of multifunctional organic molecules, remains a difficult catalytic transformation that most organic chemists prefer to avoid altogether. In other words, the best oxidation is no oxidation and most organic chemists would prefer to start at a higher oxidation state and perform a reduction or, better still, avoid changing the oxidation state. An elegant example of the latter is the use of olefin metathesis to affect what is formally an allylic oxidation which would be nigh impossible to achieve via catalytic oxidation (Fig. 1.27) [81].

1.8
Catalytic C–C Bond Formation

Another key transformation in organic synthesis is C–C bond formation and an important catalytic methodology for generating C–C bonds is carbonylation. In the bulk chemicals arena it is used, for example, for the production of acetic acid by rhodium-catalysed carbonylation of methanol [82]. Since such reactions are 100% atom efficient they are increasingly being applied to fine chemicals manufacture [83, 84]. An elegant example of this is the Hoechst-Celanese process for the manufacture of the analgesic, ibuprofen, with an annual production of several thousands tons. In this process ibuprofen is produced in two catalytic steps (hydrogenation and carbonylation) from p-isobutylactophenone (Fig. 1.28) with 100% atom efficiency [83]. This process replaced a more classical route which involved more steps and a much higher E factor.

In a process developed by Hoffmann-La Roche [55] for the anti-Parkinsonian drug, lazabemide, palladium-catalysed amidocarbonylation of 2,5-dichloropyridine replaced an original synthesis that involved eight steps, starting from 2-

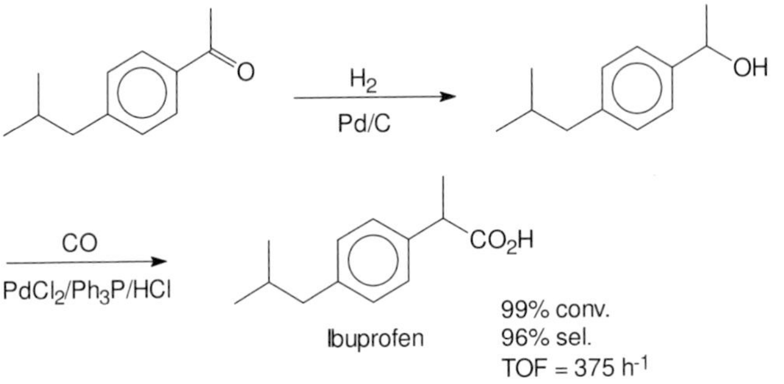

Fig. 1.28 Hoechst-Celanese process for ibuprofen.

methyl-5-ethylpyridine, and had an overall yield of 8%. The amidocarbonylation route affords lazabemide hydrochloride in 65% yield in one step, with 100% atom efficiency (Fig. 1.29).

Another elegant example, of palladium-catalysed amidocarbonylation this time, is the one-step, 100% atom efficient synthesis of α-amino acid derivatives from an aldehyde, CO and an amide (Fig. 1.30) [85]. The reaction is used, for example in the synthesis of the surfactant, N-lauroylsarcosine, from formaldehyde, CO and N-methyllauramide, replacing a classical route that generated copious amounts of salts.

Another catalytic methodology that is widely used for C–C bond formation is the Heck and related coupling reactions [86, 87]. The Heck reaction [88] involves the palladium-catalysed arylation of olefinic double bonds (Fig. 1.31) and provides an alternative to Friedel-Crafts alkylations or acylations for attaching carbon fragments to aromatic rings. The reaction has broad scope and is currently being widely applied in the pharmaceutical and fine chemical industries. For example, Albemarle has developed a new process for the synthesis of the anti-in-

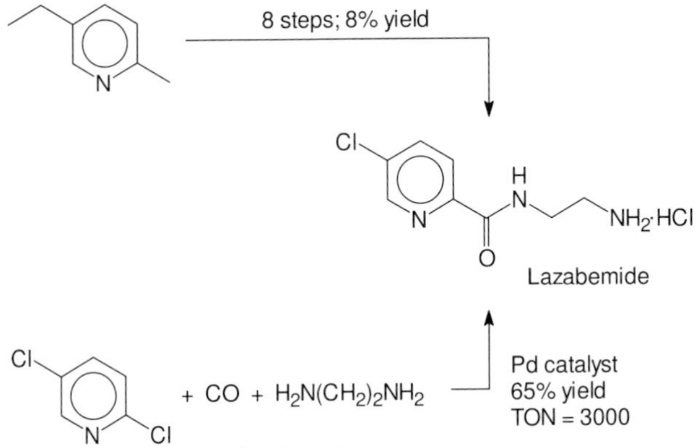

Fig. 1.29 Two routes to lazabemide.

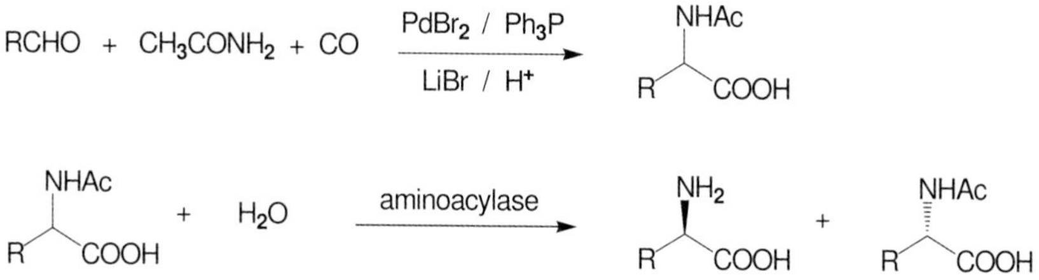

Fig. 1.30 Palladium-catalysed amidocarbonylation.

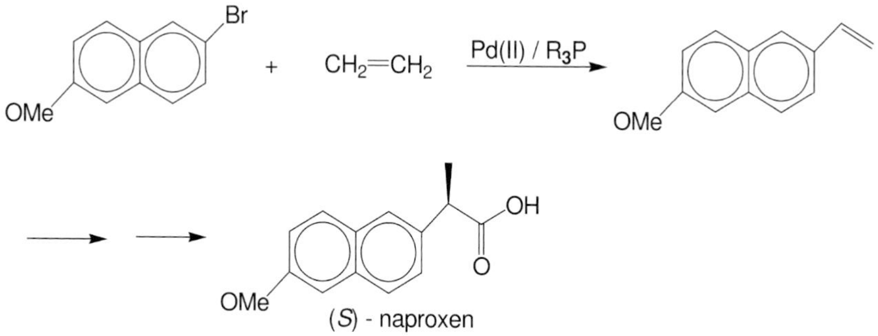

Fig. 1.31 Heck coupling reaction.

flammatory drug, naproxen, in which a key step is the Heck reaction shown in Fig. 1.31 [86].

The scope of the Heck and related coupling reactions was substantially broadened by the development, in the last few years, of palladium/ligand combinations which are effective with the cheap and readily available but less reactive aryl chlorides [86, 87] rather than the corresponding bromides or iodides. The process still generates one equivalent of chloride, however. Of interest in this context, therefore, is the report of a halide-free Heck reaction which employs an aromatic carboxylic anhydride as the arylating agent and requires no base or phosphine ligands [89].

A closely related reaction, that is currently finding wide application in the pharmaceutical industry, is the Suzuki coupling of arylboronic acids with aryl halides [90]. For example this technology was applied by Clariant scientists to the production of *o*-tolyl benzonitrile, an intermediate in the synthesis of angiotensin II antagonists, a novel class of antihypertensive drugs (Fig. 1.32) [91]. Interestingly, the reaction is performed in an aqueous biphasic system using a water soluble palladium catalyst, which forms the subject of the next section: the question of reaction media in the context of green chemistry and catalysis.

However, no section on catalytic C–C bond formation would be complete without a mention of olefin metathesis [92, 93]. It is, in many respects, the epitome of green chemistry, involving the exchange of substituents around the double bonds in the presence of certain transition metal catalysts (Mo, W, Re and Ru) as shown in Fig. 1.33. Several outcomes are possible: straight swapping

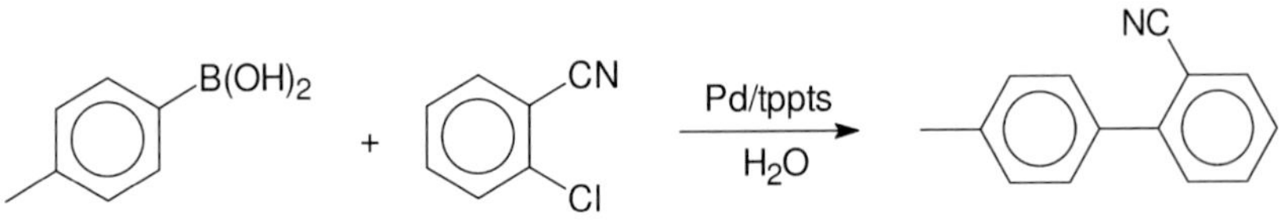

Fig. 1.32 A Suzuki coupling.

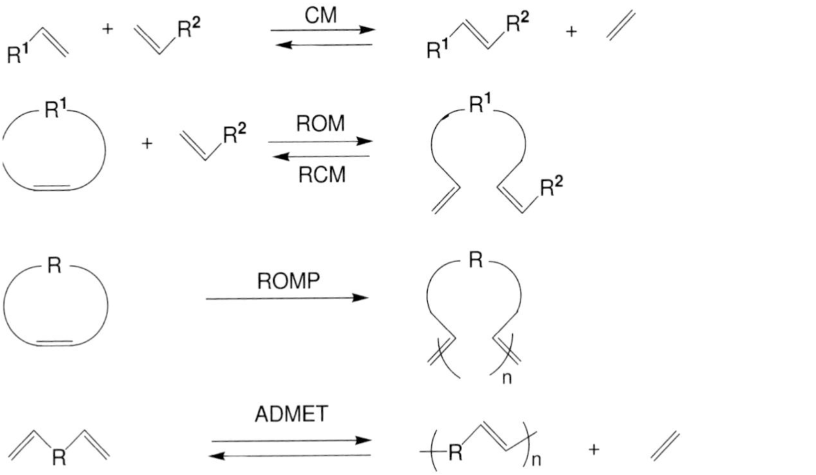

CM = Cross metathesis
ROM = Ring opening metathesis
RCM = Ring closing methatnesis

ROMP = Ring opening metathesis polymerization
ADMET = Acyclic diene metathesis

Catalysts = Mo, W, Re and Ru complexes

Fig. 1.33 Olefin metathesis reactions.

of groups between two acyclic olefins (cross metathesis, CM), closure of large rings (ring closing metathesis, RCM), diene formation from reaction of a cyclic olefin with an acyclic one (ring opening metathesis, ROM), polymerization of cyclic olefins (ring opening metathesis polymerization, ROMP) and polymerization of acyclic dienes (acyclic diene metathesis polymerisation, ADMET).

Following its discovery in the 1960s olefin metathesis was applied to bulk chemicals manufacture, a prominent example being the Shell Higher Olefins Process (SHOP) [94]. In the succeeding decades the development of catalysts, in particular the ruthenium-based ones, that function in the presence of most functional groups, paved the way for widespread application of olefin metathesis in the synthesis of complex organic molecules [92, 93]. Indeed, olefin metathesis has evolved into a pre-eminent methodology for the formation of C–C bonds under mild conditions. An illustrative example is the RCM reaction shown in Fig. 1.34 [95]. The ruthenium carbene complex catalyst functioned in undistilled protic solvents (MeOH/H_2O) in the presence of air.

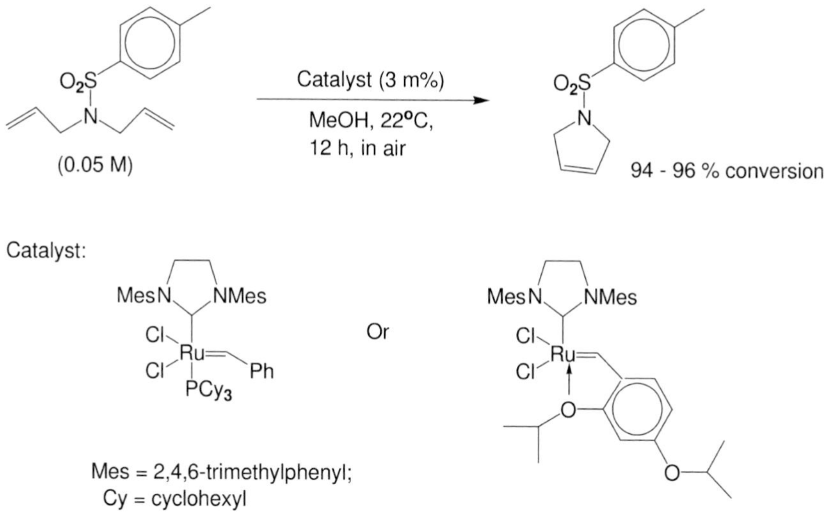

Catalyst:

Mes = 2,4,6-trimethylphenyl;
Cy = cyclohexyl

Fig. 1.34 Ru-catalysed ring closing metathesis.

1.9
The Question of Solvents: Alternative Reaction Media

Another important issue in green chemistry is the use of organic solvents. The use of chlorinated hydrocarbon solvents, traditionally the solvent of choice for a wide variety of organic reactions, has been severely curtailed. Indeed, so many of the solvents that are favored by organic chemists have been blacklisted that the whole question of solvent use requires rethinking and has become a primary focus, especially in the fine chemicals industry [96]. It has been estimated by GSK workers [97] that ca. 85% of the total mass of chemicals involved in pharmaceutical manufacture comprises solvents and recovery efficiencies are typically 50–80% [97]. It is also worth noting that in the redesign of the sertraline manufacturing process [98], for which Pfizer received a Presidential Green Chemistry Challenge Award in 2002, among other improvements a three-step sequence was streamlined by employing ethanol as the sole solvent. This eliminated the need to use, distil and recover four solvents (methylene chloride, tetrahydrofuran, toluene and hexane) employed in the original process. Similarly, impressive improvements were achieved in a redesign of the manufacturing process for sildenafil (Viagra^{TM}) [99].

The best solvent is no solvent and if a solvent (diluent) is needed then water is preferred [100]. Water is non-toxic, non-inflammable, abundantly available and inexpensive. Moreover, owing to its highly polar character one can expect novel reactivities and selectivities for organometallic catalysis in water. Furthermore, this provides an opportunity to overcome a serious shortcoming of homogeneous catalysts, namely the cumbersome recovery and recycling of the catalyst. Thus, performing the reaction in an aqueous biphasic system, whereby the catalyst resides in the water phase and the product is dissolved in the organic phase [101, 102], allows recovery and recycling of the catalyst by simple phase separation.

An example of a large scale application of this concept is the Ruhrchemie/ Rhône Poulenc process for the hydroformylation of propylene to n-butanal, which employs a water-soluble rhodium(I) complex of trisulfonated triphenyl-phosphine (tppts) as the catalyst [103]. The same complex also functions as the catalyst in the Rhône Poulenc process for the manufacture of the vitamin A intermediate, geranylacetone, via reaction of myrcene with methyl acetoacetate in an aqueous biphasic system (Fig. 1.35) [104].

Similarly, Pd/tppts was used by Hoechst [105] as the catalyst in the synthesis of phenylacetic acid by biphasic carbonylation of benzyl chloride (Fig. 1.36) as an alternative to the classical synthesis via reaction with sodium cyanide. Although the new process still produces one equivalent of sodium chloride, this is substantially less salt generation than in the original process. Moreover, sodium cyanide is about seven times more expensive per kg than carbon monoxide.

The salt production can be circumvented by performing the selective Pd/ tppts-catalysed carbonylation of benzyl alcohol in an acidic aqueous biphasic system (Fig. 1.36) [106]. This methodology was also applied to the synthesis of ibuprofen (see earlier) by biphasic carbonylation of 1-(4-isobutylphenyl)ethanol [107] and to the biphasic hydrocarboxylation of olefins [108].

As mentioned earlier (Section 1.5) another example of novel catalysis in an aqueous medium is the use of lanthanide triflates as water-tolerant Lewis acid catalysts for a variety of organic transformations in water [39].

Other non-classical reaction media [96] have, in recent years, attracted increasing attention from the viewpoint of avoiding environmentally unattractive solvents and/or facilitating catalyst recovery and recycling. Two examples, which readily come to mind, are supercritical carbon dioxide and room temperature ionic liquids. Catalytic hydrogenation in supercritical CO_2, for example, has

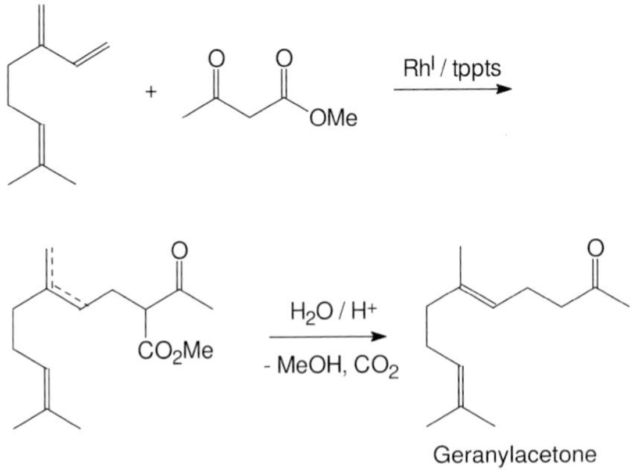

Fig. 1.35 Manufacture of n-butanal and geranylacetone in aqueous biphasic systems.

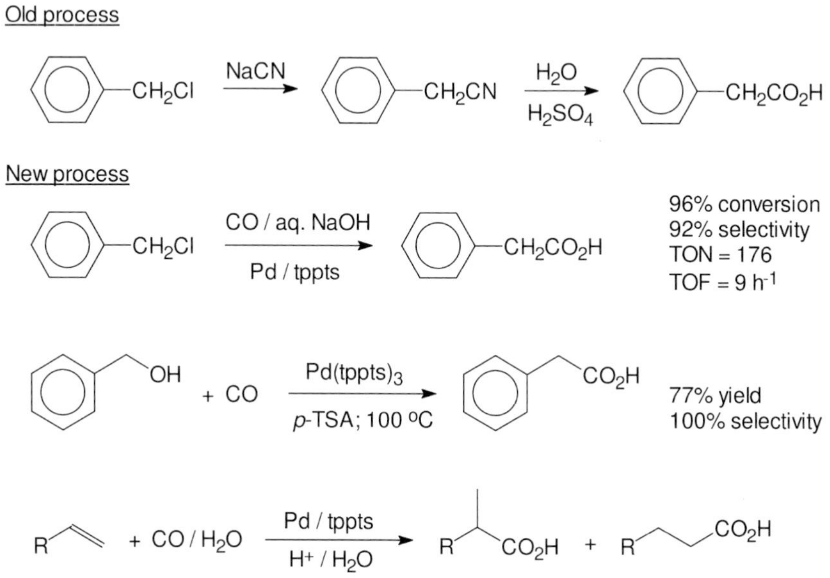

Old process

New process

96% conversion
92% selectivity
TON = 176
TOF = 9 h^{-1}

77% yield
100% selectivity

R	Selectivity (%)	
	iso-	n-
CH$_3$	43	57
C$_6$H$_5$	56	33
4-i-BuC$_6$H$_4$	82	18

Fig. 1.36 Aqueous biphasic carbonylations.

been commercialised by Thomas Swan and Co. [109]. Ionic liquids are similarly being touted as green reaction media for organic synthesis in general and catalytic reactions in particular [110–112]. They exhibit many properties which make them potentially attractive reaction media, e.g. they have essentially no vapor pressure and cannot, therefore, cause emissions to the atmosphere. These non-conventional reaction media will be treated in depth in Chapter 7.

1.10
Biocatalysis

Biocatalysis has many attractive features in the context of green chemistry: mild reaction conditions (physiological pH and temperature), an environmentally compatible catalyst (an enzyme) and solvent (often water) combined with high activities and chemo-, regio- and stereoselectivities in multifunctional molecules. Furthermore, the use of enzymes generally circumvents the need for functional group activation and avoids protection and deprotection steps required in traditional organic syntheses. This affords processes which are shorter, generate less waste and are, therefore, both environmentally and economically more attractive than conventional routes.

The time is ripe for the widespread application of biocatalysis in industrial organic synthesis and according to a recent estimate [113] more than 130 processes have been commercialised. Advances in recombinant DNA techniques have made it, in principle, possible to produce virtually any enzyme for a commercially acceptable price. Advances in protein engineering have made it possible, using techniques such as site directed mutagenesis and *in vitro* evolution, to manipulate enzymes such that they exhibit the desired substrate specificity, activity, stability, pH profile, etc. [114]. Furthermore, the development of effective immobilisation techniques has paved the way for optimising the performance and recovery and recycling of enzymes.

An illustrative example of the benefits to be gained by replacing conventional chemistry by biocatalysis is provided by the manufacture of 6-aminopenicillanic acid (6-APA), a key raw material for semi-synthetic penicillin and cephalosporin antibiotics, by hydrolysis of penicillin G [115]. Up until the mid-1980s a chemical procedure was used for this hydrolysis (Fig. 1.37). It involved the use of environmentally unattractive reagents, a chlorinated hydrocarbon solvent (CH_2Cl_2) and a reaction temperature of –40 °C. Thus, 0.6 kg Me_3SiCl, 1.2 kg PCl_5, 1.6 kg $PhNMe_2$, 0.2 kg NH_3, 8.41 kg of n-BuOH and 8.41 kg of CH_2Cl_2 were required to produce 1 kg of 6-APA [116].

In contrast, enzymatic cleavage of penicillin G (Fig. 1.37) is performed in water at 37 °C and the only reagent used is NH_3 (0.9 kg per kg of 6-APA), to adjust the pH. The enzymatic process currently accounts for the majority of the several thousand tons of 6-APA produced annually on a world-wide basis.

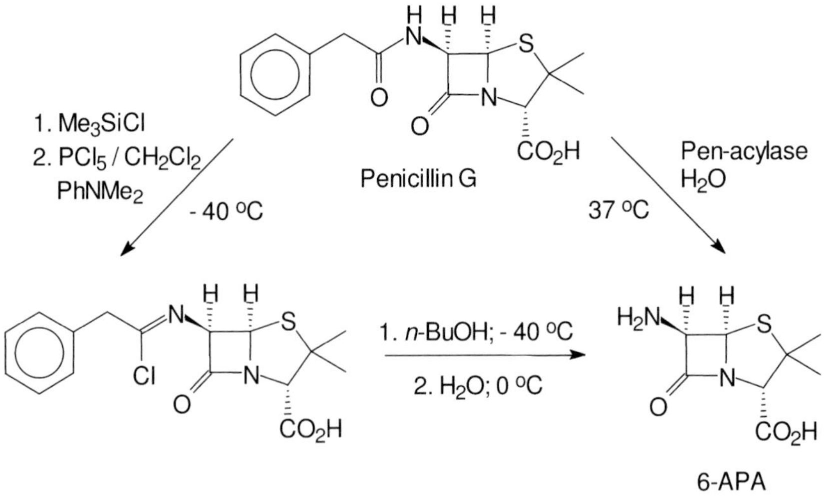

Fig. 1.37 Enzymatic versus chemical deacylation of penicillin G.

Another advantage of biocatalysis is the high degree of chemo-, regio- and stereoselectivities which are difficult or impossible to achieve by chemical means. A pertinent example is the production of the artificial sweetener, aspartame. The enzymatic process, operated by the Holland Sweetener Company (a joint venture of DSM and Tosoh) is completely regio- and enantiospecific (Fig. 1.38) [117].

The above-mentioned processes employ isolated enzymes – penicillin G acylase and thermolysin – and the key to their success was an efficient production of the enzyme. As with chemical catalysts, another key to success in biocatalytic processes is an effective method for immobilisation, providing for efficient recovery and re-use.

In some cases it is more attractive to use whole microbial cells, rather than isolated enzymes, as biocatalysts. This is the case in many oxidative biotransformations where cofactor regeneration is required and/or the enzyme has limited stability outside the cell. By performing the reaction with growing microbial cells, i.e. as a fermentation, the cofactor is continuously regenerated from an energy source, e.g. glucose. Lonza, for example, has commercialised processes for the highly chemo- and regioselective microbial ring hydroxylation and side-chain oxidation of heteroaromatics (see Fig. 1.39 for examples) [118]. Such conversions would clearly not be feasible by conventional chemical means.

DuPont has developed a process for the manufacture of glyoxylic acid, a large volume fine chemical, by aerobic oxidation of glycolic acid, mediated by resting

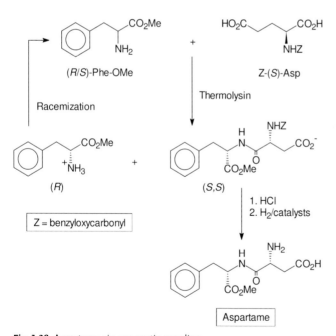

Fig. 1.38 Aspartame via enzymatic coupling.

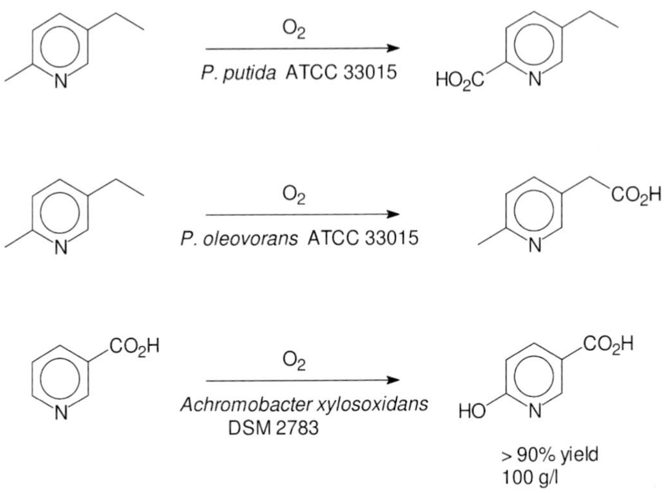

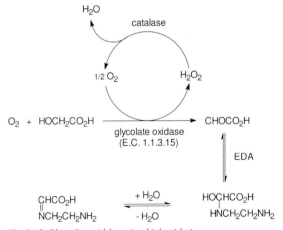

Fig. 1.39 Microbial oxidations of heteroaromatics.

whole cells of a recombinant methylotrophic yeast (Fig. 1.40) [119]. The glycolic acid is readily available from acid-catalysed carbonylation of formaldehyde. Traditionally, glyoxylic acid was produced by nitric acid oxidation of acetaldehyde or glyoxal, processes with high E factors, and more recently by ozonolysis of maleic anhydride.

The key enzyme in the above process is an oxidase which utilises dioxygen as the oxidant, producing one equivalent of hydrogen peroxide, without the need for cofactor regeneration. Another class of enzymes which catalyse the oxidation of alcohols comprises the alcohol dehydrogenases. However, in this case cofactor regeneration is required, which is an impediment to commercialisation. Re-

Fig. 1.40 Glyoxylic acid by microbial oxidation.

cently, a highly enantioselective alcohol dehydrogenase, showing broad substrate specificity and exceptionally high tolerance for organic solvents, was isolated from *Rhodococcus ruber* DSM 4451 [120]. The enzyme maintains a high activity at concentrations of up to 20% (v/v) acetone and 50% (v/v) 2-propanol. This enables the use of the enzyme, conveniently as whole microbial cells, as a catalyst for (enantioselective) Oppenauer oxidation of a broad range of alcohols, using acetone (20% v/v in phosphate buffer at pH 8) as the oxidant (Fig. 1.41), with substrate concentrations up to 1.8 mol l^{-1} (237 g l^{-1} for octan-2-ol).

Alternatively, the reaction could be performed in a reduction mode, using the ketone as substrate and up to 50% v/v isopropanol as the reductant, affording the corresponding (S)-alcohol in 99% ee at conversions ranging from 65 to 92%.

Another example in which a biocatalytic transformation has replaced a chemocatalytic one, in a very simple reaction, is the Mitsubishi Rayon process for the production of acrylamide by hydration of acrylonitrile (Fig. 1.42). Whole cells of *Rhodococcus rhodocrous*, containing a nitrile hydratase, produced acrylamide in >99.9% purity at >99.9% conversion, and in high volumetric and space time yields [121]. The process (Fig. 1.42) currently accounts for more than 100 000 tons annual production of acrylamide and replaced an existing process which employed a copper catalyst. A major advantage of the biocatalytic process is the high product purity, which is important for the main application of acrylamide as a specialty monomer.

Similarly, DuPont employs a nitrile hydratase (as whole cells of *P. chlororaphis* B23) to convert adiponitrile to 5-cyanovaleramide, a herbicide intermediate [122]. In the Lonza nitrotinamide (vitamin B6) process [123] the final step (Fig. 1.42) involves the nitrile hydratase (whole cells of *Rh. rhodocrous*) catalysed hydration of 3-cyanopyridine. Here again the very high product purity is a major advantage as conventional chemical hydrolysis affords a product contaminated with nicotinic acid, which requires expensive purification to meet the specifications of this vitamin.

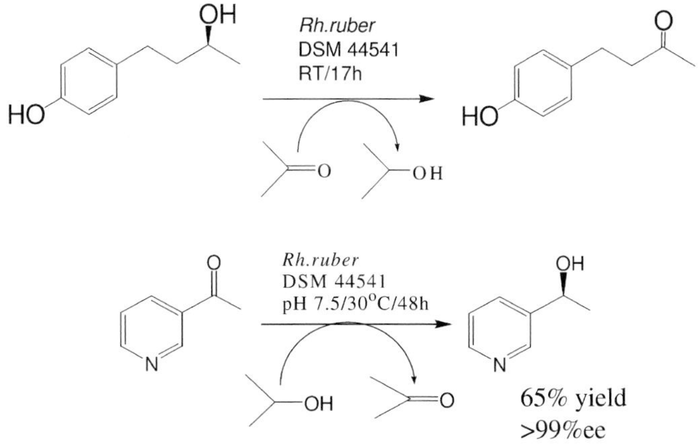

Fig. 1.41 Biocatalytic Oppenauer oxidations and MPV reductions.

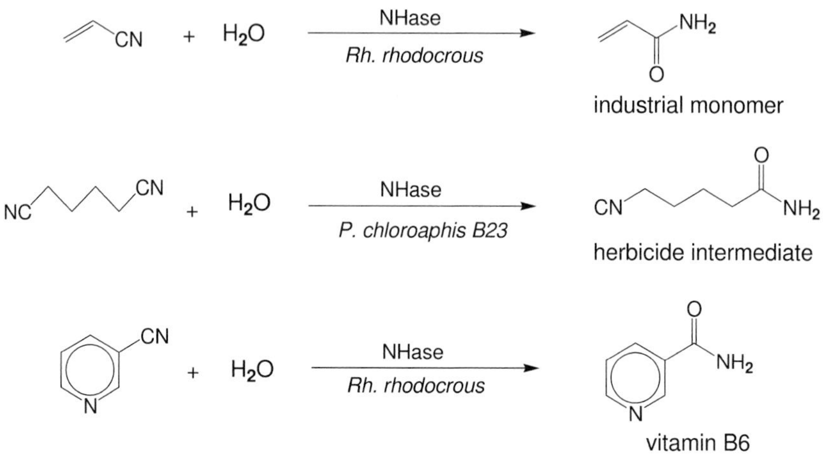

Fig. 1.42 Industrial processes employing a nitrile hydratase.

1.11
Renewable Raw Materials and White Biotechnology

Another important goal of green chemistry is the utilisation of renewable raw materials, i.e. derived from biomass, rather than crude oil. Here again, the processes used for the conversion of renewable feedstocks – mainly carbohydrates but also triglycerides and terpenes – should produce minimal waste, i.e. they should preferably be catalytic.

In the processes described in the preceding section a biocatalyst – whole microbial cells or an isolated enzyme – is used to catalyse a transformation (usually one step) of a particular substrate. When growing microbial cells are used this is referred to as a precursor fermentation. Alternatively, one can employ *de novo* fermentation to produce chemicals directly from biomass. This has become known as white biotechnology, as opposed to red biotechnology (biopharmaceuticals) and green biotechnology (genetically modified crops). White biotechnology is currently the focus of considerable attention and is perceived as the key to developing a sustainable chemical industry [124].

Metabolic pathway engineering [125] is used to optimise the production of the required product based on the amount of substrate (usually biomass-derived) consumed. A so-called biobased economy is envisaged in which commodity chemicals (including biofuels), specialty chemicals such as vitamins, flavors and fragrances and industrial monomers will be produced in biorefineries (see Chapter 8 for a more detailed discussion).

De novo fermentation has long been the method of choice for the manufacture of many natural L-amino acids, such as glutamic acid and lysine, and hydroxy acids such as lactic and citric acids. More recently, *de novo* fermentation is displacing existing multistep chemical syntheses, for example in the manufacture of vitamin B2 (riboflavin) and vitamin C. Other recent successes of white

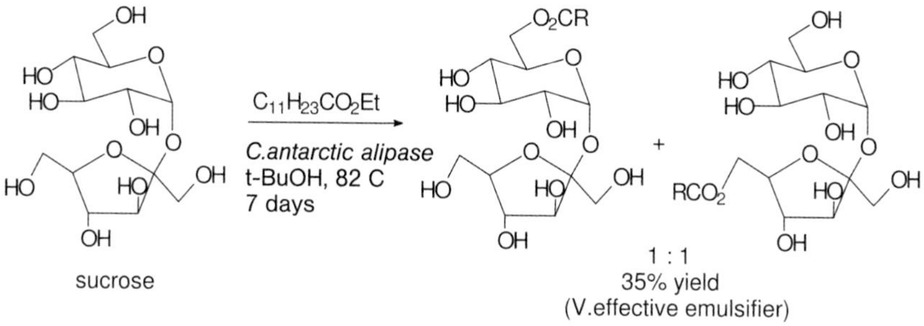

$C_{11}H_{23}CO_2Et$

C.antarctic alipase
t-BuOH, 82 C
7 days

sucrose

+

RCO$_2$

1 : 1
35% yield
(V.effective emulsifier)

Fig. 1.43 Synthesis of a bioemulsifier from renewable raw materials.

biotechnology include the biodegradable plastic, polylactate, produced by Cargill-Dow and 1,3-propanediol, a raw material for the new polyester fibre, Sorona (poly-trimethylene terephthalate) developed by DuPont/Genencor. The latter process has become a benchmark in metabolic pathway engineering [125]. Both of these processes employ corn-derived glucose as the feedstock.

Finally, an elegant example of a product derived from renewable raw materials is the bioemulsifier, marketed by Mitsubishi, which consists of a mixture of sucrose fatty acid esters. The product is prepared from two renewable raw materials – sucrose and a fatty acid – and is biodegradable. In the current process the reaction is catalysed by a mineral acid, which leads to a rather complex mixture of mono- and di-esters. Hence, a more selective enzymatic esterification (Fig. 1.43) would have obvious benefits. Lipase-catalysed acylation is possible [126] but reaction rates are very low. This is mainly owing to the fact that the reaction, for thermodynamic reasons, cannot be performed in water. On the other hand, sucrose is sparingly soluble in most organic solvents, thus necessitating a slurry process.

1.12
Enantioselective Catalysis

Another major trend in performance chemicals is towards the development of products – pharmaceuticals, pesticides and food additives, etc. – that are more targeted in their action with less undesirable side-effects. This is also an issue which is addressed by green chemistry. In the case of chiral molecules that exhibit biological activity the desired effect almost always resides in only one of the enantiomers. The other enantiomer constitutes isomeric ballast that does not contribute to the desired activity and may even exhibit undesirable side-effects. Consequently, in the last two decades there has been a marked trend towards the marketing of chiral pharmaceuticals and pesticides as enantiomerically pure compounds. This generated a demand for economical methods for the synthesis of pure enantiomers [127].

The same reasoning applies to the synthesis of pure enantiomers as to organic synthesis in general: for economic and environmental viability, processes should be atom efficient and have low E factors, that is, they should employ catalytic methodologies. This is reflected in the increasing focus of attention on enantiose-lective catalysis, using either enzymes or chiral metal complexes. Its importance was acknowledged by the award of the 2001 Nobel Prize in Chemistry to Knowles, Noyori and Sharpless for their contributions to asymmetric catalysis.

An elegant example of a highly efficient catalytic asymmetric synthesis is the Takasago process [128] for the manufacture of 1-menthol, an important flavour and fragrance product. The key step is an enantioselective catalytic isomerisa-tion of a prochiral enamine to a chiral imine (Fig. 1.44). The catalyst is a Rh-Bi-nap complex (see Fig. 1.44) and the product is obtained in 99% ee using a sub-strate/catalyst ratio of 8000; recycling of the catalyst affords total turnover num-bers of up to 300 000. The Takasago process is used to produce several thousand tons of 1-menthol on an annual basis.

An even more impressive example of catalytic efficiency is the manufacture of the optically active herbicide, (S)-metolachlor. The process, developed by No-vartis [129], involves asymmetric hydrogenation of a prochiral imine, catalysed

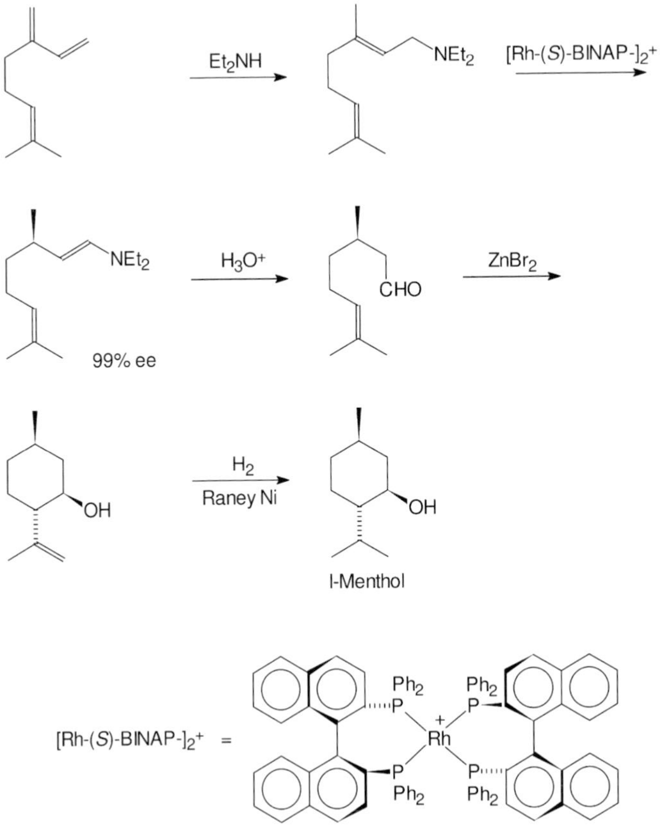

Fig. 1.44 Takasago l-menthol process.

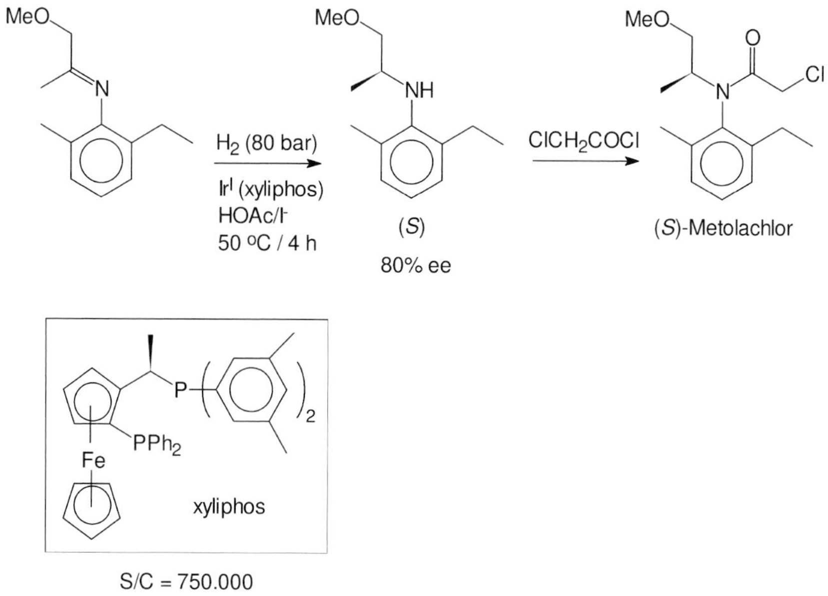

Fig. 1.45 Novartis process for (S)-metalachlor.

by an iridium complex of a chiral ferrocenyldiphosphine (Fig. 1.45). Complete conversion is achieved within 4 h at a substrate/catalyst ratio of >1 000 000 and an initial TOF exceeding 1 800 000 h^{-1}, giving a product with an ee of 80%. A higher ee can be obtained, at lower substrate/catalyst ratios, but is not actually necessary for this product. The process is used to produce several thousand tons of this optically active herbicide.

The widespread application of enantioselective catalysis, be it with chiral metal complexes or enzymes, raises another issue. These catalysts are often very expensive. Chiral metal complexes generally comprise expensive noble metals in combination with even more expensive chiral ligands. A key issue is, therefore, to minimise the cost contribution of the catalyst to the total cost price of the product; a rule of thumb is that it should not be more than ca. 5%. This can be achieved either by developing an extremely productive catalyst, as in the metachlor example, or by efficient recovery and recycling of the catalyst. Hence, much attention has been devoted in recent years to the development of effective methods for the immobilisation of metal complexes [130, 131] and enzymes [132]. This is discussed in more detail in Chapter 9.

1.13
Risky Reagents

In addition to the increasingly stringent environmental regulations with regard to the disposal of aqueous effluent and solid waste, tightened safety regulations are making the transport, storage and, hence, use of many hazardous and toxic chemicals prohibitive. The ever increasing list includes, phosgene, dimethyl sulfate, formaldehyde/hydrogen chloride (for chloromethylations), sodium azide, hydrogen fluoride, and even chlorine and bromine.

Although it will not be possible to dispense with some of these reagents entirely, their industrial use is likely to be confined to a small group of experts who are properly equipped to handle and contain these materials. In some cases, catalytic alternatives may provide an answer, such as the use of catalytic carbonylation instead of phosgene and solid acids as substitutes for hydrogen fluoride.

Chlorine-based chemistry is a case in point. In addition to the problem of salt generation, the transport of chlorine is being severely restricted. Moreover, chlorine-based routes often generate aqueous effluent containing trace levels of chlorinated organics that present a disposal problem.

Obviously, when the desired product contains a chlorine atom, the use of chlorine can be avoided only by replacing the product. However, in many cases chlorine is a reagent that does not appear in the product, and its use can be circumvented. How remarkably simple the solution can be, once the problem has been identified, is illustrated by the manufacture of a family of sulfenamides that are used as rubber additives.

Traditionally, these products were produced using a three-step, chlorine-based, oxidative coupling process (Fig. 1.46). In contrast, Monsanto scientists [133] developed a process involving one step, under mild conditions (<1 h at 70 °C). It uses molecular oxygen as the oxidant and activated charcoal as the catalyst (Fig. 1.46). The alkylaminomercaptobenzothiazole product is formed in essentially quantitative yield, and water is the coproduct. We note that activated charcoal contains various trace metals which may be the actual catalyst.

Another elegant example, also developed by Monsanto scientists [134], is the synthesis of *p*-phenylene diamine, a key raw material for aramid fibres. The traditional process involves nitration of chlorobenzene followed by reaction of the resulting *p*-nitrochlorobenzene with ammonia to give *p*-nitroaniline, which is hydrogenated to *p*-phenylenediamine. Monsanto scientists found that benzamide reacts with nitrobenzene, in the presence of a base and dioxygen, to afford 4-nitrobenzanilide. Reaction of the latter with methanolic ammonia generates *p*-nitroaniline and benzamide, which can be recycled to the first step (Fig. 1.47), resulting in an overall reaction of nitrobenzene with ammonia and dioxygen to give *p*-nitroaniline and a molecule of water. The key step in the process is the oxidation of the intermediate Meisenheimer complex by the nitrobenzene substrate, resulting in an overall oxidative nucleophilic substitution. The nitrosobenzene coproduct is reoxidised by dioxygen. Hence, a remarkable feature of the process is that no external catalyst is required; the substrate itself acts as the catalyst.

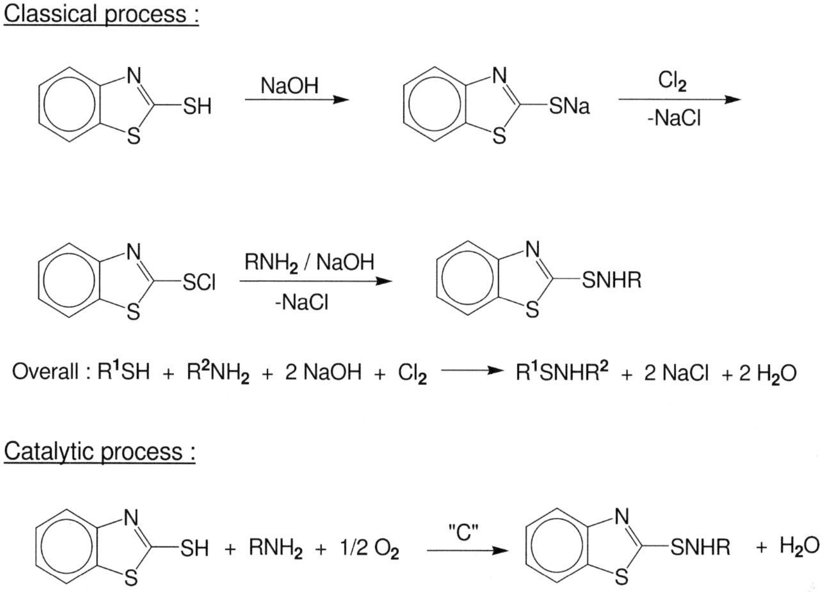
Classical process :
NaOH
Cl2
-NaCl
RNH2 / NaOH
-NaCl
Overall : R1SH + R2NH2 + 2 NaOH + Cl2 → R1SNHR2 + 2 NaCl + 2 H2O
Catalytic process :
SH + RNH2 + 1/2 O2
"C"
SNHR + H2O

Fig. 1.46 Two routes to alkylaminomercaptobenzothiazoles.

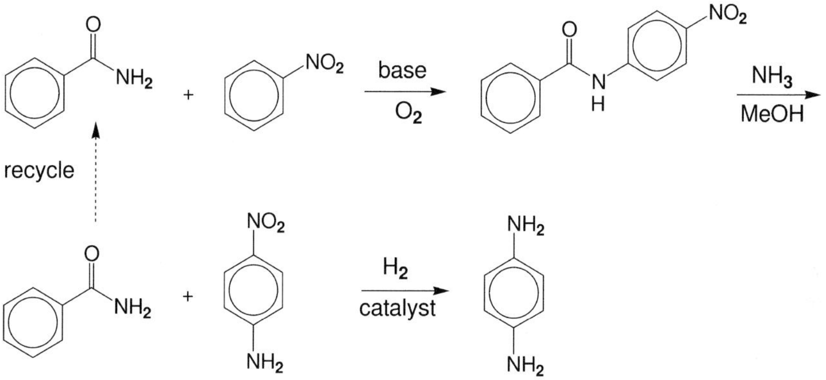
base
O2
NH3
MeOH
recycle
H2
catalyst

Fig. 1.47 Monsanto process for *p*-phenylenediamine.

1.14
Process Integration and Catalytic Cascades

The widespread application of chemo- and biocatalytic methodologies in the manufacture of fine chemicals is resulting in a gradual disappearance of the traditional barriers between the subdisciplines of homogeneous and heterogeneous catalysis and biocatalysis. An effective integration of these catalytic technologies in organic synthesis is truly the key to success.

An elegant example is the Rhodia process for the manufacture of the flavor ingredient, vanillin [30]. The process involves four steps, all performed with a heterogeneous catalyst, starting from phenol (Fig. 1.48). Overall, one equivalent of phenol, H_2O_2, CH_3OH, H_2CO and O_2 are converted to one equivalent of vanillin and three equivalents of water.

Another pertinent example is provided by the manufacture of caprolactam [135]. Current processes are based on toluene or benzene as feedstock, which can be converted to cyclohexanone via cyclohexane or phenol. More recently, Asahi Chemical [136] developed a new process via ruthenium-catalysed selective hydrogenation to cyclohexene, followed by zeolite-catalysed hydration to cyclohexanol and dehydrogenation (Fig. 1.49). The cyclohexanone is then converted to caprolactam via ammoximation with NH_3/H_2O_2 and zeolite-catalysed Beckmann rearrangement as developed by Sumitomo (see earlier).

Alternatively, caprolactam can be produced from butadiene, via homogeneous nickel-catalysed addition of HCN (DuPont process) followed by selective catalytic hydrogenation of the adiponitrile product to the amino nitrile and vapor phase hydration over an alumina catalyst (Rhodia process) as shown in Fig. 1.49 [137].

Interestingly, the by-product in the above-described hydrocyanation of butadiene, 2-methylglutaronitrile, forms the raw material for the Lonza process for nicotinamide (see earlier) [123]. Four heterogeneous catalytic steps (hydrogenation, cyclisation, dehydrogenation and ammoxidation) are followed by an enzymatic hydration of a nitrile to an amide (Fig. 1.50).

The ultimate in integration is to combine several catalytic steps into a one-pot, multi-step catalytic cascade process [138]. This is truly emulating Nature where metabolic pathways conducted in living cells involve an elegant orchestration of a series of biocatalytic steps into an exquisite multicatalyst cascade, without the need for separation of intermediates.

An example of a one-pot, three-step catalytic cascade is shown in Fig. 1.51 [139]. In the first step galactose oxidase catalyses the selective oxidation of the primary alcohol group of galactose to the corresponding aldehyde. This is fol-

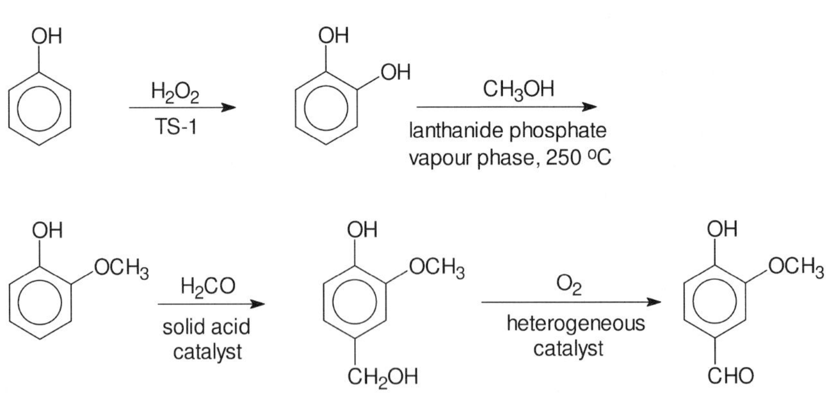

Fig. 1.48 Rhodia vanillin process.

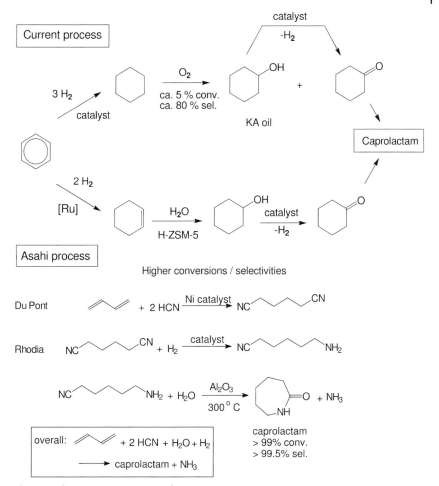

Fig. 1.49 Alternative routes to caprolactam.

lowed by L-proline catalysed elimination of water and catalytic hydrogenation, affording the corresponding deoxy sugar.

In some cases the answer may not be to emulate Nature's catalytic cascades but rather to streamline them through metabolic pathway engineering (see earlier). The elegant processes for vanillin, caprolactam and nicotinamide described above may, in the longer term, be superseded by alternative routes based on *de novo* fermentation of biomass. For many naturally occurring compounds this will represent a closing of the circle that began, more than a century ago, with the synthesis of natural products, such as dyestuffs, from raw materials derived from coal tar. It is perhaps appropriate, therefore, to close this chapter with a mention of the dyestuff indigo. Its first commercial synthesis, in the nineteenth century [140] involved classical organic chemistry and has hardly changed since

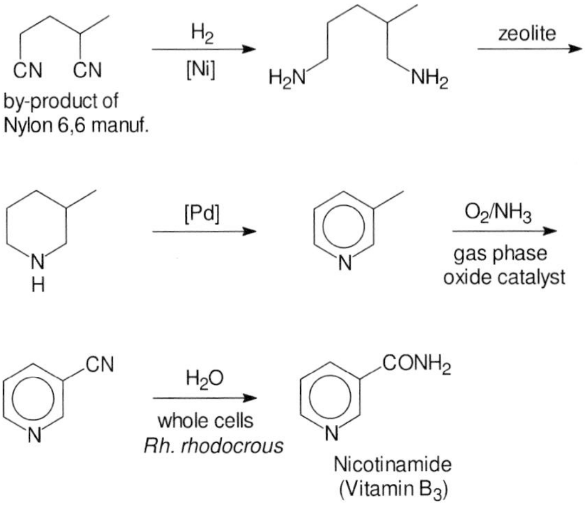

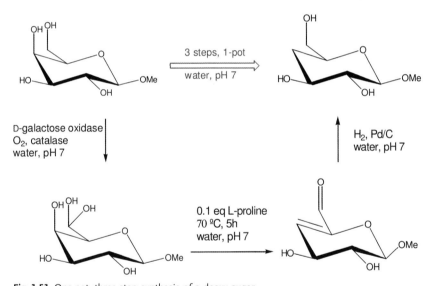

Fig. 1.50 Lonza nicotinamide process.

Fig. 1.51 One-pot, three-step synthesis of a deoxy sugar.

that time. Mitsui Toatsu reported [141] an alternative conversion of aniline to indigo in two catalytic steps. However, in the future indigo may be produced by an even greener route. Genencor [142] has developed a one-step fermenatation of glucose to indigo using a recombinant *E. coli* strain in which the tryptophan pathway has been engineered, to produce high levels of indole, and genes encoding for naphthalene dioxygenase have been incorporated. The latter enzyme catalyses the aerobic oxidation of indole to indigo. The process (see Chapter for a more detailed discussion) is not yet commercially viable, probably because of relatively low volumetric (18 g l^{-1}) and space–time yields (<1 g l^{-1} h^{-1}), but may be further optimised in the future.

References

1 P. Anastas, J.C. Warner (Eds.), *Green Chemistry: Theory and Practice*, Oxford University Press, Oxford, 1998.

2 P.T. Anastas, T.C. Williamson (Eds.), *Green Chemistry: Frontiers in Chemical Synthesis and Processes*, Oxford University Press, Oxford, 1998.

3 P.T. Anastas, M.M. Kirchhoff, *Acc. Chem. Res.* **2002**, *35*, 686-693.

4 P.T. Anastas, L.G. Heine, T.C. Williamson (Eds.), *Green Chemical Syntheses and Processes*, American Chemical Society, Washington DC, 2000.

5 P.T. Anastas, C.A. Farris (Eds.), *Benign by Design: Alternative Synthetic Design for Pollution Prevention*, ACS Symp. Ser. nr. 577, American Chemical Society, Washington DC, 1994.

6 J.H. Clark, D.J. Macquarrie, *Handbook of Green Chemistry and Technology*, Blackwell, Abingdon, 2002.

7 A.S. Matlack, *Introduction to Green Chemistry*, Marcel Dekker, New York, 2001.

8 M. Lancaster, *Green Chemistry: An Introductory Text*, Royal Society of Chemistry, Cambridge, 2002.

9 J.H. Clark (Ed.), *The Chemistry of Waste Minimization*, Blackie, London, 1995.

10 R.A. Sheldon, *C.R. Acad. Sci. Paris, IIc, Chimie/Chemistry*, **2000**, *3*, 541–551.

11 C.G. Brundtland, *Our Common Future*, The World Commission on Environmental Development, Oxford University Press, Oxford, 1987.

12 R.A. Sheldon, *Chem. Ind. (London)*, **1992**, 903–906; **1997**, 12–15.

13 R.A. Sheldon, in *Industrial Environmental Chemistry*, D.T. Sawyer, A.E. Martell (Eds.), Plenum, New York, 1992, pp. 99–119.

14 R.A. Sheldon, in *Precision Process Technology*, M.P.C. Weijnen, A.A.H. Drinkenburg (Eds.), Kluwer, Dordrecht, 1993, pp. 125–138.

15 R.A. Sheldon, *Chemtech*, March 1994, 38–47.

16 R.A. Sheldon, *J. Chem. Technol. Biotechnol.* **1997**, *68*, 381–388.

17 R.A. Sheldon, *J. Mol. Catal. A: Chemical* **1996**, *107*, 75–83.

18 R.A. Sheldon, *Pure Appl. Chem.* **2000**, *72*, 1233–1246.

19 T. Hudlicky, D.A. Frey, L. Koroniak, C.D. Claeboe, L.E. Brammer, *Green Chem.* **1999**, *1*, 57–59

20 D.J.C. Constable, A.D. Curzons, V.L. Cunningham, *Green Chem.* **2002**, *4*, 521–527; see also A.D. Curzons, D.J.C. Constable, D.N. Mortimer, V.L. Cunningham, *Green Chem.* **2001**, *3*, 1–6; D.J.C. Constable, A.D. Curzons, L.M. Freitas dos Santos, G.R. Green, R.E. Hannah, J.D. Hayler, J. Kitteringham, M.A. McGuire, J.E. Richardson, P. Smith, R.L. Webb, M. Yu, *Green Chem.* **2001**, *3*, 7–10.

21 B.M. Trost, *Science* **1991**, *254*, 1471–1477.

22 B.M. Trost, *Angew. Chem. Int. Ed.* **1995**, *34*, 259–281.

23 T. Iwata, H. Miki, Y. Fujita, in *Ullmann's Encyclopedia of Industrial Chemistry*, Vol. A19, VCH, Weinheim, 1991, p. 347.

24 P.M. Hudnall, in *Ullmann's Encyclopedia of Industrial Chemistry*, Vol. A13, VCH, Weinheim, 1991, p. 499.

25 R. Larsson (Ed.), *Perspectives in Catalysis: In Commemoration of J.J. Berzelius*, CNK Gleerup, Lund, Sweden, 1981.

26 W.H. Perkin, *J. Chem. Soc.* **1862**, 232; *Br. Pat.* **1856**, 1984.

27 K. Tanabe, W. Hölderich, *Appl. Catal. A: General* **1999**, *181*, 399–434.

28 R.A. Sheldon, R.S. Downing, *Appl. Catal. A: General* **1999**, *189*, 163–183; R.S. Downing, H. van Bekkum, R.A. Sheldon, *Cattech* **1997**, *2*, 95–109.

29 R.A. Sheldon, H. van Bekkum (Eds.), *Fine Chemicals Through Heterogeneous Catalysis*, Wiley-VCH, Weinheim, 2001, Ch. 3–7.

30 S. Ratton, *Chem. Today* **1997**, *3–4*, 33–37.

31 M. Spagnol, L. Gilbert, D. Alby, in *The Roots of Organic Development*, J. R. Desmurs, S. Ratton (Eds.), Ind. Chem. Lib., Vol. 8, Elsevier, Amsterdam, 1996, pp. 29–38.

32 H. Ichihashi, M. Kitamura, *Catal. Today* **2002**, *73*, 23–28.

33 H. Ichihashi, H. Sato, *Appl. Catal. A: General* **2001**, *221*, 359–366.

34 P. Roffia, G. Leofanti, A. Cesana, M. Mantegazza, M. Padovan, G. Petrini, S. Tonti, P. Gervasutti, *Stud. Surf. Sci. Catal.* **1990**, *55*, 43–50; G. Bellussi, C. Perego, *Cattech* **2000**, *4*, 4–16.

35 W. F. Hölderich, J. Röseler, G. Heitmann, A. T. Liebens, *Catal. Today* **1997**, *37*, 353–366.

36 P. J. Kunkeler, J. C. van der Waal, J. Bremmer, B. J. Zuurdeeg, R. S. Downing, H. van Bekkum, *Catal. Lett.* **1998**, *53*, 135–138.

37 J. A. Elings, H. E. B. Lempers, R. A. Sheldon, *Stud. Surf. Sci. Catal.* **1997**, *105*, 1165–1172.

38 M. Wegman, J. M. Elzinga, E. Neeleman, F. van Rantwijk, R. A. Sheldon, *Green Chem.* **2001**, *3*, 61–64.

39 S. Kobayashi, S. Masaharu, H. Kitagawa, L. Hidetoshi, W. L. Williams, *Chem. Rev.* **2002**, *102*, 2227–2312; W. Xe, Y. Jin, P. G. Wang, *Chemtech.* February 1999, 23–29.

40 R. A. Sheldon, H. van Bekkum (Eds.), *Fine Chemicals Through Heterogeneous Catalysis*, Wiley-VCH, Weinheim, 2001, Ch. 7.

41 B. F. Sels, D. E. De Vos, P. A. Jacobs, *Cat. Rev. Sci. Eng.* **2001**, *43*, 443–488.

42 A. Vaccari, *Catal. Today* **1998**, *41*, 53–71; A. Vaccari, *Appl. Clay Sci.* **1999**, *14*, 161–198.

43 F. Figueras, D. Tichit, M. Bennani Naciri, R. Ruiz, in *Catalysis of Organic Reactions*, F. E. Herkes (Ed.), Marcel Dekker, New York, 1998, pp. 37–49.

44 A. Corma, R. M. Martín-Aranda, *Appl. Catal. A: General* **1993**, *105*, 271–279.

45 M. J. Climent, A. Corma, S. Iborra, J. Primo, *J. Catal.* **1995**, *151* 60–66.

46 D. Brunel, A. C. Blanc, A. Galarneau, F. Fajula, *Catal. Today* **2002**, *73*, 139–152.

47 A. C. Blanc, S. Valle, G. Renard, D. Brunel, D. Macquarrie, C. R. Quinn, *Green Chem.* **2000**, *2*, 283–288.

48 D. J. Macquarrie, D. Brunel, in *Fine Chemicals Through Heterogeneous Catalysis*, R. A. Sheldon, H. van Bekkum (Eds.), Wiley-VCH, Weinheim, 2001, pp. 338–349.

49 D. Brunel, *Micropor. Mesopor. Mater.* **1999**, *27*, 329–344.

50 Y. V. Subba Rao, D. E. De Vos, P. A. Jacobs, *Angew. Chem. Int. Ed.* **1997**, *36*, 2661–2663.

51 P. N. Rylander, *Catalytic Hydrogenation over Platinum Metals*, Academic Press, New York, **1967**.

52 R. L. Augustine, *Heterogeneous Catalysis for the Synthetic Chemist*, Marcel Dekker, New York, 1996, pp. 315–343.

53 P. Sabatier, A. Mailhe, *Compt. Rend.* **1904**, *138*, 245; P. Sabatier, *Catalysis in Organic Chemistry* (translated by E. E. Reid), Van Nostrand, Princeton, 1923.

54 R. Noyori (Issue Ed.), *Adv. Synth. Catal.* **2003**, *345*, 1–324.

55 F. Roessler, *Chimia* **1996**, *50*, 106–109.

56 W. S. Knowles, R. Noyori, K. B. Sharpless, Nobel Lectures, *Angew. Chem. Int. Ed.* **2002**, *41*, 1998–2022; see also W. S. Knowles, *Adv. Synth. Catal.* **2003**, *345*, 3–13; R. Noyori, *Adv. Synth. Catal.* **2003**, *345*, 15–32.

57 H. U. Blaser, C. Malan, B. Pugin, F. Spindler, H. Steiner, M. Studer, *Adv. Synth. Catal.* **2003**, *345*, 103–151.

58 R. R. Bader, P. Baumeister, H. U. Blaser, *Chimia* **1996**, *50*, 99–105.

59 A. W. Williamson, *J. Chem. Soc.* **1852**, *4*, 106.

60 F. Fache, V. Bethmont, L. Jacquot, M. Lemaire, *Recl. Trav. Chim. Pays-Bas* **1996**, *115*, 231–238.

61 C. F. de Graauw, J. A. Peters, H. van Bekkum, J. Huskens, *Synthesis* **1994**, *10*, 1007–1017.

62 E. J. Creyghton, S. D. Ganeshie, R. S. Downing, H. van Bekkum, *J. Mol. Catal. A: Chemical* **1997**, *115*, 457–472.

63 P. J. Kunkeler, B. J. Zuurdeeg, J. C. van der Waal, J. van Bokhoven, D. C. Koningsberger, H. van Bekkum, *J. Catal.* **1998**, *180*, 234–244.

64 A. Corma, M.E. Domine, L. Nemeth, S. Valencia, *J. Am. Chem. Soc.* **2002**, *124*, 3194–3195.

65 E. Mossettig, R. Mozingo, *Org. React.* **1948**, *4*, 362.

66 T. Yokoyama, T. Setoyama, N. Fujita, M. Nakajima, T. Maki, *Appl. Catal. A: General* **1992**, *88*, 149–161.

67 R.A. Sheldon, J.K. Kochi, *Metal-Catalyzed Oxidations of Organic Compounds*, Academic Press, New York, 1981.

68 A. Chauvel, A. Delmon, W.F. Hölderich, *Appl. Catal. A: General* **1994**, *115*, 173–217.

69 A.E.J. de Nooy, A.C. Besemer, H. van Bekkum, *Synthesis* **1996**, 1153–1174.

70 B.D. Hewitt, in Ref. 2, pp. 347–360.

71 A. Dijksman, I.W.C.E. Arends, R.A. Sheldon, *Chem. Commun.* **2000**, 271–272.

72 A. Dijksman, I.W.C.E. Arends, R.A. Sheldon, *Chem. Commun.* **1999**, 1591–1592.

73 G.J. ten Brink, I.W.C.E. Arends, R.A. Sheldon, *Science* **2000**, *287*, 1636–1639.

74 B. Notari, *Stud. Surf. Sci. Catal.* **1998**, *37*, 413–425.

75 J. le Bars, J. Dakka, R.A. Sheldon, *Appl. Catal. A: General* **1996**, *136*, 69–80.

76 I.W.C.E. Arends, R.A. Sheldon, M. Wakau, U. Schuchardt, *Angew. Chem. Int. Ed. Engl.* **1997**, *36*, 1144–1163.

77 A. Corma, L.T. Nemeth, M. Renz, S. Valencia, *Nature* **2001**, *412*, 423–425.

78 A. Corma, V. Fornes, S. Iborra, M. Mifsud, M. Renz, *J. Catal.* **2004**, *221*, 67–76.

79 C. Venturello, E. Alneri, M. Ricci, *J. Org. Chem.* **1983**, *48*, 3831–3833.

80 R. Noyori, M. Aoki, K. Sato, *Chem. Commun.* **2003**, 1977–1986, and references cited therein.

81 A.K. Chatterjee, R.H. Grubbs, *Angew. Chem. Int. Ed.* **2002**, *41*, 3172–3174.

82 P.M. Maitlis, A. Haynes, G.J. Sunley, M.J. Howard, *J. Chem. Soc. Dalton Trans.* **1996**, 2187–2196; D.J. Watson, in *Catalysis of Organic Reactions*, F.E. Herkes (Ed.), Marcel Dekker, New York, 1998, pp. 369–380.

83 V. Elango, M.A. Murhpy, B.L. Smith, K.G. Davenport, G.N. Mott, G.L. Moss, US Pat. 4981995 (1991) to Hoechst-Celanese Corp.

84 M. Beller, A.F. Indolese, *Chimia* **2001**, *55*, 684–687.

85 M. Beller, M. Eckert, W.A. Moradi, H. Neumann, *Angew. Chem. Int. Ed.* **1999**, *38*, 1454–1457; D. Gördes, H. Neumann, A. Jacobi van Wangelin, C. Fischer, K. Drauz, H.P. Krimmer, M. Beller, *Adv. Synth. Catal.* **2003**, *345*, 510–516.

86 A. Zapf, M. Beller, *Top. Catal.* **2002**, *19*, 101–108.

87 C.E. Tucker, J.G. de Vries, *Top. Catal.* **2002**, *19*, 111–118.

88 R.F. Heck, *Palladium Reagents in Organic Synthesis*, Academic Press, New York, 1985, and references cited therein.

89 M.S. Stephan, A.J.J.M. Teunissen, G.K.M. Verzijl, J.G. de Vries, *Angew. Chem. Int. Ed.* **1998**, *37*, 662–664.

90 G.C. Fu, A.F. Littke, *Angew. Chem. Int. Ed.* **1998**, *37*, 3387–3388, and references cited therein.

91 W. Bernhagen, *Chim. Oggi (Chemistry Today)*, March/April 1998, 18.

92 A. Fürstner, R.H. Grubbs, R.R. Schrock (Eds.), *Olefin Metathesis Issue of Adv. Synth. Catal.* **2002**, *344* (6+7), 567–793; R.H. Grubbs (Ed.), *Handbook of Metathesis*, Wiley-VCH, Weinheim, 2003.

93 T.M. Trnka, R.H. Grubbs, *Acc. Chem. Res.* **2001**, *34*, 18–29; A. Fürstner, *Angew. Chem. Int. Ed.* **2000**, *39*, 3012–3043; S. Blechert, *Pure Appl. Chem.* **1999**, *71*, 1393–1399.

94 B. Reuben, H. Wittcoff, *J. Chem. Educ.* **1988**, *65*, 605–607.

95 S.J. Connon, M. Rivard, M. Zaja, S. Blechert, *Adv. Synth. Catal.* **2003**, *345*, 572–575.

96 W. Leitner, K.R. Seddon, P. Wasserscheid (Eds.), *Special Issue on Green Solvents for Catalysis*, *Green Chem.* **2003**, *5*, 99–284.

97 C. Jiminez-Gonzalez, A.D. Curzons, D.J.C. Constable, V.L. Cunningham, *Int. J. Life Cycle Assess.* **2004**, *9*, 115–121.

98 G.P. Taber, D.M. Pfistere, J.C. Colberg, *Org. Proc. Res. Dev.* **2004**, *8*, 385–388; A.M. Rouhi, *C&EN*, April 2002, 31–32.

99 P.J. Dunn, S. Galvin, K. Hettenbach, *Green, Chem.* **2004**, *6*, 43–48.

100 S. Kobayashi (Ed.), *Special Issue on Water, Adv. Synth. Catal.* **2002**, *344*, 219–451.

101 G. Papadogianakis, R. A. Sheldon, in *Catalysis*, Vol. 13, Specialist Periodical Report, Royal Society of Chemistry, Cambridge, 1997, pp. 114–193.

102 B. Cornils, W. A. Herrmann (Eds.), *Aqueous Phase Organometallic Catalysis. Concepts and Applications*, Wiley-VCH, Weinheim, 1998.

103 B. Cornils, E. Wiebus, *Recl. Trav. Chim. Pays-Bas* **1996**, *115*, 211.

104 C. Mercier, P. Chabardes, *Pure Appl. Chem.* **1994**, *66*, 1509.

105 C. W. Kohlpaintner, M. Beller, *J. Mol. Catal. A: Chemical* **1997**, *116*, 259–267.

106 G. Papadogianakis, L. Maat, R. A. Sheldon, *J. Mol. Catal. A: Chemical* **1997**, *116*, 179–190.

107 G. Papadogianakis, L. Maat, R. A. Sheldon, *J. Chem. Technol. Biotechnol.* **1997**, *70*, 83–91.

108 G. Papadogianakis, G. Verspui, L. Maat, R. A. Sheldon, *Catal. Lett.* **1997**, *47*, 43–46.

109 P. Licence, J. Ke, M. Sokolova, S. K. Ross, M. Poliakoff, *Green Chem.* **2003**, *5*, 99–104.

110 R. A. Sheldon, *Chem. Commun.* **2001**, 2399–2407.

111 R. D. Rogers, K. R. Seddon (Eds.), *Ionic Liquids as Green Solvents. Progress and Prospects*, ACS Symp. Ser. **2003**, *856*.

112 P. Wasserscheid, T. Welton (Eds.), *Ionic Liquids in Synthesis*, Wiley-VCH, Weinheim, 2003.

113 A. J. J. Straathof, S. Panke, A. Schmid, *Curr. Opin. Biotechnol.* **2002**, *13*, 548–556.

114 K. A. Powell, S. W. Ramer, S. B. del Cardayré, W. P. C. Stemmer, M. B. Tobin, P. F. Longchamp, G. W. Huisman, *Angew. Chem. Int. Ed.* **2001**, *40*, 3948–3959.

115 M. A. Wegman, M. H. A. Janssen, F. van Rantwijk, R. A. Sheldon, *Adv. Synth. Catal.* **2001**, *343*, 559–576; A. Bruggink, E. C. Roos, E. de Vroom, *Org. Proc. Res. Dev.* **1998**, *2*, 128–133.

116 *Ullmann's Encyclopedia of Industrial Chemistry*, 5th edn., Vol. B8, VCH, Weinheim, 1995, pp. 302–304.

117 K. Oyama, in *Chirality in Industry*, A. N. Collins, G. N. Sheldrake, J. Crosby (Eds.), Wiley, New York, 1992, pp. 237.

118 M. Petersen, A. Kiener, *Green Chem.* **1999**, *1*, 99–106.

119 J. E. Gavagan, S. K. Fager, J. E. Seip, M. S. Payne, D. L. Anton, R. DiCosimo, *J. Org. Chem.* **1995**, *60*, 3957–3963.

120 W. Stampfer, B. Kosjek, C. Moitzi, W. Kroutil, K. Faber, *Angew. Chem. Int. Ed.* **2002**, *41*, 1014–1017.

121 M. Kobayashi, T. Nagasawa, H. Yamada, *Trends Biotechnol.* **1992**, *1*, 402–408; M. Kobayashi, S. Shimizu, *Curr. Opin. Chem. Biol.* **2000**, *4*, 95–102.

122 E. C. Hann, A. Eisenberg, S. K. Fager, N. E. Perkins, F. G. Gallagher, S. M. Cooper, J. E. Gavagan, B. Stieglitz, S. M. Hennessey, R. DiCosimo, *Bioorg. Med. Chem.* **1999**, *7*, 2239–2245.

123 J. Heveling, *Chimia* **1996**, *50*, 114.

124 B. E. Dale, *J. Chem. Technol. Biotechnol.* **2003**, *78*, 1093–1103.

125 C. E. Nakamura, G. M. Whited, *Curr. Opin. Biotechnol.* **2003**, *14*, 454–459.

126 M. Woudenberg-van Oosterom, F. van Rantwijk, R. A. Sheldon, *Biotechnol. Bioeng.* **1996**, *49*, 328–333.

127 R. A. Sheldon, *Chirotechnology: The Industrial Synthesis of Optically Active Compounds*, Marcel Dekker, New York, 1993.

128 H. Komobayashi, *Recl. Trav. Chim. Pays-Bas* **1996**, *115*, 201–210.

129 H. U. Blaser, *Adv. Synth. Catal.* **2002**, *344*, 17–31; see also H. U. Blaser, W. Brieden, B. Pugin, F. Spindler, M. Studer, A. Togni, *Top. Catal.* **2002**, *19*, 3–16.

130 D. E. De Vos, I. F. J. Vankelecom, P. A. Jacobs, *Chiral Catalyst Immobilization and Recycling*, Wiley-VCH, Weinheim, 2000.

131 U. Kragl, T. Dwars, *Trend Biotechnol.* **2001**, *19*, 442–449.

132 L. Cao, L. M. van Langen, R. A. Sheldon, *Curr. Opin. Biotechnol.* **2003**, *14*, 387–394.

133 D. Riley, M. K. Stern, J. Ebner, in *The Activation of Dioxygen and Homogeneous Catalytic Oxidation*, D. H. R. Barton,

A. E. Martell, D. T. Sawyer (Eds.), Plenum Press, New York, 1993, p. 31–44.

134 M. K. Stern, F. D. Hileman, J. K. Bashkin, *J. Am. Chem. Soc.* **1992**, *114*, 9237–9238.

135 G. Dahlhoff, J. P. M. Niederer, W. F. Hoelderich, *Catal. Rev.* **2001**, *43*, 381–441.

136 H. Ishida, *Catal. Surveys Jpn.* **1997**, *1*, 241–246.

137 L. Gilbert, N. Laurain, P. Leconte, C. Nedez, *Eur. Pat. Appl.*, EP 0748797, **1996**, to Rhodia.

138 A. Bruggink, R. Schoevaart, T. Kieboom, *Org. Proc. Res. Dev.* **2003**, *7*, 622–640.

139 R. Schoevaart, T. Kieboom, *Tetrahedron Lett.* **2002**, *43*, 3399–3400.

140 A. Bayer, *Chem. Ber.* **1878**, *11*, 1296; K. Heumann, *Chem. Ber.* **1890**, *23*, 3431.

141 Y. Inoue, Y. Yamamoto, H. Suzuki, U. Takaki, *Stud. Surf. Sci. Catal.* **1994**, *82*, 615.

142 A. Berry, T. C. Dodge, M. Pepsin, W. Weyler, *J. Ind. Microbiol. Biotechnol.* **2002**, *28*, 127–133.

2
Solid Acids and Bases as Catalysts

2.1
Introduction

Processes catalyzed by acids and bases play a key role in the oil refining and petrochemical industries and in the manufacture of a wide variety of specialty chemicals such as pharmaceuticals, agrochemicals and flavors and fragrances. Examples include catalytic cracking and hydrocracking, alkylation, isomerization, oligomerization, hydration/dehydration, esterification and hydrolysis and a variety of condensation reactions, to name but a few [1, 2]. Many of these processes involve the use of traditional Brønsted acids (H_2SO_4, HF, HCl, p-toluenesulfonic acid) or Lewis acids ($AlCl_3$, $ZnCl_2$, BF_3) in liquid-phase homogeneous systems or on inorganic supports in vapor phase systems. Similarly, typical bases include NaOH, KOH, NaOMe and KOBu^t. Their subsequent neutralization leads to the generation of inorganic salts which ultimately end up in aqueous waste streams. Even though only catalytic amounts are generally, but not always, used in the oil refining and petrochemical industries, the absolute quantities of waste generated are considerable owing to the enormous production volumes involved. In contrast, in the fine and specialty chemical industries the production volumes are much smaller but acids and bases are often used in stoichiometric quantities, e.g. in Friedel-Crafts acylations and aldol and related condensations, respectively [3].

An obvious solution to this salt generation problem is the widespread replacement of traditional Brønsted and Lewis acids with recyclable solid acids and bases [3–7]. This will obviate the need for hydrolytic work-up and eliminate the costs and environmental burden associated with the neutralization and disposal of e.g. liquid acids such as H_2SO_4. The use of solid acids and bases as catalysts provides additional benefits:

- Separation and recycling is facilitated, resulting in a simpler process, which translates to lower production costs.
- Solid acids are safer and easier to handle than their liquid counterparts, e.g. H_2SO_4, HF, that are highly corrosive and require expensive construction materials.
- Contamination of the product by trace amounts of (neutralized) catalyst is generally avoided when the latter is a solid.

Green Chemistry and Catalysis. I. Arends, R. Sheldon, U. Hanefeld
Copyright © 2007 WILEY-VCH Verlag GmbH & Co. KGaA, Weinheim
ISBN: 978-3-527-30715-9

2.2
Solid Acid Catalysis

Acid-catalyzed processes constitute one of the most important areas for the application of heterogeneous catalysis. A wide variety of solid catalysts are used. They include mixed oxides such as silica–alumina and sulfated zirconia, acidic clays [8–10], zeolites [11–14], supported heteropoly acids [15], organic ion exchange resins [16, 17] and hybrid organic–inorganic materials such as mesoporous oxides containing pendant organic sulfonic acid moieties [18, 19]. For the purpose of our further discussion of this topic we can conveniently divide solid acid catalysts into three major categories: amorphous mixed oxides typified by the acidic clays, the crystalline zeolites and related materials (zeotypes), and solid acids containing surface sulfonic acid groups. The latter category embraces organic and inorganic cationic exchange resins and the above-mentioned hybrid organic–inorganic materials. Zeolites definitely occupy center stage, and will be the major focus of our discussion, but the use of (acidic) clays as catalysts is of earlier vintage so we shall begin with this interesting class of materials.

2.2.1
Acidic Clays

Clays are naturally occurring minerals that are produced in enormous quantities and find a wide variety of applications including their use as catalysts [8–10, 20, 21]. They were widely used as solid acid catalysts in oil refining from the 1930s until the mid 1960s when they were replaced by the zeolites which exhibited better activity and selectivity.

Clays are amorphous, layered (alumino)silicates in which the basic building blocks – SiO_4 tetrahedra and MO_6 octahedra ($M = Al^{3+}$, Mg^{2+}, Fe^{3+}, Fe^{2+}, etc.) – polymerize to form two-dimensional sheets [8–10, 20]. One of the most commonly used clays is montmorillonite in which each layer is composed of an octahedral sheet sandwiched between two tetrahedral silicate sheets (see Fig. 2.1). Typically, the octahedral sheet comprises oxygens attached to Al^{3+} and some lower valence cations such as Mg^{2+}. The overall layer has a net negative charge which is compensated by hydrated cations occupying the interlamellar spaces. Immersion in water results in a swelling of the clay and exposure of the intercalated cations making them accessible for cation exchange. The interlamellar cations are largely responsible for the clay's Brønsted and/or Lewis acidity. The more electronegative the cation, the stronger the acidity and both the amount and strength of Brønsted and Lewis acid sites can be enhanced by cation exchange or treatment with a mineral acid, e.g. H_2SO_4.

An Al^{3+}-exchanged montmorillonite, for example, is as active as concentrated sulfuric acid in promoting acid-catalyzed reactions. Sulfuric acid treatment of natural montmorillonite similarly affords the much more active and widely used acid catalyst known as K-10 or KSF (from Sud-Chemie or Fluka, respectively).

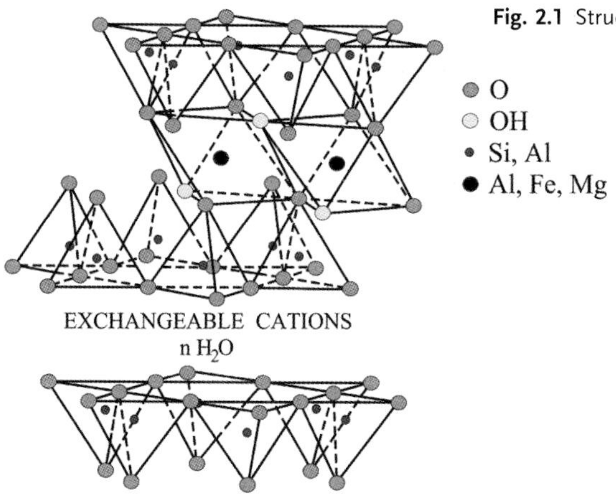

Fig. 2.1 Structure of montmorillonite clay.

○● O
○ OH
• Si, Al
● Al, Fe, Mg

EXCHANGEABLE CATIONS
n H$_2$O

For example, K-10 has been successfully used as a Friedel-Crafts alkylation catalyst (see Fig. 2.2) [22].

Clays or acid-treated clays are also effective supports for Lewis acids such as
ZnCl$_2$ or FeCl$_3$ [23]. Montmorillonite-supported zinc chloride, known as Clayzic,
has been extensively studied as a catalyst for e.g. Friedel-Crafts alkylations [24,
25] (see Fig. 2.2).

A serious shortcoming of clays, however, is their limited thermal stability. Heating of exchanged clays results in a loss of water, accompanied by collapse of the
interlamellar region, thereby dramatically decreasing the effective surface area.
This problem was addressed by developing so-called pillared clays (PILCs) [26–
28], in which the layered structure is intercalated with pillaring agents which
act as 'molecular props'. Inorganic polyoxocations such as $[Al_{13}O_4(OH)_{24}$
$(H_2O)_{12}]^{7+}$ are popular pillaring agents but a variety of organic and organometallic

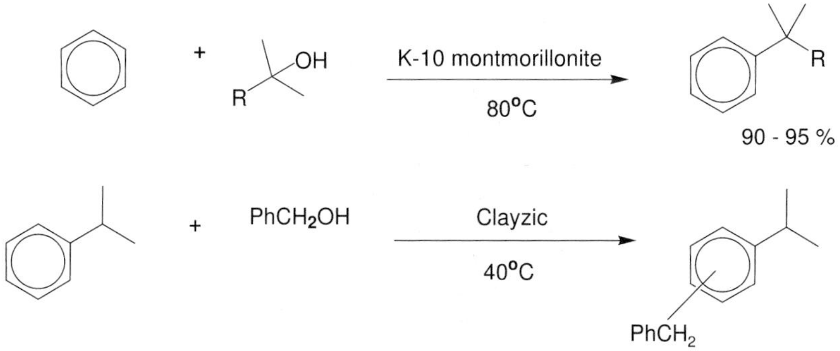

Fig. 2.2 Friedel-Crafts alkylations.

pillaring agents have also been used. Pillaring with Al_{13} provides an interlamellar space of ca. 0.8 nm, which remains after drying. The major goal of the pillaring process is to produce novel, inexpensive materials with properties (pore shape and size, acidity, etc.) complementary to zeolites (see next section).

2.2.2
Zeolites and Zeotypes: Synthesis and Structure

Zeolites are crystalline aluminosilicates comprising corner-sharing SiO_4 and AlO_4^- terahedra and consisting of a regular system of pores (channels) and cavities (cages) with diameters of molecular dimensions (0.3 to 1.4 nm) [11–14]. A large number of zeolites are known, some of which are naturally occurring, but most of which have been synthesized [29]. Analogous structures containing TO_4 tetrahedra composed of Si, Al or P as well as other main group and transition elements, e.g. B, Ga, Fe, Ge, Ti, V, Cr, Mn and Co, have also been synthesised and are generically referred to as zeotypes. They include, for example, AlPOs, SAPOs and MeAPOs [30]. As with amorphous aluminosilicates, zeolites contain an extraframework cation, usually Na^+, to maintain electroneutrality with the AlO_4^- moiety. These extraframework cations can be replaced by other cations by ion exchange. Since zeolites react with acids the proton-exchanged species (H-form) is best prepared by ion exchange with an ammonium ion followed by thermal dissociation, affording ammonia and the acid form of the zeolite. The Brønsted acid strength of the latter, which is of the same order as that of concentrated sulfuric acid, is related to the Si/Al ratio. Since an AlO_4^- moiety is unstable when attached to another AlO_4^- unit, it is necessary that they are separated by at least one SiO_4 unit. i.e. the Si/Al ratio cannot be lower than one (the H-form is depicted in Fig. 2.3). The number of proton donor hydroxy groups corresponds to the number of aluminum atoms present in the structure. The more isolated this silanol species, the stronger the acid, i.e. the acid strength increases with decreasing aluminum content or increasing Si/Al ratio (but note that complete replacement of aluminum affords a material with lower acidity).

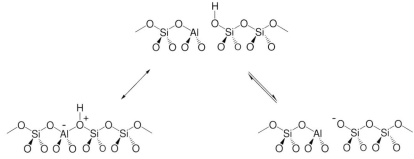

Fig. 2.3 The acid form of zeolites.

Zeolites are prepared by so-called hydrothermal synthesis a simplified scheme of which is shown in Fig. 2.4. The basic ingredients – SiO_2, Na_2SiO_3 or $Si(OR)_4$ and Al_2O_3, $NaAlO_2$ or $Al(OR)_3$ – together with a structure directing agent (template), usually an amine or tetraalkylammonium salt, are added to aqueous alkali (pH 8–12). This results in the formation of a sol–gel comprising monomeric and oligomeric silicate species. Gradual heating of this mixture up to ca. 200 °C results in dissolution of the gel to form clusters of SiO_4/AlO_4^- units which constitute the building blocks for the zeolite structure. In the presence of the template these building blocks undergo polymerization to form the zeolite which crystallizes slowly from the reaction mixture. At this point the zeolite still contains the occluded organic template. This is subsequently removed by calcination, i.e. heating in air at 400–600 °C to burn out the template and evaporate water.

Similarly, zeotype molecular sieves are synthesized by mixing the basic ingredients with the organic template, e.g. aluminophosphates are prepared from alumina and phosphoric acid. Other main group or transition elements can be incorporated into the framework by adding them to the initial sol–gel. Alternatively, different elements can be introduced by post-synthesis modification (see later), e.g. by dealumination followed by insertion of the new elements into the framework position [31].

The most important feature of zeolites (and zeotypes), in the context of catalysis, is not their range of acid–base properties, since that is also available with amorphous alumino-silicates. It is the presence of a regular structure containing

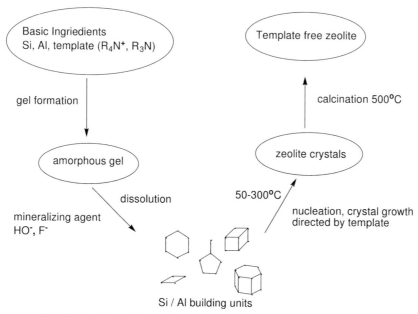

Fig. 2.4 Simplified zeolite synthesis scheme.

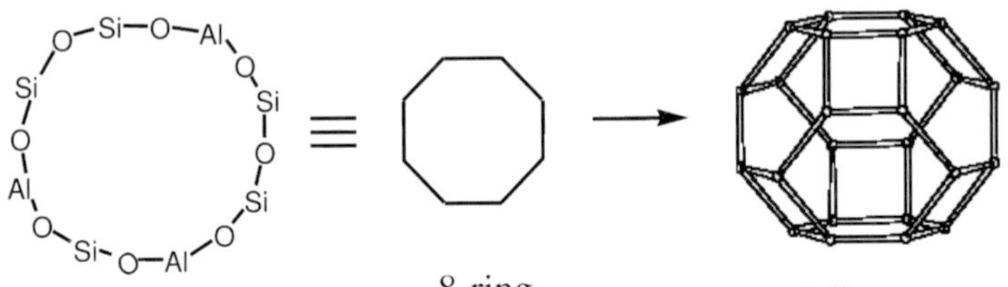

8-ring
0.35-0.45 nm

sodalite cage

Fig. 2.5 Basic units in zeolites.

molecular sized cavities and channels that make them unique as shape selective catalysts for a wide variety of organic transformations.

Representation of zeolite structures is different to that used by organic chemists. The intersection of lines represents either an SiO_4 or an AlO_4^- tetrahedron and the line itself represents an O atom joining the two tetrahedra. In Fig. 2.5 eight TO_4 octahedra are joined together in a ring, commonly referred to as the 8-ring structure as there are eight oxygen atoms in the ring. These basic ring structures are combined to form three dimensional arrangements which constitute the building blocks for the zeolite. For example, one of these building units is the sodalite cage, a truncated octahedron with four-membered rings made during truncation and six-membered rings as part of the original octahedron.

In zeolites derived from the sodalite unit these cages are joined together through extensions of either the 4-ring or the 6-ring. Zeolite A (see Fig. 2.6) is an example of the former. The center of the structure comprises a supercage, with a diameter of 1.14 nm, surrounded by eight sodalite cages. Access to these cages is via the six mutually perpendicular 8-ring openings having a diameter of 0.42 nm, enabling linear hydrocarbons to enter. The Si/Al ratio is 1–1.2 making zeolite A among the least acidic of all zeolites.

Faujasites are naturally occurring zeolites composed of sodalite cages joined through extensions of their 6-ring faces. An internal supercage of ca. 1.3 nm diameter is accessed by four 12-ring openings with a diameter of 0.74 nm. The latter provide access for relatively large aliphatic and aromatic molecules, e.g. naphthalene. The synthetic zeolites X and Y have the same crystal structure as faujasite but differ in their Si/Al ratios. In zeolites X and Y the Si/Al ratio is 1–1.5 and 1.5–3, respectively, while faujasite has a Si/Al ratio of ca. 2.2.

Mordenite is a naturally occurring zeolite with a Si/Al ratio of ca. 10 and a structure composed of 12-ring and 8-ring tunnels with diameters of 0.39 nm and ca. 0.7 nm, respectively, extending through the entire framework (Fig. 2.7). Every framework atom forms a part of the walls of these tunnels and is accessible to substrate molecules diffusing through them.

The synthetic zeolite, ZSM-5, is a highly siliceous material with a Si/Al ratio from 25 up to 2000 [32]. As shown in Fig. 2.7, it consists of a three-dimensional network of two intersecting 10-ring tunnel systems of 0.55–0.6 nm diameter. One of these resembles the 12-ring tunnels passing through mordenite (see

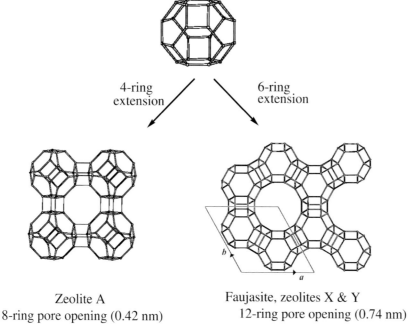

4-ring
extension

6-ring
extension

Zeolite A
8-ring pore opening (0.42 nm)

Faujasite, zeolites X & Y
12-ring pore opening (0.74 nm)

Fig. 2.6 Structures of zeolite A and faujasite.

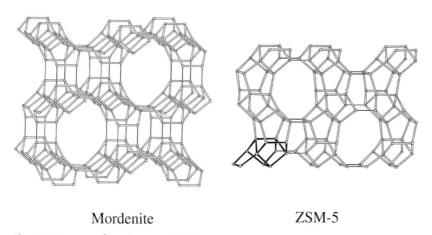

Mordenite   ZSM-5

Fig. 2.7 Structures of mordenite and ZSM-5.

above) while the other follows a sinusoidal path perpendicular to, and intersecting with, the first. The basic structural unit of this zeolite is the pentasil unit composed of fused 5-rings (Fig. 2.8).

Zeolites are conveniently divided into groups on the basis of their pore sizes – small, medium, large and ultra large – which are determined by the ring size

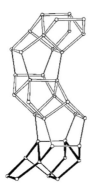

Fig. 2.8 The pentasil unit.

of the pore openings. Examples of commonly used zeolites and zeotypes are collected in Table 2.1.

Obviously the pore size determines which molecules can access the acidic sites inside the zeolite framework (molecular sieving effect) and is responsible for the shape selectivity observed with these materials (see later). The catalytic activity is also influenced by the acid strength of these sites which is determined by the Si/Al ratio (see above). The latter can be increased by post-synthesis removal of Al atoms. Dealumination can be achieved by treatment with a

Table 2.1 Pore dimensions of molecular sieves.

Ring size	Structure type	Isotopic framework structures	Pore size (nm)	Dimensionality
Small pore				
8	LTA	Zeolite A	4.1	3
Medium pore				
10	MFI	ZSM-5 silicalite	5.6×5.3	3
10	MEL	ZSM-11	5.4×5.3	3
10	AEL	AlPO-11 SAPO-11	6.3×3.9	1
Large pore				
12	FAU	X, Y	7.4	3
12	MOR	Mordenite	7.0×6.5	2
12	BEA	Beta	7.5×5.7	3
12	LTL	L	7.1	1
12	AFI	AlPO-5	7.3	1
Extra large pore				
14		AlPO-8	8.7×7.9	1
18	VFI	VPI-5	12.1	1
20		Cloverite	13.2×6.0	3
Unknown	M41S	MCM-41	40–100	1

strong acid or a chelating agent, such as EDTA, or by steaming. It results in the formation of 'silanol nests' with retention of the framework structure (Fig. 2.9).

The wide range of pore sizes available, coupled with their tunable acidity, endows the zeolites with unique properties as tailor-made (acid) catalysts. These important features of zeolites (and zeotypes) may be summarized as follows:
- regular microenvironment and uniform internal structure
- large internal surface area
- pores of molecular dimensions (shape selectivity)
- control of pore size and shape by choice of template and/or post synthesis modification
- control of pore hydrophobicity/hydrophilicity
- control of acidity by adjusting constitution (Si/Al ratio), ion exchange or post-synthesis modification
- framework substitution by transition elements
- the presence of strong electric fields within the confined space of the channels and cavities can serve to activate substrate molecules

The earliest applications of zeolites utilized the molecular sieving properties of small pore zeolites, e.g. zeolite A, in separation and purification processes such as drying and linear/branched alkane separation [33]. In 1962 Mobil Oil introduced the use of synthetic zeolite X, an FCC (fluid catalytic cracking) catalyst in oil refining. In the late sixties the W. R. Grace company introduced the "ultra-

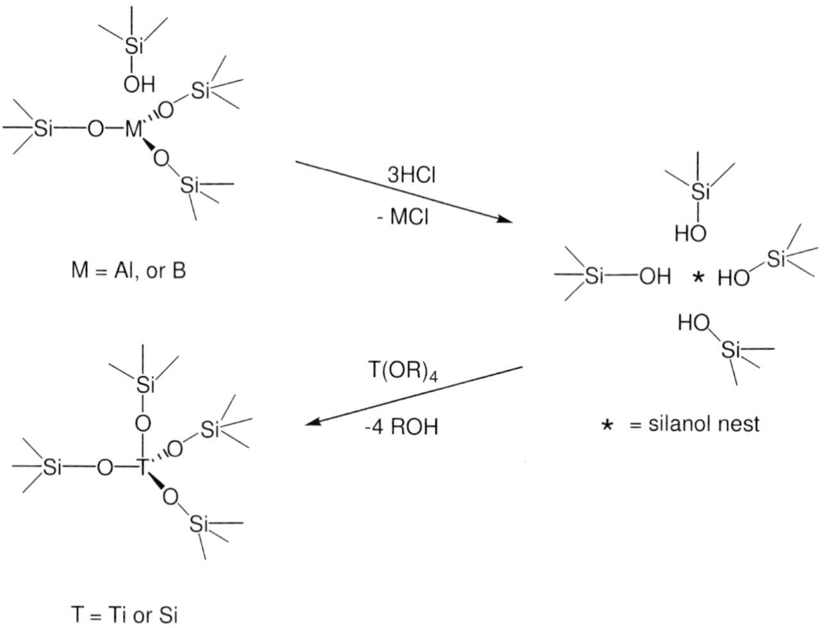

Fig. 2.9 Formation of a silanol nest by dealumination.

stable" zeolite Y (USY) as an FCC catalyst, produced by steaming of zeolite Y, which is still used today. The use of zeolite A as a replacement for phosphates in detergents was first introduced by Henkel in 1974. Although their synthesis had already been reported by Mobil in the late 1960s, extensive studies of catalytic applications of the high-silica zeolites, typified by ZSM-5 and beta, were not conducted until the 1980s and 1990s. ZSM-5 was first applied as an FCC additive to generate high octane gasoline by taking advantage of its shape selectivity in cracking linear rather than branched alkanes. ZSM-5 was subsequently applied as a shape selective acid catalyst in a wide variety of organic transformations (see later).

The 1980s also witnessed the explosive development of a wide variety of zeotypes, notably the aluminophosphate (AlPO) based family of molecular sieves [30]. The early 1990s saw the advent of a new class of mesoporous molecular sieves, the ordered mesoporous (alumino)silicates synthesized with the help of surfactant micelle templates (see Fig. 2.10) [34–37]. Exemplified by the Mobil M415 materials, of which MCM-41 is the most well-known, these micelle-templated silicas contain uniform channels with tunable diameters in the range of 1.5 to 10 nm and a greater uniformity of acid sites than other amorphous materials. According to the IUPAC definition, microporous materials are those having pores <2 nm in diameter and mesoporous solids those with pore diameters in the range 2 to 50 nm.

As noted above, one of the most important characteristics of zeolites is their ability to discriminate between molecules solely on the basis of their size. This feature, which they share with enzymes, is a consequence of them having pore dimensions close to the kinetic diameter of many simple organic molecules. Hence, zeolites and zeotypes have sometimes been referred to as mineral enzymes. This so-called shape selectivity can be conveniently divided into three categories: substrate selectivity, product selectivity and transition state selectivity. Examples of each type are shown in Fig. 2.11.

In *substrate selectivity*, access to the catalytically active site is restricted to one or more substrates present in a mixture, e.g. dehydration of a mixture of n-butanol and isobutanol over the small pore zeolite, CaA, results in dehydration of only the n-butanol [38] while the bulkier isobutanol remains unreacted. *Product*

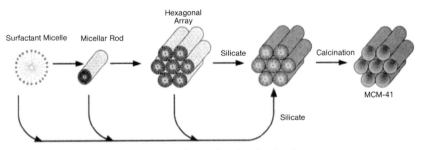

Fig. 2.10 Synthesis of mesoporous, micelle-templated molecular sieves.

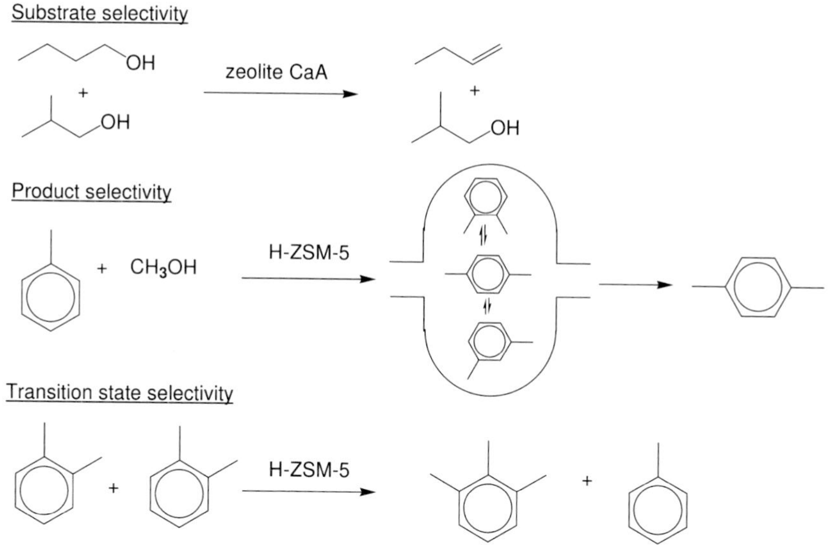

Substrate selectivity

Product selectivity

Transition state selectivity

Fig. 2.11 Shape selective catalysis with zeolites.

selectivity is a result of differences in the size of products formed, in a reversible process, inside a molecular sieve. For example, when methanol is allowed to react with toluene over H-ZSM-5, *p*-xylene is formed almost exclusively because this molecule can easily diffuse out of the molecular sieve. The bulkier *o*- and *m*-isomers, in contrast, cannot pass easily through the pores and consequently undergo isomerization, via demethylation, to the *p*-isomer. *Restricted transition state selectivity* is observed when the zeolite discriminates between two different transformations on the basis of the bulk of their transition states. In the example shown, the disproportionation of *o*-xylene to trimethylbenzene and toluene involves a bulky diaryl species in the transition state whereas isomerization to *m*- or *p*-xylene does not. Hence, disproportionation is not observed over H-ZSM-5. Most applications of zeolites as acid catalysts involve one or more of these types of shape selectivity, as we shall see in the following section.

2.2.3
Zeolite-catalyzed Reactions in Organic Synthesis

Pertinent examples of zeolite-catalyzed reactions in organic synthesis include Friedel-Crafts alkylations and acylations and other electrophilic aromatic substitutions, additions and eliminations, cyclizations, rearrangements and isomerizations, and condensations.

2.2.3.1 Electrophilic Aromatic Substitutions

Friedel-Crafts alkylations are widely used in both the bulk and fine chemical industries. For example, ethylbenzene (the raw material for styrene manufacture)
is manufactured by alkylation of benzene with ethylene (Fig. 2.12).

The original process, developed in the 1940s, involved traditional homogeneous catalysis by $AlCl_3$. This process was superseded by one employing a heterogeneous catalyst consisting of H_3PO_4 or BF_3 immobilized on a support
(UOP process). This system is highly corrosive and, because of the enormous
production volumes involved, generates substantial amounts of acidic waste. A
major breakthrough in FC alkylation technology was achieved in 1980 with the
application of the medium pore zeolite, H-ZSM-5, as a stable recyclable catalyst
for ethylbenzene manufacture (Mobil-Badger process). An added benefit of this
process is the suppression of polyalkylation owing to the shape selective properties of the catalyst.

This process marked the beginning of an era of intensive research, which
continues to this day, on the application of zeolites in the manufacture of petrochemicals and, more recently, fine chemicals.

Zeolite-based processes have gradually displaced conventional ones, involving
supported H_3PO_4 or $AlCl_3$ as catalysts, in the manufacture of cumene, the raw
material for phenol production [1, 6, 39]. A three-dimensional dealuminated
mordenite (3-DDM) catalyst was developed by Dow Chemical for this purpose
[39]. Dealumination, using a combination of acid and thermal treatments, increases the Si/Al ratio from 10–30 up to 100–1000 and, at the same time,
changes the total pore volume and pore-size distribution of the mordenite. The

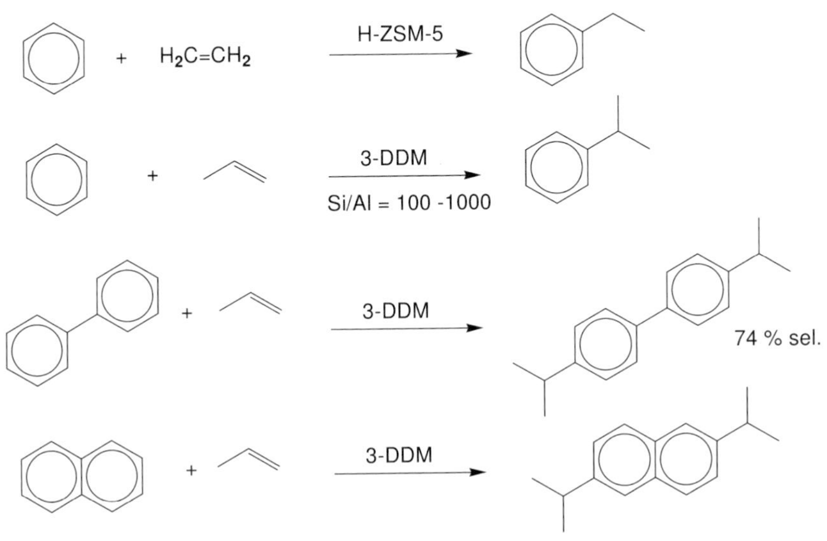

3-DDM = 3-dimensional dealuminated Mordenite (Si/Al = 100 - 1000)

Fig. 2.12 Zeolite-catalyzed Friedel-Crafts alkylations.

3-DDM catalysts have a new pore structure consisting of crystalline domains of 8- and 12-ring pores connected by mesopores (5–10 nm). The presence of the latter enhances accessibility to the micropore regions without seriously compromising the shape-selective character of the catalyst. This combination of changes in acidity and pore structure transforms synthetic mordenites into highly active, stable and selective alkylation catalysts.

Dow Chemical also pioneered the shape-selective dialkylation of polyaromatics, e.g. naphthalene and biphenyl (see Fig. 2.12), using the same type of 3-DDM catalyst [39]. The products are raw materials for the production of the corresponding dicarboxylic acids, which are important industrial monomers for a variety of high performance plastics and fibres.

Zeolites are also finding wide application as catalysts for FC alkylations and related reactions in the production of fine chemicals. For example, a H-MOR type catalyst with a Si/Al ratio of 18 was used for the hydroxyalkylation of guaiacol (Fig. 2.13) to p-hydroxymethylguaiacol, the precursor of vanillin [40]. Aromatic hydroxyalkylation with epoxides is of importance in the production of fine chemicals, e.g. the fragrance 2-phenylethanol from benzene and ethylene oxide (Fig. 2.13). Attempts to replace the conventional process, which employs stoichiometric amounts of AlCl$_3$, with a zeolite-catalyzed conversion, met with little success [6, 41]. This was largely due to the large polarity difference between the aromatic substrate and the epoxide alkylating agent which leads to unfavorable adsorption ratios between substrate and reagent. This adsorption imbalance is avoided by having the aromatic and epoxide functions in the same molecule, i.e. in an intramolecular hydroxyalkylation. For example, H-MOR and H-Beta were shown to be effective catalysts for the cyclialkylation of 4-phenyl-1-butene oxide (Fig. 2.13) [41].

The Pechmann synthesis of coumarins via condensation of phenols with β-keto esters also involves an intramolecular hydroxyalkylation, following initial

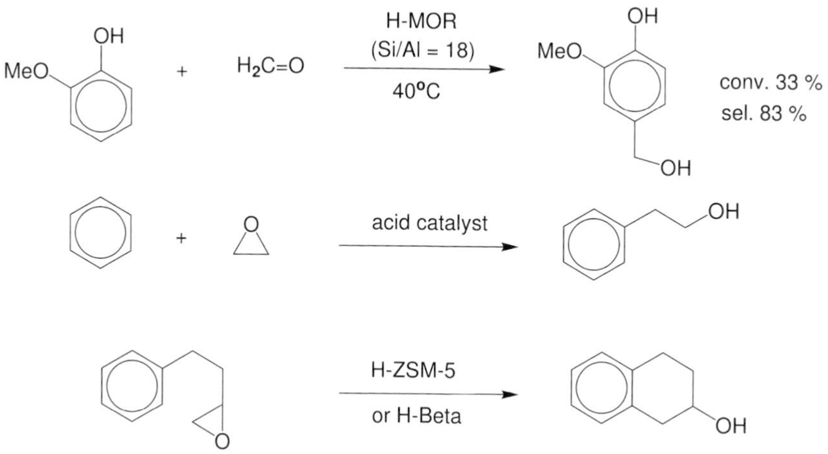

Fig. 2.13 Hydroxyalkylation of aromatics.

transesterification, and subsequent dehydration. It was found that H-Beta could successfully replace the sulfuric acid conventionally used as catalyst. For example, reaction of resorcinol with ethyl acetoacetate afforded methylumbelliferone (Fig. 2.14), a perfumery ingredient and insecticide intermediate [42]. Other examples of the synthesis of coumarin derivatives via zeolite-catalyzed intramolecular alkylations have also been described (Fig. 2.14) [6, 43].

Friedel-Crafts acylation of aromatics generally requires a stoichiometric amount of a Lewis acid such as AlCl₃. It further shares with hydroxyalkylation the problem of widely differing polarities of the substrate, acylating agent and the product, which makes it difficult to achieve a favorable adsorption ratio of substrate and reagent. Moreover, favored reagents such as the carboxylic acid or anhydride produce water and carboxylic acid, respectively, which may also be preferentially adsorbed. Hence, heterogeneous catalysis of Friedel-Crafts acylation is much more difficult than the related alkylations and poses a formidable challenge. As in the case of hydroxyalkylation, intramolecular processes such as the cycliacylation of 4-phenylbutyric acid to a-tetralone over a H-Beta catalyst, proved to be more tractable (Fig. 2.15) [44]. As already noted in Chapter 1, the commercialization of the first zeolite-catalyzed FC acylation, by Rhodia, constitutes a benchmark in this area [45, 46]. The process employs H-Beta or HY, in fixed-bed operation, for the acylation of activated aromatics, such as anisole, with acetic anhydride (Fig. 2.15). It avoids the production of HCl from acetyl chloride in the classical process in addition to circumventing the generation of stoichiometric quantities of AlCl₃ waste.

Furthermore, very high *para*-selectivities are observed as a result of the shape-selective properties of the zeolite catalyst.

Similarly, acylation of isobutylbenzene with acetic anhydride over H-Beta at 140 °C afforded *p*-acetylisobutylbenzene (Fig. 2.15), an intermediate in the syn-

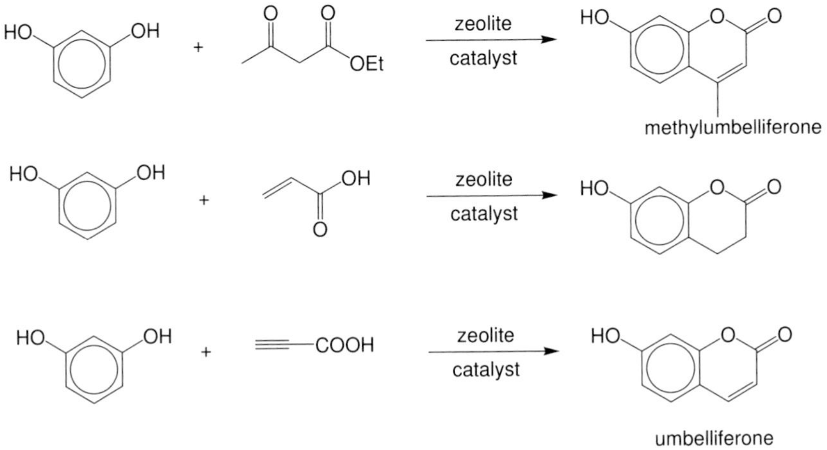

Fig. 2.14 Zeolite-catalyzed synthesis of coumarins.

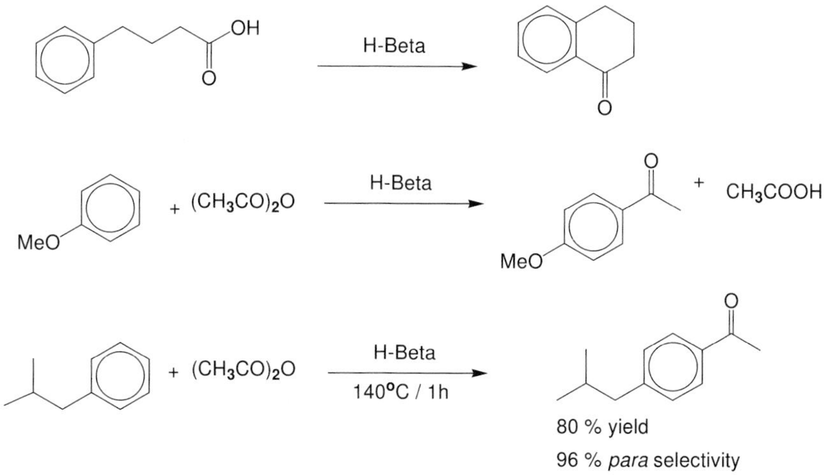

Fig. 2.15 Zeolite-catalyzed Friedel-Crafts acylations.

thesis of the antiinflammatory drug ibuprofen, in 80% yield and 96% *para*-selectivity [47].

Pioneering work in zeolite-catalyzed FC acylations was reported by Geneste and coworkers in 1986 [48]. They showed that a Ce^{3+}-exchanged zeolite Y catalyzed the acylation of toluene and xylenes with carboxylic acids (Fig. 2.16). This demonstrated that relatively mild acidity was sufficient to catalyze the reaction and that the free carboxylic acid could be used as the acylating agent. The reaction exhibited a very high *para*-selectivity. However, only the more lipophilic, higher carboxylic acids were effective, which can be ascribed to differences in preferential adsorption of substrate and acylating agent in the pores of the catalyst, i.e. the adsorption imbalance referred to above.

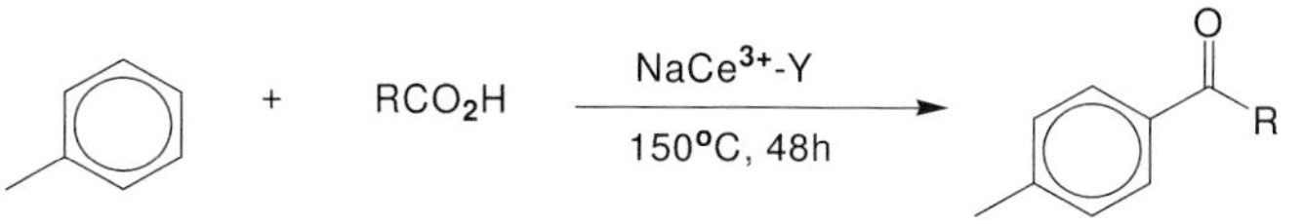

R	Yield (%)	Selectivity (%)		
		o	*m*	*p*
CH_3	trace			
CH_3CH_2	6	3	2	95
$CH_3CH_2CH_2$	20	3	2	95
$CH_3(CH_2)_4CH_2$	30	3	3	94
$CH_3(CH_2)_6CH_2$	75	3	3	94
$CH_3(CH_2)_{10}CH_2$	96	3	3	94

Fig. 2.16 Acylation of toluene over Ce^{3+}-exchanged zeolite Y.

Subsequently, Corma and coworkers [49] reported the acylation of anisole with phenacetyl chloride over H-Beta and H-Y. The FC acylation of electron-rich hetero-aromatics, such as thiophene and furan, with acetic anhydride over modified ZSM-5 catalysts (Fig. 2.17) in the gas phase [50] or liquid phase [51] was also reported.

Almost all of the studies of zeolite-catalyzed FC acylations have been conducted with electron-rich substrates. There is clearly a commercial need, therefore, for systems that are effective with electron-poor aromatics. In this context, the reports [52] on the acylation of benzene with acetic acid over H-ZSM-5 in the gas phase are particularly interesting. These results suggest that the adsorption ratios of substrate, acylating agent and product are more favorable in the gas phase than in the liquid phase.

Zeolites have also been investigated as heterogeneous catalysts for other electrophilic aromatic substitutions, e.g. nitration [53] and halogenation [54]. Aromatic nitration, for example, traditionally uses a mixture of sulfuric and nitric acids which leads to the generation of copious quantities of spent acid waste. Dealuminated mordenite (see earlier) is sufficiently robust to function effectively as a catalyst for the vapor phase nitration of benzene with 65% aqueous nitric acid [55]. The advantage of using vapor phase conditions is that the water is continuously removed. Although these results are encouraging, aromatic nitration over solid acid catalysts is still far from commercialization.

An interesting example of aromatic halogenation, in the context of green chemistry, is the production of 2,6-dichlorobenzonitrile, an agrochemical intermediate. The conventional process involves a series of reactions (see Fig. 2.18) with Cl_2, HCN, and $POCl_3$ as stoichiometric reagents, having an atom efficiency of 31% and generating substantial amounts of salts as waste. The new process, developed by Toray [56], involves vapor phase chlorination of toluene over a Ag-H-Mordenite catalyst to give a mixture of dichlorotoluene isomers. The required 2,6-isomer is separated by adsorption in faujasite or AlPO-11 and the other isomers are returned to the chlorination reactor where equilibration occurs. The 2,6-dichlorotoluene can be subsequently converted to 2,6-dichlorobenzonitrile by vapour phase ammoxidation over an oxide catalyst.

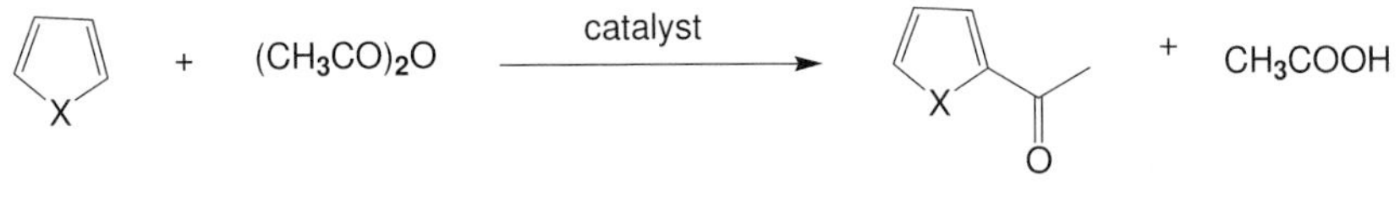

X	Catalyst	Temp ($^\circ$C)	Conv (%)	Sel. (%)
S	B-ZSM-5	250	24	99
O	Ce-ZSM-5	200	23	99
N	B-Cs-ZSM-5	150	41	98

Fig. 2.17 Acylation of heteroaromatics over zeolite catalysts.

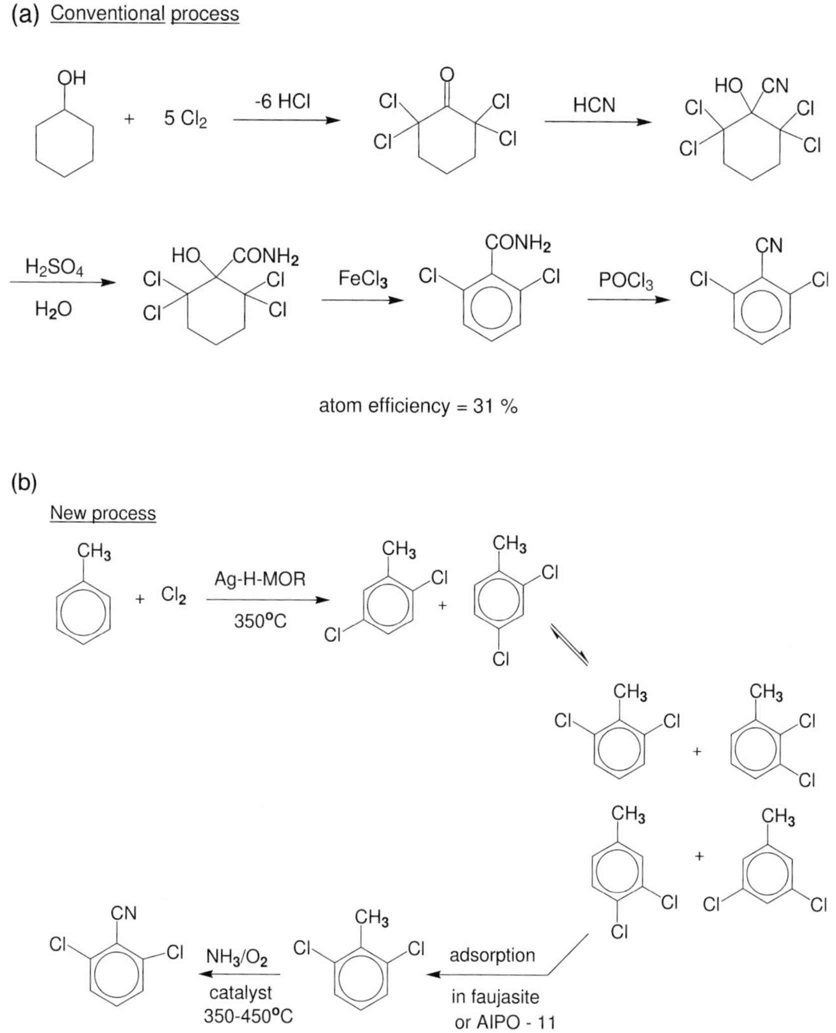

(a) Conventional process

atom efficiency = 31 %

(b)

New process

Fig. 2.18 Two processes for 2,6-dichlorobenzonitrile.

2.2.3.2 Additions and Eliminations

Zeolites have been used as (acid) catalysts in hydration/dehydration reactions. A pertinent example is the Asahi process for the hydration of cyclohexene to cyclo-hexanol over a high silica (Si/Al > 20), H-ZSM-5 type catalyst [57]. This process has been operated successfully on a 60 000 tpa scale since 1990, although many problems still remain [57] mainly due to catalyst deactivation. The hydration of cyclohexanene is a key step in an alternative route to cyclohexanone (and phenol) from benzene (see Fig. 2.19). The conventional route involves hydrogenation to cyclohexane followed by autoxidation to a mixture of cyclohexanol and

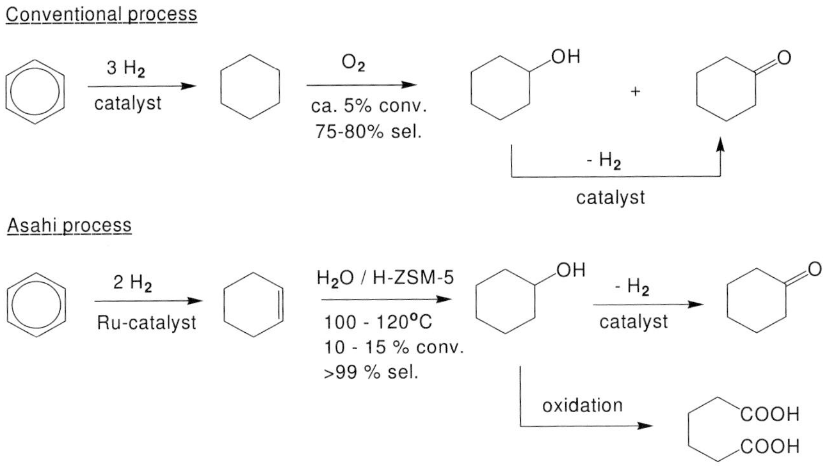

Conventional process

Asahi process

Fig. 2.19 Two processes for cyclohexanone.

cyclohexanone and subsequent dehydrogenation of the former. A serious disadvantage of this process is that the autoxidation gives reasonable selectivities (75–80%) only at very low conversions (ca. 5%), thus necessitating the recycle of enormous quantities of cyclohexane. The Asahi process involves initial partial hydrogenation of benzene to cyclohexene over a heterogeneous ruthenium catalyst. The cyclohexane byproduct is recycled by dehydrogenation back to benzene. The cyclohexene hydration proceeds with >99% selectivity at 10–15% conversion. If the ultimate product is adipic acid the cyclohexanone can be by-passed by subjecting cyclohexanol to oxidation (see Fig. 2.19) [57].

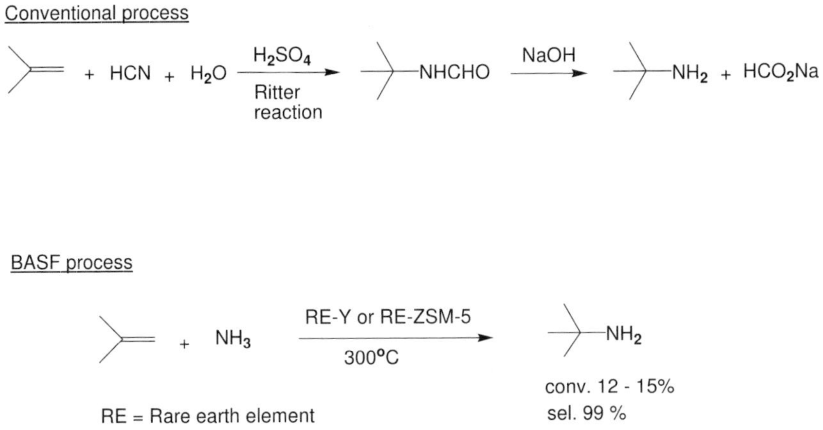

Conventional process

BASF process

RE = Rare earth element

Fig. 2.20 Two processes for *tert*-butylamine.

Similarly, zeolites can catalyze the addition of ammonia to an olefinic double bond, as is exemplified by the BASF process for the production of *tert*-butylamine by reaction of isobutene with ammonia, in the vapor phase, over a rare earth exchanged ZSM-5 or Y zeolite (Fig. 2.20) [58, 59]. This process has an atom efficiency of 100% and replaced a conventional synthesis via a Ritter reaction, which employs HCN and sulfuric acid and generates formate as a coproduct.

2.2.3.3 Rearrangements and Isomerizations

As already discussed in Chapter 1, the commercialization, by Sumitomo [60–64], of a vapor phase *Beckmann rearrangement* of cyclohexanone oxime to caprolactam over a high-silica MFI (ZSM-5 type) zeolite (Fig. 2.21) is another benchmark in zeolite catalysis. The process, which currently operates on a 90 000 tpa scale, replaces a conventional one employing stoichiometric quantities of sulfuric acid and producing ca. 2 kg of ammonium sulfate per kg of caprolactam.

Interestingly, the activity of the catalyst was proportional to its external surface area. This, together with the fact that caprolactam does not fit easily into the pores of the zeolite, strongly suggests that the reaction takes place on the external surface, possibly at the pore openings. Ichihashi and coworkers [61] proposed that the reaction takes place in a silanol nest resulting from dealumination (see earlier), as was also suggested by Hoelderich and coworkers [65].

The Friedel-Crafts acylation of phenols proceeds via initial esterification followed by *Fries rearrangement* of the resulting aryl ester to afford the hydroxyaryl

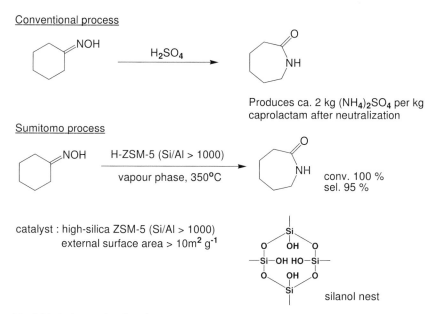

Fig. 2.21 Zeolite-catalyzed Beckmann rearrangement.

ketone. For example, phenylacetate affords a mixture of o- and p-hydroxyaceto-phenone (see Fig. 2.22). The latter is a key intermediate in the Hoechst Cela-nese process for the manufacture of the analgesic, paracetamol.

Traditionally, the Fries rearrangement is conducted with stoichiometric amounts of mineral (H_2SO_4, HF) or Lewis acids ($AlCl_3$, $ZnCl_2$) and generates large amounts of inorganic salts as by-products, i.e. it suffers from the same disadvantages as standard Friedel-Crafts acylations. Substitution of these corro-sive, polluting and non-regenerable catalysts by regenerable solid acid catalysts has obvious environmental benefits. Consequently, a variety of solid acids, par-ticularly zeolites, have been studied as catalysts for the Fries rearrangement, both in the vapor and liquid phases [66]. Most studies were performed with phenyl acetate as the substrate, prepared ex situ or formed in situ from phenol and acetic anhydride or acetic acid. A variety of zeolites have been used, e.g. H-Y, H-ZSM-5 and H-beta, generally affording o-hydroxyacetophenone as the ma-jor product, often in very high selectivity [67]. None of these processes has been brought to commercialization, presumably because they produce the commer-cially less interesting isomer as the product.

Another example of commercial interest is the Fries rearrangement of the benzoate ester of resorcinol to afford 2,4-dihydroxybenzophenone, the precursor of the UV-absorbent 4-O-octyl-2-hydroxybenzophenone. Reaction of benzoic acid with one equivalent of resorcinol (see Fig. 2.22), over various solid catalysts, in chlorobenzene as solvent, with continuous removal of water, was investigated by Hoefnagel and van Bekkum [68]. H-Beta was slightly less active than the ion-ex-

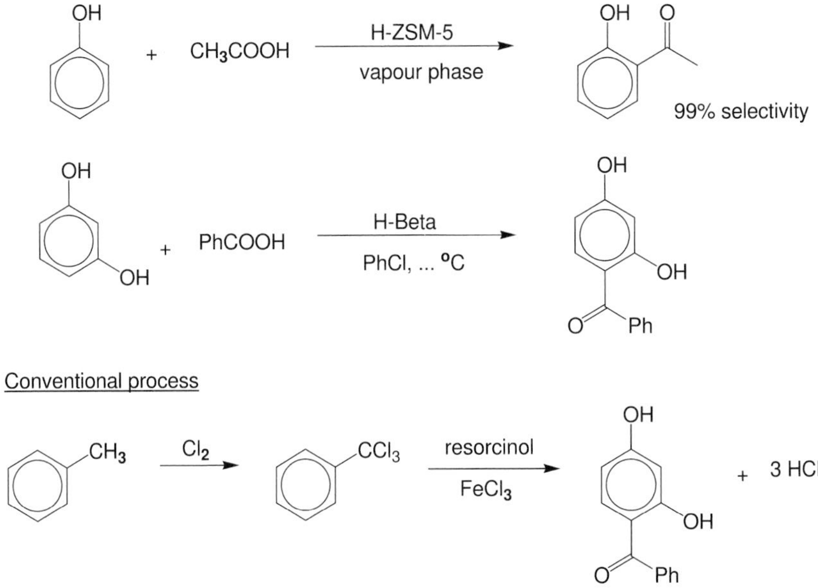

Fig. 2.22 Fries rearrangement over zeolites.

change resin, Amberlyst 15, but gave less by-product formation and has the advantage of being regenerable by simple air burn-off. It would appear to have environmental benefits compared to the existing industrial process which generates substantial amounts of chloride waste (see Fig. 2.22).

Epoxide rearrangements are key steps in the manufacture of numerous synthetic intermediates in the fine chemical industry. They are generally performed with conventional Brønsted or Lewis acids or strong bases as catalysts, often in stoichiometric amounts. Here again, replacement by recyclable solid catalysts has obvious benefits [69].

For example, rearrangement of α-pinene oxide produces, among the ten or so major products, campholenic aldehyde, the precursor of the sandalwood fragrance santalol. The conventional process employs stoichiometric quantities of zinc chloride but excellent results have been obtained with a variety of solid acid catalysts (see Fig. 2.23), including a modified H-USY [70] and the Lewis acid Ti-Beta [71]. The latter afforded campholenic aldehyde in selectivities up to 89% in the liquid phase and 94% in the vapor phase.

Similarly, the rearrangement of substituted styrene oxides to the corresponding phenylacetaldehydes (see Fig. 2.23) affords valuable intermediates for fragrances, pharmaceuticals and agrochemicals. Good results were obtained using zeolites, e.g. H-ZSM-5, in either the liquid or vapor phase [69]. The solid Lewis acid, titanium silicalite-1 (TS-1) also gave excellent results, e.g. styrene oxide was converted to phenylacetaldehyde in 98% selectivity at 100% conversion in 1–2 h at 70 °C in acetone as solvent. Other noteworthy epoxide rearrangements

Catalyst	conv (%)	sel. (%)
US-Y	98	76
Ti-Beta	100	89 (liquid phase)
	100	94 (gas phase)

	conv (%)	sel. (%)
H-ZSM-5	99	97 - 100
Ti-silicate-1	100	98

98% yield

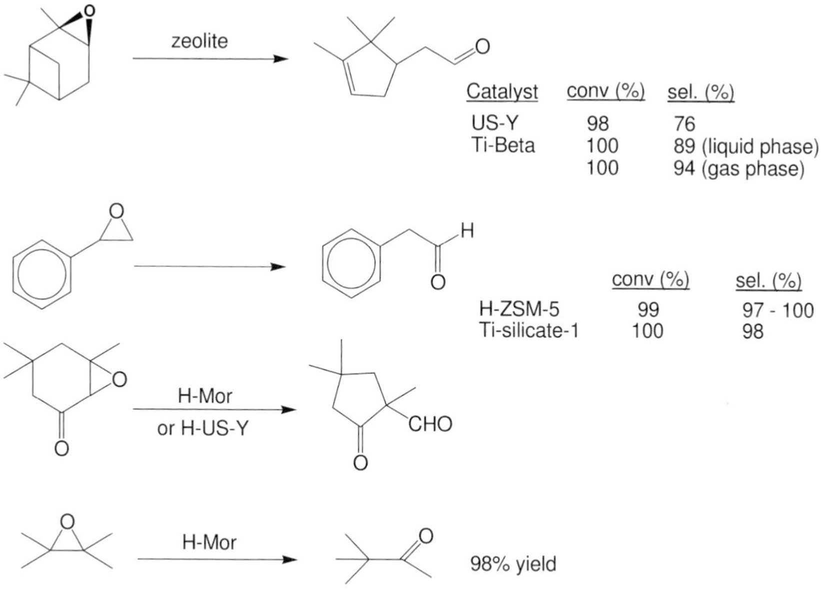

Fig. 2.23 Zeolite-catalyzed epoxide rearrangements.

are the conversion of isophorone oxide to the keto aldehyde (see Fig. 2.23), over H-mordenite [72] or H-US-Y (Si/Al=48) [69], and 2,3-dimethyl-2-butene oxide to pinacolone [73]. The mechanistically related pinacol rearrangement, e.g. of pinacol to pinacolone, has also been conducted over solid acid catalysts including acidic clays and zeolites [74].

2.2.3.4 Cyclizations

Zeolites have also been shown to catalyze a variety of acid-promoted cyclizations. Many of these involve the formation of N-heterocycles via intramolecular amination reactions [75–77]. Some examples are shown in Fig. 2.24.

Reaction of ammonia with various combinations of aldehydes, over solid acid catalysts in the vapor phase, is a convenient route for producing pyridines [77]. For example, amination of a formaldehyde/acetaldehyde mixture affords pyridine and 3-picoline (Fig. 2.25). Mobil scientists found that MFI zeolites such as H-ZSM-5 were particularly effective for these reactions.

Alternatively, 3-picoline is produced by vapor phase cyclization of 2-methyl-pentane-1,5-diamine (Fig. 2.25) over, for example, H-ZSM-5 followed by palladium-catalyzed dehydrogenation [78]. This diamine is a by-product of the manufacture of hexamethylenediamine, the raw material for nylon 6,6, and these two reactions are key steps in the Lonza process for nicotinamide production (see Chapter 1) [79].

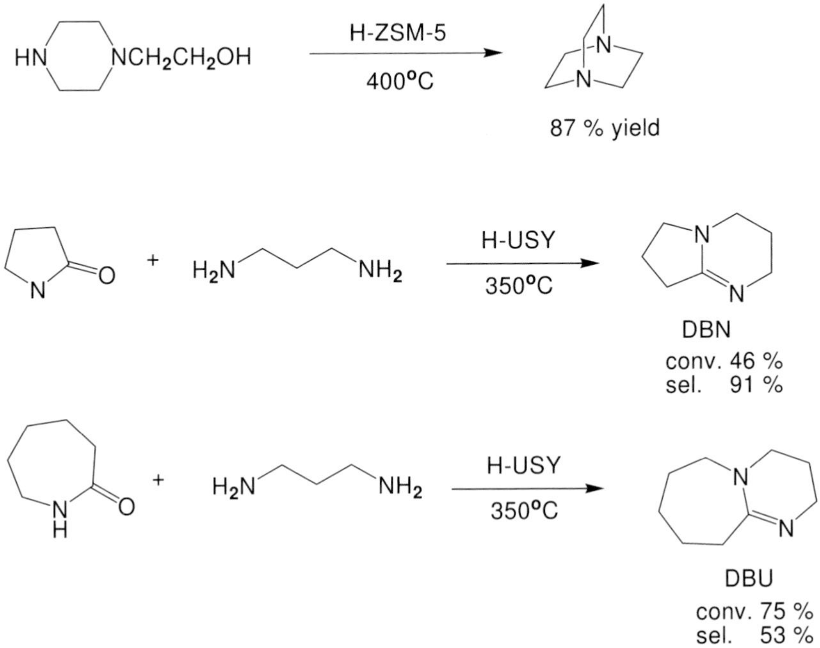

Fig. 2.24 Zeolite-catalyzed cyclizations.

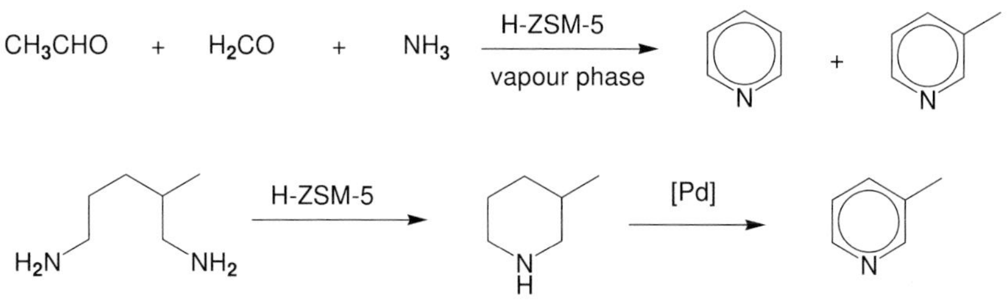

Fig. 2.25 Synthesis of pyridines via zeolite-catalyzed cyclizations.

Other examples of zeolite-catalyzed cyclizations include the Fischer indole synthesis [80] and Diels-Alder reactions [81].

2.2.4
Solid Acids Containing Surface SO₃H Functionality

This category encompasses a variety of solid acids with the common feature of a sulfonic acid moiety attached to the surface, i.e. they are heterogeneous equivalents of the popular homogeneous catalysts, *p*-toluenesulfonic and methanesulfonic acids. The cross-linked polystyrene-based, macroreticular ion exchange resins, such as Amberlyst®-15, are probably the most familiar examples of this class. They are the catalysts of choice in many industrial processes and in laboratory scale organic syntheses, e.g. esterifications, etherifications and acetalizations [2, 16, 17]. However, a serious shortcoming of conventional, polystyrene-based resins is their limited thermal stability. Reactions requiring elevated temperatures can be conducted with the more thermally stable Nafion® resins, which consist of a perfluorinated, Teflon-like polymeric backbone functionalized with terminal SO₃H groups [2, 82–85]. These materials (see Fig. 2.26 for structure) were originally developed by DuPont for application as membranes in electrochemical processes, based on their inertness to corrosive environments. Nafion is commercially available as Nafion NR50 (DuPont) in the

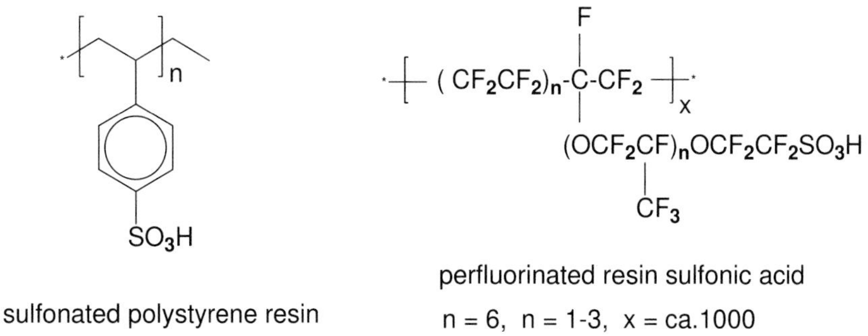

perfluorinated resin sulfonic acid

sulfonated polystyrene resin   n = 6, n = 1-3, x = ca.1000

Fig. 2.26 Structures of polystyrene- and perfluorocarbon-based ion exchange resins.

form of relatively large (2–3 mm) beads or as a 5% solution in a mixture of a lower alcohol and water. Other major differences with the polystyrene-based resins are their superior acid strength and higher number of acid sites. As a consequence of their highly electron-withdrawing perfluoroalkyl backbone, Nafion-H has an acidity comparable with that of 100% sulfuric acid, i.e. an acidity function (H_o) of –11 to –13 compared with –2.2 for Amberlyst-15. It also has five times as many acid sites as the latter. This superacidity coupled with thermal stability (up to 280 °C) make Nafion-H an attractive catalyst for a variety of processes [2, 16, 17]. Thus, reactions not requiring especially strong acidity and/ or high temperatures are usually conducted with the less expensive polystyrene-based resins while those that require higher acidity and/or temperatures can be performed better with the more robust Nafion-H.

However, Nafion-H NR50 beads have one serious drawback: they have a very low surface area (< 0.02 m^2 g^{-1}). To overcome this disadvantage DuPont researchers developed Nafion-silica composites, consisting of small (20–60 nm) Nafion resin particles embedded in a porous silica matrix [86]. The microstructure can be regarded as a porous silica network containing a large number of 'pockets' of very strong acid sites (the Nafion polymer) in domains of ca. 10–20 nm. The surface area is increased by several orders of magnitude and the catalytic activity per unit weight is up to 1000 times that of the pure Nafion. They are prepared by a sol–gel technique and are available as SAC 13, which contains 13% (w/w) Nafion. The improved accessibility to the active sites makes this material a particularly attractive solid acid catalyst [2, 84] for a variety of reactions such as Friedel-Crafts alkylations [86, 87] and Friedel-Crafts acylations [2]. The latter are effective only with electron-rich aromatics such as anisole. The Nafion-silica composite was compared with the pure polymer and Amberlyst-15 in the alkylation of p-cresol with isobutene [2]. The results are shown in Table 2.2.

Nafion-silica composites were compared with zeolites such as H-Beta, H-USY and H-ZSM-5, in the Fries rearrangement of phenyl acetate (see earlier). The highest conversion was observed with H-Beta [88].

Another class of thermally stable polymers containing surface SO$_3$H functionalities comprises the sulfonated polysiloxanes (see Fig. 2.27) [89]. They are best prepared by a sol–gel technique involving copolymerization of functionalized and non-functionalized silanes, e.g. 3-(tris-hydroxysilyl)propyl sulfonic acid with tetraethoxysilane. This sol–gel process affords material with a high surface area (300–600 m^2 g^{-1}), high porosity and large mesopores (> 20 nm). Although they have excellent chemical and thermal stability, and have been marketed under the trade name Deloxan®, industrial applications have not yet been forthcoming, probably owing to their relatively high price.

Similarly, perfluoroalkylsulfonic acid moieties can be covalently attached to a silica matrix by grafting a silica surface with (EtO)$_3$Si(CH$_2$)$_3$(CF$_2$)$_2$O(CF$_2$)$_2$SO$_3$H (see Fig. 2.27) or by using the latter in a sol–gel synthesis [18, 90].

More recently, attention has shifted to the preparation of hybrid organic–inorganic ordered mesoporous silicas, e.g. of the M41S type (see earlier), bearing pendant alkylsulfonic acid moieties [18, 91–96]. These materials combine a high

Table 2.2 Acid catalyzed alkylation of p-cresol.

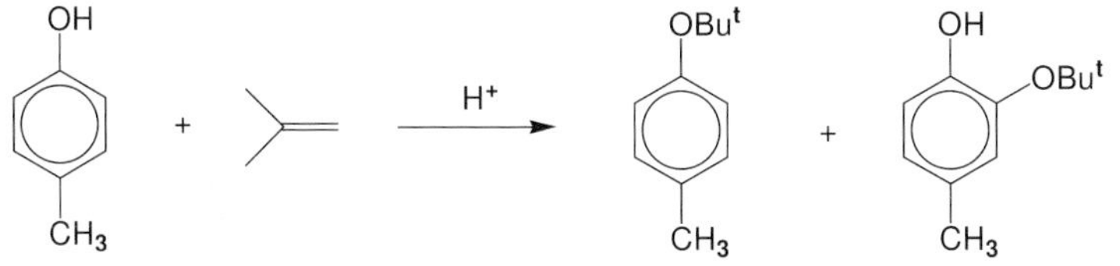

Catalyst	Conv. (%)	Selectivity (%)		Rate (mMg^{-1} cat·h^{-1})
		Ether	Alkylates	
13% Nafion/SiO$_2$	82.6	0.6	99.4	581.0
Nafion NR50	19.5	28.2	71.8	54.8
Amberlyst-15	62.4	14.5	85.5	171.0

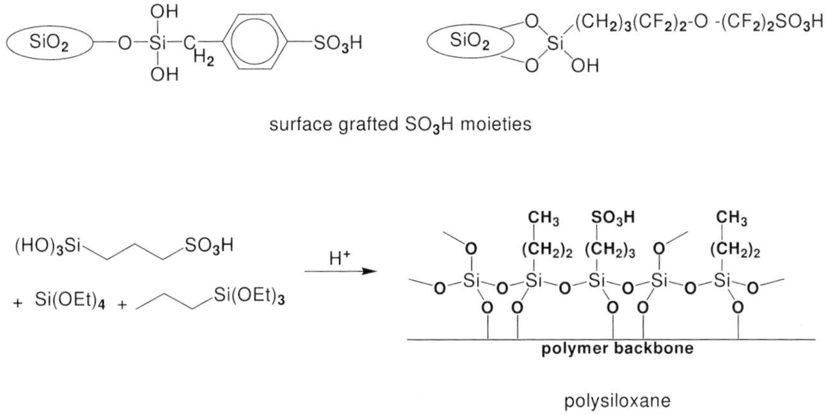

surface grafted SO$_3$H moieties

polysiloxane

Fig. 2.27 Synthesis of SO$_3$H-functionalized silicas.

surface area, high loading and acid strength with excellent site accessibility and regular mesopores of narrow size distribution. This unique combination of properties makes them ideal solid acid catalysts for, in particular, reactions involving relatively bulky molecules as reactants and/or products. They are prepared by grafting of preformed MCM-41 with, for example, 3-mercaptopropyltrimethoxysilane (MPTS), followed by oxidation of the pendant thiol groups to SO$_3$H with hydrogen peroxide (Fig. 2.28) [18]. Alternatively, direct co-condensation of tetraethoxysilane (TEOS) with MPTS in the presence of a surfactant templating agent, such as cetyltrimethylammonium bromide or dodecylamine, affords functionalized MCM-41 or HMS (hexagonal mesoporous silica), respectively [18, 91–96]. Non-ionic surfactants, e.g. the poly(ethylene oxide)–poly(propylene oxide) block copolymer, commercially available as Pluronic 123, have

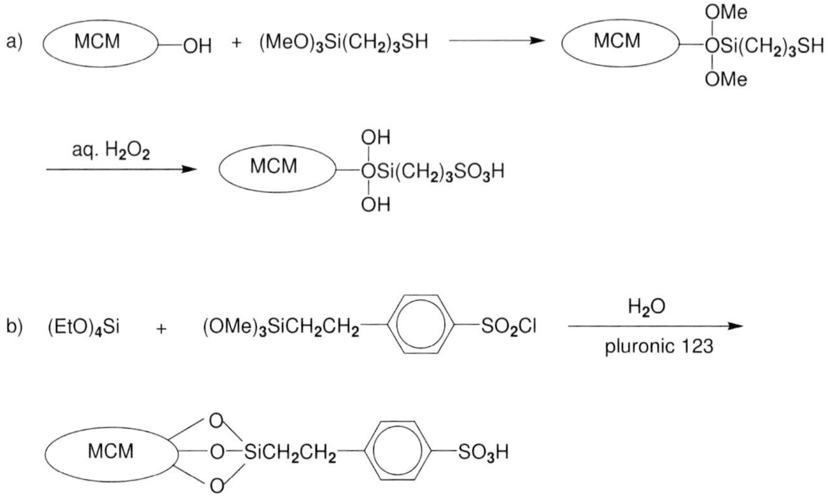

a) MCM—OH + (MeO)$_3$Si(CH$_2$)$_3$SH $\longrightarrow$ MCM—OSi(CH$_2$)$_3$SH (with OMe groups)

$\xrightarrow{\text{aq. H}_2\text{O}_2}$ MCM—OSi(CH$_2$)$_3$SO$_3$H (with OH groups)

b) (EtO)$_4$Si + (OMe)$_3$SiCH$_2$CH$_2$—⟨ring⟩—SO$_2$Cl $\xrightarrow[\text{pluronic 123}]{\text{H}_2\text{O}}$

MCM—O—SiCH$_2$CH$_2$—⟨ring⟩—SO$_3$H

pluronic 123 = ethylene oxide / propylene oxide block copolymer

Fig. 2.28 Synthesis of ordered mesoporous silicas containing pendant SO$_3$H groups.

been used under acidic conditions [91, 92, 96]. When hydrogen peroxide is added to the sol–gel ingredients the SH groups are oxidized *in situ*, affording the SO$_3$H-functionalized mesoporous silica in one step [91, 92]. Alternatively, co-condensation of TEOS with 2-(4-chlorosulfonylphenyl)ethyltrimethoxysilane, in the presence of Pluronic 123 as template, afforded a mesoporous material containing pendant arenesulfonic acid moieties (see Fig. 2.28) [91, 92].

These mesoporous SO$_3$H-functionalized silicas are attractive, recyclable acid catalysts for reactions involving molecules that are too large to access the smaller pores of conventional molecular sieves or those affording bulky products. The additional option of modifying their hydrophobicity/hydrophilicity balance, by introducing varying amounts of alkyltrialkoxysilanes, e.g. CH$_3$CH$_2$CH$_2$Si(OEt)$_3$, into the sol–gel synthesis provides an opportunity to design tailor-made solid acid catalysts [94]. These organic–inorganic materials have been used as acid catalysts in, for example, esterifications of polyols with fatty acids [18, 94, 95], Friedel-Crafts acylations [97] and bisphenol-A synthesis [93]. The latter is an important raw material for epoxy resins, and is manufactured by reaction of phenol with acetone (see Fig. 2.29) using ion exchange resins such as Amberlyst®-15 as catalyst. However, thermal stability and fouling are major problems with these catalysts and there is a definite need for thermally stable, recyclable solid acid catalysts. A sulfonic acid functionalized MCM-41, produced by grafting of preformed MCM-41 with MPTS and subsequent H$_2$O$_2$ oxidation, exhibited a superior activity and selectivity compared with various zeolites. Unfortunately, the crucial comparison with Amberlyst-15 was not reported.

Finally, the functionalization of mesoporous silicas with perfluoroalkylsulfonic acid groups, by grafting with 1,2,2-trifluoro-2-hydroxy-1-trifluoromethylethane

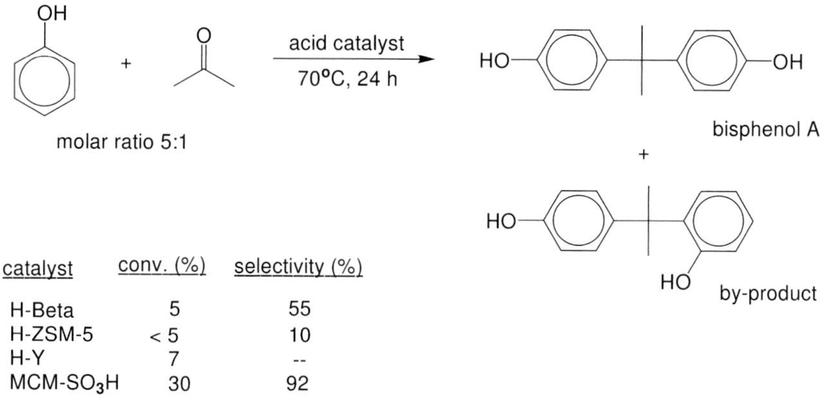

molar ratio 5:1

catalyst	conv. (%)	selectivity (%)
H-Beta	5	55
H-ZSM-5	< 5	10
H-Y	7	--
MCM-SO$_3$H	30	92

Fig. 2.29 Bisphenol-A synthesis over MCM-SO$_3$H.

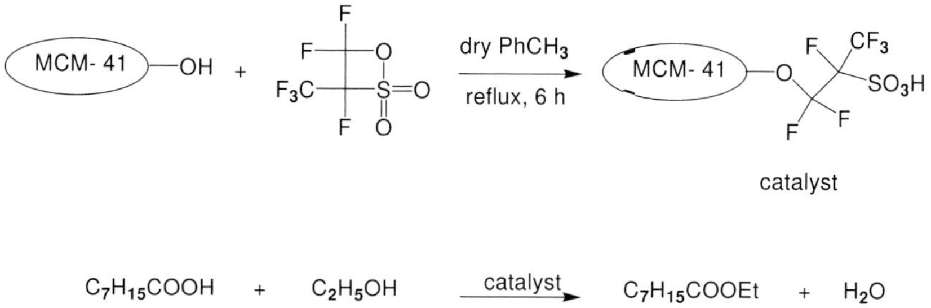

$$C_7H_{15}COOH \quad + \quad C_2H_5OH \xrightarrow{\text{catalyst}} C_7H_{15}COOEt \quad + \quad H_2O$$

Fig. 2.30 Mesoporous silica grafted with perfluoroalkylsulfonic acid groups.

sulfonic acid β-sultane (see Fig. 2.30) was recently reported [98]. The resulting materials displayed a higher activity than commercial Nafion® silica composites (see earlier) in the esterification of ethanol with octanoic acid [98].

2.2.5
Heteropoly Acids

Heteropoly acids (HPAs) are mixed oxides composed of a central ion or 'heteroatom' generally P, As, Si or Ge, bonded to an appropriate number of oxygen atoms and surrounded by a shell of octahedral MO$_6$ units, usually where M = Mo, W or V. HPAs, having the so-called Keggin structure, are quite common and consist of a central tetrahedron, XO$_4$ (X = P, As, Si, Ge, etc.) surrounded by 12 MO$_6$ octahedra (M = Mo, W, V) arranged in four groups of three edge-sharing M$_3$O$_{13}$ units (see Fig. 2.31).

Many HPAs exhibit superacidity. For example, H$_3$PW$_{12}$O$_{40}$ has a higher acid strength ($H_o = -13.16$) than CF$_3$SO$_3$H or H$_2$SO$_4$. Despite their rather daunting formulae they are easy to prepare, simply by mixing phosphate and tungstate in the required amounts at the appropriate pH [99]. An inherent drawback of

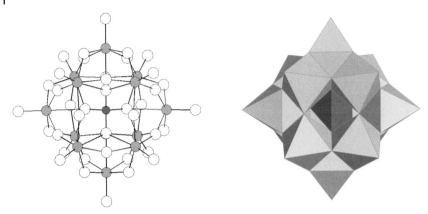

Fig. 2.31 Keggin structure of $[PW_{12}O_{40}]^{3-}$
● P-atoms; ● W-atoms; ○ O-atoms.

HPAs, however, is their solubility in polar solvents or reactants, such as water or ethanol, which severely limits their application as recyclable solid acid catalysts in the liquid phase. Nonetheless, they exhibit high thermal stability and have been applied in a variety of vapor phase processes for the production of petrochemicals, e.g. olefin hydration and reaction of acetic acid with ethylene [100, 101]. In order to overcome the problem of solubility in polar media, HPAs have been immobilized by occlusion in a silica matrix using the sol–gel technique [101]. For example, silica-occluded $H_3PW_{12}O_{40}$ was used as an insoluble solid acid catalyst in several liquid phase reactions such as ester hydrolysis, esterification, hydration and Friedel-Crafts alkylations [101]. HPAs have also been widely applied as catalysts in organic synthesis [102].

2.3
Solid Base Catalysis

Examples of the application of recyclable solid base catalysts are far fewer than for solid acids [103]. This is probably because acid-catalyzed reactions are much more common in the production of commodity chemicals. The various categories of solid bases that have been reported are analogous to the solid acids described in the preceding sections and include anionic clays, basic zeolites and mesoporous silicas grafted with pendant organic bases.

2.3.1
Anionic Clays: Hydrotalcites

Anionic clays are natural or synthetic lamellar mixed hydroxides with interlayer spaces containing exchangeable anions [10, 104]. The generic terms, layered double hydroxides (LDHs) or hydrotalcites are widely used, the latter because extensive characterization has been performed on hydrotalcite (a Mg/Al hydroxy-carbonate) which is both inexpensive and easy to synthesize.

Hydrotalcite is a natural mineral of ideal formula $Mg_6Al_2(OH)_{16}CO_3 \cdot 4H_2O$, having a structure similar to brucite, $Mg(OH)_2$. In hydrotalcite the Mg cations are partially replaced with Al^{3+} and the resulting positive charge is compensated by anions, typically carbonate, in the interlamellar space between the brucite-like sheets. When hydrotalcite is calcined at ca. 500 °C it is decarbonated and dehydrated to afford a strongly basic mixed Mg/Al oxide. Rehydration restores the original hydrotalcite structure and creates Brønsted base sites (OH^-) in the interlamellar space.

Activated hydrotalcites prepared in this way have been used as solid base catalysts in, for example, the aldol and related reactions [105]. The aldol condensation is a key step in the production of several commodity chemicals, including the solvent methylisobutylketone (MIBK) and 2-ethylhexanol, the precursor of the PVC plasticizer, dioctylphthalate (see Fig. 2.32). More than 1 million tons of these chemicals are produced worldwide on an annual basis using homogeneous bases such as 30% aqueous caustic (NaOH) [106]. Major problems are associated with the safe handling of the 30% caustic and treatment of caustic contaminated waste streams. About 1 kg of spent catalyst is generated per 10 kg of product and it has been estimated that 30% of the selling price is related to product recovery, purification and waste treatment. Hence, there is a definite incentive to replace these homogeneous bases with recyclable solid bases and hydrotalcites have been used to catalyze the aldol condensation of n-butyraldehyde in the liquid phase and acetone in the vapor phase to give 2-ethylhexenal and mesityl oxide, respectively [105]. Hydrogenation of the aldol condensation products yields 2-ethylhexanol and MIBK, respectively (Fig. 2.32). Impregnation of

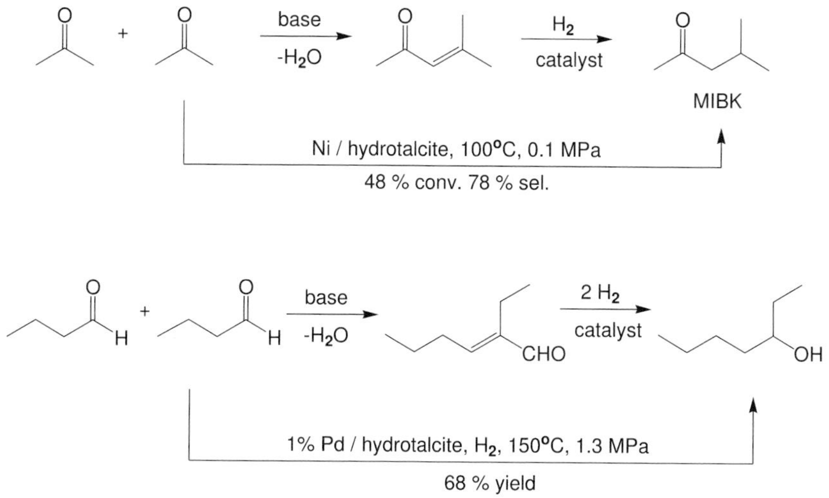

Fig. 2.32 Synthesis of MIBK and 2-ethylhexanol via aldol condensations.

the hydrotalcite with Pd or Ni affords bifunctional catalysts which are able to mediate the one-pot reductive aldol condensation of n-butyraldehyde or acetone to 2-ethylhexanol or MIBK, respectively [105].

Aldol and related condensation reactions such as Knoevenagel and Claisen-Schmidt condensations are also widely used in the fine chemicals and specialty chemicals, e.g. flavors and fragrances, industries. Activated hydrotalcites have been employed as solid bases in many of these syntheses. Pertinent examples include the aldol condensation of acetone and citral [107, 108], the first step in the synthesis of ionones, and the Claisen-Schmidt condensation of substituted 2-hydroxyacetophenones with substituted benzaldehydes [109], the synthetic

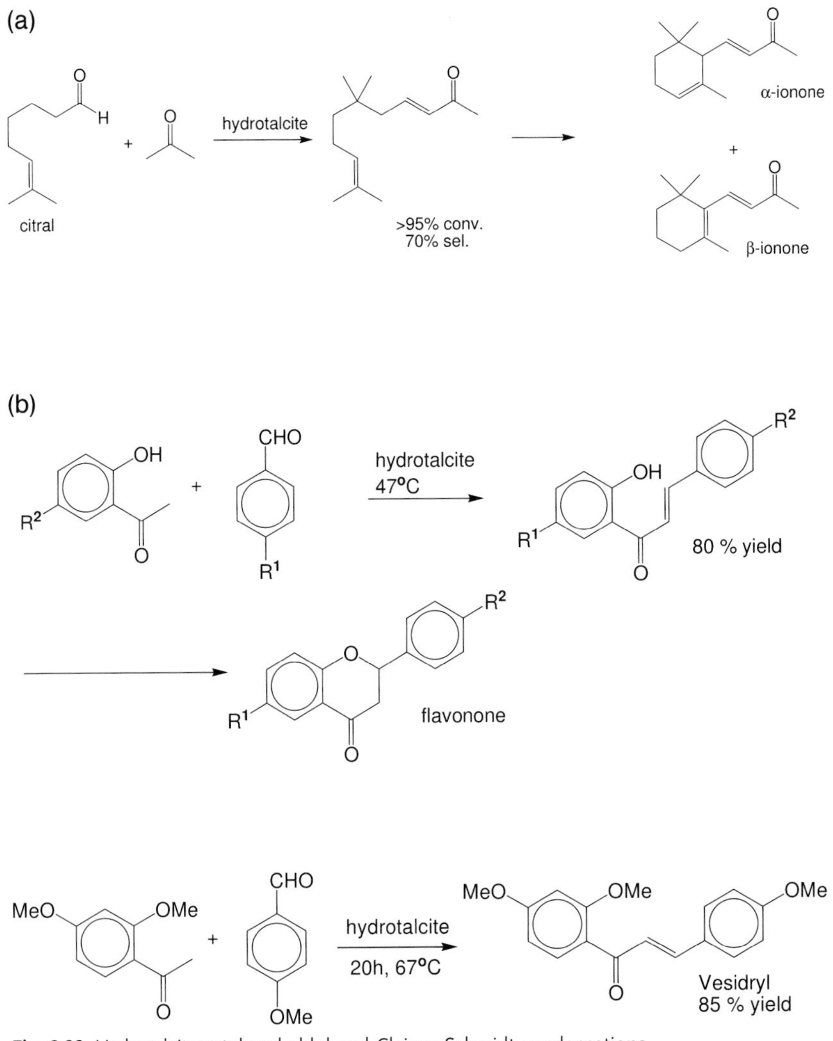

(a)

(b)

Fig. 2.33 Hydrotalcite-catalyzed aldol and Claisen-Schmidt condensations

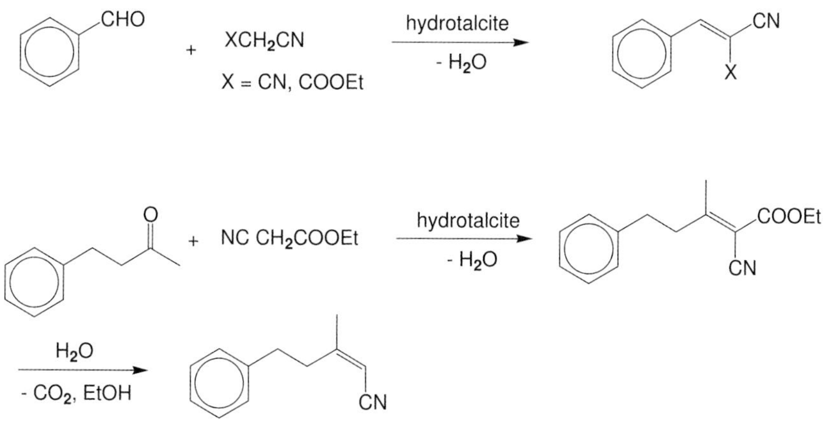

Fig. 2.34 Hydrotalcite-catalyzed Knoevenagel condensations.

route to flavonoids (see Fig. 2.33). The diuretic drug, Vesidryl, was similarly synthesized by condensation of 2,4-dimethoxyacetophenone with *p*-anisaldehyde (see Fig. 2.33) [109].

Excellent results were also obtained using activated hydrotalcite as a solid base catalyst in the Knoevenagel condensation of benzaldehyde with ethylcyanoacetate [110], ethylacetoacetate [111] or malononitrile [112] (see Fig. 2.34). Similarly, citronitrile, a perfumery compound with a citrus-like odor, was synthesized by hydrotalcite-catalyzed condensation of benzylacetone with ethyl-cyanoacetate, followed by hydrolysis and decarboxylation (Fig. 2.34) [113].

Other reactions which have been shown to be catalyzed by hydrotalcites include Michael additions, cyanoethylations and alkylations of, e.g. 1,3-dicarbonyl

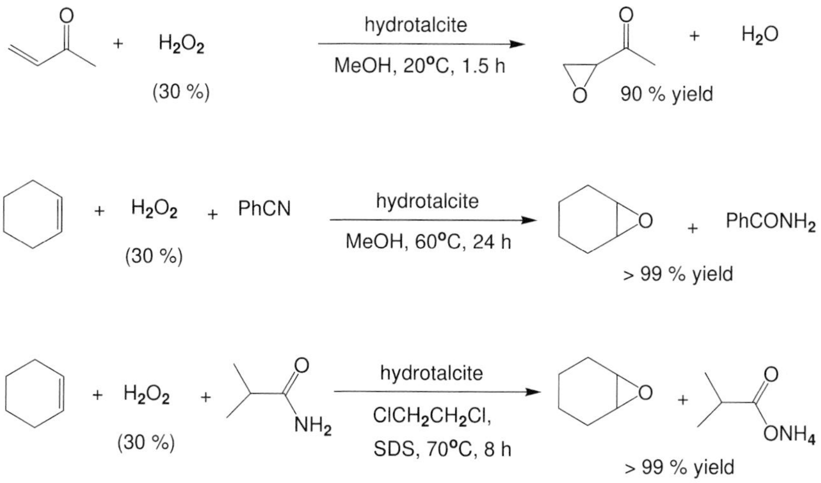

Fig. 2.35 Hydrotalcite-catalyzed oxidations with H_2O_2.

compounds [105, 114]. Another interesting application is as a solid base catalyst in oxidations (Fig. 2.35) involving the hydroperoxide anion, HO_2^- [115], e.g. the epoxidation of electron deficient olefins with H_2O_2 [116] and the epoxidation of electron-rich olefins with $PhCN/H_2O_2$ [117]. The latter reaction involves the peroxycarboximidate (Payne reagent) as the active oxidant, producing one equivalent of the amide as a co-product. Subsequent studies showed that carboxylic amides also act as catalysts in this reaction (Fig. 2.35), isobutyramide being particularly effective [118]. In this case, one equivalent of the ammonium carboxylate is formed as the co-product.

Interesting recent developments are the use of hydrotalcite supported on carbon nanofibers [119], to facilitate recovery of the catalyst by filtration, and the use of synthetic hydroxyapatite, $Ca_{10}(PO_4)_6(OH)_2$ as a solid base catalyst in a variety of reactions including Michael additions [120]. The supported hydrotalcite exhibited higher activities and selectivities than the conventional unsupported material in the aldol condensation of citral with acetone [119].

2.3.2
Basic Zeolites

In contrast with the widespread application of zeolites as solid acid catalysts (see earlier), their use as solid base catalysts received scant attention until fairly recently [121]. This is probably because acid-catalyzed processes are much more common in the oil refining and petrochemical industries. Nonetheless, basic zeolites and related mesoporous molecular sieves can catalyze a variety of reactions, such as Knoevenagel condensations and Michael additions, which are key steps in the manufacture of flavors and fragrances, pharmaceuticals and other specialty chemicals [121]. Indeed, the Knoevenagel reaction of benzaldehyde with ethyl cyanoacetate (Fig. 2.36) has become a standard test reaction for solid base catalysts [121].

Two approaches have been used to prepare basic zeolites by post-synthesis modification: (i) ion exchange of protons by alkali metal or rare earth cations and (ii) generation of nanoparticles of alkali metal or alkaline earth metal oxides within the zeolite channels and cavities (see later). In zeolites, Lewis basicity is associated with the negatively charged framework oxygens and, as expected, increases with the size of the counter cation, i.e. Li < Na < Ka < Cs. Alkali-exchanged zeolites contain a large number of relatively weak basic sites capable of abstracting a proton from molecules having a pK_a in the range 9–11 and a few more basic ones (up to $pK_a = 13.3$) [121]. This corresponds with the basic

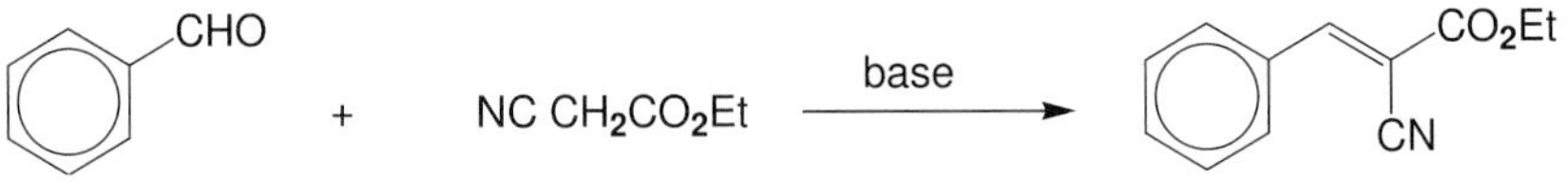

Fig. 2.36 Knoevenagel reaction.

strength required to catalyze many synthetically useful reactions such as the Knoevenagel and Michael reactions referred to above. The use of the ion-exchanged zeolites as base catalysts has the advantage that they are stable towards reaction with moisture and/or carbon dioxide.

Cesium-exchanged zeolite X was used as a solid base catalyst in the Knoevenagel condensation of benzaldehyde or benzyl acetone with ethyl cyanoacetate [121]. The latter reaction is a key step in the synthesis of the fragrance molecule, citronitrile (see Fig. 2.37). However, reactivities were substantially lower than those observed with the more strongly basic hydrotalcite (see earlier). Similarly, Na-Y and Na-Beta catalyzed a variety of Michael additions [122] and K-Y and Cs-X were effective catalysts for the methylation of aniline and phenylacetonitrile with dimethyl carbonate or methanol, respectively (Fig. 2.37) [123]. These procedures constitute interesting green alternatives to classical alkylations using methyl halides or dimethyl sulfate in the presence of stoichiometric quantities of conventional bases such as caustic soda.

Alkali-exchanged mesoporous molecular sieves are suitable solid base catalysts for the conversion of bulky molecules which cannot access the pores of zeolites. For example, Na- and Cs-exchanged MCM-41 were active catalysts for the Knoevenagel condensation of benzaldehyde with ethyl cyanoacetate ($pK_a = 10.7$) but low conversions were observed with the less acidic diethyl malonate ($pK_a = 13.3$) [123]. Similarly, Na-MCM-41 catalyzed the aldol condensation of several bulky ketones with benzaldehyde, including the example depicted in Fig. 2.38, in which a flavonone is obtained by subsequent intramolecular Michael-type addition [123].

As noted above, the basic sites generated by alkali metal exchange in zeolites are primarily weak to moderate in strength. Alkali and alkaline earth metal oxides, in contrast, are strong bases and they can be generated within the pores and cavities of zeolites by over-exchanging them with an appropriate metal salt, e.g. an acetate, followed by thermal decomposition of the excess metal salt, to afford highly dispersed basic oxides occluded in the pores and cavities. For example, cesium oxide loaded faujasites, exhibiting super-basicities, were prepared by impregnating CsNa-X or CsNa-Y with cesium acetate and subsequent thermal decomposition [124, 125]. They were shown to catalyze, inter alia, Knoevenagel condensations, e.g. of benzaldehyde with ethyl cyanoacetate [126]. A simi-

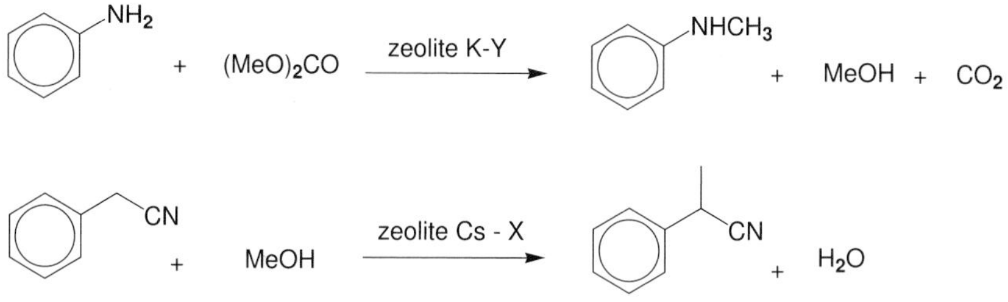

Fig. 2.37 Alkylations catalyzed by basic zeolites.

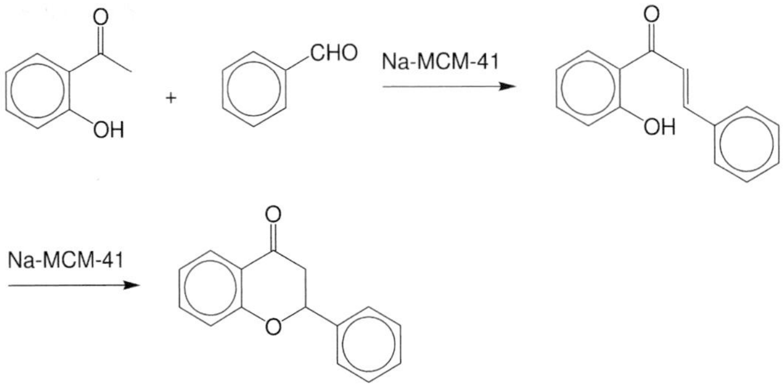

Fig. 2.38 Na-MCM-41 catalyzed flavonone synthesis.

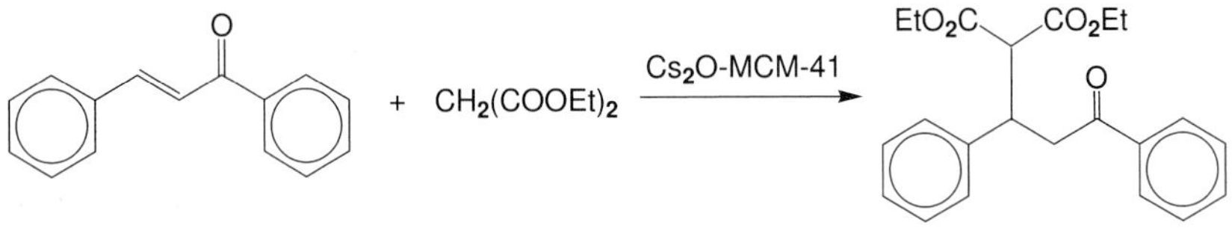

Fig. 2.39 Cesium oxide loaded MCM-41 as a catalyst for the Michael addition.

larly prepared Cs oxide loaded MCM-41 was an active catalyst for the Michael addition of diethylmalonate to chalcone (Fig. 2.39) [127].

A serious drawback of these alkali metal oxide loaded zeolites and mesoporous molecular sieves is their susceptibility towards deactivation by moisture and/or carbon dioxide, which severely limits their range of applications.

2.3.3
Organic Bases Attached to Mesoporous Silicas

Analogous to the attachment of organic alkyl(aryl)sulfonic acid groups to the surface of organic or inorganic polymers (see earlier), recyclable solid base catalysts can be prepared by grafting, e.g. amine moieties, to the same supports. Anion exchange resins, such as Amberlyst, are composed of amine or tetraalkylammonium hydroxide functionalities grafted to cross-linked polystyrene resins and they are widely used as recyclable basic catalysts [128]. As with the cation exchange resins, a serious limitation of these materials is their lack of thermal stability under reaction conditions, in this case a strongly alkaline medium. This problem can be circumvented by grafting basic moieties to inorganic polymers that are thermally stable under reaction conditions.

The use of silica-grafted primary and tertiary amines as solid base catalysts in Knoevenagel condensations was reported already in 1988. Yields were high but activities were rather low, owing to the relatively low loadings (<1 mmol g^{-1}). A decade later the groups of Brunel [129–131], Macquarrie [132–135] and Jacobs [136] obtained much higher loadings (up to 5 mmol g^{-1}), and activities, by at-

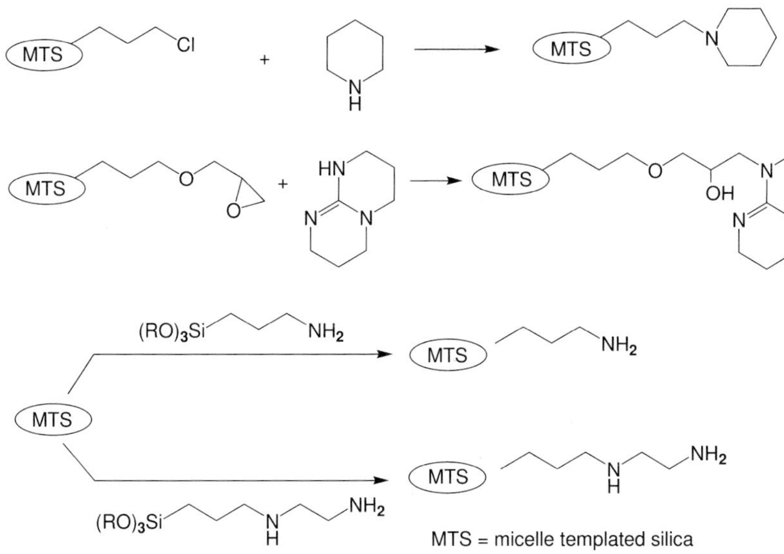

Fig. 2.40 Methods for attachment of organic bases to micelle templated silicas.

taching amine functionalities to mesoporous silicas, thus taking advantage of their high surface areas (ca. $1000 \text{ m}^2 \text{ g}^{-1}$). Two strategies were used for their preparation: (i) grafting of pre-formed micelle templated silicas (MTS) with amine functionalities or with other functionalities, e.g. halide, epoxide, which can be subsequently converted to amines by nucleophilic displacement (see Fig. 2.40) and (ii) direct incorporation by co-condensation of $(EtO)_4Si$ with an appropriately functionalized silane in a sol–gel synthesis.

For example, Brunel and coworkers [137] anchored primary amine groups to the surface of mesoporous silica by grafting with 3-aminopropyltriethoxysilane. Anchoring of tertiary amine functionality was achieved by grafting with chloro- or iodopropyltriethoxysilane followed by nucleophilic displacement of halide with piperidine. Both materials were active catalysts for the Knoevenagel condensation of benzaldehyde with ethyl cyanoacetate at $80\,^\circ$C in dimethyl sulfoxide. The primary amine was more active, which was explained by invoking a different mechanism: imine formation with the aldehyde group rather than classical base activation of the methylene group in the case of the tertiary amine.

Stronger solid base catalysts can be prepared by grafting guanidine bases to mesoporous silicas. For example, the functionalization of MCM-41 with 1,5,7-triazabicyclo[4,4,0]dec-5-ene (TBD), as shown in Fig. 2.40, afforded a material (MCM-TBD) that was an effective catalyst for Michael additions with ethylcyanoacetate or diethylmalonate (Fig. 2.41) [136].

In a variation on this theme, a bulky guanidine derivative, N,N,N''-tricyclohexylguanidine was encapsulated by assembly within the supercages of hydrophobic zeolite Y. The resulting 'ship-in-a-bottle' catalyst was active in the aldol reaction of

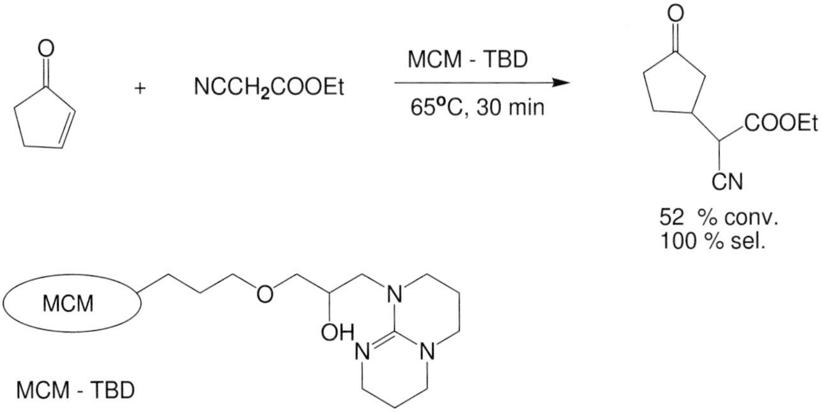

Fig. 2.41 Michael reactions catalyzed by MCM-TBD.

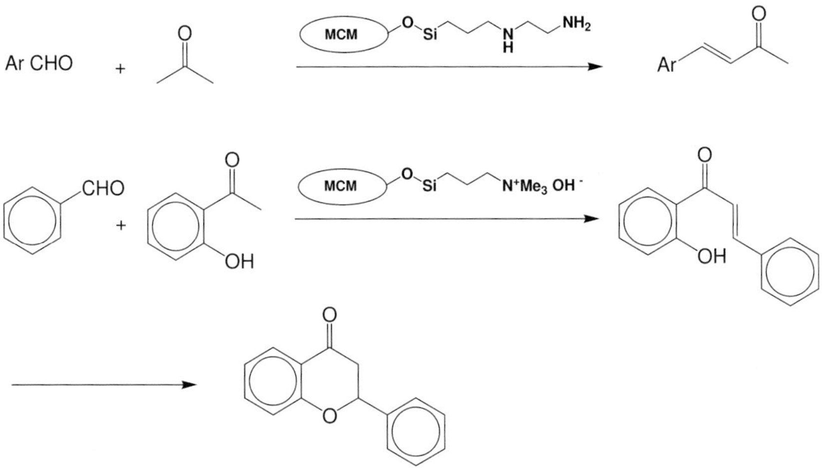

Fig. 2.42 Aldol reactions catalyzed by various solid bases.

acetone with benzaldehyde, giving 4-phenyl-4-hydroxybutan-2-one (Fig. 2.42) [138]. The catalyst was stable and recyclable but activities were relatively low, presumably owing to diffusion restrictions. Grafting of the same guanidine onto MCM-41 afforded a more active catalyst but closer inspection revealed that leaching of the base from the surface occurred. Similarly, diamine functionalized MCM-41 [139] and a tetraalkylammonium hydroxide functionalized MCM-41 [140] were shown to catalyze aldol condensations (Fig. 2.42).

2.4
Other Approaches

In Section 2.1 we discussed the use of solid Brønsted acids as recyclable alternatives to classical homogeneous Brønsted and Lewis acids. In the case of Lewis acids alternative strategies for avoiding the problems associated with their use can be envisaged. The need for greener alternatives for conventional Lewis acids such as $AlCl_3$ and $ZnCl_2$ derives from the necessity for decomposing the acid–base adduct formed between the catalyst and the product. This is generally achieved by adding water to the reaction mixture. Unfortunately, this also leads to hydrolysis of the Lewis acid, thus prohibiting its re-use and generating aqueous effluent containing copious amounts of inorganic salts. Since the product is often more basic than the substrate, e.g. in Friedel-Crafts acylations (see earlier), Lewis acid catalyzed processes often require stoichiometric amounts of catalyst.

One approach is to incorporate Lewis acids into, for example, zeolites or mesoporous silicas [141]. For example, incorporation of Sn(IV) into the framework of zeolite beta afforded a heterogeneous water-tolerant Lewis acid [142]. It proved to be an effective catalyst for the intramolecular carbonyl-ene reaction of citronellal to isopulegol [143] (Fig. 2.43) in batch or fixed bed operation. Hydrogenation of the latter affords menthol (Fig. 2.43).

Another approach is to design homogeneous Lewis acids which are water-compatible. For example, triflates of Sc, Y and lanthanides can be prepared in water and are resistant to hydrolysis. Their use as Lewis acid catalysts in aqueous media was pioneered by Kobayashi and coworkers [144–146]. The catalytic activity is dependent on the hydrolysis constant (K_h) and water exchange rate constant (WERC) for substitution of inner sphere water ligands of the metal cation [145]. Active catalysts were found to have pK_h values in the range 4–10. Cations having a pK_h of less than 4 are easily hydrolyzed while those with a pK_h greater than 10 display only weak Lewis acidity.

$Sc(OTf)_3$ and lanthanide triflates, particularly $Yb(TOTf)_3$ have been shown to catalyze a variety of reactions in aqueous/organic cosolvent mixtures [145, 146]. For example, they catalyze the nitration of aromatics with 69% aqueous nitric acid, the only by-product being water [147].

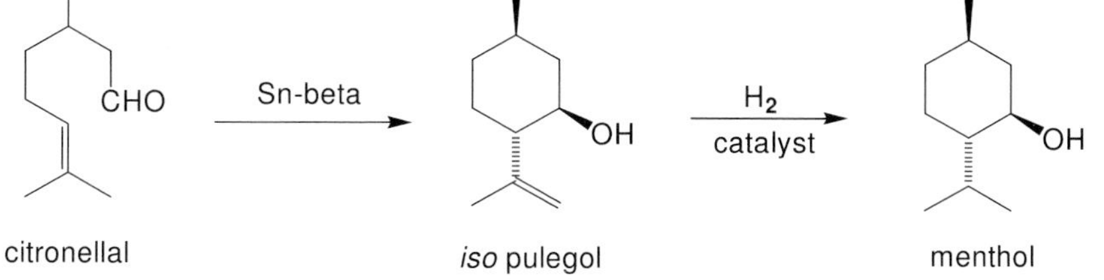

citronellal   *iso* pulegol   menthol

Fig. 2.43 Sn-Beta catalyzed carbonyl-ene reaction.

The lanthanide triflate remains in the aqueous phase and can be re-used after concentration. From a green chemistry viewpoint it would be more attractive to perform the reactions in water as the only solvent. This was achieved by adding the surfactant sodium dodecyl sulfate (SDS; 20 mol%) to the aqueous solution of e.g. $Sc(OTf)_3$ (10 mol%) [145]. A further extension of this concept resulted in the development of lanthanide salts of dodecyl sulfate, so-called Lewis acid–surfactant combined catalysts (LASC) which combine the Lewis acidity of the cation with the surfactant properties of the anion [148]. These LASCs, e.g. $Sc(DS)_3$, exhibited much higher activities in water than in organic solvents. They were shown to catalyze a variety of reactions, such as Michael additions and a three component α-aminophosphonate synthesis (see Fig. 2.44) in water [145].

Another variation on this theme is the use of a scandium salt of a hydrophobic polystyrene-supported sulfonic acid ($PS-SO_3H$) as an effective heterogeneous Lewis acid catalyst in aqueous media [149].

Ishihara and Yamamoto and coworkers [150] reported the use of 1 mol% of $ZrOCl_2 \cdot 8H_2O$ and $HfOCl_2 \cdot 8SH_2O$ as water-tolerant, reusable homogeneous catalysts for esterification.

Finally, it should be noted that Lewis acids and bases can also be used in other non-conventional media, as described in Chapter 7, e.g. fluorous solvents, supercritical carbon dioxide and ionic liquids by designing the catalyst, e.g. for solubility in a fluorous solvent or an ionic liquid, to facilitate its recovery and reuse. For example, the use of the ionic liquid butylmethylimidazolium hydroxide, [bmim][OH], as both a catalyst and reaction medium for Michael additions (Fig. 2.45) has been recently reported [151].

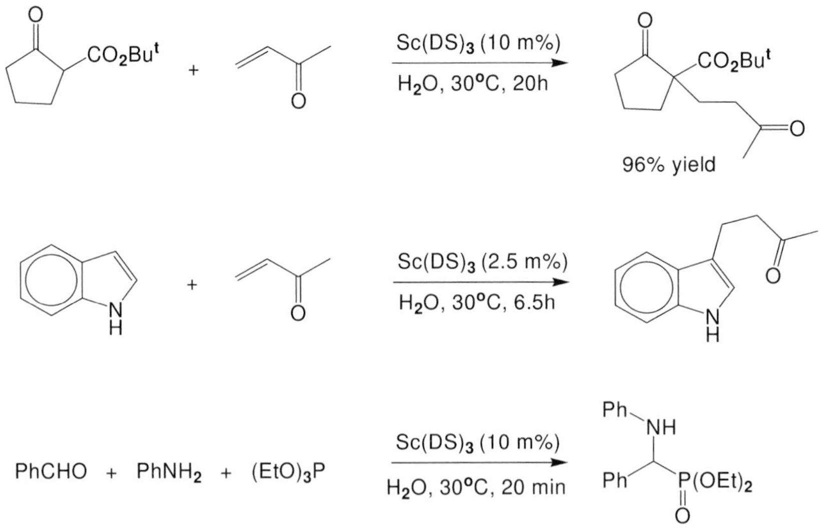

Fig. 2.44 $Sc(DS)_3$ catalyzed reactions in water.

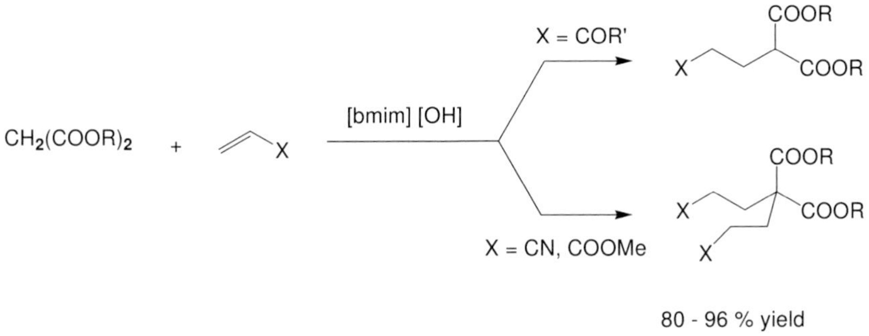

Fig. 2.45 Michael additions catalyzed by [bmim][OH].

It is clear that there are many possibilities for avoiding the use of classical acid and base catalysts in a wide variety of chemical reactions. Their application will result in the development of more sustainable processes with a substantial reduction in the inorganic waste produced by the chemical industry. Particularly noteworthy in this context is the use of chemically modified expanded corn starches, containing pendant SO_3H or NH_2 groups, as solid acid or base catalysts, respectively [152]. In addition to being recyclable these catalysts are biodegradable and derived from renewable raw materials (see Chapter 8).

References

1 K. Tanabe, W. F. Hölderich, *Appl. Catal. A: General*, **1999**, *181*, 399.
2 M. Harmer, in *Handbook of Green Chemistry and Technology*, J. Clark, D. Macquarrie (Eds.), Blackwell, Oxford, **2002**, pp. 86–119.
3 R. A. Sheldon, H. van Bekkum (Eds.), *Fine Chemicals through Heterogeneous Catalysis*, Wiley-VCH, Weinheim, **2001**.
4 W. Hölderich, *Catal. Today*, **2000**, *62*, 115.
5 A. Corma, H. Garcia, *Chem. Rev.*, **2003**, *103*, 4307.
6 R. A. Sheldon, R. S. Downing, *Appl. Catal. A: General*, **1999**, *189*, 163.
7 A. Mitsutani, *Catal. Today*, **2002**, *73*, 57.
8 M. Campanati, A. Vaccari, in Ref. [3], pp. 61–79.
9 A. Vaccari, *Catal. Today*, **1998**, *41*, 53.
10 A. Vaccari, *Appl. Clay Sci.*, **1999**, *14*, 161.
11 H. van Bekkum, E. M. Flanigen, J. C. Jansen (Eds.), *Introduction to Zeolite*

Science and Practice, Elsevier, Amsterdam, **1991**.
12 A. Corma, in Ref. [3], pp. 80–91.
13 A. Corma, *J. Catal.*, **2003**, *216*, 298.
14 A. Corma, *Chem. Rev.*, **1995**, *95*, 559.
15 Y. Izumi, in Ref. [3], pp. 100–105.
16 M. M. Sharma, *React. Funct. Polym.*, **1995**, *26*, 3.
17 A. Chakrabarti, M. M. Sharma, *React. Funct. Polym.*, **1993**, *20*, 1.
18 W. van Rhijn, D. De Vos, P. A. Jacobs, in Ref. [3], pp. 106–113.
19 M. H. Valkenberg, W. F. Hölderich, *Catal. Rev.*, **2002**, *44* (2), 321.
20 Y. Izumi, K. Urabe, M. Onaka, *Zeolite, Clay and Heteropoly Acids in Organic Reactions*, VCH, Weinheim, **1992**.
21 In Ref. [20], p. 53.
22 O. Sieskind, P. Albrecht, *Tetrahedron Lett.*, **1993** *34*, 1197.
23 A. Cornelis, P. Laszlo, *SynLett.*, **1994**, *3*, 155.

24 J. H. Clark, A. P. Kybett, D. J. Macquarrie, S. J. Barlow, *J. Chem. Soc., Chem. Commun.*, **1989**, 1353.

25 S. J. Barlow, T. W. Bastock, J. H. Clark, S. R. Cullen, *J. Chem. Soc., Perkin Trans.*, **1994**, *2*, 411.

26 T. J. Pinnavaia, *Science*, **1983**, *220*, 4595.

27 F. Figueras, *Catal. Rev. Sci. Eng.*, **1988**, *30*, 457.

28 J. Sterte, *Catal. Today*, **1988**, *2*, 219.

29 S. L. Suib, *Chem. Rev.*, **1993**, *93*, 803.

30 S. T. Wilson, in Ref. [11], pp. 137–151.

31 R. Szostak, in Ref. [11], pp. 153–199.

32 G. T. Kokotailo, S. L. Lawton, D. H. Olsen, W. M. Meijer, *Nature*, **1978**, *272*, 437.

33 For a historical account see E. M. Flanigen, in Ref. [11], pp. 31–34.

34 A. Corma, *Chem. Rev.*, **1997**, *97*, 2373.

35 C. T. Kresge, M. E. Leonowicz, W. J. Roth, J. C. Vartuli, J. S. Beck, *Nature*, **1992**, *359*, 710.

36 J. S. Beck, J. C. Vartuli, W. J. Roth, M. E. Leonowicz et al., *J. Am. Chem. Soc.*, **1992**, *114*, 10834.

37 D. Macquarrie, in *Handbook of Green Chemistry and Technology*, J. Clark, D. Macquarrie (Eds.), Blackwell, Oxford, **2002**, pp. 120–147.

38 P. B. Weisz, *Pure Appl. Chem.*, **1980**, *52*, 2091.

39 G. R. Meima, G. S. Lee, J. M. Garces, in Ref. [3], pp. 151–160.

40 P. Métivier, in Ref. [3], pp. 173–177.

41 J. A. Elings, R. S. Downing, R. A. Sheldon, *Stud. Surf. Sci. Catal.*, **1997**, *105*, 1125.

42 E. A. Gunnewegh, A. J. Hoefnagel, R. S. Downing, H. van Bekkum, *Recl. Trav. Chim. Pays-Bas*, **1996**, *115*, 226.

43 E. A. Gunnewegh, A. J. Hoefnagel, H. van Bekkum, *J. Mol. Catal.*, **1995**, *100*, 87.

44 E. A. Gunnewegh, R. S. Downing, H. van Bekkum, *Stud. Surf. Sci. Catal.*, **1995**, *97*, 447.

45 M. Spagnol, L. Gilbert, D. Alby, in *The Roots of Organic Development*, J. R. Desmurs, S. Ratton (Eds.), Ind. Chem. Lib., Vol. 8, Elsevier, Amsterdam, **1996**, pp. 29–38; S. Ratton, *Chem. Today*, **1997**, *3–4*, 33.

46 P. Métivier, in Ref. [3], pp. 161–172.

47 A. Vogt, A. Pfenninger, EP0701987A1, **1996**, to Uetikon AG.

48 B. Chiche, A. Finiels, C. Gauthier, P. Geneste, J. Graille, D. Pioch, *J. Org. Chem.*, **1986**, *51*, 2128.

49 A. Corma, M. J. Ciment, H. Garcia, J. Primo, *Appl. Catal.*, **1989**, *49*, 109.

50 W. Hölderich, M. Hesse, F. Naumann, *Angew. Chem.*, **1988**, *27*, 226.

51 A. Finiels, A. Calmettes, P. Geneste, P. Moreau, *Stud. Surf. Sci. Catal.*, **1993**, *78*, 595; J. J. Fripiat, Y. Isaev, *J. Catal.*, **1999**, *182*, 257.

52 A. K. Pandey, A. P. Singh, *Catal. Lett.*, **1997**, *44*, 129; R. Sreekumar, R. Padmakumar, *Synth. Commun.*, **1997**, *27*, 777.

53 A. Kogelbauer, H. W. Kouwenhoven, in Ref. [3], pp. 123–132.

54 P. Ratnasamy, A. P. Singh, in Ref. [3], pp. 133–150.

55 L. Bertea, H. W. Kouwenhoven, R. Prins, *Appl. Catal. A: General*, **1995**, *129*, 229.

56 K. Iwayama, S. Yamakawa, M. Kato, H. Okino, Eur. Pat. Appl. EP948988, **1999**, to Toray.

57 H. Ishida, *Catal. Surveys Jpn.*, **1997**, *1*, 241.

58 W. F. Hölderich, *Stud. Surf. Sci. Catal.*, **1993**, *75*, 127; W. F. Hölderich, in *Comprehensive Supramolecular Chemistry*, J.-M. Lehn et al. (Eds.), **1996**, Vol. 7, pp. 671–692.

59 A. Chauvel, B. Delmon, W. F. Hölderich, *Appl. Catal. A*, **1994**, *115*, 173.

60 H. Ichihashi, M. Ishida, A. Shiga, M. Kitamura, T. Suzuki, K. Suenobu, K. Sugita, *Catal. Surveys Asia*, **2003**, *7* (4), 261.

61 H. Ichihashi, H. Sato, *Appl. Catal. A: General*, **2001**, *221*, 359; H. Ichihashi, M. Kitamura, *Catal. Today*, **2002**, *73*, 23.

62 H. Sato, *Catal. Rev. Sci. Eng.*, **1997**, *39*, 395.

63 T. Tatsumi, in Ref. [3], pp. 185–204.

64 For a discussion of by-product free routes to caprolactam see: G. Dahlhoff, J. P. M. Niederer, W. F. Hölderich, *Catal. Rev.*, **2001**, *43*, 381.

65 G. P. Heitmann, G. Dahlhoff, W. F. Hölderich, *J. Catal.*, **1999**, *186*, 12.

66 M. Guisnet, G. Perot, in Ref. [3], pp. 211–216, and references cited therein.

67 I. Nicolai, A. Aguilo, US 4652683, **1987**; B. B. G. Gupta, US 4668826, **1987**, to Hoechst-Celanese.

68 A. J. Hoefnagel, H. van Bekkum, *Appl. Catal. A: General*, **1993**, *97*, 87.

69 W. F. Hölderich, U. Barsnick, in Ref. [3], pp. 217–231.

70 W. F. Hölderich, J. Röseler, G. Heitman, A. T. Liebens, *Catal. Today*, **1997**, *37*, 353.

71 P. J. Kunkeler, J. C. van der Waal, J. Bremmer, B. J. Zuurdeeg, R. S. Downing, H. van Bekkum, *Catal. Lett.*, **1998**, *53*, 135.

72 J. A. Elings, H. E. B. Lempers, R. A. Sheldon, *Eur. J. Org. Chem.*, **2000**, 1905.

73 R. A. Sheldon, J. A. Elings, S. K. Lee, H. E. B. Lempers, R. S. Downing, *J. Mol. Catal. A*, **1998**, *134*, 129.

74 A. Molnar, in Ref. [3], pp. 232–239.

75 R. A. Budnik, M. R. Sanders, Eur. Pat. Appl. EP158310, **1985**, to Union Carbide.

76 M. Hesse, W. Hölderich, E. Gallei, *Chem. Ing. Tech.*, **1991**, *63*, 1001.

77 C. H. McAteer, E. F. V. Scriven, in Ref. [3], pp. 275–283.

78 C. D. Chang, W. H. Lang, US 4220783, **1980**, to Mobil Oil Corp.

79 J. Heveling, *Chimia*, **1996**, *50*, 114.

80 R. S. Downing, P. J. Kunkeler, in Ref. [3], pp. 178–183, and references cited therein.

81 G. J. Meuzelaar, R. A. Sheldon, in Ref. [3], pp. 284–294.

82 M. A. Harmer, W. E. Farneth, Q. Sun, *Adv. Mater.*, **2001**, *221*, 45.

83 M. A. Harmer, Q. Sun, *Appl. Catal. A: General*, **2001**, *221*, 45.

84 A. E. W. Beers, T. A. Nijhuis, F. Kapteijn, in Ref. [3], pp. 116–121.

85 G. A. Olah, P. S. Iyer, G. K. Prakash, *Synthesis*, **1986**, *7*, 513.

86 M. A. Harmer, W. E. Farneth, Q. Sun, *J. Am. Chem. Soc.*, **1996**, *118*, 7708.

87 M. A. Harmer, Q. Sun, A. J. Vega, W. E. Farneth, A. Heidekum, W. F. Hoelderich, *Green Chem.*, **2000**, *6*, 7.

88 A. Heidekum, M. A. Harmer, W. F. Hölderich, *J. Catal.*, **1998**, *176*, 260.

89 S. Wieland, P. Panster, in Ref. [3], pp. 92–99.

90 M. A. Harmer, Q. Sun, M. J. Michalczyk, Z. Yang, *Chem. Commun.*, **1997**, 1803.

91 D. Margolese, J. A. Melero, S. C. Christiansen, B. F. Chmelka, G. D. Stucky, *Chem. Mater.*, **2000**, *12*, 2448.

92 J. A. Melero, G. D. Stucky, R. van Grieken, G. Morales, *J. Mater. Chem.*, **2002**, *12*, 1664.

93 D. Das, J. F. Lee, S. Cheng, *Chem. Commun.*, **2001**, 2178.

94 I. Diaz, C. Marquez-Alvarez, F. Mohino, J. Perez-Pariente, E. Sastre, *Micropor. Mesopor. Mater.*, **2001**, *44–45*, 295.

95 W. M. van Rhijn, D. E. De Vos, B. F. Sels, W. D. Bossaert, P. A. Jacobs, *Chem. Commun.*, **1998**, 317; W. D. Bossaert, D. E. De Vos, W. M. van Rhijn, J. Bullen, P. J. Grobet, P. A. Jacobs, *J. Catal.*, **1999**, *182*, 156.

96 R. Richer, L. Mercier, *Chem. Commun.*, **1998**, 1775.

97 J. A. Melero, R. van Grieken, G. Morales, V. Nuno, *Catal. Commun.*, **2004**, *5*, 131.

98 M. Alvaro, A. Corma, D. Das, V. Fornes, H. Garcia, *Chem. Commun.*, **2004**, 956.

99 M. T. Pope, *Heteropoly and Isopoly Oxometalates*, Springer, Berlin, 1983.

100 N. Mizuno, M. Misono, *Chem. Rev.*, **1998**, *98*, 199.

101 Y. Izumi, in Ref. [3], pp. 100–105.

102 For reviews see: I. V. Kozhevnikov, *Catal. Rev. Sci. Eng.*, **1995**, *37*, 311; I. V. Kozhevnikov, *Chem. Rev.*, **1998**, *98*, 171; T. Okuhara, N. Mizuno, M. Misono, *Adv. Catal.*, **1996**, *41*, 113; M. Misono, I. Ono, G. Koyano, A. Aoshima, *Pure Appl. Chem.*, **2000**, *72*, 1305.

103 H. Hattori, *Chem. Rev.*, **1995**, *95*, 537.

104 F. Cavani, F. Trifiro, A. Vaccari, *Catal. Today*, **1991**, *11*, 173.

105 B. F. Sels, D. E. De Vos, P. A. Jacobs, *Catal. Rev.*, **2001**, *43*, 392.

106 G. J. Kelly, F. King , M. Kett, *Green Chem.*, **2002**, *4*, 392.

107 F. Figueras, J. Lopez, in Ref. [3], pp. 327–337.

108 J. C. Roelofs, A. J. van Dillen, K. P. de Jong, *Catal. Today*, **2000**, *60*, 297.

109 M. J. Climent, A. Corma, S. Iborra, J. Primo, *J. Catal.*, **1995**, *151*, 60.

110 F. Figueras, D. Tichit, M. Bennani Naciri, R. Ruiz, in *Catalysis of Organic Reactions*, F. E. Herkes (Ed.), Marcel Dekker, New York, 1998, p. 37.

111 A. Corma, V. Fornes, R. M. Martin-Aranda, F. Rey, *J. Catal.*, **1992**, *134*, 58.

112 A. Corma, R. M. Martin-Aranda, *Appl. Catal. A: General*, **1993**, *105*, 271.

113 A. Corma, S. Iborra, J. Primo, F. Rey, *Appl. Catal. A: General*, **1994**, *114*, 215.

114 C. Cativiela, F. Figueras, J. I. Garcia, J. A. Mayoral, M. M. Zurbano, *Synth. Commun.*, **1995**, *25*, 1745.

115 K. Kaneda, K. Yamaguchi, K. Mori, T. Mizugaki, K. Ebitani, *Catal. Surveys Japan*, **2000**, *4*, 31; K. Kaneda, S. Ueno, K. Ebitani, *Curr. Top. Catal.*, **1997**, *1*, 91.

116 C. Cativiela, F. Figueras, J. M. Fraile, J. I. Garcia, J. A. Mayoral, *Tetrahedron Lett.*, **1995**, *36*, 4125.

117 S. Ueno, K. Yamaguchi, K. Yoshida, K. Ebitani, K. Kaneda, *Chem. Commun.*, **1998**, 295.

118 K. Yamaguchi, K. Ebitani, K. Kaneda, *J. Org. Chem.*, **1999**, *64*, 2966.

119 F. Winter, A. J. van Dillen, K. P. de Jong, *Chem. Commun.*, **2005**, 3977.

120 M. Zahouily, Y. Abrouki, B. Bahlaouan, A. Rayadh, S. Sebti, *Catal. Commun.*, **2003**, *4*, 521.

121 A. Corma, S. Iborra, in Ref. [3], pp. 309–326.

122 R. Streekumar, P. Rugmini, R. Padmakumar, *Tetrahedron Lett.*, **1997**, *38*, 6557.

123 Y. Ono, *CATTECH*, March **1997**, 31.

124 K. R. Kloetstra, H. van Bekkum, *Chem. Soc. Chem. Commun.*, **1995**, 1005.

125 P. A. Hathaway, M. E. Davis, *J. Catal.*, **1989**, *116*, 263 and 279.

126 I. Rodriguez, H. Cambon, D. Brunel, M. Lasperas, *J. Mol. Catal. A: Chemical*, **1998**, *130*, 95.

127 K. R. Kloetstra, H. van Bekkum, *Stud. Surf. Sci. Catal.*, **1997**, *105*, 431.

128 E. Angeletti, C. Canepa, G. Martinetti, P. Venturello, *Tetrahedron Lett.*, **1988**, *18*, 2261; see also E. Angeletti, C. Canepa, G. Martinelli, P. Venturello, *J. Chem. Soc. Perkin Trans.*, **1989**, *1*, 105.

129 M. Lasperas, T. Llorett, L. Chaves, I. Rodriguez, A. Cauvel, D. Brunel, *Stud. Surf. Sci. Catal.*, **1997**, *108*, 75.

130 A. Cauvel, G. Renard, D. Brunel, *J. Org. Chem.*, **1997**, *62*, 749.

131 D. Brunel, *Micropor. Mesopor. Mat.*, **1999**, 329.

132 D. J. Macquarrie, D. Brunel, in Ref. [3], pp. 338–348.

133 D. J. Macquarrie, in Ref. [2], pp. 120–149.

134 D. B. Jackson, D. J. Macquarrie, *Chem. Commun.*, **1997**, 1781.

135 D. J. Macquarrie, *Green Chem.*, **1999**, *1*, 195.

136 Y. V. Subba Rao, D. E. De Vos, P. A. Jacobs, *Angew. Chem. Int. Ed.*, **1997**, *36*, 2661.

137 M. Lasperas, T. Lloret, L. Chaves, I. Rodriguez, A. Cauvel, D. Brunel, *Stud. Surf. Sci. Catal.*, **1997**, *108*, 75.

138 R. Sercheli, A. L. B. Ferreira, M. C. Guerreiro, R. M. Vargas, R. A. Sheldon, U. Schuchardt, *Tetrahedron Lett.*, **1997**, *38*, 1325.

139 B. M. Choudry, M. Lakshmi-Kantam, P. Sreekanth, T. Bandopadhyay, F. Figueras, A. Tuel, *J. Mol. Catal.*, **1999**, *142*, 361.

140 I. Rodriguez, S. I. Iborra, A. Corma, F. Rey, J. J. Jorda, *Chem. Commun.*, **1999**, 593.

141 A. Corma, H. Garcia, *Chem. Rev.*, **2002**, *102*, 3837; A. Corma, H. Garcia, *Chem. Rev.*, **2003**, *103*, 4307.

142 A. Corma, L. T. Nemeth, M. Renz, S. Valencia, *Nature*, **2001**, *412*, 423.

143 A. Corma, M. Renz, *Chem. Commun.*, **2004**, 550.

144 S. Kobayashi, *Chem. Lett.*, **1991**, 2187.

145 S. Kobayashi, K. Manabe, *Acc. Chem. Res.*, **2002**, *35*, 209.

146 S. Kobayashi, M. Sugiura, H. Kitagawa, W. W. L. Lam, *Chem. Rev.*, **2002**, *102*, 2227.

147 F. J. Waller, A. G. M. Barrett, D. C. Braddock, D. Ramprasad, *Chem. Commun.*, **1997**, 613.

148 S. Kobayashi, T. Wakabayashi, *Tetrahedron Lett.*, **1998**, *39*, 5389.

149 S. Imura, K. Manabe, S. Kobayashi, *Tetrahedron*, **2004**, *60*, 7673.

150 M. Nakayama, A. Sato, K. Ishihara, H. Yamamoto, *Adv. Synth. Catal.*, **2004**, *346*, 1275.

151 B. C. Ranu, S. Banerjee, *Org. Lett.*, **2005**, *7*, 3049.

152 S. Doi, J. H. Clark, D. J. Macquarrie, K. Milkowski, *Chem. Commun.* **2002**, 2632.

3
Catalytic Reductions

3.1
Introduction

Catalytic hydrogenation – using hydrogen gas and heterogeneous catalysts – can be considered as the most important catalytic method in synthetic organic chemistry on both laboratory and production scales. Hydrogen is, without doubt, the cleanest reducing agent and heterogeneous robust catalysts have been routinely employed. Key advantages of this technique are (i) its broad scope, many functional groups can be hydrogenated with high selectivity; (ii) high conversions are usually obtained under relatively mild conditions in the liquid phase; (iii) the large body of experience with this technique makes it possible to predict the catalyst of choice for a particular problem and (iv) the process technology is well established and scale-up is therefore usually straightforward. The field of hydrogenation is also the area where catalysis was first widely applied in the fine chemical industry. Standard hydrogenations of olefins and ketones, and reductive aminations, using heterogeneous catalysts, have been routinely performed for more than two decades [1]. These reactions are usually fast and catalyst separation is easy. Catalysts consist of supported noble metals, Raney nickel, and supported Ni or Cu. However, once enantioselectivity is called for homogeneous catalysis and biocatalysis generally become the methods of choice. The field of homogeneous asymmetric catalysis had a major breakthrough with the discovery of ligands such as BINAP, DIPAMP and DIOP (for structures see later), designed by the pioneers in this field, Noyori, Knowles and Kagan [2–4]. These ligands endow rhodium, ruthenium and iridium with the unique properties which allow us nowadays to perform enantioselective reduction of a large number of compounds using their homogeneous metal complexes. Besides homogeneous and heterogeneous catalysis, the potential of biocatalytic reduction is still growing and has already delivered some interesting applications [5]. Reduction of many secondary carbonyl compounds can be performed using enzymes as catalysts, yielding chiral compounds with high enantioselectivities [6]. Use of new genetic engineering techniques is rapidly expanding the range of substrates to be handled by enzymes. The issue of cofactor-recycling, often denoted as the key problem in the use of biocatalytic reductions, is nowadays mainly a technological issue which can be solved by applying a

Green Chemistry and Catalysis. I. Arends, R. Sheldon, U. Hanefeld
Copyright © 2007 WILEY-VCH Verlag GmbH & Co. KGaA, Weinheim
ISBN: 978-3-527-30715-9

substrate or enzyme coupled approach in combination with novel reactor concepts. In this chapter, all three catalytic methods: heterogeneous, homogeneous and biocatalytic reductions, will be separately described and illustrated with industrial examples. Enantioselective hydrogenation applications, which are mainly the domain of homogeneous and enzymatic catalysts, will constitute a main part of this chapter.

3.2
Heterogeneous Reduction Catalysts

3.2.1
General Properties

The classical hydrogenation catalysts for preparative hydrogenation are supported noble metals, Raney nickel and supported Ni and Cu, all of which are able to activate hydrogen under mild conditions [1]. Because only surface atoms are active, the metal is present as very small particles in order to give a high specific surface area. It is generally accepted that the catalytic addition of hydrogen to an X=Y bond does not occur in a concerted manner but stepwise. In other words, the H–H bond has to be cleaved first giving intermediate M–H species, which are then added stepwise to the X=Y bond. This is depicted schematically in Fig. 3.1 for a metallic surface for the hydrogenation of ethylene [7]. The first step is called dissociative adsorption and the newly formed M–H bonds deliver the energetic driving force for the cleavage of the strong H–H bond. The metal surface also forms complexes with the X=Y, most probably via a π-bond, thereby activating the second reactant and placing it close to the M–H fragments, allowing the addition to take place. There is consensus that the H is added to the complexed or adsorbed X=Y from the metal side, leading to an overall cis-addition [7]. The outcome of a diastereoselective hydrogenation can thus also be rationalized. For substrate coordination to the surface, preferential adsorption of the less hindered face will take place preferentially, due to repulsive interactions. Furthermore if anchoring groups are present in the molecule, such as OH, sulfide or amine, repulsive interactions can be overruled and the opposite stereoisomer will be obtained.

It is important to realize that even today it is not possible to adequately characterize a preparative heterogeneous catalyst on an atomic level. Most research in heterogeneous hydrogenation took place in the 1970s to 1980s, and the results thereof can be found in the pioneering monographs of Rylander [1], Augustine [8], and Smith [9]. Catalysts are still chosen on an empirical basis by trial and error and it is rarely understood why a given catalyst is superior to another one. The factors which influence the reactivity and selectivity are: (i) *Type of metal:* noble metals (Pd, Pt, Rh, Ru) versus base metals (Ni and Cu). Bimetallic catalysts are also applied. (ii) *Type of support:* charcoal, versus aluminas or silicas. (iii) *Type of catalyst:* supported on carrier, fine powders, Ni as Raney nickel

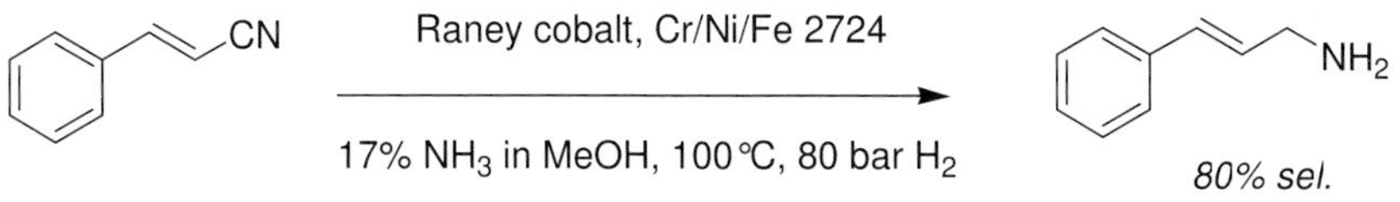

Fig. 3.7 Selective cinnamonitrile hydrogenation to 3-phenylallylamine.

The field of arene hydrogenation is almost 100 years old. The initial work of Sabatier on the interaction of finely divided nickel with ethylene and hydrogen gas led to the development of the first active catalyst for the hydrogenation of benzene [29]. In the last two decades, arene hydrogenation has also become important for the area of fine chemistry. Traditionally it is accomplished by use of a Group VIII metal, with the rate of hydrogenation depending on the metal used, i.e. Rh > Ru > Pt > Ni > Pd > Co [1]. Hydrogenation of multiply substituted aromatic rings can lead to the formation of a variety of stereoisomers. Hydrogenation of functionalized arenes usually leads predominantly to the cis-substituted product. The nature of carrier, type of metal, solvent, temperature and pressure determine the exact amount of stereochemical induction. In Fig. 3.8, the use of Ru/C enables hydrogenation of tri-substituted benzoic acid into the corresponding cis-product with high stereoselectivity [30].

The partial hydrogenation of an arene to its cyclohexene derivative is difficult to achieve, because complete hydrogenation to cyclohexane tends to occur. Often the use of a Ru/C catalyst can solve this problem because Ru is not very effective at hydrogenating olefinic double bonds. Alternatively Pt/C and Rh/C can be used [31]. The Asahi Corporation has developed a benzene-to-cyclohexene process involving a liquid–liquid two-phase system (benzene–water) with a solid ruthenium catalyst dispersed in the aqueous phase. The low solubility of cyclohexene in water promotes rapid transfer towards the organic phase. An 80000 tons/year plant using this process is in operation [32]. Another way to scavenge the intermediate cyclohexene is to support the metal hydrogenation catalysts on an acidic carrier (e.g. silica–alumina). On such a bifunctional catalyst the cyclohexene enters the catalytic alkylation of benzene to yield cyclohexylbenzene [33]. The latter can be converted with high selectivity, by oxidation and rearrangement reactions, into phenol and cyclohexanone [34]. The complete reaction cycle is shown in Fig. 3.9.

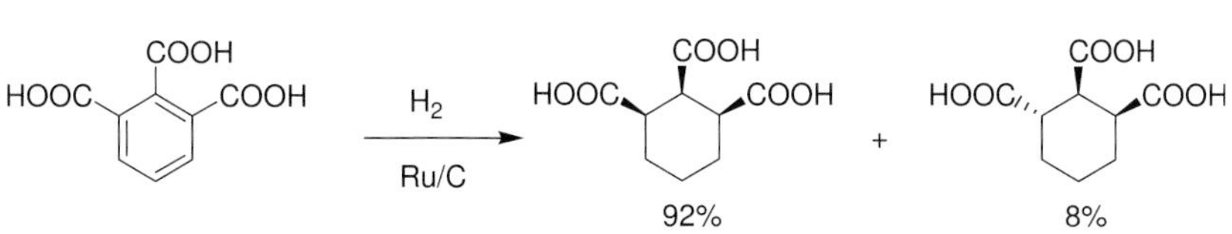

Fig. 3.8 Selective hydrogenation of a tri-substituted aromatic ring.

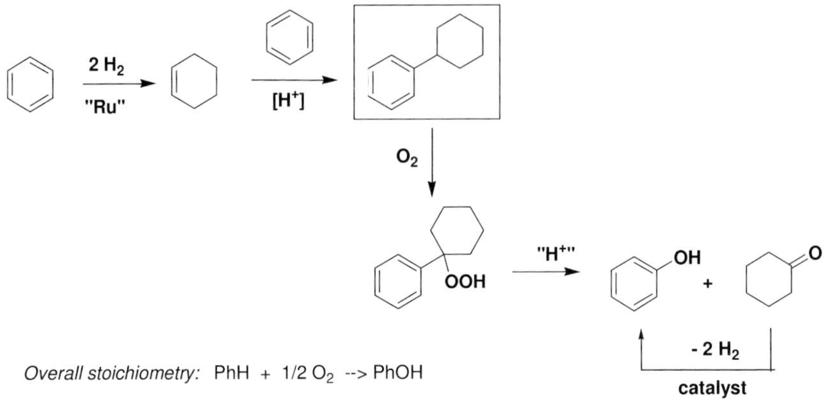

Overall stoichiometry: PhH + 1/2 O$_2$ --> PhOH

Fig. 3.9 Partial hydrogenation of benzene and production of phenol.

Another example of catalytic hydrogenation, which demonstrates the flexibility of this method, is the hydrogenation of nitrogen-containing aromatic ring systems. For example isoquinoline ring saturation can be directed towards either or both of the two rings (Fig. 3.10). When a platinum catalyst was used in acetic acid, the hydrogenation of isoquinoline results in saturation of the nitrogen-containing aromatic ring. Changing the solvent to methanolic hydrogen chloride results in hydrogenation of the other aromatic ring, and use of ethanol in combination with sulfuric acid results in saturation of both aromatic rings. Under the latter reaction conditions, especially when using Ru catalysts, the cisproduct is formed preferentially [32].

Additional reactions which need to be highlighted are the reductive alkylation of alcohols and amines with aldehydes leading to the green synthesis of ethers and amines. These reactions are generally catalyzed by palladium [35]. This reaction can replace the classical Williamson's synthesis of ethers which requires an alcohol and an alkyl halide together with a base, and always results in the concomitant production of salt. The choice of Pd/C as catalyst is due to the low efficiency of this metal for the competitive carbonyl reduction. Analysis of the

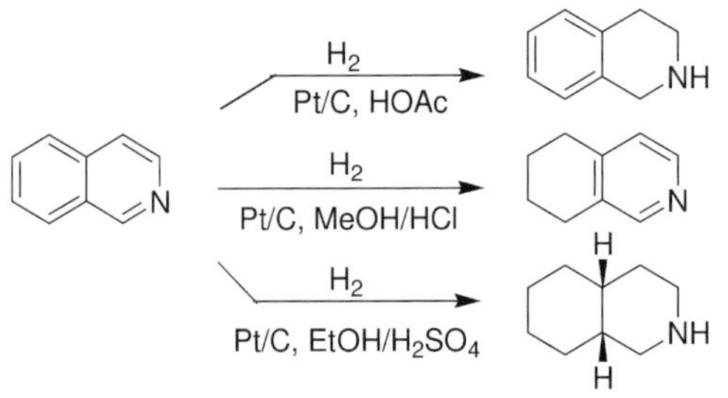

Fig. 3.10 Influence of solvents on the Pt/C catalyzed selective hydrogenation of isoquinoline.

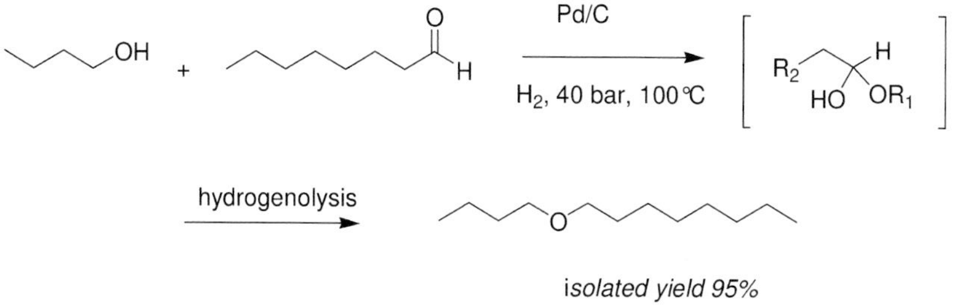

Fig. 3.11 Catalytic alternative to the Williamson's synthesis of ethers.

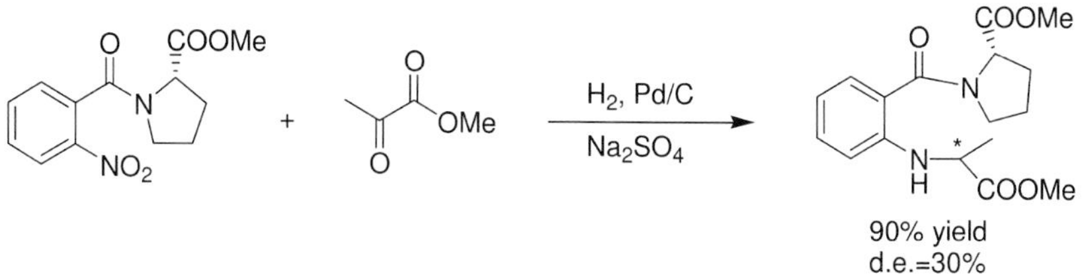

Fig. 3.12 Reductive amination of nitro derivatives leading to non-natural amino acid derivatives.

supposed reaction mechanism indicates that the first step of the reaction is the formation of the hemiketal which is favored by the use of one of the reactants (the alcohol or the aldehyde) as solvent (Fig. 3.11). After hydrogenolysis under H$_2$ pressure, ethers are produced in high yields after simple filtration of the catalyst and evaporation of the solvent.

In reductive amination, the alkylation is performed under hydrogen in the presence of a catalyst and an aldehyde or ketone as the alkylating agent. The method was developed for anilines and extended to amide N-alkylation [36]. A notable example is the use of nitro derivatives as aniline precursors in a one-pot reduction of the nitro group and subsequent reductive alkylation of the resulting aniline (Fig. 3.12).

N- and O-benzyl groups are among the most useful protective groups in synthetic organic chemistry and the method of choice for their removal is catalytic hydrogenolysis [37]. Recently the most important reaction conditions were identified [38]: Usually 5–20% Pd/C; the best solvents are alcoholic solvents or acetic acid; acids promote debenzylation, whereas amines can both promote and hinder hydrogenolysis. Chemoselectivity can mainly be influenced by modifying the classical Pd/C catalysts.

3.2.2
Transfer Hydrogenation Using Heterogeneous Catalysts

On a small scale it can be advantageous to perform a transfer hydrogenation, in which an alcohol, such as isopropanol, serves as hydrogen donor. The advantage of this technique is that no pressure equipment is needed to handle hydrogen, and that the reactions can be performed under mild conditions, without the risk of reducing other functional groups. The so-called Meerwein-Ponndorf-Verley (MPV) reduction of aldehydes and ketones, is a hydrogen-transfer reaction using easily oxidizable alcohols as reducing agents. Industrial applications of the MPV-reactions are found in the fragrance and pharmaceutical industries. MPV reactions are usually performed by metal alkoxides such as Al(O-iPr)₃. The activity of these catalysts is related to their Lewis-acidic character in combination with ligand exchangeability. Naturally, a heterogeneous catalyst would offer the advantage of easy separation from the liquid reaction mixture. Many examples of heterogeneously catalyzed MPV reactions have now been reported [39]. The catalysts comprise (modified) metal oxides which have either Lewis acid or base properties. The mechanism involves the formation of an alkoxide species in the first step, and in the second step a cyclic six-membered transition state (see Fig. 3.13). Examples are the use of alumina, ZrO_2 and immobilized zirconium complexes, magnesium oxides and phosphates and Mg–Al hydrotalcites.

Especially worth mentioning in a green context is the use of mesoporous materials and zeolites, as stable and recyclable catalysts for MPV reductions. High activity was obtained by using zeolite-beta catalysts. Beta zeolites have a large pore three-dimensional structure with pores of size 7.6×6.4 Å² which makes them suitable for a large range of substrates. Al, Ti- and Sn-beta zeolite have all been used as catalysts for the selective reduction of cyclohexanones [40–42]. The

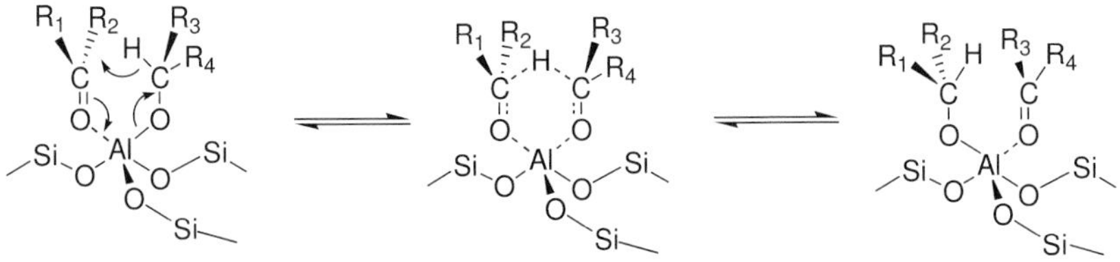

Fig. 3.13 Mechanism of MPV reduction catalyzed by Al-beta zeolite.

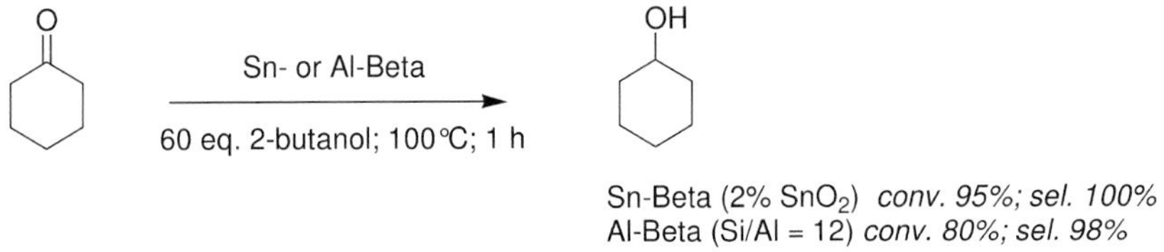

Sn-Beta (2% SnO₂) *conv. 95%; sel. 100%*
Al-Beta (Si/Al = 12) *conv. 80%; sel. 98%*

Fig. 3.14 MPV reductions catalyzed by Sn- and Al-zeolite beta.

reaction is shown in Fig. 3.14. In terms of both activity and selectivity Sn-beta (containing 2% SnO_2) seems to be superior. This can be ascribed to the interaction of the carbonyl group with the Sn-center, which is stronger than with Ti and more selective than with Al-centers [42]. Additionally, shape-selectivity effects can be observed for all three zeolites. When using 4-*tert*-butylcyclohexanone as the substrate, the selectivity for the cis-isomer easily reaches 99%. This is clear proof that the reaction occurs within the pores of the zeolite and that the active tin centers are not at the external surface of the zeolite or in solution.

3.2.3
Chiral Heterogeneous Reduction Catalysts

Despite the fact that enantioselective hydrogenation is largely the domain of homogeneous catalysts, some classical examples of heterogeneous reduction catalysts need to be mentioned. In this case the metal surface is modified by a (natural) chiral additive. The first successful attempts were published about 60 years ago and despite a large effort in this field, only three examples have shown success: the Raney nickel system for β-functionalized ketones, Pt catalysts modified with cinchona alkaloids for α-functionalized ketones, and Pd catalysts modified with cinchona alkaloids for selected activated C=C bonds [43, 44]. The Raney-Ni–tartaric acid–NaBr catalyst system, known as the Izumi system [45], affords good to high enantioselectivity in the hydrogenation of β-functionalized ketones and reasonable results have also been obtained for unfunctionalized ketones. Besides Raney nickel, Ni powder and supported nickel are almost as good precursors. In the optimized preparation procedure, RaNi undergoes ultrasonication in water followed by modification with tartaric acid and NaBr at 100 °C and pH 3.2. The modification procedure is highly corrosive and produces large amounts of nickel- and bromide-containing waste. This, together with its low activity (typical reaction time is 48 h), hampers its industrial application. Examples are given in Fig. 3.15 [43].

For the hydrogenation of α-functionalized ketones, the Pt on alumina system, modified with cinchonidine or its simple derivative 10,11-dihydro-O-methylcinchonidine, is the best catalyst [43, 44, 46]. This so-called Orito system [47] is

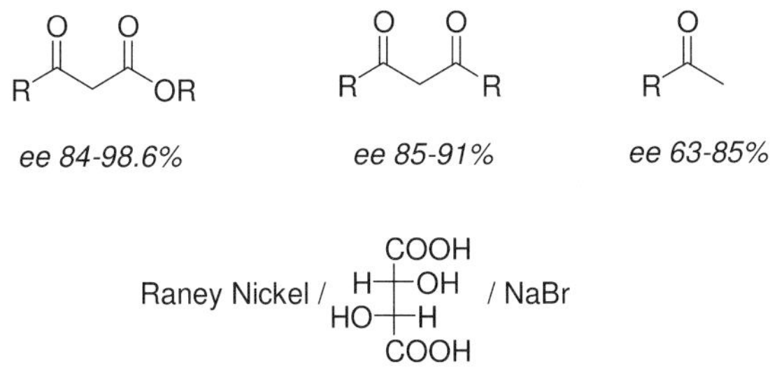

ee 84-98.6%   ee 85-91%   ee 63-85%

Raney Nickel / [tartaric acid structure: COOH, H–OH, HO–H, COOH] / NaBr

Fig. 3.15 Best results for the enantioselective hydrogenation of β-functionalized ketones and non-functionalized ketones using the Izumi system.

most well known for the hydrogenation of a-ketoesters. Acetic acid or toluene as solvent, close to ambient temperature and medium to high pressure (10–70 bar) are sufficient to ensure high enantioselectivities of 95 to 97.5% (see Fig. 3.16). The highest *ee* has been obtained after reductive heat treatment of Pt/Al$_2$O$_3$ and subsequent sonochemical pretreatment at room temperature [48]. In recent years the substrate range of cinchona-modified Pt has been extended to the hydrogenation of selected activated ketones, including ketopantolactone, a-keto acids, linear and cyclic a-keto amides, a-keto acetals and trifluoroacetophenone [43, 44]. Examples are shown in Fig. 3.16.

Up to 72% *ee* has been achieved in the hydrogenation of a diphenyl-substituted reactant, (*trans*)-a-phenylcinnamic acid, with a Pd/TiO$_2$ catalyst and cinchonidine at 1 bar in strongly polar solvent mixtures [49]. For aliphatic a,β-unsaturated acids the enantioselectivities that can be attained are much lower. Therefore, for these type of substrates, homogeneous metal-catalysts are preferred.

Another approach towards asymmetric heterogeneous catalysts is the immobilization of chiral homogeneous complexes via different methods. In this way the advantages of homogeneous catalysts (high activity and selectivity) and heterogeneous catalysts (easy recovery) can be combined. For a complete overview of this active research field the reader is referred to several reviews on this topic [50, 51]. The practical applicability of these catalysts is hampered by the fact that severe demands of recyclability and stability need to be obeyed. In certain cases promising results have been obtained as outlined here.

1. The use of solid and soluble catalysts with covalently attached ligands:
The covalent anchoring of ligands to the surface or to a polymer is a conventional approach which often requires extensive modification of the already expensive ligands. The advantage is that most catalysts can be heterogenized by this approach and various supports can be used. An illustrative example is the use of the BINAP ligand that has been functionalized using several different

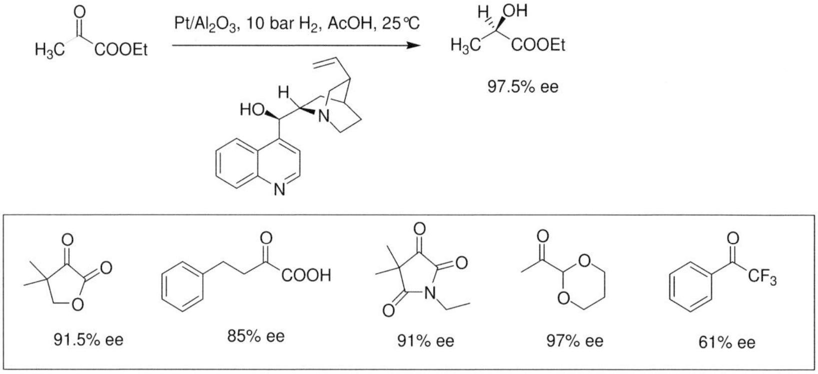

Fig. 3.16 The Pt/cinchona alkaloid system for the enantio-selective hydrogenation of a-functionalized ketones.

methods, followed by attachment to supports such as polystyrene (see Fig. 3.17) [52] or polyethyleneglycol [53]. Alternatively it was rendered insoluble by oligo-merization [54] or co-polymerized with a chiral monomer [55]. The resulting cat-alysts have catalytic performances that are comparable to their soluble analogs. The latter soluble catalyst (see Fig. 3.17) achieved even better activities than the parent Ru-BINAP, which was attributed to a cooperative effect of the polymeric backbone. Alternatively, the ligand can be covalently attached to a "smart" poly-mer which is soluble at the reaction temperature and which precipitates by cool-ing or heating the reaction mixture (see Chapter 7). In this way the reaction can be conducted homogeneously under optimum mass transport conditions and phase separation can be induced by changing the temperature [56].

2. Catalysts immobilized on support via ionic interaction:
This is an attractive method since it does not require the functionalization of the ligand. Augustine et al. [57] developed heteropolyacids (HPA), notably phos-photungstates, as a "magic glue" to attach cationic Rh and Ru complexes to var-ious surfaces. The interaction of the heteropolyacid is thought to occur directly

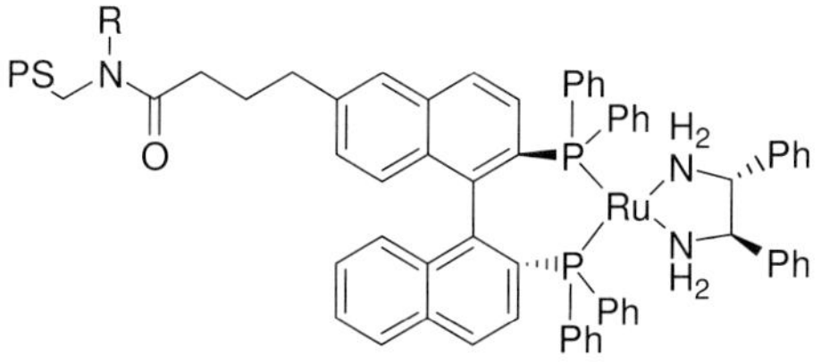

solid catalyst

acetophenone: ee 98%; TOF 390 h^{-1}

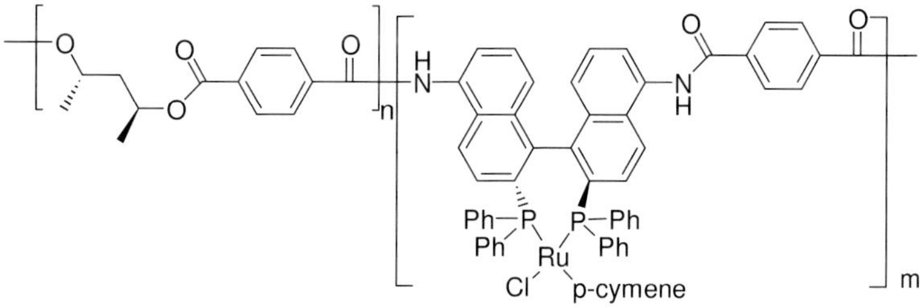

soluble polymeric catalyst

cinnamic acid: ee 96%; TOF 50 h^{-1}

Fig. 3.17 Heterogenization of Ru-BINAP catalysts by covalent modification according to Refs. [52, 55].

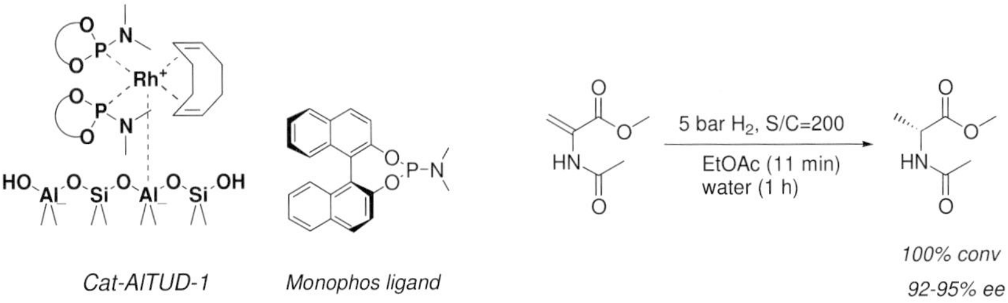

Fig. 3.18 Immobilization through ionic interaction: Rh-Monophos on mesoporous Al-TUD1.

with the metal ion. The most promising catalyst seems to be Rh-DIPAMP (for structure of ligand see Fig. 3.23 below) attached to HPA-clay, due to the high stability of the DIPAMP ligand. This catalyst hydrogenated methyl acetamidoacrylic acid (MAA) with 97% *ee* (TON 270, TOF up to 400 h^{-1}). Recently a mesoporous Al-containing material was used to attach various cationic chiral rhodium complexes to its anionic surface. In this way a stable material was obtained, which showed excellent recyclability over five cycles and which could even be applied in water [58]. Best results were obtained by using Rh-monophos (see Fig. 3.18).

3. Catalysts entrapped or occluded in polymer matrices:
While the entrapment in rigid matrices such as zeolites usually leads to a significant rate decrease, the occlusion in polyvinylalcohol (PVA) or polydimethylsilane (PDMS) looks more promising but is still far from being practically useful. Leaching of the metal complex is often observed in solvents which swell the organic matrix. Rh-DUPHOS (see Fig. 3.23, below) was occluded in PVA or PDMS and was applied for the hydrogenation of methyl acetamidoacrylic acid. Activities and selectivities in this case (TON 140, TOF ca. 15 h^{-1}, *ee* 96%) [59], were significantly lower than for the homogeneous complex (*ee* 99%, TOF 480 h^{-1}).

3.3
Homogeneous Reduction Catalysts

3.3.1
Wilkinson Catalyst

In 1965 Wilkinson invented the rhodium-tris(triphenylphosphine) catalyst as a hydrogenation catalyst [60]. It still forms the basis for many of the chiral hydrogenations performed today. The most effective homogeneous hydrogenation catalysts are complexes consisting of a central metal ion, one or more (chiral) ligands and anions which are able to activate molecular hydrogen and to add the two H atoms to an acceptor substrate. Experience has shown that low-valent Ru,

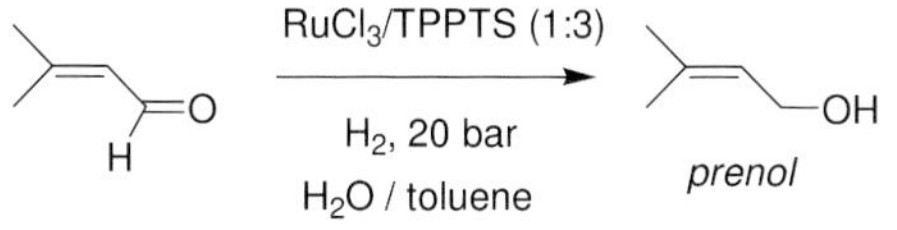

RuCl₃/TPPTS (1:3)

H₂, 20 bar

H₂O / toluene

prenol

Fig. 3.19 Chemoselective hydrogenation of 3-methyl-2-buten-1-al.

Rh and Ir complexes stabilized by tertiary (chiral) phosphorus ligands are the
most active and the most versatile catalysts. Although standard hydrogenations
of olefins, ketones and reductive aminations are best performed using heterogeneous catalysts (see above), homogeneous catalysis becomes the method of
choice once selectivity is called for. An example is the chemoselective hydrogenation of α,β-unsaturated aldehydes which is a severe test for the selectivity of
catalysts.

The use of a tris(triphenyl) ruthenium catalyst resulted in a high selectivity to
the desired unsaturated alcohol – prenol – in the hydrogenation of 3-methyl-2-
buten-1-al (Fig. 3.19) [61]. Thus the double C=C bond stays intact while the aldehyde group is hydrogenated. The latter product is an intermediate for the production of citral. In this case RuCl₃ was used in combination with water-soluble
tris-sulfonated triphenylphosphine (TPPTS) ligands, and in this way the whole
reaction could be carried out in a two-phase aqueous/organic system (see Chapter 7). This allowed easy recycling of the catalyst by phase separation [61]. The
catalytic cycle for the Wilkinson catalyst according to Halpern is depicted in
Fig. 3.20 [62].

The catalytic cycle starts with the dissociation of one ligand P which is replaced e.g. by a solvent molecule. An oxidative addition reaction of dihydrogen

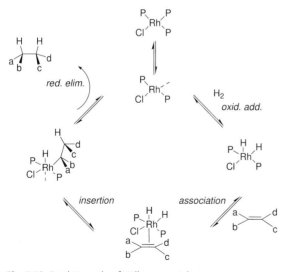

Fig. 3.20 Catalytic cycle of Wilkinson catalyst.

then takes place. This occurs in a cis fashion and can be promoted by the sub-stitution of more electron-rich phosphines on the rhodium complex. The next step is the migration of hydride forming the alkyl group. Reductive elimination of alkane completes the cycle. Obviously the rate of this step can be increased by using electron-withdrawing ligands. The function of the ligand, besides influ-encing the electronic properties of the metal, is to determine the geometry around the metal core. This forms the basis for the enantioselective properties of these catalysts (see below).

3.3.2
Chiral Homogeneous Hydrogenation Catalysts and Reduction of the C=C Double Bond

The foundation for the development of catalysts for asymmetric hydrogenation was the concept of replacing the triphenylphosphane ligand of the Wilkinson cat-alyst with a chiral ligand. This was demonstrated in the work of the early pioneers Horner [63] and Knowles [64]. With these new catalysts, it turned out to be possi-ble to hydrogenate prochiral olefins. Important breakthroughs were provided by the respective contributions of Kagan and Dang [65], and the Monsanto group headed by Knowles [66]. Chelating chiral phosphorus atoms played a central role in this selectivity, and this culminated in the development of the PAMP and later its corresponding dimer the bidentate DIPAMP ligand [67]. The DIPAMP ligand led to high enantioselectivities in the rhodium catalyzed hydrogenation of pro-tected dehydroaminoacids. On an industrial scale this catalyst is used for the mul-ti-ton scale production of the anti-Parkinson drug L-DOPA [68] (Fig. 3.21).

It must be noted that the enantioselectivity reached by the catalyst does not have to be absolute for industrial production. In this case 100% enantioselectiv-ity of the product was obtained by crystallization. In the past, Selke in the for-mer GDR independently developed a sugar based bis-phosphinite as a ligand which formed the basis for the L-DOPA process by the company VEB-Isis [69,

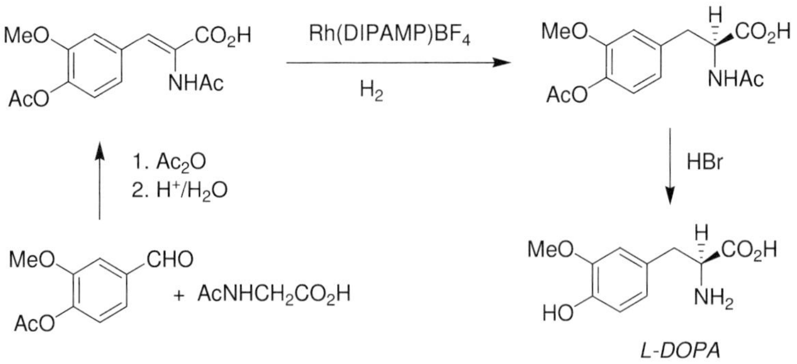

Fig. 3.21 Monsanto's L-DOPA process.

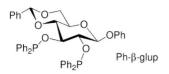

Fig. 3.22 "Selke's" ligand for the production of L-DOPA.

Ph-β-glup

70], see Fig. 3.22. The process was introduced in 1985 and it ended in 1990, one year after the collapse of the socialist system.

A breakthrough in ligand design was achieved by the work of Kagan, who demonstrated that phosphorus ligands with chirality solely in the backbone, which are much easier to synthesize, led to even better enantioselectivities [4]. This was the beginning in the 1990s of a still-ongoing period in which many new bisphosphines were invented and tested in a variety of enantioselective hydrogenations [71]. Selected examples of chiral bisphosphine ligands have been collected in Fig. 3.23.

Ligands based on ferrocenes have been developed extensively by Hayashi, Togni and the Ciba-Geigy catalysis group (now operating as part of Solvias AG) [72]. An example is found in Lonza's new biotin process (Fig. 3.24) in which Rh catalyzed hydrogenation of the olefinic double bond of the substrate takes place with 99% *de* [73, 74]. Another example of the use of this class of ferrocene-derived ligands is the use of Josiphos ligand with Ru for the production of (+) *cis*-methyl dihydrojasmonate (Fig. 3.25) [75].

Especially worth mentioning is the use of bis-phospholane DUPHOS as a ligand [76]. A wide variety of substrates, notably enamides, vinylacetic acid derivatives and enol acetates, can be reduced with high enantioselectivities. A recent application is the rhodium catalyzed enantioselective hydrogenation of the α,β-unsaturated carboxylic ester in Fig. 3.26, which is an intermediate for Pfizer's Candoxatril, a new drug for the treatment of hypertension and congestive heart failure [77]. Other examples are the diastereoselective hydrogenation leading to Pharmacia & Upjohn's Tipranavir [78] (Fig. 3.27), an HIV protease inhibitor, and the manufacture of α-amino acid derivatives [79]. Both processes were developed by Chirotech (now part of Dow). For the production of α-amino acids, the example of *N*-Boc-(*S*)-3-fluorophenylalanine is given in Fig. 3.28. However, many derivatives have been produced using the same methodology. It is noteworthy that the use of Rh-MeDUPHOS catalysts is accompanied by using a biocatalytic deprotection of the resulting amide using acylase enzymes (see Fig. 3.28).

One of the drawbacks of the use of bisphosphines is the elaborate syntheses necessary for their preparation. Many efforts have been directed towards the development of bis-phosphonites and bis-phosphites. However, surprisingly monodentate phosphinates, phosphates and phosphoramidates recently emerged as effective alternatives for bidentate phosphines (Fig. 3.29). This constitutes an important breakthrough in this area as these can be synthesized in one or two steps, and the cost of these ligands is an order of magnitude lower. Monophos can be made in a single step from BINOL and HMPT.

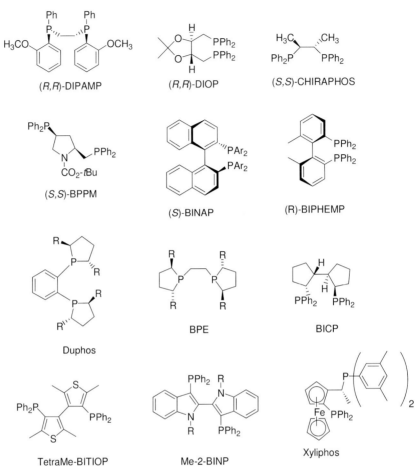

Fig. 3.23 Selected examples of chiral ligands used in homogeneous hydrogenation catalysts.

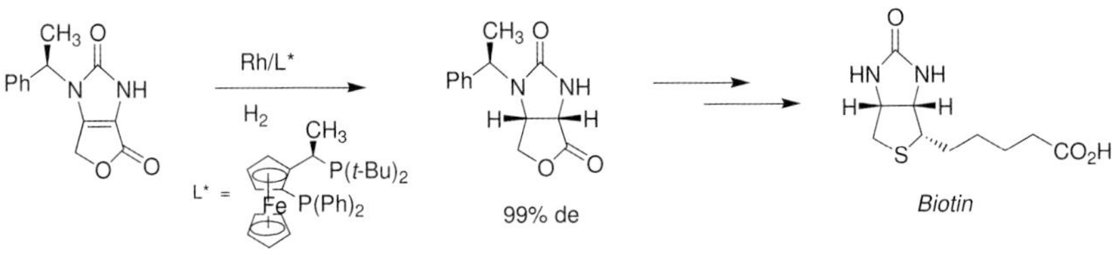

Fig. 3.24 Lonza's biotin process using Rh-Xyliphos-type ligands.

All of the above examples involve an extra coordinating group such as enamide, acid, or ester in the substrate. This is necessary for optimum coordination to the metal. Asymmetric hydrogenation of olefins without functional groups is an emerging area [80].

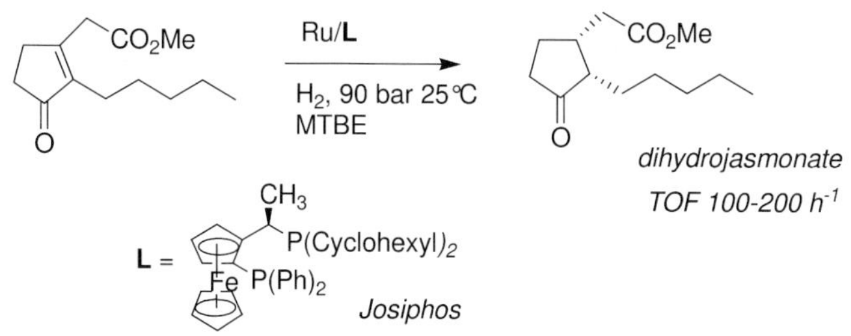

Fig. 3.25 Production of (+) *cis*-methyl dihydrojasmonate using Ru-Josiphos catalyzed hydrogenation.

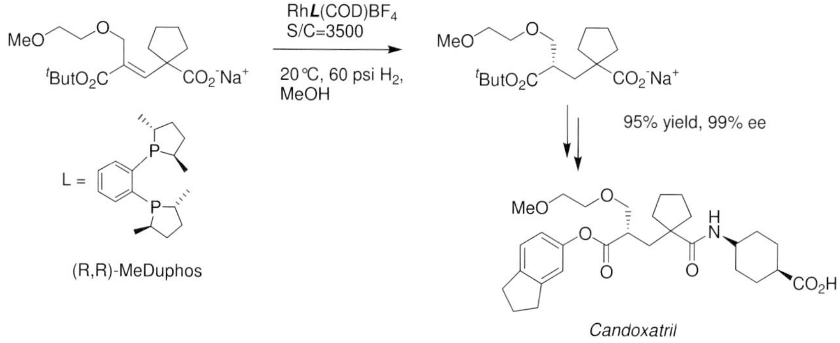

Fig. 3.26 Rh-MeDUPHOS catalyzed hydrogenation in the production of Candoxatril.

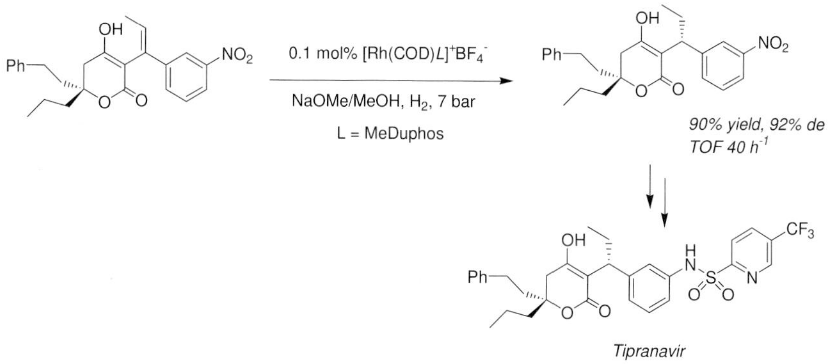

Fig. 3.27 Rh-MeDUPHOS for production of Tipranavir.

According to the mechanism of the enantioselective reduction, a concept has been evolved, termed the "quadrant rule" [81, 82]. This model addresses how the chiral ligand influences the preferential addition of hydrogen either to the *re-* or *si*-face of a C=X bond. As visualized in Fig. 3.30, upon coordination of the ligands to the metal atom, the substituents are oriented in such a way that

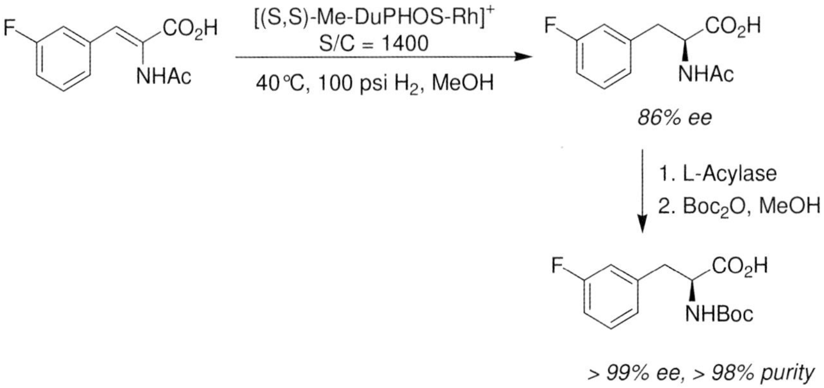

Fig. 3.28 Production of *N*-Boc-(*S*)-3-fluorophenylalanine using Chirotech's Rh-MeDUPHOS technology.

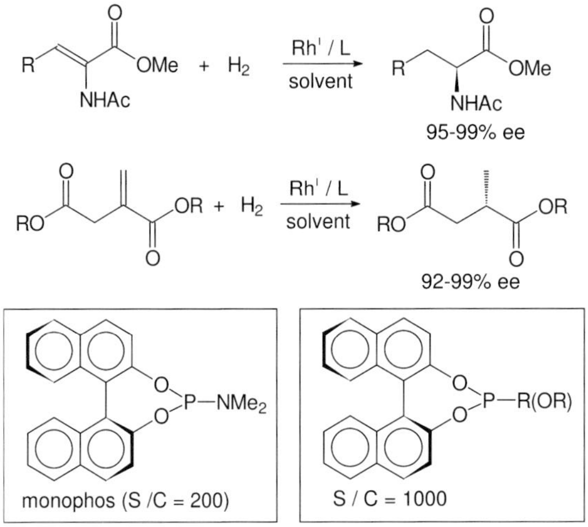

Fig. 3.29 Use of monodentate ligands for the reduction of prochiral olefins.

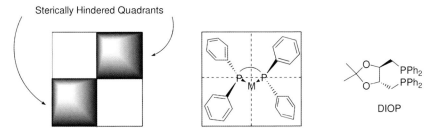

Fig. 3.30 The quadrant model used to predict enantioselectivity in homogeneous hydrogenation.

a chiral array is formed where two diagonal quadrants are blocked by bulky sub-stituents. The situation is visualized for DIOP, a diphosphine with two PAr$_2$ moieties attached to a flexible chiral backbone. When the substrate coordinates to the metal atom, it will orient in such a way that steric repulsion is minimal. In many cases, this simple model is able to predict the sense of induction. However it will not be able to predict more complex situations [83].

3.3.3
Chiral Homogeneous Catalysts and Ketone Hydrogenation

Enantiopure alcohols can be produced using chiral hydrogenation catalysts for the reduction of ketones. A major breakthrough in this area was achieved in the mid-1980s by Takaya and Noyori, following the initial work of Ikariya's group [84], on the development of the BINAP ligand for Ru-catalyzed hydrogenations [85]. The ruthenium-BINAP ligand is famous for its broad reaction scope: many different classes of ketones can be hydrogenated with very high enantioselectiv-ities. BINAP is a chiral atropisomeric ligand, and its demanding synthesis has been optimized [86]. The Japanese company Takasago has commercialized var-ious Ru-BINAP processes [87]. Drawbacks of the Ru-BINAP procedure are the relatively high pressures and temperatures in combination with long reaction times. This can be circumvented by adding small amounts of strong acids [88] or turning to Rh-DUPHOS type complexes. Furthermore, unlike rhodium com-plexes, most ruthenium bisphosphines have to be preformed for catalysis.

An example of the hydrogenation of a β-ketoester is the dynamic kinetic reso-lution of racemic 2-substituted acetoacetates [89]. In this process one of the two enantiomeric acetoacetates is hydrogenated with very high diastereoselective preference and in high enantioselectivity. At the same time the undesired acet-oacetate undergoes a continuous racemization. This has found application in the hydrogenation of the intermediate for the carbapenem antibiotic intermedi-ate Imipenem on a scale of 120 ton/year [90]. The reaction is shown in Fig. 3.31. Ru-BINAP is also suitable for the enantioselective hydrogenation of α-substituted ketones. In Fig. 3.32, the reduction of acetol leading to the (R)-1,2-propane-diol is denoted. This diol is an intermediate for the antibiotic Levofloxa-cin [87, 90].

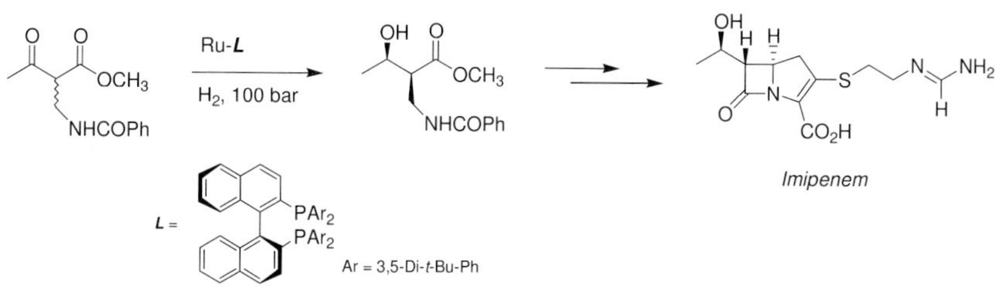

Fig. 3.31 Hydrogenation of a β-ketoester for the production of Imipenem.

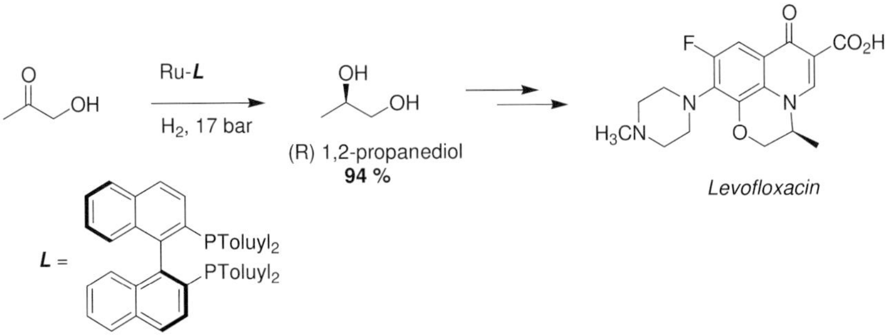

Fig. 3.32 Enantioselective hydrogenation of acetol, intermediate for Levofloxacin.

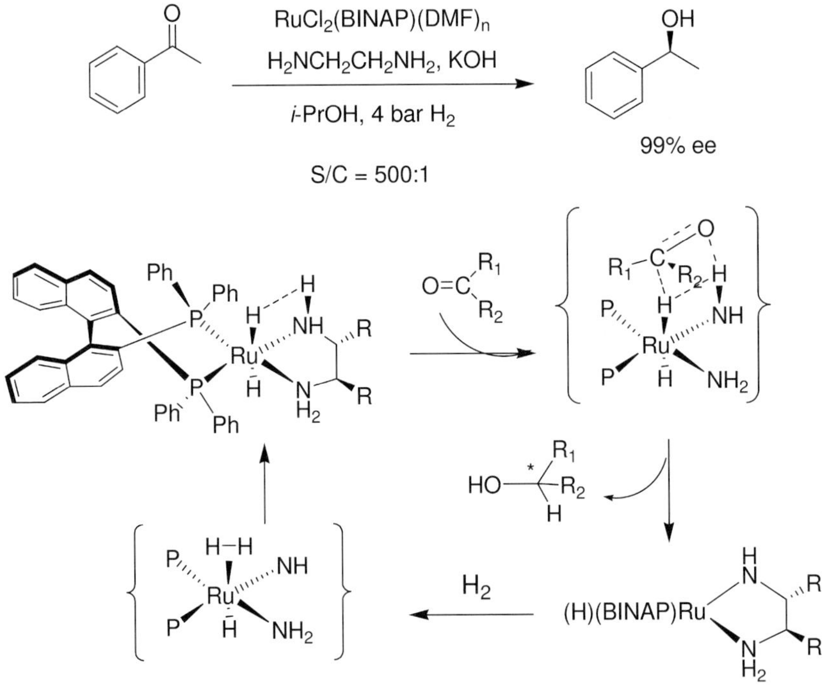

Fig. 3.33 Bifunctional Ru-BINAP(diamine) complexes for enantioselective hydrogenation of simple ketones.

Another breakthrough was made by Noyori for the enantioselective hydrogenation of aromatic ketones by introducing a new class of Ru-BINAP (diamine) complexes [91]. In this case hydrogen transfer is facilitated by ligand assistance. The company Takasago used this catalyst for the production of (R)-1-phenylethanol in 99% *ee* using only 4 bar of hydrogen [86] (see Fig. 3.33). (R)-1-phenylethanol, as its acetate ester, is sold as a fragrance and has a floral, fresh, green note.

There is consensus that the transfer of the two H atoms occurs in a concerted manner as depicted in Fig. 3.33 [91–94]. This hypothesis explains the need for an N–H moiety in the ligand.

3.3.4
Imine Hydrogenation

The asymmetric hydrogenation of imines is an important application because it gives access to enantiopure amines. For a long period the development of asymmetric imine hydrogenation lagged behind the impressive progress made in asymmetric olefin and ketone hydrogenation [95, 96]. It was difficult to achieve good results both in terms of rate and selectivity. The best results were obtained with rhodium and bisphosphines. However rhodium-based catalysts required 70 bar of hydrogen, which hampered their industrial application. The largest asymmetric catalytic process nowadays, however, involves imine hydrogenation as the key step (see Fig. 3.34). Ciba-Geigy (now Syngenta) produces the herbicide (S)-metolachlor on a scale of 10 000 ton/year. The use of iridium, instead of rhodium, finally resulted in the required activity: TOFs with this catalyst are in excess of 100 000 h^{-1} and TONs of more than 1×10^6 were reached [97]. As a ligand xyliphos, a ferrocenyl-type bisphosphine, was used in combination with iodide and acetic acid as promoters. The enantioselectivity for this process is ca. 80%, which can be easily increased by lowering the substrate/catalyst ratio. However, this is not necessary as 80% *ee* is sufficient. Compared to the first generation process which produced racemic metolachlor, an environmental burden reduction of 89% could be realized.

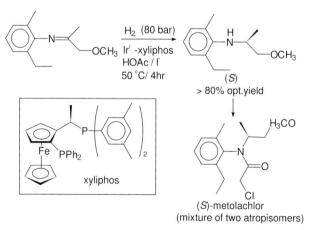

Fig. 3.34 Ir-Xyliphos based process for the production of (S)-metolachlor.

3.3.5
Transfer Hydrogenation using Homogeneous Catalysts

For small scale production, the use of hydrogen at elevated pressure imposes practical problems related to reactor design and safety issues. An attractive alternative is the use of alcohols or formate as a reductant (see also below for biocatalytic reductions). The Meerwein–Ponndorf–Verley (MPV) reaction has already been mentioned above and is traditionally performed using stoichiometric amounts of aluminim salts or zeolites [98]. In particular, for enantioselective transformations, a wide variety of Ru-, Rh- and Ir-based homogeneous catalysts are now known and the technique of transfer hydrogenation seems to find application in industry [99–101]. Whereas complexes containing chiral phosphine ligands are the catalysts of choice for hydrogenation reactions with H_2, Ru, Rh and Ir complexes with chiral NN or NO ligands have been shown to be very effective for asymmetric transfer hydrogenations (Fig. 3.35). These complexes are not able to activate molecular hydrogen. Especially in the case of aryl ketones and ketimines, transfer hydrogenation can be potentially interesting, because the ruthenium hydrogenation technology in this case is limited by medium TONs and TOFs [10].

A plethora of ligands has been reported in the literature but the most effective ones are 1,2-amino alcohols, monotosylated diamines and selected phosphine-oxazoline ligands. The active structures of the complexes reported are half-sand-

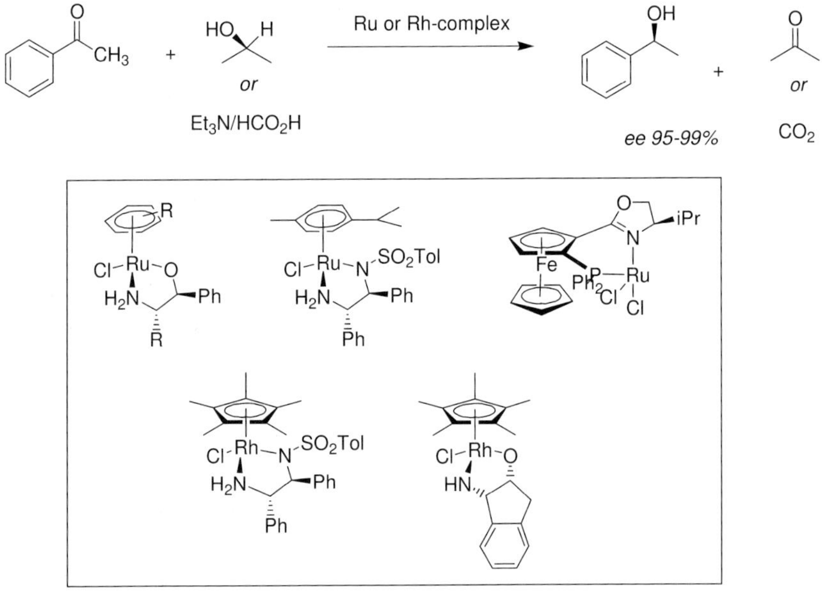

Fig. 3.35 Structure of the most effective Ru and Rh
precursors used in transfer hydrogenations.

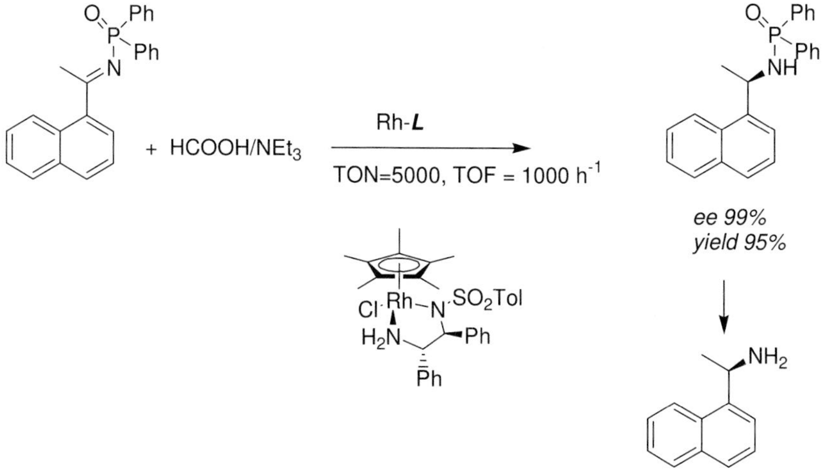

Fig. 3.36 Transfer hydrogenation of imine, as developed by Avecia.

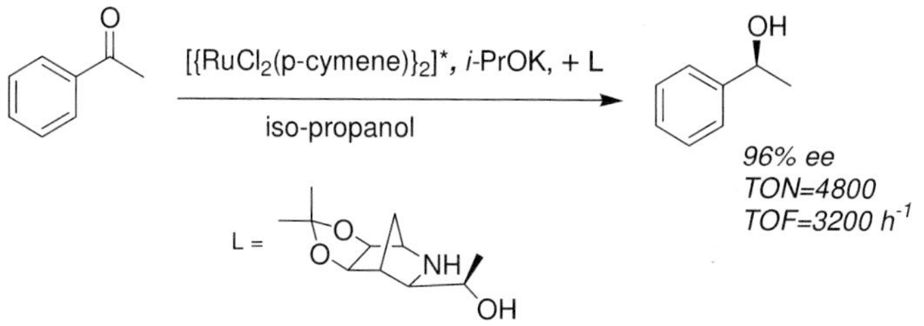

Fig. 3.37 Azanorbornane-based ligands form complexes with
Ru, which are the most effective complexes reported today for
transfer hydrogenation.

wich π-complexes, Ru-arene and Rh (or Ir)-cyclopentadiene complexes (see
Fig. 3.35) [10]. The introduction of these catalysts in 1995 by Noyori [100] was a
breakthrough because it led to a great improvement in terms of both reaction
rate and enantioselectivity. For a discussion of the bifunctional mechanism operating in this case see the discussion above and Ref. [94]. These catalysts have
been developed with acetophenone as model substrate. However many functionalized acetophenones, aryl ketones, acetylenic ketones as well as cyclic aryl ketones give very good results. Imine reduction can also be performed highly efficiently using this technique. An example is shown in Fig. 3.36 where a TOF of
$1000 \ h^{-1}$ was reached [102].

Furthermore, Avecia has developed the Rh-cyclopentadienyl complexes in
Fig. 3.36 for the large scale production of 1-tetralol and substituted 1-phenyl-
ethanol [102]. The challenge in this area is to increase the activity of the catalyst.
Recently, Andersson and co-workers reported an azanorbornane-based ligand
which can reach a TOF up to $3680 \ h^{-1}$ for acetophenone hydrogenation (see

Fig. 3.37) [103]. The combined introduction of a dioxolane in the backbone and a methyl group in the α-position to the OH group proved essential to reach this activity. The ligand without these functionalities, resulted in 10 times lower TOFs.

3.4
Biocatalytic Reductions

3.4.1
Introduction

Oxidoreductases play a central role in the metabolism and energy conversion of living cells. About 25% of the presently known enzymes are oxido reductases [104]. The classification of oxidoreductases is presented in Fig. 3.38. The groups of oxidases, monooxygenases and peroxidases – dealing with oxidations – will be described in Chapter 4.

For industrial applications the group of alcohol dehydrogenases, otherwise known as carbonyl-reductases, is of prime interest. The natural substrates of the enzymes are alcohols such as ethanol, lactate, glycerol, etc. and the corresponding carbonyl compounds. However, unnatural ketones can also be reduced enantioselectively. In this way chiral alcohols, hydroxy acids and their esters or amino acids, respectively, can be easily generated. An excellent overview of this field up to 2004 can be found in Refs. [105–108]. The advantages of biocatalytic reductions are that the reactions can be performed under mild conditions leading to excellent enantioselectivities. A vast reservoir of wild-type microorganisms has been screened for new enzymes, and tested with numerous substrates. A number of alcohol dehydrogenases are commercially available. These include baker's yeast and the alcohol dehydrogenases from Baker's yeast, *Thermoanaerobium brockii*, horse liver and the hydroxysteroid dehydrogenase from *Pseudomenas testosterone* and *Bacillus spherisus* [105, 106]. Table 3.1 gives a selection of well known alcohol dehydrogenases. Genetic engineering tools can be applied to

1. Dehydrogenases

$$SH_2 + D \longrightarrow S + DH_2$$

2. Oxidases
(Oxidative dehydrogenation)

$$SH_2 + O_2 \longrightarrow S + H_2O_2$$

3. Oxygenases (oxygen insertion)
mono- :

$$S + DH_2 + O_2 \longrightarrow SO + D + H_2O$$

di- :

$$SH + O_2 \longrightarrow SO_2H$$

4. Peroxidases

$$SH_2 + H_2O_2 \longrightarrow S + 2 H_2O$$

$$S + H_2O_2 \longrightarrow SO + H_2O$$

S, SH, SH_2 = substrate
S, SO, SO_2, SO_2H = oxidized substrate
D, DH_2 = cofactor e.g. NAD / NADH$_2$

Fig. 3.38 Classification of oxidoreductases.

Table 3.1 Examples of commonly used alcohol dehydrogenases.

Dehydrogenase	Specificity	Cofactor
Yeast-ADH	Prelog	NADH
Horse liver-ADH	Prelog	NADH
Thermoanaerobium brockii-ADH	Prelog	NADPH
Hydroxysteroid-DH	Prelog	NADH
Candida parapsilosis	Prelog	NADH
Rhodococcus erythropolis	Prelog	NADH
Lactobacillus kefir-ADH	Anti-Prelog	NADPH
Mucor javanicus-ADH [a]	Anti-Prelog	NADPH
Pseudomenas sp.-ADH [a]	Anti-Prelog	NADH

a) Not commercially available.

make enzymes available in larger quantities and at lower costs in recombinant hosts. Furthermore, by directed evolution, tailor-made catalysts can be generated. The scientific knowledge in this field is increasing rapidly and in the future more tailor-made enzymes will be used in industrial processes [109, 110].

The vast majority of alcohol dehydrogenases require nicotanimide cofactors, such as nicotinamide adenine dinucleotide (NADH) and its respective phosphate NADPH. The structure of NAD/NADP is shown in Fig. 3.39. Hydrogen and two electrons are transferred from the reduced nicotinamide to the carbonyl group to effect a reduction of the substrate (see Fig. 3.39).

The mechanism of horse liver dehydrogenase has been described in great detail. In this case the alcohol is coordinated by a Zn ion and Serine. Both the alcohol and the cofactor are held in position via a three-point attachment and efficient hydrogen transfer takes place [111, 112]. The alcohol dehydrogenases can be used to reduce a variety of ketones to alcohols with high enantioselectivity under mild conditions. In Fig. 3.40 a number of examples is shown using *Lactobacillus kefir* [113]. During the course of the reaction, the enzyme delivers the hydride preferentially either from the *si* or the *re*-side of the ketone to give, for

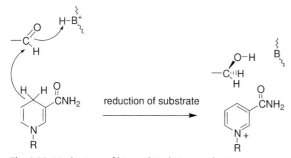

Fig. 3.39 Mechanism of biocatalytic ketone reduction.

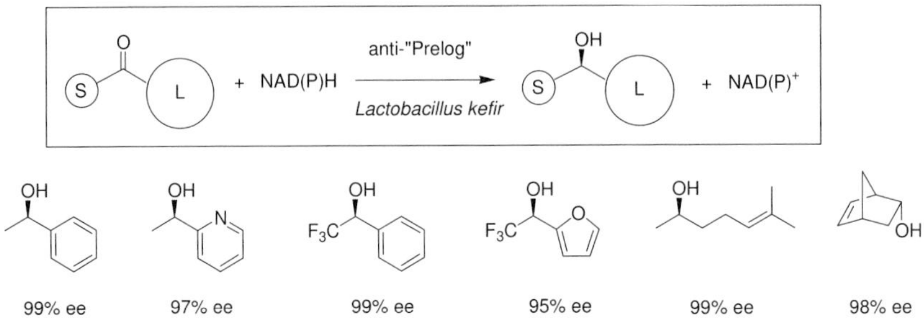

Fig. 3.40 Examples of ketone reduction and the Prelog rule.

simple systems, the (*R*)- or (*S*)-alcohols respectively. For most cases, the stereo-chemical course of the reaction, which is mainly dependent on the steric re-quirements of the substrate, may be predicted from a simple model which is generally referred to as the "Prelog" rule [106, 114]. As shown in Fig. 3.40, *Lactobacillus kefir* leads to anti-Prelog selectivity. Obviously the enzymatic reduction is most efficient when the prochiral center is adjoined by small and large groups on either side (Fig. 3.40).

The cofactors are relatively unstable molecules and expensive if used in stoichiometric amounts. In addition, they cannot be substituted by less expensive simple molecules. Since it is only the oxidation state of the cofactor which changes during the reaction it may be regenerated *in situ* by using a second redox reaction to allow it to re-enter the reaction cycle. Thus, the expensive cofactor is needed only in catalytic amounts, leading to a drastic reduction in costs [115]. Much research effort in the field of alcohol dehydrogenases is therefore directed towards cofactor recycling (see below). Cofactor recycling is no problem when whole microbial cells are used as biocatalysts for redox reactions. Many examples have been described using "fermenting yeast" as a reducing agent [116]. In this case cheap sources of redox equivalents, such as carbohydrates, can be used since the microorganism possesses all the enzymes and cofactors which are required for the metabolism. In Fig. 3.41, an example of the production of a pharmaceutically relevant prochiral ketone is given using fermentation technology [117].

However from the standpoint of green chemistry, the use of isolated enzymes (or dead whole cells) is highly preferred because it avoids the generation of copious amounts of biomass. It must be emphasized that the productivity of microbial conversions is usually low, since non-natural substrates are only tolerated at concentrations as low as 0.1–0.3% [106]. The large amount of biomass present in the reaction medium causes low overall yields and makes product recovery troublesome. Therefore the E-factors for whole cell processes can be extremely high. Moreover the use of wild-type cells often causes problems because an array of enzymes is present which can interfere in the reduction of a specific ketone (giving opposite selectivities). The use of recombinant techniques, however, which only express the desired enzyme can overcome this problem [108].

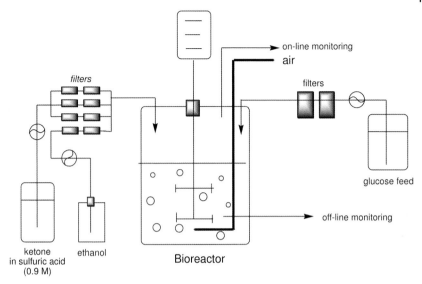

Fig. 3.41 Whole cell technology for biocatalytic reduction, example adapted from Ref. [117].

Much higher productivities can be obtained using isolated enzymes or cell extracts [118]. This approach is therefore highly preferred. Because of the importance of whole cell technology for biocatalytic reduction a few examples will be given. However the main part of this chapter will be devoted to industrial examples of bioreduction involving isolated enzymes and cofactor recycling.

3.4.2
Enzyme Technology in Biocatalytic Reduction

To keep the amount of biomass involved to a minimum, the use of isolated enzymes or cell extracts is highly preferable. In this case, for their efficient recycling, two strategies are feasible [106]. These are depicted in Fig. 3.42. In the first case, the substrate-coupled approach, only one enzyme is needed and additional alcohol is needed to complete the reduction. In terms of overall reaction, these coupled substrate-systems bear a strong resemblance to transition-metal transfer hydrogenation, the so-called Meerwein-Ponndorf-Verley reaction, see above. In the second approach [106–108], a large variety of reducing agents can be used, such as formate, glucose, H_2 or phosphite, in combination with a second enzyme, namely formate dehydrogenase, glucose dehydrogenase, hydrogenase [119] or phosphite dehydrogenase [120], respectively. Industrial processes (see below) commonly use glucose or formate as coreductants.

In the first approach, one enzyme is needed to convert both substrate (ketone) and co-substrate (alcohol). The second alcohol needs to afford a ketone which can be easily removed by distillation, e.g. acetone, to drive the reduction to completion. For most alcohol-dehydrogenases, however, the enzyme can only

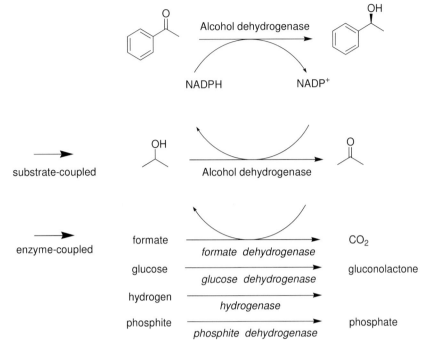

Fig. 3.42 Cofactor recycling during bioreduction processes.

tolerate certain levels of alcohols and aldehydes and therefore these coupled-substrate systems are commonly impeded by cosubstrate-inhibition [121]. Engineering solutions performed by Liese and coworkers were found to address this problem: *in situ* product removal accompanied by constant addition of reducing alcohol can improve the yields using isopropanol as the reductant.

To push the productivity limits of coupled-substrate systems even further, the maximum tolerable concentration of isopropanol applied for cofactor-regeneration with different alcohol dehydrogenases (as isolated enzymes or in preparations) has been constantly increased in the period 2000–2002 by using even more chemostable enzyme preparations [108]. In general, bacterial alcohol dehydrogenases dominate the stability ranking and the best results up till now were reached using *Rhodococcus ruber* DSM 44541 [122]. Elevated concentrations of isopropanol, up to 50% (v/v), using whole cells, not only shift the equilibrium toward the product-side, but also enhance the solubility of lipophilic substrates in the aqueous/organic medium (Fig. 3.43). The alcohol dehydrogenases from *Rhodococcus ruber*, employed as a whole cell preparation, tolerate substrate concentrations of approximately 1–2 M. Using this system a variety of methyl ketones, ethyl ketones and chloromethyl ketones could be reduced with Prelog preference. It is also worth noting that this system has been used in a reduction as well as an oxidation mode (see Chapter 4).

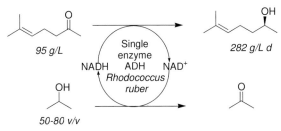

Fig. 3.43 6-Methyl-hept-5-en-2-one reduction and simultaneous NAD⁺ regeneration using alcohol dehydrogenase from *Rhodococcus rubber*.

To circumvent the drawback of thermodymanic equilibrium limitations and cosubstrate-inhibition, cofactor recycling can be performed completely independently, by using a second irreversible reaction, and another enzyme. This is the so-called coupled-enzyme approach. In this case a variety of reducing agents has been used. The most advanced regeneration systems make use of innocuous hydrogen as reductant. An example has been demonstrated where hydrogenase from *Pyrococcus furiosus* was used to recycle the NADP⁺ cofactor, showing a turnover frequency up to 44 h⁻¹ [119, 123]. In nature this enzyme – which operates under thermophilic conditions – is able to reversibly cleave the heterolytic cleavage of molecular hydrogen. It contains both Ni and Fe in the active site. Unfortunately, the enhanced activity of the enzyme at elevated temperature (maximum 80 °C) could not be used due to the thermal instability of NADPH and reactions were performed at 40 °C. This system, despite the oxygen sensitivity of hydrogenase I, represents an elegant example of green chemistry and bears significant potential (see Fig. 3.44).

One of the prominent industrial bioreduction processes, run by cofactor regeneration, is performed by leucine dehydrogenase. This enzyme can catalyze the reductive amination of trimethylpyruvic acid, using ammonia (see Fig. 3.45). For this process the cofactor regeneration takes place by using formate as the reductant and formate dehydrogenase as the second enzyme. The advantage of

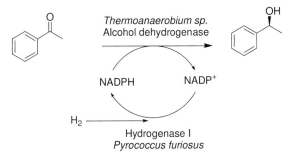

Fig. 3.44 Cofactor recycling using Hydrogenase I from *Pyrococcus furiosus*.

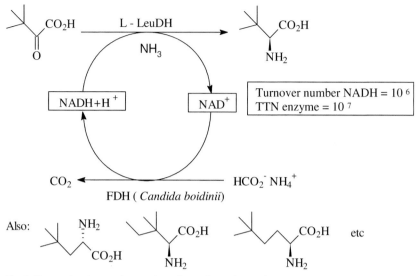

Fig. 3.45 L-tert-leucine production by cofactor-dependent reductive amination.

this concept is that the cofactor-regeneration is practically irreversible since carbon dioxide is produced, which can be easily removed. Although the substrate scope of this enzyme is not too good, a number of other substrates have been screened [124, 125].

Another industrial example, albeit on a lower scale, of the use of formate as formal reductant is represented by the synthesis of (R)-3-(4-fluorophenyl)-2-hydroxy propionic acid, which is a building block for the synthesis of *Rupintrivir*, a rhinovirus protease inhibitor currently in human clinical trials to treat the common cold [126]. The chiral 2-hydroxy-acid **A** can, in principle, be readily prepared by asymmetric reduction of the α-keto acid salt **B** (see Fig. 3.46) [127]. Chemocatalysts, such as Ru(II)-BINAP and Rh(I)-NORPHOS, gave unsatisfactory results for this substrate. However using D-lactate dehydrogenase (D-LDH), the keto acid salt **B** could be stereoselectively reduced to the corresponding α-hydroxy acid by NADH with high *ee* and high conversion. For scaling up, an enzyme coupled approach was chosen where formate was used as the stoichiometric reductant. The process is depicted in Fig. 3.46. A relatively simple continuous membrane reactor was found to satisfactorily produce α-hydroxy acid with a productivity of $560 \text{ g L}^{-1} \text{ d}^{-1}$. Enzyme deactivation was a key factor in determining the overall cost of the process. In a period of nine days, without adding fresh D-LDH and FDH, the rate of reaction decreased slowly and the enzymes lost their activity at a rate of only 1% per day. Therefore the reactor was charged with new enzymes periodically to maintain a high conversion of above 90%.

An example of glucose coupled ketone reduction is the continuous mode reduction using whole (dead) cells of *Lactobacillus kefir* for the enantioselective reduction of 2,5-hexanedione to (2R,5R)-hexanediol – a popular chiral ligand for

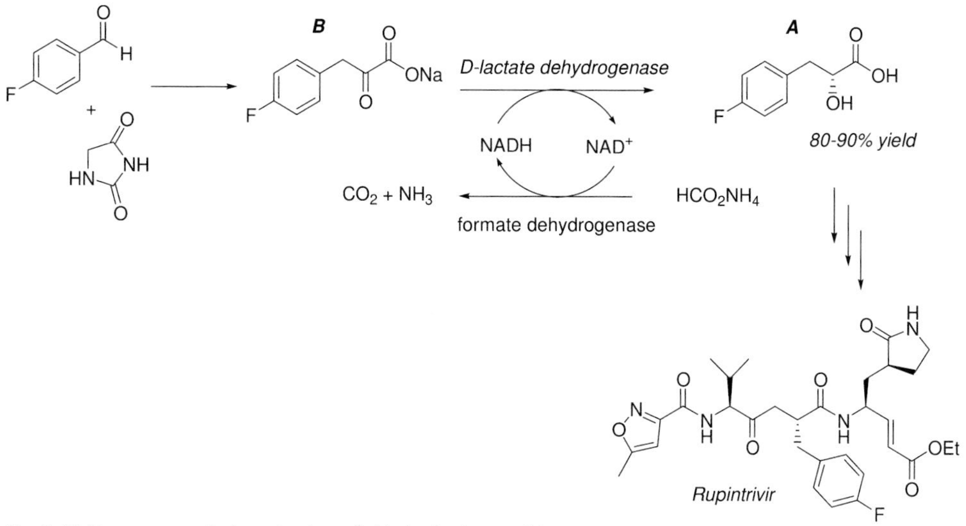

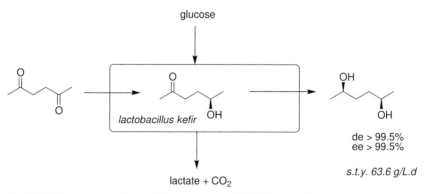

Fig. 3.46 Formate coupled production of chiral α-hydroxy acid.

transition metal catalyzed asymmetric hydrogenations. In this case the biocatalytic process is superior to its chemical counterpart because the *de* > 99.5% and *ee* > 99.5% are difficult to achieve with chemical catalysts [128]. While substrate and glucose for cofactor regeneration were constantly fed to a stirred tank reactor, the product was removed *in situ*. Using this technique, the productivity could be increased to 64 g L^{-1} d^{-1} (Fig. 3.47) [129].

Another example for biocatalytic reduction using glucose as the reductant, is the production of (R)-ethyl-4,4,4-trifluoro-3-hydroxybutanoate by Lonza [130]. It is a building block for pharmaceuticals such as *Befloxatone*, an anti-depressant monoamine oxidase-A inhibitor from Synthelabo. The process uses whole cells of *Escheria coli* that contain two plasmids. One carries an aldehyde reductase

Fig. 3.47 Glucose coupled biocatalytic production of (2R,5R)-hexanediol.

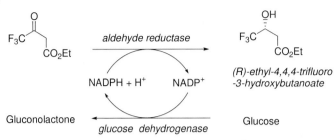

Fig. 3.48 Stereoselective reduction of (R)-ethyl-4,4,4-trifluoro-acetoacetate by aldehyde reductase.

gene from the yeast *Sporobolomyces salmonicolor*, which catalyzes the reduction of ethyl-4,4,4-trifluoroacetoacetate, and the second carries a glucose dehydrogenase gene from *Bacillus megaterium* to generate NADPH from NADP⁺ (Fig. 3.48). By this "co-expression" approach the desired enzymes can be produced together in one fermentation. The cells can be stored frozen before use in the biotransformation. The Lonza process is carried out in a water/butyl acetate two-phase system to avoid inhibition of the reductase by the substrate and product. An advantage of the two-phase system is that the cells are permeabilized, allowing the transfer of NADP⁺ and NADHP through the cell wall. This whole-cell biocatalyst was originally constructed for the stereoselective reduction of 4-chloro-3-oxobutanoate (a precursor for L-carnitine), in which case productivities of up to 300 g L⁻¹, and *ee* values of up to 92% were reported [131].

The enantioselective reduction of alkyl-4-chloroacetoacetates is an important area, because it leads to an intermediate for the side chain of statins. Statins are cholesterol-lowering drugs, and the market thereof is the largest in the pharmaceutical sector. In 2003 revenues of US$ 9.2 billion and US$ 6.1 billion were recorded for astorvastatin and simvastatin respectively [132]. The established strategy of enantioselective reduction of alkyl-4-chloroacetoacetates has been optimized: Alcohol dehydrogenase from *Candida magnoliae* and glucose dehydrogenase from *Bacillus megatorium* were used in a biphasic organic solvent aqueous buffer system for the production of ethyl (S)-4-chloro-3-hydroxybutano-

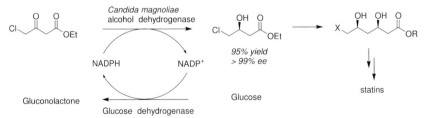

Fig. 3.49 Biocatalytic production of ethyl (S)-4-chloro-3-hydroxybutanoate using alcohol dehydrogenase from *Candida magnoliae*.

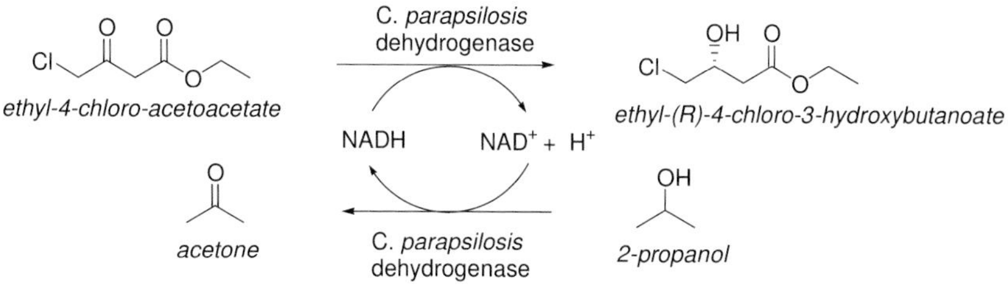

C. parapsilosis
dehydrogenase

CI ethyl-4-chloro-acetoacetate

NADH   NAD⁺ + H⁺

ethyl-(R)-4-chloro-3-hydroxybutanoate

acetone

C. parapsilosis
dehydrogenase

2-propanol

Fig. 3.50 Preparation of ethyl (R)-4-chloro-3-hydroxybutanoate by using alcohol dehdyrogenase from *Candida parapsilosis*.

ate in enantiopure form (see Fig. 3.49). Product concentrations of 63 g L^{-1} were observed in the organic phase and the product was isolated in 95% yield [133].

When the enantioselective reduction of ethyl 4-chloroacetoacetate was carried out with alcohol dehydrogenase from *Candida parapsilosis*, the other enantiomer was produced: ethyl (R)-4-chloro-3-hydroxybutanoate [134]. This product is a key intermediate in a synthesis of (R)-carnitine. In this case a substrate coupled approach was chosen. The enzyme also has a strong oxidation activity for 2-propanol, which was therefore selected as the cosubstrate. The situation is depicted in Fig. 3.50. Under optimized conditions, the yield of (R)-ethyl-4-chloro-3-hydroxy-butanoate reached 36.6 g L^{-1} (>99% *ee*, 95% yield) on a 30 L scale.

3.4.3
Whole Cell Technology for Biocatalytic Reduction

For small-scale reductions on a laboratory scale the use of (dead cells of) baker's yeast is a cheap alternative. It results in Prelog face reduction of a large variety of aliphatic and aromatic ketones to give the (S)-alcohols in good optical purities [106]. Examples were given above. On a larger scale the use of fermenting yeast can also be attractive [116]. An example is the production of a natural "green note" flavor compound [104]. Natural flavors have high consumer appeal and are desired in the food industry. Products obtained by fermentation or enzymatic catalysis usually qualify for the label "natural" and can be sold at higher prices than nature identical flavors derived by chemical synthesis. The green notes are a mixture of hexenal, hexan-1-ol, *trans*-2-hexenal, *trans*-2-hexen-1-ol and *cis*-3-hexen-1-ol. The latter dominates the impact of freshness. The flavor company Firmenich SA, Geneva, CH [135], developed a process starting from polyunsaturated fatty acids, which are converted to six-carbon hexenals and hex-anal by a combined action of lipoxygenase and hydroperoxide lyase. In the second conversion, the aldehydes were reduced employing a 20% baker's yeast suspension. This process is operated on the ton scale. Separation in this case is relatively facile because the flavor compounds are sufficiently volatile to be isolated by distillation. The yeast is discarded after the biotransformation.

Another example is the microbial reduction of 3,4-methylene-dioxyphenylace-tone to the corresponding (S) alcohol at Lilly Research Laboratories [136] (see

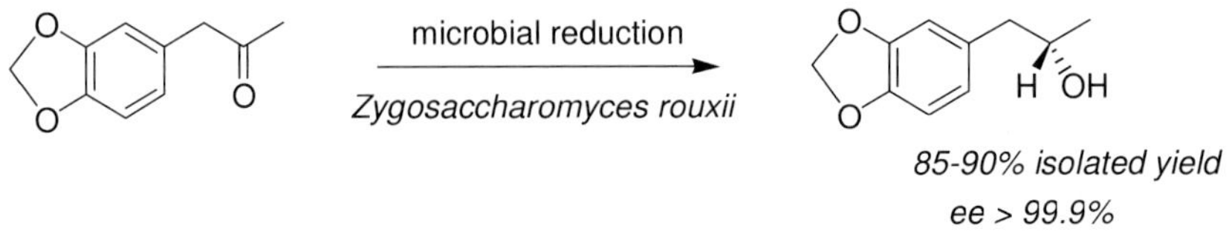

Fig. 3.51 Asymmetric reduction of 3,4-methylene-dioxyphenylacetone.

Fig. 3.51). The chiral alcohol is a key intermediate in the synthesis of an anti-convulsant drug. The yeast *Zygosaccharomyces rouxii* employed for this process is hampered by substrate and product toxicity at levels of >6 g L^{-1}. This was solved by using *in situ* adsorption on Amberlite XAD-7. In this way *in situ* product removal could be achieved. At the end of the process (8–12 h) 75–80 g of the alcohol is found on the resin and about 2 g L^{-1} remains in the aqueous phase. Finally an isolated yield of 85–90% could be obtained with an *ee* $>99.9\%$.

Screening for the novel enzyme, although the classical method, is still one of the most powerful tools for finding biocatalytic reduction systems [108]. Enzyme sources used for the screening can be soil samples, commercial enzymes, culture sources, a clone bank, etc. Their origin can be microorganisms, animals or plants. In most of the examples mentioned above, the biocatalyst was chosen after an extensive screening program. A third example of a fermenting process, which illustrates the importance of screening, is shown in Fig. 3.52. The key intermediate in the synthesis of Montelukast, an anti-asthma drug, was prepared from the corresponding ketone by microbial transformation [137]. After screening 80 microorganisms, the biotransforming organism *Microbacterium campoquemadoensis* (MB5614) was identified, which was capable of reducing the complex ketone to the (S) alcohol.

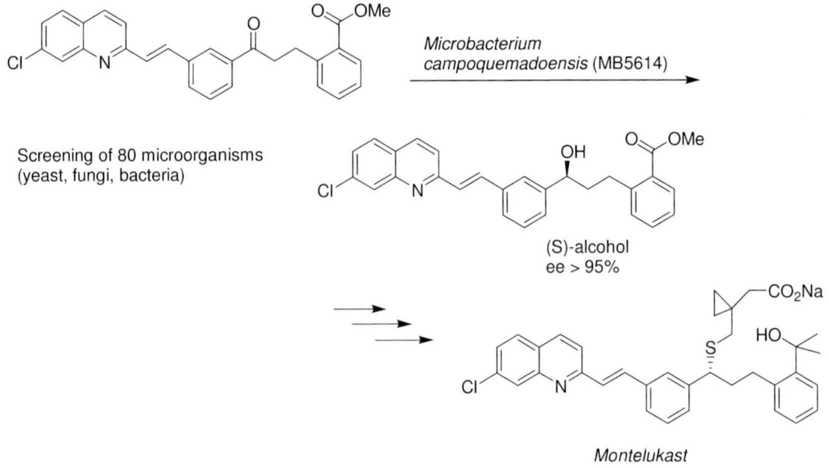

Fig. 3.52 Fermenting process for the production of Montelukast.

3.5
Conclusions

Summarizing, it is evident that catalytic reduction is an important technology that is widely applied for the production of fine chemicals. It is a key example of green technology, due to the low amounts of catalysts required, in combination with the use of hydrogen (100% atom efficient!) as the reductant. In general, if chirality is not required, heterogeneous supported catalysts can be used in combination with hydrogen. Apart from the standard reductions, such as ketone and nitro reductions, reductive aminations etc., the Pd-catalyzed reductive alkylation of alcohols and amines with aldehydes, leading to the green synthesis of ethers and amines, needs to be mentioned. Once selectivity and chirality is called for, homogeneous catalysts and biocatalysts are required. The use and application of chiral Ru, Rh and Ir catalysts has become a mature technology. It allows a large variety of transformations: imines and functionalized ketones and alkenes can be converted with high selectivity in most cases. The enantioselective reduction of non-functionalized and non-aromatic alkenes is still an area of development.

Biocatalysis is developing rapidly. Especially in the area of ketone reduction, biocatalytic reduction is often the method of choice, due to the high purities of product that can be achieved (>99%). In these cases alcohol, formate or glucose are used as the reductants. Significant progress can be expected in this area due to the advancement in enzyme production technologies and the possibility of tailor-made enzymes. In the future more applications of catalytic reduction will certainly come forward, for new chemical entities as well as for the replacement of current less-green technologies.

References

1 P. N. Rylander, *Hydrogenation Methods*, Academic Press, San Diego, 1990 and preceding volumes.
2 W. S. Knowles, *Adv. Synth. Catal.*, **2003**, *345*, 3.
3 R. Noyori, *Adv. Synth. Catal.*, **2003**, *345*, 15.
4 H. B. Kagan, in *Asymmetric Synthesis*, Vol 5, Chiral Catalysis, J. D. Morrison (Ed.), Academic Press, New York, 1985.
5 A. J. J. Straathof, S. Panke, A. Schmid, *Curr. Opin. Biotechnol.*, **2002**, *13*, 548.
6 K. Nakamura, R. Yamanaka, T. Matsuda and T. Harada, *Tetrahedron Asym.*, **2003**, *14*, 2659.
7 G. A. Somorjai, K. McCrea, *Appl. Catal. A: General*, **2001**, *222*, 3; G. A. Somorjai, *Top. Catal.*, **2002**, *18*, 157.
8 R. L. Augustine, *Heterogeneous Catalysis for the Synthetic Chemist*, CRC Press, New York, 1995.
9 G. V. Smith, F. Notheisz, *Heterogeneous Catalysis in Organic Chemistry*, Academic Press, San Diego, 1999.
10 H.-U. Blaser, C. Malan, B. Pugin, F. Spindler, H. Steiner, M. Studer, *Adv. Synth. Catal.*, **2003**, *345*, 103.
11 E. N. Marvell, T. Li, *Synthesis*, **1973**, 457.
12 J. Sobczak, T. Boleslawska, M. Pawlowska, W. Palczewska, in *Heterogeneous Catalysis and Fine Chemicals*, M. Guisnet et al. (Eds.), Elsevier, Amsterdam, 1988, p. 197.

13 R. A. Raphael, in *Acetylenic Compounds in Organic Synthesis*, Butterworths, London, 1955, p. 26.

14 H. Lindlar, *Helv. Chim. Acta*, **1952**, *35*, 446.

15 C. A. Drake, US Pat. 4596783 and 4605797, 1996, to Philips Petroleum Company.

16 L. A. Sarandeses, J. L. Mascareñas, L. Castedo, A. Mouriño, *Tetrahedron Lett.*, **1992**, *33*, 5445.

17 S. Bailey, F. King, in *Fine Chemicals through Heterogeneous Catalysis*, R. A. Sheldon, H. van Bekkum (Eds.), Wiley-VCH, Weinheim, 2001, p. 351.

18 A. Giroir-Fendler, D. Richard, P. Gallezot, *Stud. Surf. Sci. Catal.*, **1988**, *41*, 171.

19 A. Giroir-Fendler, D. Richard, P. Gallezot, *Catal. Lett.*, **1990**, *5*, 175.

20 S. Ratton, *Chem. Today*, March/April **198**, 33–37.

21 T. Yokohama, T. Setoyama, N. Fujita, M. Nakajima, T. Maki, *Appl. Catal.*, **1992**, *88*, 149.

22 T. Yokoyama, T. Setoyama, in *Fine Chemicals through Heterogeneous Catalysis*, R. A. Sheldon, H. van Bekkum (Eds.), Wiley-VCH, Weinheim, 2001, p. 370.

23 U. Siegrist, P. Baumeister, H.-U. Blaser, M. Studer, *Chem. Ind.*, **1998**, *75*, 207.

24 C. De Bellefon, P. Fouilloux, *Catal. Rev.*, **1994**, *36*, 459.

25 S. Gomez, J. A. Peters, T. Maschmeyer, *Adv. Synth. Catal.*, **2002**, *344*, 1037.

26 A. M. Allgeier, M. W. Duch, *Chem. Ind.*, **2001**, *82*, 229.

27 S. N. Thomas-Pryor, T. A. Manz, Z. Liu. T. A. Koch, S. K. Sengupta, W. N. Delgass, *Chem. Ind.*, **1998**, *75*, 195.

28 P. Kukula, M. Studer, H.-U. Blaser, *Adv. Synth. Catal.*, **2004**, *346*, 1487.

29 P. Sabatier, *Ind. Eng. Chem.*, **1926**, *18*, 1004.

30 S. N. Balasubrahmanyam, N. Balasubrahmanyam, *Tetrahedron*, **1973**, *29*, 683.

31 H. van Bekkum, H. M. A. Buurmans, G. van Minnen-Pathuis, B. M. Wepster, *Recl. Trav. Chim. Pays-Bas*, **1969**, *88*, 779.

32 J. G. Donkervoort, E. G. M. Kuijpers, in *Fine Chemicals through Heterogeneous Catalysis*, R. A. Sheldon, H. van Bekkum (Eds.), Wiley-VCH, Weinheim, 2001, p. 407.

33 L. H. Slaugh, J. A. Leonard, *J. Catal.*, **1969**, *13*, 385.

34 I. W. C. E. Arends, M. Sasidharan, A. Kühnle, M. Duda, C. Jost, R. A. Sheldon, *Tetrahedron*, **2002**, *58*, 9055.

35 J. Muzart, *Tetrahedron*, **2005**, *61*, 5955.

36 F. Fache, F. Valot, M. Lemaire, in *Fine Chemicals through Heterogeneous Catalysis*, R. A. Sheldon, H. van Bekkum (Eds.), Wiley-VCH, Weinheim, 2001, p. 461.

37 T. W. Greene, P. G. M. Wuts, in *Protective Groups in Organic Synthesis*, 2nd edn., Wiley, New York, 1991.

38 H.-U. Blaser, H. Steiner, M. Studer, in *Transition Metals for Organic Synthesis*, Vol. 2, C. Bolm, M. Beller (Eds.), Wiley-VCH, Weinheim, 1998, p. 97.

39 E. J. Creyghton, J. C. van der Waal, in *Fine Chemicals through Heterogeneous Catalysis*, R. A. Sheldon, H. van Bekkum (Eds.), Wiley-VCH, Weinheim, 2001, p. 438.

40 E. J. Creyghton, S. D. Ganeshie, R. S. Downing, H. van Bekkum, *J. Chem. Soc., Chem. Commun.*, **1995**, 1859; E. J. Creyghton, S. D. Ganeshie, R. S. Downing, H. van Bekkum, *J. Mol. Catal. A: Chemical*, **1997**, *115*, 457.

41 J. C. van der Waal, K. Tan, H. van Bekkum, *Catal. Lett.*, **1996**, *41*, 63.

42 A. Corma, M. E. Domine, L. Nemeth, S. Valencia, *J. Am. Chem. Soc.*, **2002**, *124*, 3194.

43 M. Studer, H.-U. Blaser, C. Exner, *Adv. Synth. Catal.*, **2003**, *346*, 45; H.-U. Blaser, H. P. Jalett, M. Müller, M. Studer, *Catal. Today*, **1997**, *37*, 441.

44 T. Mallat, A. Baiker, in *Fine Chemicals through Heterogeneous Catalysis*, R. A. Sheldon, H. van Bekkum (Eds.), Wiley-VCH, Weinheim, 2001, p. 449.

45 Y. Izumi, *Adv. Catal.*, **1983**, *32*, 215.

46 A. Baiker, *J. Mol. Catal. A: Chemical*, **1997**, *115*, 473.

47 Y. Orito, S. Imai, S. Niwa, *J. Chem. Soc. Jpn.*, **1979**, 1118.

48 B. Török, K. Balàzsik, G. Szöllösi, K. Felföldi, M. Bartók, *Chirality*, **1999**, *11*, 470.

49 Y. Nitta, K. Kobiro, *Chem. Lett.*, **1996**, 897.

50 H.-U. Blaser, B. Pugin, M. Studer, in *Chiral Catalyst Immobilization and Recycling*, D. E. De Vos, I. F. J. Vankelecom, P. A. Jacobs (Eds.), Wiley-VCH, Weinheim, 2000, p. 1.

51 K. Fodor, S. G. A. Kolmschot, R. A. Sheldon, *Enantiomer*, **1999**, *4*, 497, and references cited therein.

52 T. Ohkuma, H. Takeno, Y. Honda, R. Noyori, *Adv. Synth. Catal.*, **2001**, *343*, 369.

53 P. Guerreiro, V. Rato-Velomanana-Viadl, J.-P. Genet, P. Dellis, *Tetrahedron Lett.*, **2001**, *42*, 3423.

54 T. Lamouille, C. Saluzzo, R. ter Halle, F. Le Guyader, M. Lemaire, *Tetrahedron Lett.*, **2001**, *42*, 663.

55 Q. Fan, C. Ren, C. Yeung, W. Hu, A. S. C. Chan, *J. Am. Chem. Soc.*, **1999**, *121*, 7407.

56 D. E. Bergbreiter, *Chem. Rev.*, **2002**, *102*, 3345.

57 R. L. Augustine, S. K. Tanielyan, S. Anderson, H. Yang, Y. Gao, *Chem. Ind.*, **2001**, *82*, 497.

58 C. Simons, U. Hanefeld, I. W. C. E. Arends, R. A. Sheldon, T. Maschmeyer, *Chem. Eur. J.*, **2004**, 5839; C. Simons, U. Hanefeld, I. W. C. E. Arends, R. A. Sheldon, T. Maschmeyer, *Chem. Commun.*, **2004**, 2830.

59 A. Wolfson, S. Geresh, M. Gottlieb, M. Herskowitz, *Tetrahedron: Asym.*, **2002**, *13*, 465.

60 F. H. Jardine, J. A. Osborn, G. Wilkinson, J. F. Young, *Chem. Ind.*, **1965**, 560; *J. Chem. Soc. (A)*, **1966**, 1711.

61 J. M. Grosselin, C. Mercier, G. Allmang, F. Grass, *Organometallics*, **1991**, *10*, 2126.

62 J. Halpern, *Pure Appl. Chem.*, **2001**, *73*, 209.

63 L. Horner, H. Buthe, H. Siegel, *Tetrahedron Lett.*, **1968**, 4923.

64 W. S. Knowles, M. J. Sabacky, *J. Chem. Soc., Chem. Commun.*, **1968**, 1445.

65 T. P. Dang, H. B. Kagan, *J. Am. Chem. Soc.*, **1972**, *94*, 6429.

66 W. S. Knowles, M. J. Sabacky, B. D. Vineyard, *J. Chem. Soc., Chem. Commun.*, **1972**, 10.

67 W. S. Knowles, *Acc. Chem. Res.*, **1983**, *16*, 106.

68 W. S. Knowles, in *Asymmetric Catalysis on an Industrial Scale*, H.-U. Blaser, E. Schmidt (Eds.), Wiley-VCH, Weinheim, 2004, p. 23.

69 R. Selke, H. Pracejus, *J. Mol. Catal.*, **1986**, *37*, 213.

70 R. Selke, in *Asymmetric Catalysis on an Industrial Scale*, H.-U. Blaser, E. Schmidt (Eds.), Wiley-VCH, Weinheim, 2004, p. 39.

71 For a recent overview of chiral ligands see e.g. Ref. [10].

72 R. R. Bader, P. Baumeister, H.-U. Blaser, *Chimia*, **1996**, *50*, 86.

73 J. McGarrity, F. Spindler, R. Fux , M. Eyer, EP 624587, 1994, to Lonza.

74 R. Imwinkelried, *Chimia*, **1997**, *51*, 300.

75 D. A. Dobbs, K. P. M. Vanhessche, E. Brazi, V. Rautenstrauch, J.-Y. Lenoir, J.-P. Genêt, J. Wiles, S. H. Bergens, *Angew. Chem. Int. Ed.*, **2000**, *39*, 1992.

76 M. J. Burk, F. Bienewald, in *Applied Homogeneous Catalysis with Organometallic Compounds*, Vol. 2, B. Cornils, W. A. Herrmann (Eds.), VCH, Weinheim, 1996, p. 13.

77 M. J. Burk, F. Bienewald, S. Challenger, A. Derrick, J. A. Ramsden, *J. Org. Chem.*, **1999**, *64*, 3290.

78 H.-U. Blaser, F. Spindler, M. Studer, *Appl. Catal. A: General*, **2001**, *221*, 119.

79 C. J. Cobley, N. B. Johnson, I. C. Lennon, R. McCague, J. A. Ramsden, A. Zanotti-Gerosa, in *Asymmetric Catalysis on an Industrial Scale*, H.-U. Blaser, E. Schmidt (Eds.), Wiley-VCH, Weinheim, 2004, p. 269.

80 R. L. Balterman, in *Comprehensive Asymmetric Catalysis*, Vol. 1, E. N. Jacobsen, A. Pfaltz, H. Yamamoto (Eds.), Springer, Berlin, 1999, p. 183.

81 K. E. Koenig, M. J. Sabacky, G. L. Bachman, W. C. Christopfel, H. D. Barnstoff, R. B. Friedman, W. S. Knowles, B. R. Stults, B. D. Vineyard, D. J. Weinkauff, *Ann. N.Y. Acad. Sci.*, **1980**, *333*, 16.

82 H. Brunner, A. Winter, J. Breu, *J. Organomet. Chem.*, **1998**, *553*, 285.

83 S. Feldgus, C. R. Landis, *J. Am. Chem. Soc.*, **2000**, *122*, 12714.

84 T. Ikarya, Y. Ishii, H. Kawano, T. Arai, M. Saburi, S. Yoshikawa, S. Akutagawa, *J. Chem. Soc., Chem. Commun.*, **1985**, 922.

85 R. Noyori, M. Ohta, Y. Hsiao,
M. Kitamura, T. Ohta, H. Takaya,
J. Am. Chem. Soc., **1986**, *108*, 7117;
R. Noyori, H. Takaya, *Acc. Chem. Res.*,
1990, *23*, 345.

86 H. Kumobayashi, T. Miura, N. Sayo,
T. Saito, X. Zhang, *Synlett*, **2001**, 1055.

87 H. Kumobayashi, *Recl. Trav. Chim.
Pays-Bas*, **1996**, *115*, 201.

88 S. A. King, A. S. Thompson, A. O. King,
T. R. Verhoeven, *J. Org. Chem.*, **1992**,
57, 6689.

89 R. Noyori, T. Ikeda, T. Ohkuma,
M. Widhalm, M. Kitamura, H. Takaya,
S. Akutagawa, N. Sayo, T. Saito,
T. Taketomi, H. Kumobayashi, *J. Am.
Chem. Soc.*, **1989**, *111*, 9134.

90 R. Noyori, *Acta Chem. Scand.*, **1996**, *50*,
380.

91 R. Noyori, T. Ohkuma, *Angew. Chem.
Int. Ed.*, **2001**, *40*, 41.

92 R. Hartmann, P. Chen, *Angew. Chem.
Int. Ed.*, **2001**, *40*, 3581.

93 K. Abdur-Rashid, M. Faatz, A. J. Lough,
R. H. Morris, *J. Am. Chem. Soc.*, **2001**,
123, 7473.

94 M. Yamakawa, H. Ito, R. Noyori,
J. Am. Chem. Soc., **2000**, *122*, 1466.

95 H.-U. Blaser, F. Spindler, in *Compre-
hensive Asymmetric Catalysis*, Vol. 1,
E. N. Jacobsen, A. Pfaltz, H. Yamamoto
(Eds.), Springer, Berlin, 1999, p. 248.

96 T. Ohkuma, M. Kitamura, R. Noyori,
in *Catalytic Asymmetric Synthesis*, 2nd
edn., I. Ojima (Ed.), VCH Publishers,
New York, 2000, p. 1.

97 H-U. Blaser, R. Hanreich, H.-D.
Schneider, F. Spindler, B. Steinacher,
in *Asymmetric Catalysis on an Industrial
Scale*,
H.-U. Blaser, E. Schmidt (Eds.),
Wiley-VCH, Weinheim, 2004, p. 55.

98 C. F. de Graauw, J. A. Peters, H. van
Bekkum, J. Huskens, *Synthesis*, **1994**,
1007.

99 G. Zassinovich, G. Mestroni, S. Gladia-
li, *Chem. Rev.*, **1992**, *92*, 1051.

100 R. Noyori, S. Hashiguchi, *Acc. Chem.
Res.*, **1997**, *30*, 97.

101 M. Wills, *Tetrahedron: Asymmetry*, **1999**,
10, 2045.

102 J. Blacker, J. Martin, in *Asymmetric
Catalysis on an Industrial Scale*,

H.-U. Blaser, E. Schmidt (Eds.), Wiley-
VCH, Weinheim, 2004, p. 201.

103 S. J. M. Nordin, P. Roth, T. Tarnai,
D. A. Alonso, P. Brandt, P. G. Andersson,
Chem. Eur. J., **2001**, *7*, 1431.

104 *Enzyme Nomenclature*, Academic Press,
New York, 1992.

105 M.-R. Kula, U. Kragl, in *Stereoselective
Biocatalysis*, R. N. Patel (Ed.), Marcel
Dekker, New York, 2000, p. 839.

106 K. Faber, *Biotransformations in Organic
Chemistry*, 4th edn., Springer, Berlin,
2000.

107 W. Kroutil, H. Mang, K. Edegger,
K. Faber, *Curr. Opin. Chem. Biol.*, **2004**,
8, 120.

108 K. Nakamura, R. Yamanaka, T. Matsuda,
T. Harada, *Tetrahedron: Asymmetry*, **2003**,
14, 2659.

109 For an introduction to the area of
directed evolution see K. A. Powell,
S. W. Ramer, S. B. del Cardayré,
W. P. C. Stemmer, M. B. Todin,
P. F. Longchamp, G. W. Huisman,
Angew. Chem. Int. Ed., **2001**, *40*, 3948.

110 M. T. Reetz, *Angew. Chem. Int. Ed.*,
2001, *40*, 284.

111 A. Fersht, *Structure and Mechanism in
Protein Science*, Freeman and Company,
New York, 1999.

112 J. K. Rubach, B. V. Plapp, *Biochemistry*,
2003, *42*, 2907, and references cited
therein.

113 C. W. Bradshaw, H. Hummel,
C. H. Wong, *J. Org. Chem.*, **1992**, *57*,
1532.

114 V. Prelog, *Pure Appl. Chem.*, **1964**, *9*,
119.

115 U. Kragl, D. Vasic-Racki, C. Wandrey,
Bioproc. Eng., **1996**, *14*, 291.

116 S. Servi, in *Biotechnology*, 2nd edn.,
Vol. 8a, H. J. Rehm, G. Reed (Eds.),
Wiley-VCH, Weinheim, 1998, p. 363.

117 M. Chartrain, R. Greasham, J. Moore,
P. Reider, D. Robinson, B. Buckland,
J. Mol. Catal. B: Enzymatic, **2001**, *11*,
503.

118 For selected examples, see Ref. [5].

119 R. Mertens, L. Greiner, E. C. D. van den
Ban, H. B. C. M. Haaker, A. Liese, *J.
Mol. Catal. B Enzymatic*, **2003**, *24--25*,
39.

120 J. M. Vrtis, A. K. White, W. M. Metcalf, W. A. van der Donk, *Angew. Chem. Int. Ed.*, **2002**, *41*, 3257.

121 T. Stillger, M. Bönitz, M. Villela Filho, A. Liese, *Chem. Ing. Technol.*, **2002**, *74*, 1035.

122 W. Stampfer, B. Kosjek, C. Moitzi, W. Kroutil, K. Faber *Angew. Chem. Int. Ed.*, **2002**, *41*, 1014.

123 L. Greiner, D. H. Müller, E. C. D. van den Ban, J. Wöltinger, C. Wandrey, A. Liese, *Adv. Synth. Catal.*, **2003**, *345*, 679.

124 G. Krix, A. S. Bommarius, K. Drauz, M. Kottenhan, M. Schwarm, M.-R. Kula, *J. Biotechnol.*, **1997**, *53*, 29.

125 H. Groger, K. Drauz, in *Asymmetric Catalysis on an Industrial Scale*, H.-U. Blaser, E. Schmidt (Eds.), Wiley-VCH, Weinheim, 2004, p. 131

126 J. Tao, K. McGee, in *Asymmetric Catalysis on an Industrial Scale*, H.-U. Blaser, E. Schmidt (Eds.), Wiley-VCH, Weinheim, 2004, p. 323.

127 M.-J. Kim, G. M. Whitesides, *J. Am. Chem. Soc.*, **1988**, *110*, 2959.

128 J. Haberland, A. Kriegesmann, E. Wolfram, W. Hummel, A. Liese, *Appl. Microb. Biotechnol.*, **2002**, *58*, 595.

129 J. Haberland, W. Hummel, T. Dausman, A. Liese, *Org. Proc. Res. Dev.*, **2002**, *6*, 458.

130 N. M. Shaw, K. T. Robbins, A. Kiener, *Adv. Synth. Catal.*, **2003**, *345*, 425.

131 S. Shimizu, M. Kataoka, A. Morishita, M. Katoh, T. Morikawa, T. Miyoshi, H. Yamada, *Biotechnol. Lett.*, **1990**, *12*, 593.

132 M. Müller, *Angew. Chem. Int. Ed.*, **2005**, *44*, 362.

133 Y. Yasohara, N. Kizaki, J. Hasegawa, M. Wada, M. Kataoka, S. Shimizu, *Tetrahedron: Asym.*, **2001**, *12*, 1713; N. Kizaki, Y. Yasohara, J. Hasegawa, M. Wada, M. Kataoka, S. Shimizu, *Appl. Microbiol. Biotechnol.*, **2001**, *55*, 590.

134 H. Yamamoto, A. Matsuyama, Y. Kobayashi, *Biosci. Biotechnol. Biochem.*, **2002**, *66*, 925.

135 B. L. Muller, C. Dean, I. M. Whitehead, Symposium: *Plant Enzymes in the Food Industry*, 1994, 26, Lausanne.

136 B. A. Anderson, M. M. Hansen, A. R. Harkness, C. L. Henry, J. T. Vicenzi, M. J. Zmijewski, *J. Am. Chem. Soc.* **1995**, *117*, 12358; M. J. Zmijewski, J. Vicenzi, B. E. Landen, W. Muth, P. Marler, B. Anderson, *Appl. Microbiol. Biotechnol.* **1997**, *47*, 162; J. T. Vicenzi, M. J. Zmijewski, M. R. Reinhard, B. E. Landen, W. L. Muth, P. G. Marler, *Enzyme Microbial. Technol.* **1997**, *20*, 494.

137 A. Shafiee, H. Motamedi, A. King, *Appl. Microbiol. Biotechnol.* **1998**, *49*, 709.

4
Catalytic Oxidations

4.1
Introduction

The controlled partial oxidation of hydrocarbons, comprising alkanes, alkenes and (alkyl)aromatics, is the single most important technology for the conversion of oil- and natural gas-based feedstocks to industrial organic chemicals [1–3]. For economic reasons, these processes predominantly involve the use of molecular oxygen (dioxygen) as the primary oxidant. Their success depends largely on the use of metal catalysts to promote both the rate of reaction and the selectivity to partial oxidation products. Both gas phase and liquid phase oxidations, employing heterogeneous and homogeneous catalysts, respectively, are used industrially, in a ca. 50/50 ratio (see Table 4.1).

The pressure of increasingly stringent environmental regulation is also providing a stimulus for the deployment of catalytic oxidations in the manufacture of fine chemicals. Traditionally, the production of many fine chemicals involved oxidations with stoichiometric quantities of, for example, permanganate or dichromate, resulting in the concomitant generation of copious amounts of inorganic salt-containing effluent. Currently, there is considerable pressure, therefore, to replace these antiquated technologies with cleaner, catalytic alternatives [3, 4]. In practice this implies an implementation of catalytic technologies which allow the use of oxygen and hydrogen peroxide (which produces water as the side product) as stoichiometric oxidants. Therefore this chapter will focus on catalysts which use either O_2 or H_2O_2 as oxidants. In some cases, e.g. in the case of stereoselective conversion where a highly added benefit of the product prevails, the use of other oxidants will be considered as well.

In principle, homogeneous as well as heterogeneous and bio-catalysts can be deployed in liquid phase oxidations but, in practice, the overwhelming majority of processes are homogeneous, i.e. they involve the use of soluble metal salts or complexes as the catalyst. Pivotal examples of the potential of selective catalyzed oxidations have already been mentioned in Chapter 1. For example, in the oxidation of alcohols and epoxidation of olefins enormous progress has been achieved over the last decade towards green methods, notably using homogeneous metal complexes. However also in the field of biocatalytic transformations, industrial applications start to appear, notably in the field of aromatic side

Green Chemistry and Catalysis. I. Arends, R. Sheldon, U. Hanefeld
Copyright © 2007 WILEY-VCH Verlag GmbH & Co. KGaA, Weinheim
ISBN: 978-3-527-30715-9

Table 4.1 Bulk chemicals via catalytic oxidation.

Product	Feedstock	Oxidant/Process [a]
Styrene	Benzene/ethene	None/G (O_2/L) [b]
Terephthalic acid	p-Xylene	O_2/L
Formaldehyde	Methanol	O_2/G
Ethene oxide	Ethene	O_2/G
Phenol	a. Benzene/propene	O_2/L
	b. Toluene	
Acetic acid	a. n-Butane	O_2/L
	b. Ethene	
Propene oxide	Propene	RO_2H/L
Acrylonitrile	Propene	O_2/G
Vinyl acetate	Ethene	O_2/L; G
Benzoic acid	Toluene	O_2/L
Adipic acid	Benzene	O_2/L
ε-Caprolactam	Benzene	O_2/L
Phthalic anhydride	o-Xylene	O_2/G
Acrylic acid	Propene	O_2/G
Methyl methacrylate	Isobutene	O_2/G
Maleic anhydride	n-Butane	O_2/G

a) L = liquid phase; G = gas phase.
b) Styrene from propene epoxidation with EBHP.

chain oxidation of heteroaromatics [5]. In the next section, the basics of mechanisms in metal-catalyzed oxidations – homolytic versus heterolytic – will be presented. The fundamentals of redox catalysis by enzymes will also be given. The green methodologies for converting different classes of substrates will then be treated consecutively.

4.2
Mechanisms of Metal-catalyzed Oxidations: General Considerations

The ground state of dioxygen is a triplet containing two unpaired electrons with parallel spins. The direct reaction of $^3O_2$ with singlet organic molecules to give singlet products is a spin forbidden process with a very low rate. Fortunately, this precludes the spontaneous combustion of living matter, a thermodynamically very favorable process.

One way of circumventing this activation energy barrier involves a free radical pathway in which a singlet molecule reacts with $^3O_2$ to form two doublets (free radicals) in a spin-allowed process (Fig. 4.1, Reaction (1)). This process is, however, highly endothermic (up to 50 kcal mol^{-1}) and is observed at moderate temperatures only with very reactive molecules that afford resonance stabilized radicals, e.g. reduced flavins (Fig. 4.1, Reaction (2)). It is no coincidence, therefore,

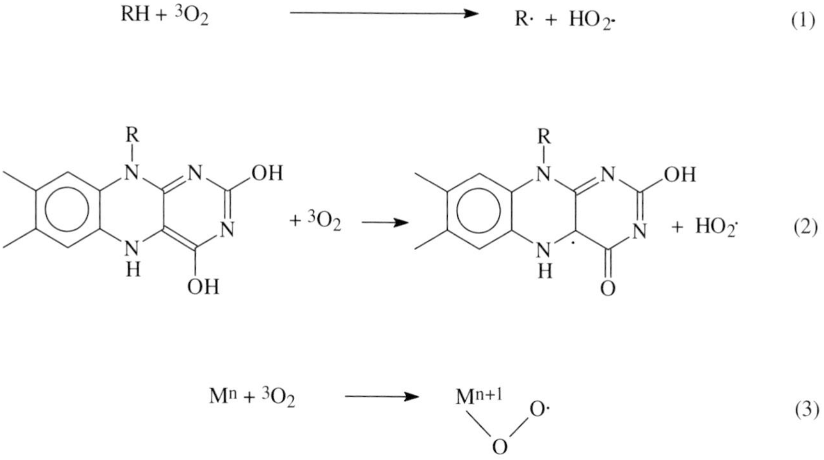

$$RH + {}^3O_2 \longrightarrow R\cdot + HO_2\cdot \qquad (1)$$

$$+ {}^3O_2 \longrightarrow + HO_2\cdot \qquad (2)$$

$$M^n + {}^3O_2 \longrightarrow M^{n+1} \qquad (3)$$

Fig. 4.1 Reactions of triplet oxygen.

that this is the key step in the activation of dioxygen by flavin-dependent oxygenases.

A second way to overcome this spin conservation obstacle is via reaction of $^3O_2$ with a paramagnetic (transition) metal ion, affording a superoxometal complex (Fig. 4.1, Reaction (3)). Subsequent inter- or intramolecular electron-transfer processes can lead to the formation of a variety of metal–oxygen species (Fig. 4.2) which may play a role in the oxidation of organic substrates.

Basically, all (catalytic) oxidations, with dioxygen or peroxide reagents, either under homogeneous or heterogeneous conditions, can be divided into two types on the basis of their mechanism: homolytic and heterolytic. The former involve free radicals as reactive intermediates. Such reactions can occur with most organic substrates and dioxygen, in the presence or absence of metal catalysts. This ubiquity of free radical processes in dioxygen chemistry renders mechanistic interpretation more difficult than in the case of hydrogenations or carbonylations where there is no reaction in the absence of the catalyst.

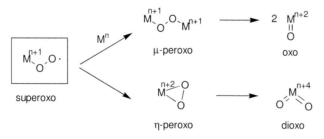

Fig. 4.2 Metal–oxygen species.

Heterolytic oxidations generally involve the (metal-mediated) oxidation of a substrate by an active oxygen compound, e.g. H_2O_2 or RO_2H. Alternatively, stoichiometric oxidation of a substrate by a metal ion or complex is coupled with the reoxidation of the reduced metal species by the primary oxidant (e.g. O_2 or H_2O_2).

4.2.1
Homolytic Mechanisms

As noted above, dioxygen reacts with organic molecules, e.g. hydrocarbons, via a free radical pathway. The corresponding hydroperoxide is formed in a free radical chain process (Fig. 4.3). The reaction is autocatalytic, i.e. the alkyl hydroperoxide accelerates the reaction by undergoing homolysis to chain initiating radicals, and such processes are referred to as autoxidations [1].

The susceptibility of any particular substrate to autoxidation is determined by the ratio $k_p/(2k_t)^{1/2}$, which is usually referred to as its oxidizability [6]. The oxidizabilities of some typical organic substrates are collected in Table 4.2.

The reaction can be started by adding an initiator which undergoes homolytic thermolysis at the reaction temperature to produce chain-initiating radicals. The initiator could be the alkyl hydroperoxide product although relatively high temperatures (>100°C) are required for thermolysis of hydroperoxides. Alternatively, chain-initiating radicals can be generated by the reaction of trace amounts of hydroperoxides with variable valence metals, e.g. cobalt, manganese, iron, cerium etc. The corresponding alkoxy and alkylperoxy radicals are produced in one-electron transfer processes (Fig. 4.4).

In such processes the metal ion acts (in combination with ROOH) as an initiator rather than a catalyst. It is important to note that homolytic decomposition of alkyl hydroperoxides via one-electron transfer processes is generally a com-

Initiation:

$$In_2 \xrightarrow{R_i} 2\ In\cdot$$

$$In\cdot + RH \longrightarrow InH + R\cdot$$

Propagation:

$$R\cdot + O_2 \xrightarrow{\text{very fast}} RO_2\cdot$$

$$RO_2\cdot + RH \xrightarrow{k_p} RO_2H + R\cdot$$

Termination:

$$RO_2\cdot + RO_2\cdot \xrightarrow{k_t} RO_4R \longrightarrow \text{nonradical products}$$

Fig. 4.3 Mechanism of autoxidation.

Table 4.2 Oxidizabilities of organic compounds at 30 °C [a].

Substrate	$k_p/(2k_t)^{1/2} \times 10^3$ $(M^{-1/2}\ s^{-1/2})$
Indene	28.4
Cyclohexene	2.3
1-Octene	0.06
Cumene	1.5
Ethylbenzene	0.21
Toluene	0.01
p-Xylene	0.05
Benzaldehyde [b]	290
Benzyl alcohol	0.85
Dibenzyl ether	7.1

a) Data taken from Ref. [6].
b) At 0 °C.

$$RO_2H + Co^{II} \longrightarrow RO^{\bullet} + Co^{III}OH$$

$$RO_2H + Co^{III} \longrightarrow RO_2^{\bullet} + Co^{II} + H^+$$

Net reaction: $2\,RO_2H \xrightarrow{Co^{II}/Co^{III}} RO^{\bullet} + RO_2^{\bullet} + H_2O$

Fig. 4.4 Metal initiated and mediated autoxidation.

peting process even with metal ions that catalyze heterolytic processes with hydroperoxides (see above). Since dioxygen can be regenerated via subsequent chain decomposition of the alkyl hydroperoxide this can lead to competing free radical autoxidation of the substrate. Generally speaking, this has not been recognized by many authors and can lead to a misinterpretation of results.

4.2.1.1 Direct Homolytic Oxidation of Organic Substrates

Another class of metal catalyzed autoxidations involves the direct one-electron oxidation of the substrate by the oxidized form of the metal catalyst. For example, the autoxidation of alkylaromatics in acetic acid in the presence of relatively high concentrations (~ 0.1 M) of cobalt(III) acetate involves rate-limiting one electron transfer oxidation of the alkylbenzene to the corresponding cation radical (Fig. 4.5). This is followed by elimination of a proton to afford the corresponding benzylic radical, which subsequently forms a benzylperoxy radical by reaction with dioxygen. The primary products from substituted toluenes are the corresponding aldehydes, formed by reaction of benzylperoxy radicals with cobalt(II) (which simultaneously regenerates the cobalt(III) oxidant). The usual reaction of alkylperoxy radicals with the toluene substrate (see Fig. 4.3) is largely

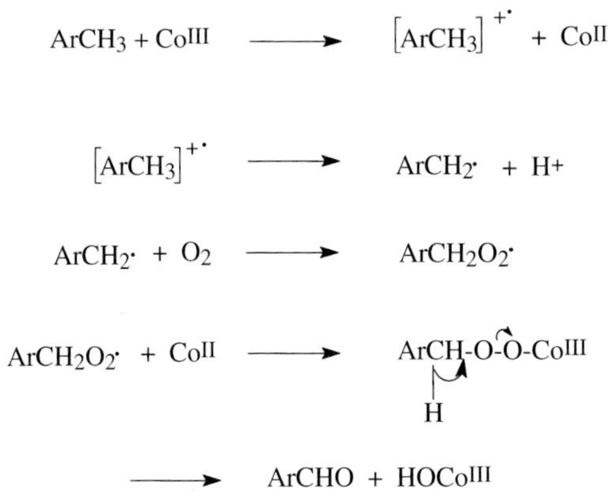

$$ArCH_3 + Co^{III} \longrightarrow \left[ArCH_3\right]^{+\bullet} + Co^{II}$$

$$\left[ArCH_3\right]^{+\bullet} \longrightarrow ArCH_2^{\bullet} + H^+$$

$$ArCH_2^{\bullet} + O_2 \longrightarrow ArCH_2O_2^{\bullet}$$

$$ArCH_2O_2^{\bullet} + Co^{II} \longrightarrow ArCH\text{-}O\text{-}O\text{-}Co^{III}$$
$$\underset{H}{|}$$

$$\longrightarrow ArCHO + HOCo^{III}$$

Fig. 4.5 Direct homolytic oxidation of benzylic compounds.

circumvented by the efficient trapping of the benzylperoxy radical with the relatively high concentration of cobalt(II) present. The aldehyde product undergoes facile autoxidation to the corresponding carboxylic acid and metal-catalyzed autoxidation of methylaromatics is a widely used method for the production of carboxylic acids (see below). Since cobalt is usually added as cobalt(II), reactive substrates such as aldehydes or ketones are often added as promoters to generate the high concentrations of cobalt(III) necessary for initiation of the reaction.

4.2.2
Heterolytic Mechanisms

Catalytic oxidations with dioxygen can also proceed via heterolytic pathways which do not involve free radicals as intermediates. They generally involve a two-electron oxidation of a (coordinated) substrate by a metal ion. The oxidized form of the metal is subsequently regenerated by reaction of the reduced form with dioxygen. Typical examples are the palladium(II)-catalyzed oxidation of alkenes (Wacker process) and oxidative dehydrogenation of alcohols (Fig. 4.6).

In a variation on this theme, which pertains mainly to gas phase oxidations, an oxometal species oxidizes the substrate and the reduced form is subsequently re-oxidized by dioxygen (Fig. 4.7). This is generally referred to as the Mars-van Krevelen mechanism [7].

A wide variety of oxidations mediated by monooxygenase enzymes are similarly thought to involve oxygen transfer from a high-valent oxoiron intermediate to the substrate (although the mechanistic details are still controversial) [8–11]. However, in this case a stoichiometric cofactor is necessary to regenerate the reduced form of the enzyme resulting in the overall stoichiometry shown in Fig. 4.8.

Indeed, the holy grail in oxidation chemistry is to design a 'suprabiotic' catalyst capable of mediating the transfer of both oxygen atoms of dioxygen to or-

$$H_2C=CH_2 + Pd^{II} + H_2O \longrightarrow CH_3CHO + Pd^0 + 2H^+$$

$$Pd^0 + 2H^+ + {}^1/_2 O_2 \xrightarrow{\text{Cu}^{II}} Pd^{II} + H_2O$$

$$R_2CHOH + Pd^{II} \longrightarrow R_2C=O + Pd^0 + H_2O$$

Fig. 4.6 Wacker oxidation and oxidative dehydrogenation of alcohols.

$$M=O + S \longrightarrow M + SO$$

$$M + {}^1/_2 O_2 \longrightarrow M=O$$

Fig. 4.7 Mars-van Krevelen mechanism.

$$RH + O_2 + DH_2 \xrightarrow{\text{monooxygenase}} ROH + D + H_2O$$

D/DH_2 = cofactor

Fig. 4.8 Stoichiometry in oxidations mediated by monooxygenases.

ganic substrates, such as alkenes and alkanes [12]. This would obviate the need for a stoichiometric cofactor as a sacrificial reductant, i.e. it would amount to a Mars-van Krevelen mechanism in the liquid phase.

4.2.2.1 Catalytic Oxygen Transfer

Another way to avoid the need for a sacrificial reductant is to use a reduced form of dioxygen, e.g. H_2O_2 or RO_2H, as a single oxygen donor. Such a reaction is referred to as a catalytic oxygen transfer and can be described by the general equation shown in Fig. 4.9.

Catalytic oxygen transfer processes are widely applicable in organic synthesis. Virtually all of the transition metals and several main group elements are known to catalyze oxygen transfer processes [13]. A variety of single oxygen donors can be used (Table 4.3) in addition to H_2O_2 or RO_2H. Next to price and

$$S + XOY \xrightarrow{\text{catalyst}} SO + XY$$

S = substrate; SO = oxidized substrate

XOY = H_2O_2, RO_2H, R_3NO, NaOCl, KHSO$_5$, etc.

Fig. 4.9 Catalytic oxygen transfer.

Table 4.3 Oxygen donors.

Donor	% Active Oxygen	Coproduct
H_2O_2	47.0 $(14.1)^{a)}$	H_2O
N_2O	36.4	N_2
O_3	33.3	O_2
CH_3CO_3H	21.1	CH_3CO_2H
tert-BuO_2H	17.8	tert-BuOH
HNO_3	25.4	NO_x
NaOCl	21.6	NaCl
NaO_2Cl	35.6	NaCl
NaOBr	13.4	NaBr
$C_5H_{11}NO_2{}^{b)}$	13.7	$C_5H_{11}NO$
$KHSO_5$	10.5	$KHSO_4$
$NaIO_4$	7.5	$NaIO_3$
PhIO	7.3	PhI

a) Figure in parentheses refers to 30% aq. H_2O_2.
b) N-Methylmorpholine-N-oxide (NMO).

ease of handling, two important considerations which influence the choice of oxygen donor are the nature of the co-product and the weight percentage of available oxygen. The former is important in the context of environmental acceptability and the latter bears directly on the volumetric productivity (kg product per unit reactor volume per unit time). With these criteria in mind it is readily apparent that hydrogen peroxide, which affords water as the co-product, is generally the preferred oxidant. The co-product from organic oxidants, such as RO_2H and amine oxides, can be recycled via reaction with H_2O_2. The overall process produces water as the co-product but requires one extra step compared to the corresponding reaction with H_2O_2. With inorganic oxygen donors environmental considerations are relative. Sodium chloride and potassium bisulfate are obviously preferable to the heavy metal salts (Cr, Mn, etc.) produced in classical stoichiometric oxidations. Generally speaking, inorganic oxidants are more difficult to recycle, in an economic manner, than organic ones. Indeed, the ease of recycling may govern the choice of oxidant, e.g. NaOBr may be preferred over NaOCl because NaBr can, in principle, be re-oxidized with H_2O_2. As noted earlier, a disadvantage of H_2O_2 and RO_2H as oxygen donors is possible competition from metal-catalyzed homolytic decomposition leading to free radical oxidation pathways and/or low selectivities based on the oxidant.

Heterolytic oxygen transfer processes can be divided into two categories based on the nature of the active oxidant: an oxometal or a peroxometal species (Fig. 4.10). Catalysis by early transition metals (Mo, W, Re, V, Ti, Zr) generally involves high-valent peroxometal complexes whereas later transition metals (Ru, Os), particularly first row elements (Cr, Mn, Fe) mediate oxygen transfer via oxometal species. Some elements, e.g. vanadium, can operate via either mechanism, depending on the substrate. Although the pathways outlined in Fig. 4.10

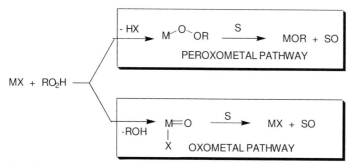

Fig. 4.10 Peroxometal versus oxometal pathways.

pertain to peroxidic reagents analogous schemes involving M=O or MOX (X=ClO, IO$_4$, HSO$_5$, R$_3$NO etc.) as the active oxidant, can be envisaged for other oxygen donors. Reactions that typically involve peroxometal pathways are alkene epoxidations, alcohol oxidations and heteroatom (N and S) oxidations. Oxometal species, on the other hand, are intermediates in the oxidation of al- kane, benzylic and allylic C–H bonds and the dihydroxylation and oxidative cleavage of olefins, in addition to the above-mentioned transformations.

For the sake of completeness we also note that oxygen transfer processes can be mediated by organic catalysts which can be categorized on the same basis as metal catalysts. For example, ketones catalyze a variety of oxidations with mono- peroxysulfate (KHSO$_5$) [14]. The active oxidant is the corresponding dioxirane and, hence, the reaction can be construed as involving a 'peroxometal' pathway. Similarly, TEMPO-catalyzed oxidations of alcohols with hypochlorite [15, 16] in- volve an oxoammonium salt as the active oxidant, i.e. an 'oxometal' pathway.

4.2.3
Ligand Design in Oxidation Catalysis

In the majority of catalytic oxidations simple metal salts are used as the cata- lysts. In contrast, oxidations mediated by redox enzymes involve metal ions co- ordinated to complex ligands: amino acid residues in a protein or a prosthetic group, e.g. a porphyrin ligand in heme-dependent enzymes. Indeed, many of the major challenges in oxidation chemistry involve demanding oxidations, such as the selective oxidation of unactivated C–H bonds, which require powerful oxi- dants. This presents a dilemma: if an oxidant is powerful enough to oxidize an unactivated C–H bond then, by the same token, it will readily oxidize most li- gands, which contain C–H bonds that are more active than the C–H bonds of the substrate. The low operational stability of, for example, heme-dependent en- zymes is a direct consequence of the facile oxidative destruction of the porphyr- in ring. Nature solves this problem *in vivo* by synthesizing fresh enzyme to re- place that destroyed. *In vitro* this is not a viable option. In this context it is worth bearing in mind that many simple metal complexes that are used as catalysts in oxidation reactions contain ligands, e.g. acetylacetonate, that are rapidly destroyed under oxidizing conditions. This fact is often not sufficiently recognized by authors of publications on catalytic oxidations.

Collins [17] has addressed the problem of ligand design in oxidation catalysis in some detail and developed guidelines for the rational design of oxidatively robust ligands. Although progress has been achieved in understanding ligand sensitivity to oxidation the ultimate goal of designing metal complexes that are both stable and exhibit the desired catalytic properties remains largely elusive. We note, in this context, that an additional requirement has to be fulfilled: the desired catalytic pathway should compete effectively with the ubiquitous free radical autoxidation pathways.

One category of oxidations deserves special mention in the context of ligand design: enantioselective oxidations. It is difficult to imagine enantioselective oxidations without the requirement for (chiral) organic ligands. Here again, the task is to design ligands that endow the catalyst with the desired activity and (enantio-)selectivity and, at the same time, are (reasonably) stable.

In the following sections oxidative transformations of a variety of functional groups will be discussed from both a mechanistic and a synthetic viewpoint.

4.2.4
Enzyme Catalyzed Oxidations

Oxidations by enzymes are performed by the group of oxidoreductases. An overview of the classification of oxidoreductases is given in Fig. 4.11.

The alcohol *dehydrogenases* were already described in Chapter 3. These enzymes are cofactor dependent and in the active site hydrogen transfer takes place from NADH or NADPH. In the reverse way they can, however, be applied for the oxidation of alcohols in some cases (see below). *Oxidases* are very appealing for biocatalytic purposes, because they use oxygen as the only oxidant without the need for a cofactor. Oxidases usually have flavins (glucose oxidase, alcohol oxidase) or copper (examples; galactose oxidase, laccase and tyrosinase) in the active site [18]. The mechanism for glucose oxidase (GOD) is denoted in

1. Dehydrogenases

$$SH_2 + D \longrightarrow S + DH_2$$

2. Oxidases
(Oxidative dehydrogenation)

$$SH_2 + O_2 \longrightarrow S + H_2O_2$$

3. Oxygenases (oxygen insertion)
mono- :

$$S + DH_2 + O_2 \longrightarrow SO + D + H_2O$$

di- :

$$SH + O_2 \longrightarrow SO_2H$$

4. Peroxidases

$$SH_2 + H_2O_2 \longrightarrow S + 2 H_2O$$

$$S + H_2O_2 \longrightarrow SO + H_2O$$

S, SH, SH$_2$ = substrate
S, SO, SO$_2$, SO$_2$H = oxidized substrate
D, DH$_2$ = cofactor e.g. NAD / NADH$_2$

Fig. 4.11 Classes of oxidoreductases.

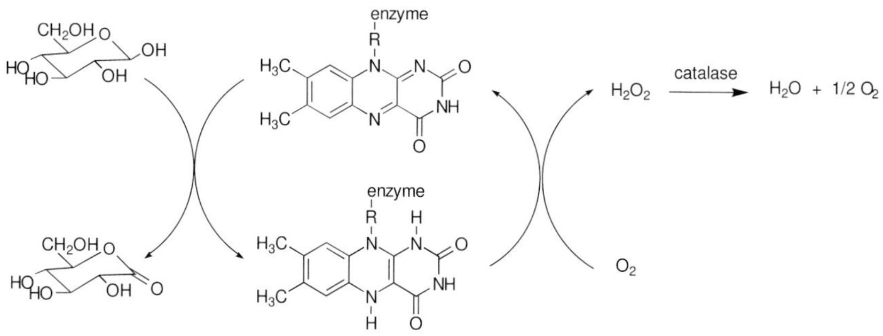

Fig. 4.12 Aerobic oxidation of D-glucose catalyzed by flavin-dependent glucose oxidase.

Fig. 4.12. GOD oxidases the oxidation of β-D-glucose to D-glucone-δ-lactone, a reaction which has attracted the attention of generations of analytical scientists due to its possible applicability in glucose sensors for diabetes control.

Galactose oxidase exhibits a mononuclear copper-active site, which is flanked by a tyrosinyl radical. In a single step, a two-electron oxidation of alcohols can be performed by this enzyme, where one electron is extracted by copper and the other by the tyrosine residue [19].

In the case of oxygenases, catalytic oxygen transfer has to take place, and many of the principles described above are operative here as well. A combination of oxygen and NADH or NADPH is supplied to the active site which results in net transfer of "O" and H_2O as the by-product. Thus NAD(P)H is required as sacrificial reductant and only one of the two oxygen atoms in oxygen ends up in the product. An exception is the dioxygenases where both oxygen atoms can end up in the product. *Oxygenases* can be classified, depending on their redox-active cofactor (Table 4.4). The active site commonly contains either a heme-iron, non-heme-iron, flavin or copper (mono- or binuclear) as redox-active group.

Table 4.4 Classification of oxygenases.

Type	Examples	Substrates
Heme-iron aromatics	Cytochrome-P450	alkanes, alkenes
Non-heme-iron	Methane monooxygenase	alkanes, alkenes
	Rieske dioxygenases	aromatics, alkenes
	Lipoxygenase	alkenes
Flavin	Hydroxybiphenyl monooxygenase	aromatics
	Styrene monooxygenase	styrenes
	Cyclohexane monooxygenase	Baeyer-Villiger
Copper	Tyrosinase	phenolic
	Dopamine β-monooxygenase	benzylic C–H

A pivotal class of oxygenases are the so-called iron heme-proteins from the Cytochrome P450 family, in which the iron is ligated by a porphyrin moiety [11, 20, 21]. In the first case, an oxometal-type mechanism is operative, where a putative $P^{+•}Fe(IV)=O$ (P = porphyrin) species transfers oxygen directly to the substrate (see Fig. 4.13). The exact identity is still a matter of controversy. Generally, a Fe(IV)=O species stabilized by a cationic radical porphyrin moiety seems to be favored instead of a formally Fe(V)=O species. Especially intriguing is the potential of these types of enzymes to perform a stereoselective hydroxylation of non-activated alkanes – a reaction which is still very rare in the case of metal-catalysed oxidations. A so-called rebound mechanism has been proposed in this case which involves a [Fe–OH R$^•$] transient species (see Fig. 4.14). Iron-heme-type enzymes are the enzymes which have the broadest substrate spectrum. They are capable of performing oxygenation on a wide range of compounds, from alkanes and fatty acids, to alkenes and alcohols. Due to the instability of the heme group, and the requirement for a cofactor, these enzymes are commonly employed under microbial whole cell conditions [22].

Another important class of oxygenases are the flavin-dependent oxygenases, notably in Baeyer-Villiger oxidation and epoxidation [23, 24]. The mechanism for Baeyer-Villiger conversion of cyclic ketones into esters by the enzyme cyclohexane monooxygenase is shown in Fig. 4.15 [25]. The prosthetic group FAD, which is tightly bound to the active site of the enzyme, is reduced to $FADH_2$ by NADPH. Rapid reaction between $FADH_2$ and molecular oxygen (see Fig. 4.1, Reaction (2)) affords 4a-hydroperoxyflavin which is a potent oxidizing agent. Addition of 4a-hydroperoxyflavin to the carbonyl group of cyclohexanone creates a tetrahedral intermediate, which rearranges to give FAD-4a-OH and ε-caprolactone. Finally a water molecule is eliminated from FAD-4a-OH to regenerate

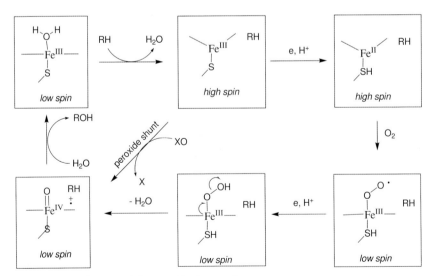

Fig. 4.13 Fe-porphyrins and the peroxide shunt pathway.

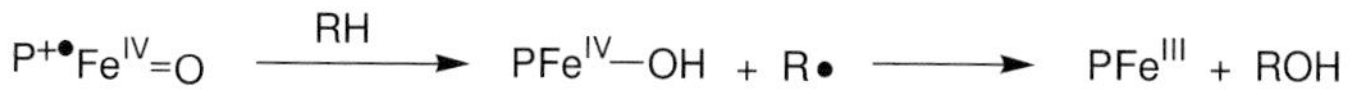

$$P^{+\bullet}Fe^{IV}=O \xrightarrow{\text{RH}} PFe^{IV}-OH + R\bullet \longrightarrow PFe^{III} + ROH$$

Fig. 4.14 Fe-porphyrins and the oxygen-rebound mechanism.

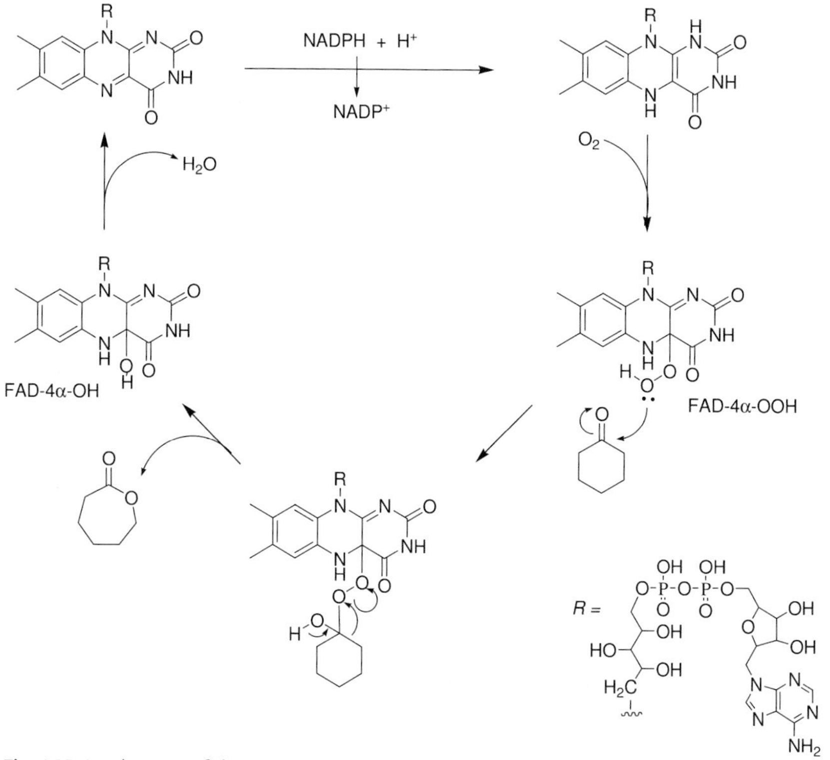

Fig. 4.15 Mechanism of the Baeyer-Villiger oxidation of cyclohexanone using flavin-dependent monooxygenase.

FAD ready for a subsequent catalytic cycle. This enzyme catalyzed Baeyer-Villiger oxidation bears great resemblance to the analogous chemical reaction performed by peroxides or peracids, which act as nucleophiles. Globally these flavin-enzymes can perform the same reactions as peracids, i.e. epoxidations, Baeyer-Villiger-reactions and nucleophilic heteroatom oxidation [26–28].

Within the group of non-heme-iron monooxygenases, two notable classes can be identified: The binuclear non-heme monooxygenases and the mononuclear dioxygenases [29, 30]. The most important example of the first group, methane monooxygenase, is capable of mediating the oxidation of a broad range of substrates including methane, at ambient temperature [31]. In this enzyme a non-heme di-iron core is present in which the two-irons are connected through two carboxylate bridges, and coordinated by histidine residues (see Fig. 4.16) [32]. Also in this case Fe(IV)=O species have been proposed as the active intermediates.

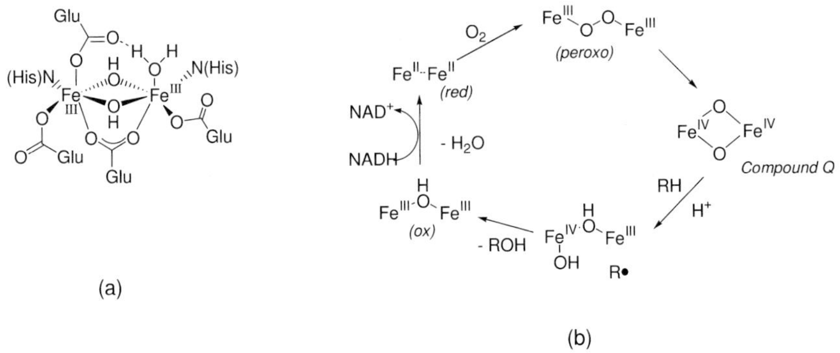

(a)

(b)

Fig. 4.16 Structure of the MMO-iron site and its catalytic cyle.

The class of mononuclear dioxygenases [30] can e.g. perform hydroperoxidation of lipids, the cleavage of catechol and dihydroxylation of aromatics. A prominent example is naphthalene dioxygenase, which was the first identified by its crystal structure. It contains iron and a Rieske (2Fe–2S) cluster and is commonly referred to as a Rieske-type dioxygenase [33]. The iron in this case is flanked by two histidines and one aspartic acid residue. Among the mononuclear iron enzymes, the 2-His-1-carboxylate is a common motif, which flanks one-side of the iron in a triangle and plays an important role in dioxygen activation [34] (Fig. 4.17).

Alternatively copper is capable of H-atom abstraction leading to aliphatic hydroxylation. The enzyme dopamine β-monooxygenase is an example thereof and the reaction is exemplified in Fig. 4.18 [35]. Recently, it has been disclosed that the two copper centers in the enzyme are far apart and not coupled by any magnetic interaction [36]. The binding and activation of dioxygen thus takes place at a single Cu-center, which can provide one electron to generate a superoxo Cu^{II}–O_2^- intermediate. This superoxo intermediate is supposed to be capable of H-atom abstraction.

All the above examples of monooxygenases are dependent on cofactors. The enzymes are usually composed of several subunits which, in a series of cascades, provide the reducing equivalents to the iron-active site. The need for a co-

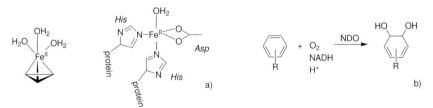

Fig. 4.17 (a) The 2-His-1-carboxylate facial triad in naphthalene dioxygenase (NDO). (b) Reaction catalyzed by NDO.

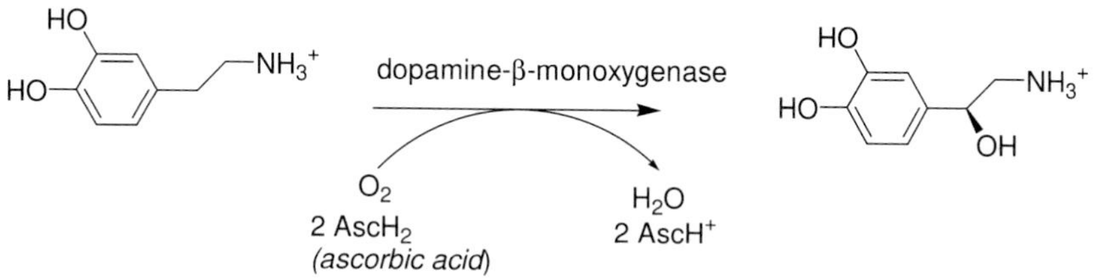

Fig. 4.18 Hydroxylation of dopamine by the non-coupled binuclear copper center in dopamine-β-monooxygenase.

factor is circumvented in the case of *peroxidases* where H_2O_2 acts as the oxidant. In this case the addition of hydrogen peroxide results in a shunt pathway, which directly recovers the [Fe(IV)=O][P$^{+\bullet}$] species from Fe(III) (see Fig. 4.13). Besides iron-heme-dependent peroxidases [37], vanadate-dependent peroxidases are also known which can catalyze sulfide oxidations and which exhibit much higher stabilities [38].

4.3
Alkenes

Various combinations of metal catalyst and single oxygen donor have been used to effect different oxidative transformations of olefins: epoxidation, dihydroxylation, oxidative cleavage, ketonization and allylic oxidation. The most extensively studied example is undoubtedly olefin epoxidation [39].

4.3.1
Epoxidation

The epoxidation of propene with *tert*-butylhydroperoxide (TBHP) or ethylbenzene hydroperoxide (EBHP), for example, accounts for more than one million tons of propene oxide production on an annual basis (Fig. 4.19).

The reaction in Fig. 4.19 is catalyzed by compounds of high-valent, early transition metals such as Mo(VI), W(VI), V(V) and Ti(IV). Molybdenum compounds are particularly effective homogeneous catalysts and are used in the ARCO process in combination with TBHP or EBHP. In the Shell process, on the other hand, a heterogeneous Ti(IV)/SiO$_2$ catalyst is used with EBHP in a continuous,

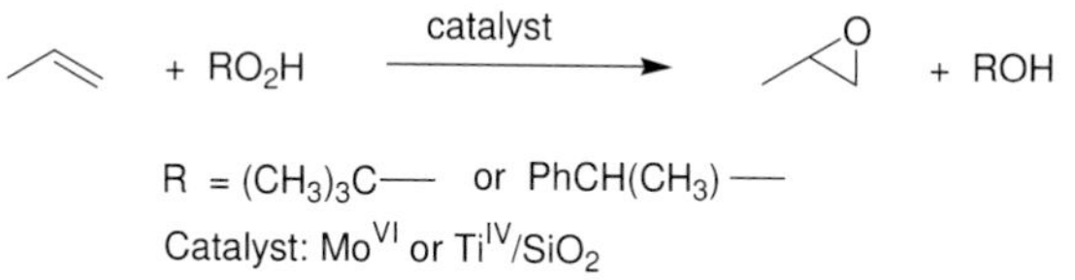

R = (CH$_3$)$_3$C— or PhCH(CH$_3$) —
Catalyst: Mo^{VI} or Ti^{IV}/SiO$_2$

Fig. 4.19 Epoxidation of propylene using RO$_2$H as oxidant.

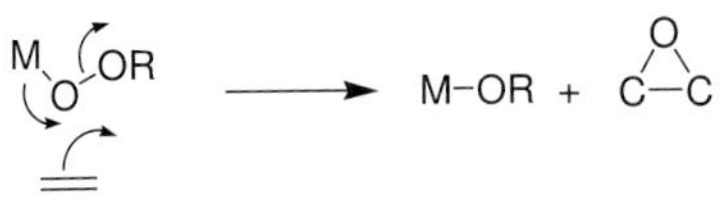

Fig. 4.20 Peroxometal mechanism in the epoxidation.

fixed-bed operation. Alkyl hydroperoxides in combination with homogeneous (Mo, W, V, Ti) or heterogeneous (Ti(IV)/SiO$_2$) catalysts can be used for the selective epoxidation of a wide variety of olefins [40]. Chiral titanium complexes are used as catalysts for enantioselective epoxidations with alkyl hydroperoxides (see below).

The epoxidation of olefins with RO$_2$H or H$_2$O$_2$ (see above) catalyzed by early transition elements involves, as would be expected, a peroxometal mechanism in which the rate-limiting step is oxygen transfer from an electrophilic (alkyl) peroxometal species to the nucleophilic olefin (Fig. 4.20). The metal center does not undergo any change in oxidation state during the catalytic cycle. It functions as a Lewis acid by withdrawing electrons from the O–O bond and thus increased the electrophilic character of the coordinated peroxide. Active catalysts are metals that are strong Lewis acids and relatively weak oxidants (to avoid one electron oxidation of the peroxide) in their highest oxidation state.

Neither the homogeneous Mo nor the heterogeneous Ti(IV)/SiO$_2$ catalysts are effective with hydrogen peroxide as the oxygen donor. Indeed, they are severely inhibited by strongly coordinating molecules such as alcohols, and particularly water. Because of the strong interest in the use of H$_2$O$_2$ as the primary oxidant, particularly in the context of fine chemicals production, much effort has been devoted to developing epoxidation catalysts that are effective with aqueous hydrogen peroxide.

In the mid-1980s two approaches were followed to achieve this goal. Enichem scientists developed a titanium(IV)-silicalite catalyst (TS-1) that is extremely effective for a variety of synthetically useful oxidations, including epoxidation with

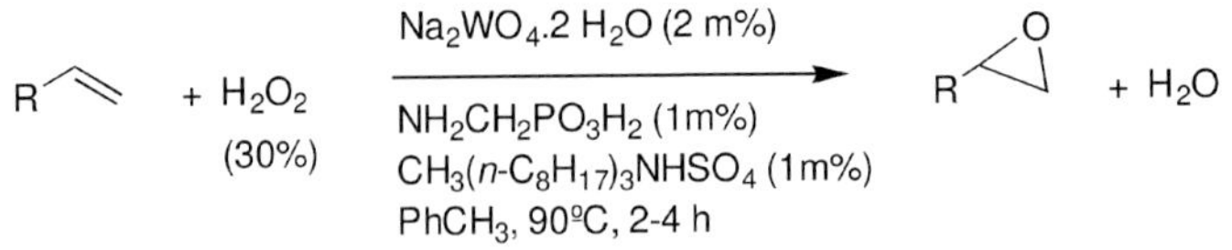

Olefin	Conv (%)	Yield (%)
1-octene	96	94
1-decene	99	99
1-dodecene	98	97
2-octene	99	99
styrene	52	3

Fig. 4.21 Epoxidation catalyzed by TS-1.

30% aqueous H_2O_2 (Fig. 4.21) [41]. The unique activity of TS-1 derives from the fact that silicalite is a hydrophobic molecular sieve, in contrast to Ti(IV)/SiO$_2$ which is hydrophilic. Consequently, hydrophobic substrates are preferentially adsorbed by TS-1 thus precluding the strong inhibition by water observed with Ti(IV)/SiO$_2$. However, a serious limitation of TS-1 in organic synthesis is that its scope is restricted to relatively small molecules, e.g. linear olefins, which are able to access the micropores (5.3×5.5 Å^2). This provoked a flurry of activity aimed at developing titanium-substituted molecular sieves with larger pores which, as yet, has not produced catalysts with comparable activity to TS-1 [42].

4.3.1.1 Tungsten Catalysts

At the same time Venturello and coworkers [43] followed a different approach. They showed that a mixture of tungstate and phosphate in the presence of a tetraalkylammonium salt as a phase transfer agent catalyzed epoxidations with H_2O_2 in a two-phase dichloroethane/water medium. Since its discovery in 1983 this system has been extensively studied [44, 45] in particular with regard to the exact nature of the active phosphotungstate catalyst [46]. More recently, Noyori and coworkers reported [47] a significant improvement of the original system. An appropriate choice of phase transfer catalyst, containing a sufficiently lipophilic tetraalkylammonium cation and a bisulfate (HSO_4^-) anion, in combination with catalytic amounts of $H_2NCH_2PO_3H_2$ and sodium tungstate produced an effective system for the epoxidation of olefins with H_2O_2 in toluene/water or in the absence of an organic solvent (Fig. 4.22). The type of ammonium salt used with phosphate/tungstic acid catalyst is important. For example n-octylammonium hydrogen sulfate is critical in Noyori's work; chloride causes deactivation and other ammonium hydrogen sulfates produce catalysts that are not as effective or are even completely inactive.

Subsequently, the Noyori-system was shown [48] to be an effective system for the oxidative cleavage of cyclic olefins to dicarboxylic acids (Fig. 4.23) using ≥4 equivalents of H_2O_2, via the intermediate formation of the epoxide. For example, cyclohexene afforded adipic acid in 93% isolated yield, thus providing a 'green route' to adipic acid. Although the economics may be prohibitive for adipic acid manufacture, owing to the consumption of 4 equivalents of H_2O_2, the method has general utility for the selective conversion of a variety of cyclic olefins to the corresponding dicarboxylic or keto-carboxylic acids.

The success of tungstate as catalyst has stimulated a wealth of research in this area. Recently it was shown that a silicotungstate compound, synthesized

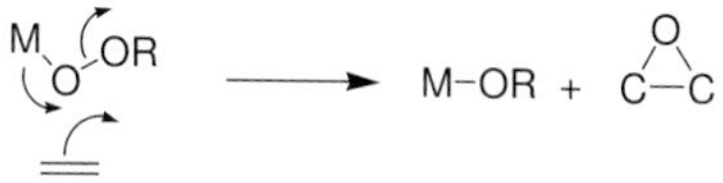

Fig. 4.22 Halide- and halogenated solvent-free biphasic epoxidation with 30% aqueous H_2O_2 using tungstate catalysts.

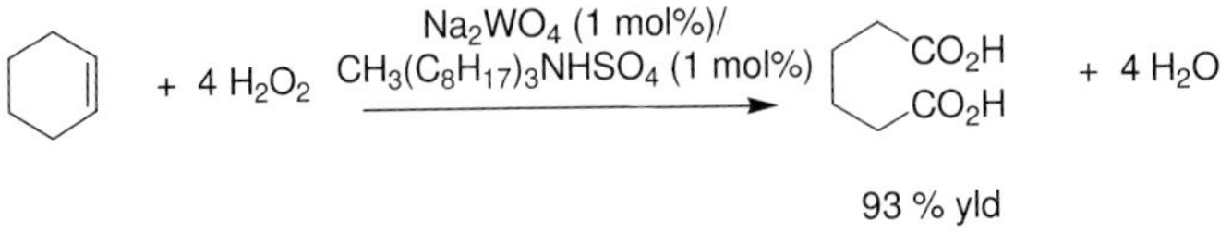

$$\text{cyclohexene} + 4\ H_2O_2 \xrightarrow[\text{CH}_3(\text{C}_8\text{H}_{17})_3\text{NHSO}_4\ (1\ \text{mol\%})]{\text{Na}_2\text{WO}_4\ (1\ \text{mol\%})/} \text{adipic acid (CO}_2\text{H, CO}_2\text{H)} + 4\ H_2O$$

93 % yld

Fig. 4.23 Oxidative cleavage of cyclohexene using tungstate catalyst.

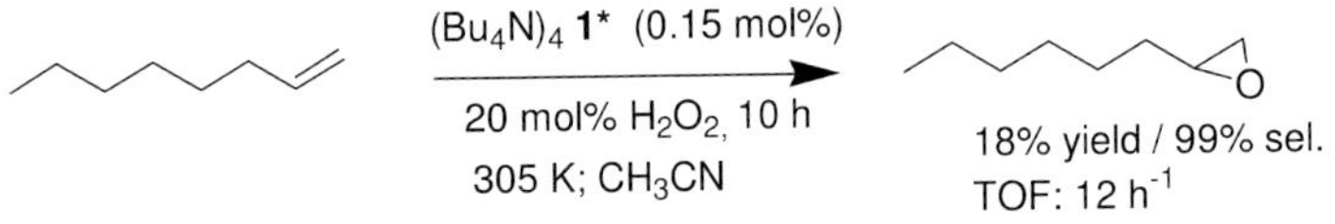

$$\text{1-octene} \xrightarrow[\substack{20\ \text{mol\%}\ H_2O_2,\ 10\ h \\ 305\ K;\ CH_3CN}]{(\text{Bu}_4\text{N})_4\ \mathbf{1}^*\ (0.15\ \text{mol\%})} \text{1,2-epoxyoctane}$$

18% yield / 99% sel.
TOF: 12 h^{-1}

Fig. 4.24 Silicotungstate as recyclable catalyst for epoxidation.

by protonation of a divalent, lacunary Keggin-type compound, exhibits high catalytic performance for epoxidations with H_2O_2. For example, propene could be
oxidized to propylene oxide with >99% selectivity at 305 K in acetonitrile using
0.15 mol% catalyst (1.5 mol% W) in 10 h (Fig. 4.24) [49]. This catalyst could be
recovered and reused at least five times.

"Green" versions of the Venturello system have been developed: A smart re-
coverable phosphate/tungstic acid was obtained by using tributylphosphate as a
cosolvent, in combination with a quaternary ammonium cetylpyridinium cation:
π-$[C_5H_5NC_{16}H_{33}]_3[PO_4(WO_3)_4]$ [50]. This soluble material oxidizes propene with
30% H_2O_2 at 35 °C, and the reduced catalyst precipitates out once hydrogen peroxide is consumed. Heterogeneous versions of the Venturello-catalyst [51] have
been obtained by direct support on ion-exchange resins [52]. Another approach
is to heterogenize the quaternary ammonium functions that charge-balance the
anionic W-catalysts as in PW$_4$-amberlite [52]. Other peroxotungstates can also be
immobilized: cetylpyridium-dodecatungstates were immobilized on fluoroapatite
and employed under anhydrous solvent-free conditions [53]. Furthermore perox-
otungstate was immobilized on ionic liquid-modified silica and used as a catalyst in acetonitrile [54]. In general these catalysts are subject to low productivities. A much more active heterogeneous W-system was obtained by combining
a $[WO_4]^{2-}$-exchanged LDH catalyst with NH$_4$Br [55]. In this case W catalyzes the
oxidation of the bromide ions. The thus formed bromohydrin is converted into
the epoxide. Interesting activities could be observed for e.g. α-methylstyrene
with H_2O_2 in the presence of NH$_4$Br at 40 °C, which produces almost 50 g of
epoxide per g $[WO_4]^{2-}$-LDH catalyst within 5 h with 95% epoxide selectivity [55].

4.3.1.2 Rhenium Catalysts

In the early 1990s Herrmann and coworkers [56] reported the use of methyltrioxorhenium (MTO) as a catalyst for epoxidation with anhydrous H_2O_2 in *tert*-
butanol. In the initial publication cyclohexene oxide was obtained in 90% yield
using 1 mol% MTO at 10 °C for 5 h. At elevated temperatures (82 °C) the corre-

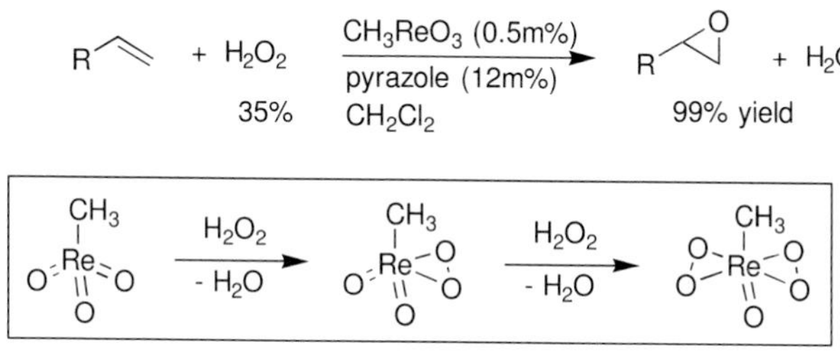

Fig. 4.25 MTO-Catalyzed epoxidations with H_2O_2.

sponding *trans*-diol was obtained (97% yield) owing to the acidity of the system promoting ring opening of the epoxide. Subsequently, the groups of Herrmann [57] and Sharpless [58] significantly improved the synthetic utility of MTO by performing reactions in the presence of 10–12 mol% of heterocyclic bases, e.g. pyridine [58], 3-cyanopyridine [58] and pyrazole [57] in a dichloromethane/water mixture. For example, using the MTO-pyrazole system high activities and epoxide selectivities were obtained [57] with a variety of olefins using 35% aqueous H_2O_2 and 0.5 mol% MTO. The use of 2,2,2-trifluoroethanol further enhances the catalytic performance of the MTO/base/hydrogen peroxide system by a three- to ten-fold increase in TOF [59, 60]: In the epoxidation of cyclohexene turnover numbers of over 10 000 could be achieved by slow addition of the substrate. Notably, the use of perfluorinated alcohols in the absence of catalysts already leads to significant epoxidation [60, 61]. This observation, that fluorinated alcohols are able to activate hydrogen peroxide, is tentatively attributed to the characteristic feature of highly fluorinated alcohols: owing to the strong electron-withdrawing effect of the perfluoroalkyl groups the hydroxy group is very electron poor and unable to accept a hydrogen bond but, at the same time, easily donates a hydrogen bond. In this way the perfluorinated alcohol acts as a Lewis acid and increases the electrophilicity of the hydrogen peroxide.

MTO-catalyzed epoxidations proceed via a peroxometal pathway involving a diperoxorhenium(VII) complex as the active oxidant (see Fig. 4.25). Major disadvantages of MTO are its limited stability under basic H_2O_2 conditions [62] and its rather difficult and, hence, expensive synthesis.

4.3.1.3 Ruthenium Catalysts

The systems described above all involve peroxometal species as the active oxidant. In contrast, ruthenium catalysts involve a ruthenium-oxo complex as the active oxidant [1]. Until recently, no Ru-catalysts were known that were able to activate H_2O_2 rather then to decompose it. However in 2005 Beller and co-workers recognized the potential of the Ru(terpyridine)(2,6-pyridinedicarboxylate) catalyst [63] for the epoxidation of olefins with H_2O_2 [64]. The result is a very efficient method for the epoxidation of a wide range of alkyl substituted or allylic alkenes using as little as 0.5 mol% Ru. In Fig. 4.26 details are given. Terminal

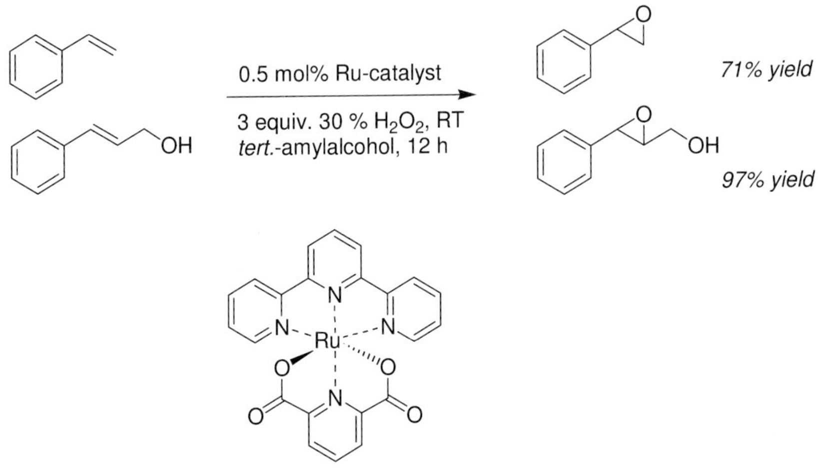

Fig. 4.26 Ru(terpyridine)(2.6-pyridinedicarboxylate) catalysed epoxidations.

olefins are not very active under these conditions. This system can be operated under neutral conditions, which makes it suitable for the synthesis of acid-sensitive epoxides as well.

4.3.1.4 Manganese Catalysts

Manganese-based catalysts probably involve an oxomanganese(V) complex as the active oxidant. Montanari and coworkers [65] described the use of a manganese(III) complex of a halogenated porphyrin, in the presence of hexylimidazole and benzoic acid, for epoxidations with 30% H_2O_2 in dichloromethane–water. More recently, Hage and coworkers [66] showed that a manganese(II) complex of trimethyl-1,4,7-triazacyclononane (tmtacn), originally developed as a bleach activator for application in detergents, is a highly active catalyst for epoxidations with H_2O_2 in aqueous methanol. In the original publication a vast excess of H_2O_2 was needed owing to competing manganese-catalyzed decomposition of the H_2O_2. Subsequent detailed investigations of this system by Jacobs and De Vos and coworkers [67] culminated in the development of an optimized system, comprising Mn(II)-tmtacn (0.1 mol%) in the presence of oxalate buffer (0.3 mol%) with 35% H_2O_2 (1.5 eq.) in acetonitrile at 5 °C. This system is an extremely active catalyst for the epoxidation of even relatively unreactive olefins. Mechanistic details are uncertain but it would seem likely that the active oxidant is an oxomanganese(IV) or (V) [68] complex containing one tmtacn and one oxalate ligand (see Fig. 4.27).

Recently it has been shown that simple manganese sulfate in the presence of sodium bicarbonate is reasonably effective in promoting the epoxidation of alkenes with aqueous H_2O_2 using DMF or t-BuOH as solvents [69]. In this system peroxocarbonate is formed *in situ*, thus minimizing the catalase activity of the Mn salt. Following this discovery, Chan and coworkers introduced an imidazole-based ionic

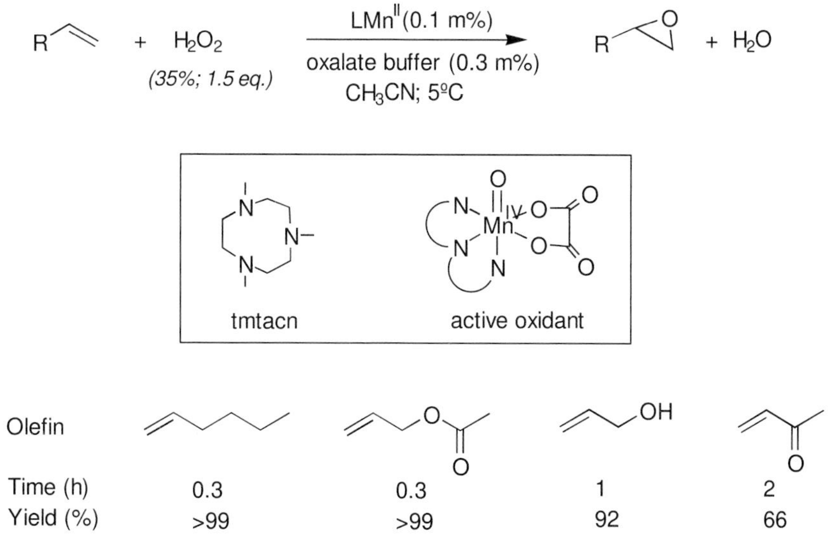

Fig. 4.27 Mn-tmtacn catalyzed epoxidation with H_2O_2.

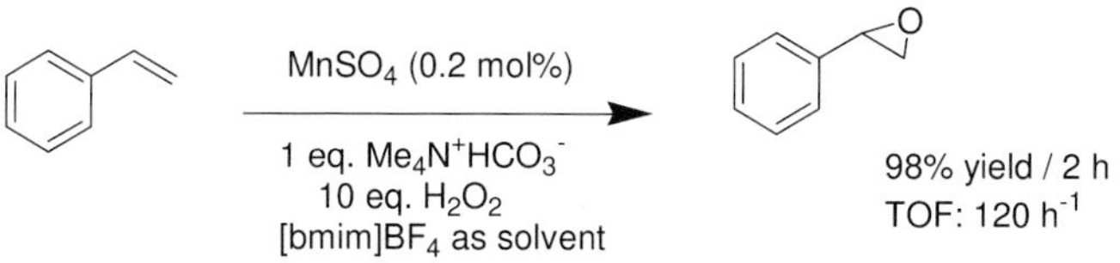

Olefin				
Time (h)	0.3	0.3	1	2
Yield (%)	>99	>99	92	66

Fig. 4.28 Mn/bicarbonate system for epoxidation in ionic liquid.

liquid in the Mn/bicarbonate system in order to overcome the requirements of volatile organic cosolvents (see Fig. 4.28) [70]. A major disadvantage of these Mn salt systems is that 10 equivalents of H_2O_2 are still required to reach substantial conversions. Moreover simple terminal olefins cannot be oxidized in this way.

In addition, recently it was found that by using peracetic acid as the oxidant $Mn^{II}(bipy)_2$ becomes an extremely active catalyst. A turnover frequency over $200\,000\ h^{-1}$ was found for 1-octene using 2 equivalents of peracetic acid in acetonitrile [71]. In spite of this high activity, the system above with Mn-tacn and H_2O_2 would be preferred because of the safety issues associated with peracetic acid and the coproduction of acetic acid.

4.3.1.5 Iron Catalysts

Mechanistically related to Mn, is the use of Fe as an epoxidation catalyst. Recently, iron complexes with a tetradentate amine core were reported, that were capable of activating H_2O_2 without the involvement of hydroxyl radicals [72]. For a variety of substituted as well as terminal alkenes, effective epoxidation

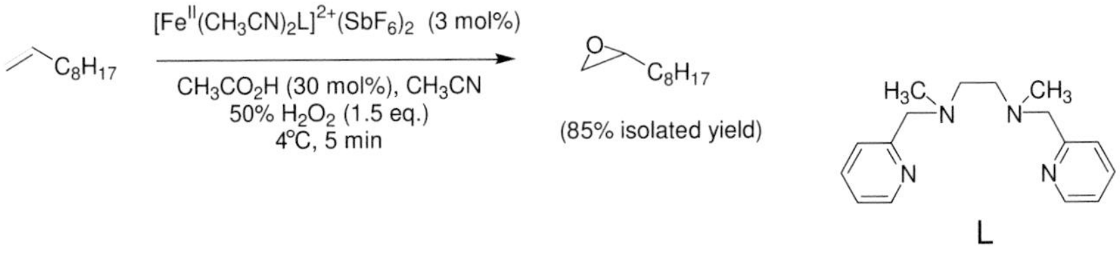

$[Fe^{II}(CH_3CN)_2L]^{2+}(SbF_6)_2$ (3 mol%)

CH_3C_8H_17 →

CH_3CO_2H (30 mol%), CH_3CN
50% H_2O_2 (1.5 eq.)
4°C, 5 min

O (epoxide) C_8H_17

(85% isolated yield)

H_3C—N N—CH_3

L

Fig. 4.29 Fe-tetradentate amine complexes as catalysts in epoxidation with H_2O_2/acetic acid.

was described using 3 mol% of Fe, ≤ 30 mol% acetic acid and 1.5 eq. of 50% H_2O_2 (see Fig. 4.29) [73]. For a variety of substituted as well as terminal alkenes, complete conversion of the alkene was obtained leading to 77–85% isolated yields. Acetic acid was essential to obtain high epoxide yields. Investigations showed that under reaction conditions with the acid, peracetic acid is formed and that the iron(II) complex self-assembled into a μ-oxo, carboxylate bridged diiron(III) complex resembling the diiron(III) core found in the active site of the oxidized enzyme methane monooxygenase (MMO). The results obtained by Jacobsen are amongst the best obtained so far in epoxidation using iron catalysts and hydrogen peroxide. At the same time, Que showed that analogs of this iron complex led to asymmetric cis-dihydroxylation of olefins (see below) [74]. In addition it was found that, *in situ* prepared iron catalysts from ferric perchlorate or ferric nitrate and phenanthroline ligands, are very active catalysts with peracetic acid as the oxidant and acetonitrile as solvent [75]. TOFs of the order of 5000 h^{-1} were obtained for a variety of terminal and internal olefins. A major disadvantage is of course the requirement for peracetic acid (see above for Mn).

4.3.1.6 Selenium and Organocatalysts

Finally, organic compounds and arylseleninic acids have been described as catalysts for epoxidations with H_2O_2. Perfluoroheptadecan-9-one was shown [76] to be a recyclable catalyst for the epoxidation of relatively reactive olefins with 60% aqueous H_2O_2 in 2,2,2-trifluoroethanol. The discovery that hydrogen peroxide can be used in conjunction with catalytic amounts of peroxyseleninic acids, dates from 1978 [77]. Recently 3,5-bis(trifluoromethyl)benzeneseleninic acid was used as the catalyst, and even sensitive epoxides could be formed in nearly quantitative yields, using S/C ratios as high as 200 (see Fig. 4.30 and Table 4.5) [78]. In these reactions bis(3,5-bis(trifluoromethyl)phenyl)-diselenide is the starting compound, which under reaction conditions gives the seleninic acid (see Fig. 4.30). Under these optimised conditions turnover frequencies of over 1000 h^{-1} were observed at room temperature.

It is clear from the preceding discussion that several systems are now available for epoxidations with aqueous H_2O_2. The key features are compared in Table 4.5 for epoxidation of cyclohexene and a terminal olefin. They all have their limitations regarding substrate scope, e.g. TS-1 is restricted to linear olefins and

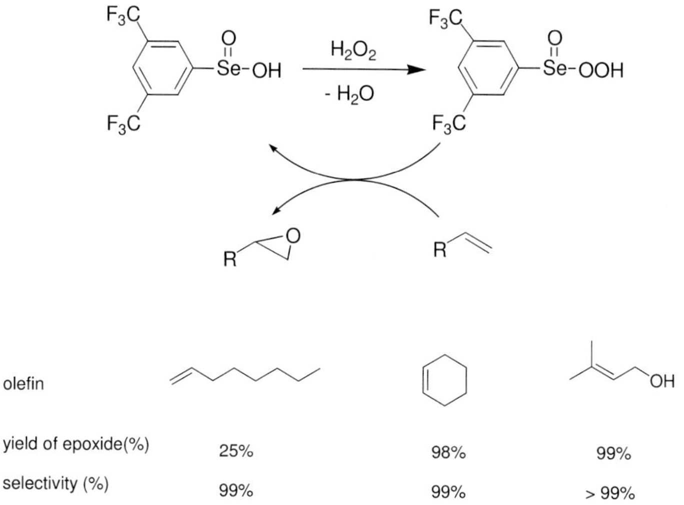

Fig. 4.30 Aromatic seleninic acids as catalysts for epoxidation.

Table 4.5 Catalytic epoxidation with H_2O_2: state of the art.

Catalyst	Mn[a]	W[b]	Re[c]	Re[d]	Se[e]	Ru[f]	Ti[g]
Ref.	67	48	57	59	78	64	41
S/C	666	50	200	1000	200	200	100
Solvent	CH_3CN	$PhCH_3$	CH_2Cl_2	CF_3CH_2OH	CF_3CH_2OH	TAA[h]	CH_3OH
Temp. (°C)	5	90	25	25	20	25	25
1-Alkene							
Yield (%)	99	81	99	99	25	–	74
TOF (h^{-1})[i]	2000	12	14	48	13	–	108
Cyclohexene							
Yield (%)	83	[j]	99	99	98	84[k]	1
TOF (h^{-1})[i]	550	[j]	198	2000	1000	14	1

a) Mn(tmtacn)/oxalate buffer.
b) $WO_4^{2-}/H_2NCH_2PO_3H_2/(C_8H_{17})_3MeN^+HSO_4^-$.
c) CH_3ReO_3/pyrazole (12 mol%).
d) CH_3ReO_3/pyrazole (10 mol%).
e) 0.25 mol% bis(3,5-bis(trifluoromethyl) phenyldiselenide.
f) Ru(terpyridine)(pyridinedicarboxylate).
g) TS-1 (=titanosilicalite).
h) TAA = *tert.*-amylalcohol.
i) mol product/mol cat/h.
j) Gives adipic acid via ring opening.
k) Activity for 1-methyl-cyclohexene. This substrate is more nucleophilic, and therefore more reactive [l]. Essentially no
reaction.

W gives ring opening with acid-sensitive epoxides, or the solvent required for good results, e.g. dichloromethane (MTO) and acetonitrile (Mn-tmtacn). Hence, the quest for even better systems continues.

4.3.1.7 Hydrotalcite and Alumina Systems

Relatively simple solid materials like hydrotalcites and clays can also act as catalysts for epoxidation in the presence of H_2O_2. In the case of anionic clays like $Mg_{10}Al_2(OH)_{24}CO_3$, these materials are basic enough to promote nucleophilic epoxidations [79]. For "normal" electrophilic epoxidation nitrile and amide additives are required [80]. It must, however, be recognized that these materials are polynuclear-alumina clays and simple alumina will also catalyze this reactions without additives [81]. A variety of alumina-materials can absorb hydrogen peroxide onto that surface forming an active oxidant (alumina-OOH). Reactions could be carried out using ethylacetate as the solvent, which is a cheap and environmentally attractive solvent. The major limitation of this system is the low tolerance towards water and to overcome this problem, reactions need to be carried out under Dean-Stark conditions. Furthermore selectivities are usually in the 70–90% range. Notably, the epoxide of α-pinene, which is very unstable, was also obtained in a reasonable 69% yield [81].

4.3.1.8 Biocatalytic Systems

A variety of monooxygenases (see above) can perform epoxidations. Some biocatalytic methods come into sight, which will become attractive for industrial use. In these cases chiral epoxides are the targeted products, and therefore this subject will be dealt with in Section 4.6.

4.3.2
Vicinal Dihydroxylation

The osmium-catalyzed vicinal dihydroxylation of olefins with single oxygen donors, typically TBHP or N-methylmorpholine-N-oxide (NMO), has been known for more than two decades [82] and forms the basis for the Sharpless asymmetric dihydroxylation of olefins (see below). The reaction involves an oxometal mechanism in which it is generally accepted that oxoosmium(VIII) undergoes an initial 2+2 cycloaddition to the olefin to give an oxametallocycle which subsequently rearranges to an osmium(VI)-diol complex (Fig. 4.31) [83]. This is followed by rate-limiting reaction with the oxidant to afford the diol product with concomitant regeneration of osmium tetroxide. Recently, the scope of this method was extended to electron-deficient olefins such as α,β-unsaturated amides by performing the reaction at acidic pH. Citric acid (25 mol%) was identified as the additive of choice and 1.1 equivalent of NMO was required as oxidant [84].

Beller and coworkers [85] showed that osmium-catalyzed dihydroxylations can be performed successfully with dioxygen as the primary oxidant (Fig. 4.32).

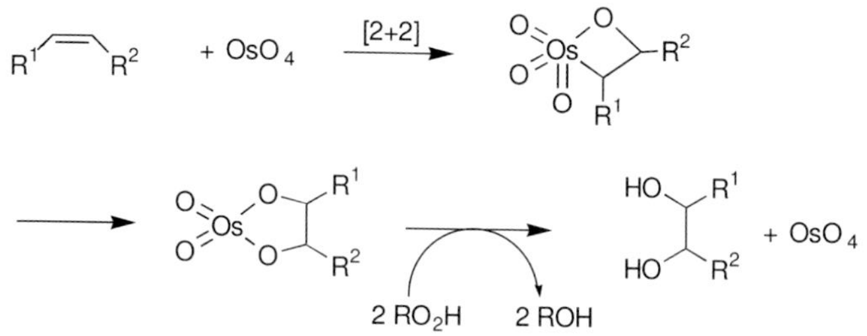

R¹—R² + OsO₄ —[2+2]→ (osmium cyclic intermediate with R², R¹)

→ (osmium dioxolane with R¹, R²) + 2 RO₂H → 2 ROH → HO–R¹ / HO–R² (vic-diol) + OsO₄

2 RO₂H   2 ROH

Fig. 4.31 Mechanism of osmium-catalyzed vicinal dihydroxylation of olefins.

R¹/R² olefin + ½ O₂ + H₂O

$$K_2OsO_2(OH)_4/L \ (0.5\text{-}2 \ mol\%)$$

tert-BuOH/H₂O
pH 10.4, 50 °C, 16-24 h

→ R¹/R² diol with OH, OH

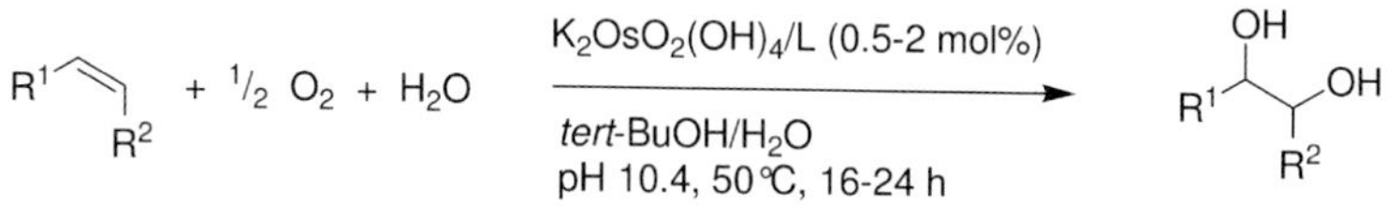

L = quinuclidine, DABCO

Olefin	C_6H_{13}	Ph	Ph	(cyclohexene)
Conv. (%)	99	66	100	90
Sel. (%)	96	77	97	75

Fig. 4.32 Osmium-catalyzed dihydroxylations using air as the oxidant.

Using potassium osmate (0.5–2 mol%) in the presence of an organic base, e.g. quinuclidine, in aqueous tert-butanol at pH 10.4, a variety of olefins was converted to the corresponding vic-diols in high yield. Apparently reoxidation of osmium(VI) to osmium(VIII) with dioxygen is possible under alkaline conditions. When chiral bases were used asymmetric dihydroxylation was observed (see Section 4.7) albeit with moderate enantioselectivities.

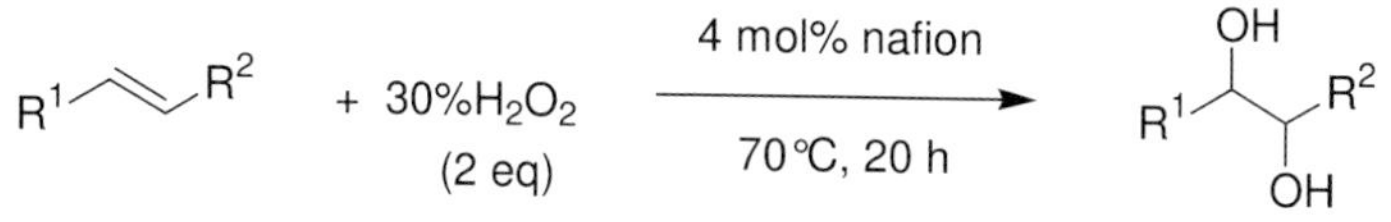

R¹/R² olefin + 30% H₂O₂
(2 eq)

4 mol% nafion

70 °C, 20 h

→ R¹/R² diol with OH, OH

Olefin	(hex-1-ene)	(hex-2-ene)	(cyclohexene)
Yield towards diol (%)	40	85	98

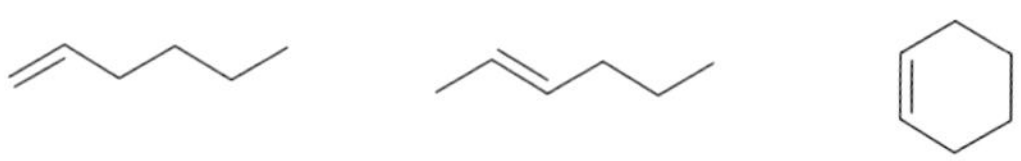

Fig. 4.33 Heterogeneous acid-catalyzed dihydroxylation of olefins.

Osmium is unrivalled as catalyst for the asymmetric cis-dihydroxylation of olefins. However, Sato and coworkers reported that the perfluorosulfonic acid resin, Nafion® (see Chapter 2) is an effective catalyst for the trans-dihydroxylation of olefins with H_2O_2 [86]. The method is organic solvent-free and the catalyst can be easily recycled (see Fig. 4.33). The first step of this reaction is epoxidation which is probably carried out by resin-supported peroxysulfonic acid formed *in situ*. This is followed by acid-catalyzed epoxide-ring opening.

4.3.3
Oxidative Cleavage of Olefins

It has long been known that oxidative cleavage of olefinic double bonds occurs with stoichiometric amounts of the powerful oxidant rutheniumtetroxide [87]. Ruthenium-catalyzed oxidative cleavage has been described with various oxygen donors, e.g. $NaIO_4$ [88, 89] and NaOCl [90] as the primary oxidant. The presumed catalytic cycle, in which RuO_4 is the active oxidant, is shown in Fig. 4.34. Various attempts have been made to introduce a green element in this reaction, i.e. recyclable heterogenous Ru-nanoparticles [91], and environmentally acceptable solvents [92], but the main problem of the oxidant has not been solved.

From an economic viewpoint it would obviously be advantageous if hydrogen peroxide or even dioxygen could be used as the primary oxidant. Unfortunately, ruthenium compounds generally catalyze rapid, non-productive decomposition of hydrogen peroxide [93]. However, a bimetallic system, comprising MoO_3/$RuCl_3$, for the oxidative cleavage of olefins with H_2O_2 in *tert*-butanol has been described [94]. For optimum performance a small amount of a carboxylic acid, which could be the product of the oxidative cleavage, was added. Presumably, the reaction involves initial molybdenum-catalyzed conversion to the *vic*-diol via the epoxide, followed by ruthenium-catalyzed oxidative cleavage of the diol.

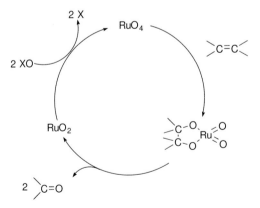

XO = NaOCl, $NaIO_4$, RCO_3H etc.

Fig. 4.34 A catalytic cycle for ruthenium-catalyzed oxidative cleavage of olefins.

Methodologies for the oxidative cleavage of olefins still, in our opinion, leave something to be desired. Bearing in mind the osmium-catalyzed aerobic dihydroxylation of olefins (see Section 4.3.2) and the ruthenium-catalyzed aerobic oxidations of alcohols and cleavage of diols (see below) one cannot help wondering if conditions cannot be designed for the ruthenium-catalyzed aerobic oxidative cleavage of olefins. Indeed for compounds like styrene and stilbene up to 80% yield can be obtained for oxidative cleavage at 130 °C using 0.004 mol% of a RuCl$_2$–PNNP complex [95]. Alternatively, a combination of osmium and ruthenium might be expected to have this capability.

Another possibility is the use of tungsten, which has led to excellent results for the conversion of cyclohexene to adipic acid (see Fig. 4.23) [48]. For linear olefins using the Venturello-system, oxidative cleavage products can be obtained in around 80% yield [96].

4.3.4
Oxidative Ketonization

The selective aerobic oxidations of ethene to acetaldehyde and terminal alkenes to the corresponding methyl ketones, in the presence of a PdCl$_2$/CuCl$_2$ catalyst are collectively referred to as Wacker oxidations [97]. The reaction involves hydroxypalladation of the olefin, via nucleophilic attack of coordinating hydroxide on a palladium–olefin π complex followed by a β-hydride shift to give the acetaldehyde (or methyl lketone) product and palladium(0) (Fig. 4.35). The latter undergoes copper-catalyzed reoxidation with dioxygen. The presence of chloride ions at acidic pH is necessary for a reasonable rate of reoxidation and to prevent the formation of palladium clusters.

A major disadvantage of the Wacker system is the extremely corrosive nature of (acidic) aqueous solutions of PdCl$_2$ and CuCl$_2$, which necessitates the use of costly titanium alloys as materials of construction. With higher olefins oxidation rates are lower and complex mixtures are often formed as a result of competing palladium-catalyzed isomerization of the olefin. Research has been focused on developing acid-free PdCl$_2$/CuCl$_2$ catalyzed Wacker oxidations using ionic liquid [98], ionic liquid/ScCO$_2$ [99], and heterogeneous Pd montmorillonite [100].

However, all these systems suffer from high concentrations of chloride ion, so that substantial amounts of chlorinated by-products are formed. For these reasons there is a definite need for chloride- and copper-free systems for Wacker oxidations. One such system has been recently described, viz., the aerobic oxidation of terminal olefins in an aqueous biphasic system (no additional solvent)

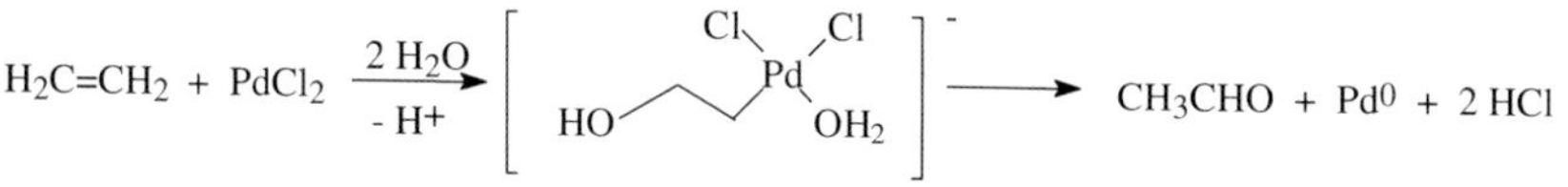

Fig. 4.35 PdCl$_2$/CuCl$_2$ catalyzed Wacker oxidation.

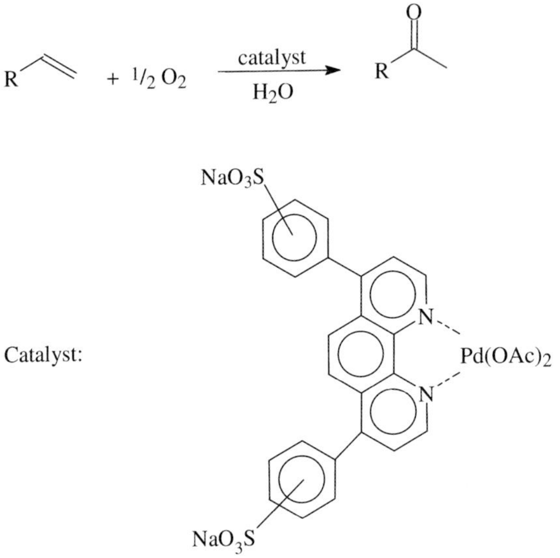

Fig. 4.36 Water-soluble palladium complexes for Wacker oxidation.

catalyzed by water-soluble palladium complexes of bidentate amines such as sulfonated bathophenanthroline (see Fig. 4.36) [101].

Moreover, it was disclosed that PdCl$_2$ in combination with N,N-dimethylacetamide (DMA) solvent could offer a simple and efficient catalyst system for acid- and Cu-free Wacker oxidation [102]. The reaction is illustrated in Fig. 4.37. A wide range of terminal olefins could be oxidized to form the corresponding methyl ketones in high yields, reaching a TOF up to 17 h^{-1}. The Pd-DMA catalyst layer could be recycled. Furthermore this system is also capable of per-

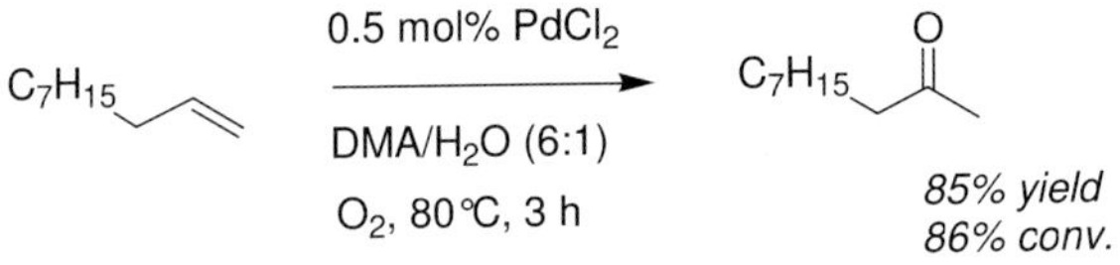

Fig. 4.37 PdCl$_2$/N,N-dimethylacetamide system for Wacker oxidation of olefins.

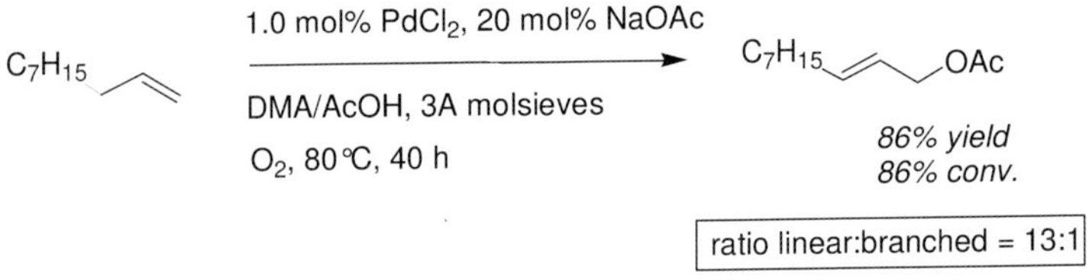

Fig. 4.38 PdCl$_2$/N,N-dimethylacetamide system for regioselective acetoxylation of terminal olefins.

forming regioselective acetoxylation of terminal olefins to linear allylic acetates (see Fig. 4.38). This system shows a strong resemblance to the Bäckvall system (see below and Fig. 4.41) [103].

In this context it is also worth mentioning that Showa Denko has developed a new process for the direct oxidation of ethene to acetic acid using a combination of palladium(II) and a heteropoly acid [104]. However, the reaction probably involves heteropoly acid-catalyzed hydration followed by palladium-catalyzed aerobic oxidation of ethanol to acetic acid rather than a classical Wacker mechanism.

4.3.5
Allylic Oxidations

Classical (metal-catalyzed) autoxidation of olefins is facile but not synthetically useful owing to competing oxidation of allylic C–H bonds and the olefinic double bond, leading to complex product mixtures [105]. Nonetheless, the synthetic chemist has a number of different tools for the allylic oxidation of olefins available.

Stoichiometric oxidation with SeO_2 became more attractive after Sharpless showed that the reaction could be carried out with catalytic amounts of SeO_2 and TBHP as the (re)oxidant [106]. The reaction involves an oxometal mechanism (see Fig. 4.39). The use of fluorous seleninic acids with iodoxybenzene as oxidant introduces the possibility of recycling the catalyst [107].

Allylic oxidation (acyloxylation) can also be achieved with copper catalysts and stoichiometric amounts of peresters or an alkylhydroperoxide in a carboxylic acid as solvent [108], via a free radical mechanism (Fig. 4.40). The use of water-soluble ligands [109] or fluorous solvents [110] allows recycling of the copper catalyst. In view of the oxidants required, this reaction is economically viable only when valuable (chiral) products are obtained using asymmetric copper catalysts [111–113]. The scope of the reaction is rather limited however.

Arguably the best, or at least the most versatile, allylic oxidation method is based on Pd [114]. Since the intermediates are palladium-allyl complexes rather than free radicals the number of by-products is limited compared to the preceding examples (Fig. 4.41). Furthermore, a large number of nucleophiles (amines, alcohols, stabilized carbanions, carboxylates or halides) may attack the palladium-allyl complex, giving a wide variety of products.

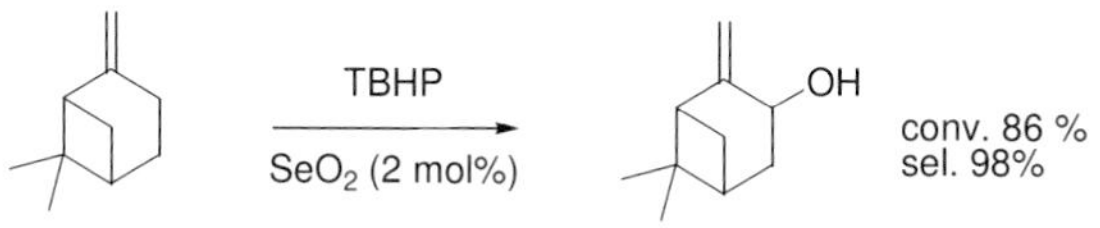

Fig. 4.39 Selenium-catalyzed allylic oxidation of olefins.

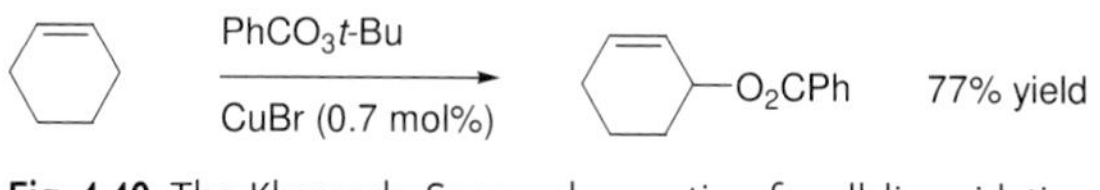

Fig. 4.40 The Kharasch–Sosnovsky reaction for allylic oxidation of olefins.

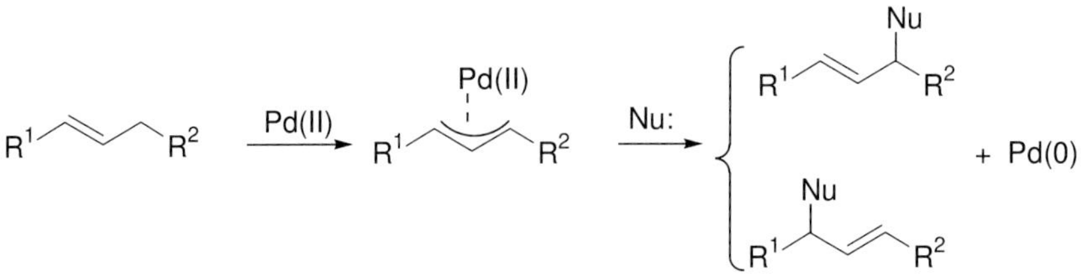

Fig. 4.41 Pd-catalyzed allylic oxidation of olefins.

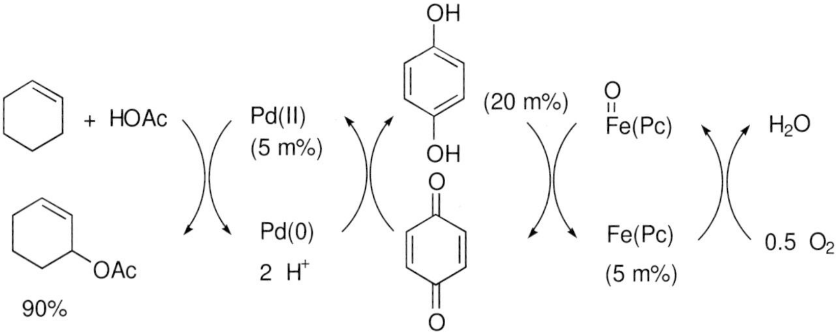

Fig. 4.42 The Pd/mediator approach for allylic oxidation of olefins.

In the group of Bäckvall a method was developed involving palladium and benzoquinone as cocatalyst (Fig. 4.42) [103]. The difficulty of the catalytic reaction lies in the problematic reoxidation of Pd(0) which cannot be achieved by dioxygen directly (see also Wacker process). To overcome this a number of electron mediators have been developed, such as benzoquinone in combination with metal macrocycles, heteropolyacids or other metal salts (see Fig. 4.42). Alternatively a bimetallic palladium(II) air oxidation system, involving bridging phosphines, can be used which does not require additional mediators [115]. This approach would also allow the development of asymmetric Pd-catalyzed allylic oxidation.

4.4
Alkanes and Alkylaromatics

One of the remaining "holy grails" in catalysis is undoubtedly the selective oxidation of unfunctionalized hydrocarbons, such as alkanes [116]. Generally speaking the difficulty lies in the activation of the poorly reactive C–H bonds and stabilization of a product which is often more reactive than the starting material. There are few notable exceptions e.g. the gas phase oxidation of butane to maleic anhydride (a highly stable product) and isobutane to tert-butylhydroperoxide (a highly reactive tertiary C–H bond). The aerobic oxidation of cyclohexane to cyclohexanol/cyclohexanone – the starting material for adipic acid used on a million tonnes scale for the manufacture of nylon-6,6 – is more challenging [117]. Since the substrate lacks reactive C–H bonds and the product cannot be stabilized, the reaction is stopped at <10% conversion to prevent further oxidation, affording a roughly 1:1 mixture of cyclohexanol/cyclohexanone in ca. 80% selectivity.

As noted in Section 4.2 the selective oxidation of unreactive C–H bonds with dioxygen is fraught with many problems, e.g. the oxidative destruction of organic ligands and competition with free radical oxidation pathways. For this reason most biomimetic systems employ sacrificial reductants or a reduced form of dioxygen, e.g. H_2O_2 [118]. One extensively studied class of biomimetic catalysts comprises the so-called Gif-systems [119], which were believed to involve direct insertion of high-valent oxoiron(V) species into C–H bonds. However, more recent results suggest that classical free radical mechanisms may be involved [120].

4.4.1
Oxidation of Alkanes

Classical autoxidation of tertiary C–H bonds in alkanes can afford the corresponding hydroperoxides in high selectivities. This is applied industrially in the conversion of pinane to the corresponding hydroperoxide, an intermediate in the manufacture of pinanol (Fig. 4.43).

More reactive hydroperoxides can be converted selectively to alcohols via the method of Bashkirov (Fig. 4.44), where a boric acid ester protects the product from further oxidation and thus increases the selectivity [121]. The method is used to convert C_{10}–C_{20} paraffins to alcohols which are used as detergents and surfactants, for the oxidation of cyclohexane (see elsewhere) and cyclododecane to cyclododecanol (cyclododecanone) for the manufacture of nylon-12.

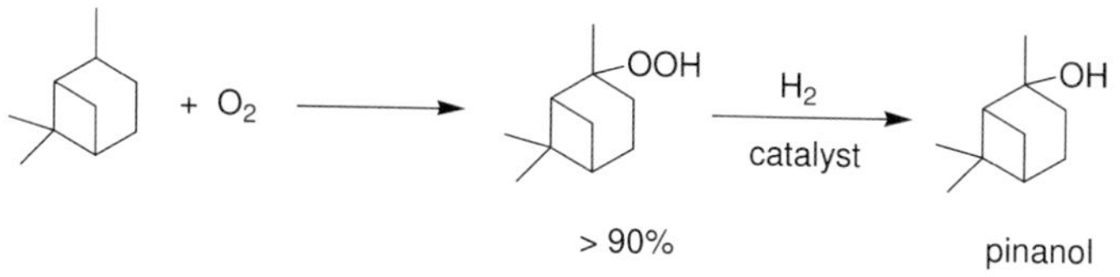

Fig. 4.43 Selective oxidation of pinane.

$$6 \ RH + 3 \ O_2 + B_2O_3 \xrightarrow[\Delta T]{\text{Initiator}} 2 \ (RO)_3B + 3 \ H_2O$$

$$(RO)_3B \xrightarrow{H_3O^+} 3 \ ROH + H_3BO_3$$

recycle

Fig. 4.44 Bashkirov method for conversion of alkanes to alcohols.

For less reactive hydrocarbons more drastic measures are required. High valent metal compounds, especially ruthenium compounds [122], can react with hydrocarbons but usually more than stoichiometric amounts of peracids or peroxides are required to reach the high oxidation state of the metal (e.g. Ru^{VI}). One of the very few examples in which a clean oxidant (H_2O_2 or O_2) is used, was reported by Drago et al. using a *cis*-dioxoruthenium complex (Fig. 4.45 a). In both cases a free radical mechanism could not be excluded, however [123]. In a variation on this theme Catalytica researchers described the direct oxidation of methane to methanol, using a platinum-bipyrimidine complex in concentrated sulfuric acid (Fig. 4.45 b) [124]. Analogous to the Bashkirov method the sulfuric acid protects the methanol, via esterification, from overoxidation to (eventually) CO_2 and H_2O, which is a highly thermodynamically favorable process.

Recently the Co/Mn/N-hydroxyphthalimide (NHPI) systems of Ishii have been added to the list of aerobic oxidations of hydrocarbons, including both aromatic side chains and alkanes. For example, toluene was oxidized to benzoic acid at 25 °C [125] and cyclohexane afforded adipic acid in 73% selectivity at 73% conversion [126], see Fig. 4.46. A related system, employing N-hydroxysaccharine, instead of NHPI was reported for the selective oxidation of large ring cycloalkanes [127].

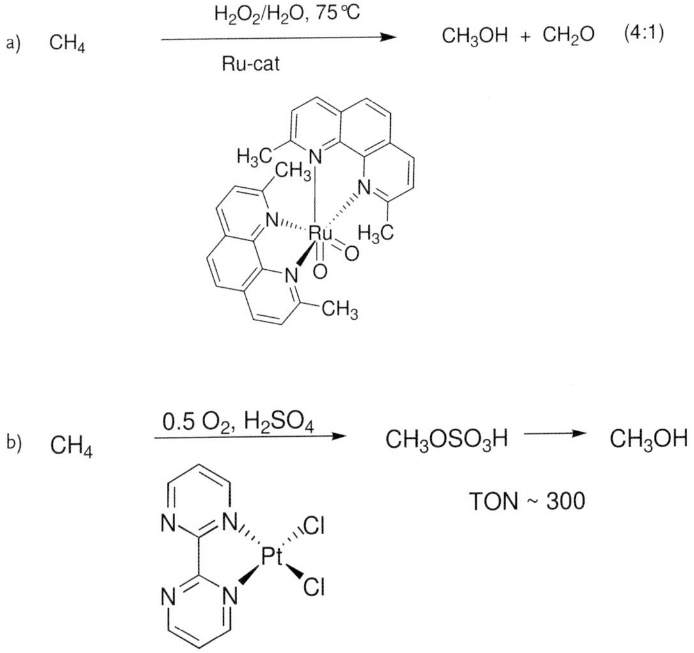

Fig. 4.45 Approaches for selective conversion of methane to methanol using O_2 or H_2O_2 as oxidants.

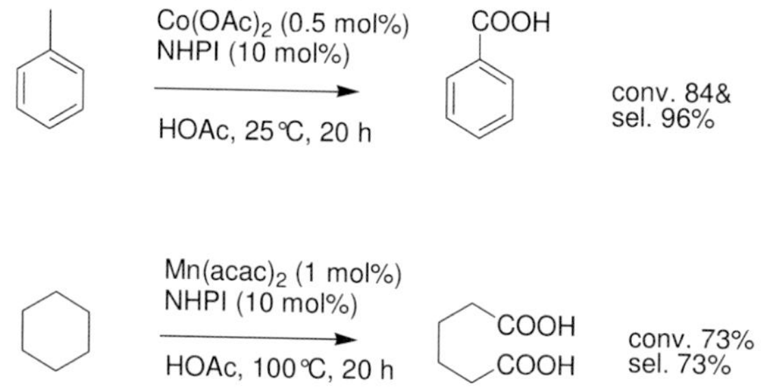

Co(OAc)$_2$ (0.5 mol%)
NHPI (10 mol%)
HOAc, 25 °C, 20 h

COOH

conv. 84&
sel. 96%

Mn(acac)$_2$ (1 mol%)
NHPI (10 mol%)
HOAc, 100 °C, 20 h

COOH
COOH

conv. 73%
sel. 73%

Fig. 4.46 NHPI as cocatalyst for aerobic oxidation of toluene and cyclohexane.

$$ArCH_3 + Br\cdot \longrightarrow ArCH_2\cdot + HBr$$

$$Co^{III} + Br^- \longrightarrow Co^{II} + Br\cdot$$

Fig. 4.47 Synergistic effect of bromide ions and NHPI compared.

The role of NHPI seems to be analogous to that of bromide ions during aut-oxidation processes: Bromide ion has a pronounced synergistic effect on metal-catalyzed autoxidations [128–130]. This results from a change in the chain-propagation step from hydrogen abstraction by alkylperoxy radicals to the energetically more favorable hydrogen abstraction by bromine atoms (Fig. 4.47). The bromine atoms are formed by one-electron oxidation of bromide ions by, for example, cobalt(III) or manganese(III), generally a more favorable process than one-electron oxidation of the hydrocarbon substrate.

4.4.2
Oxidation of Aromatic Side Chains

The power of 'green chemistry' is nicely illustrated by reference to the production of aromatic acids. Classical methods using chlorine or nitric acid have been largely displaced by catalytic oxidations with dioxygen (see Fig. 4.48). This leads to high atom utilization, low-salt technology, no chloro- or nitro-compounds as by-products and the use of a very cheap oxidant.

Oxidation of hydrocarbons with dioxygen is more facile when the C–H bond is activated through aromatic or vinylic groups adjacent to it. The homolytic C–H bond dissociation energy decreases from ca. 100 kcal mol^{-1} (alkyl C–H) to ca. 85 kcal mol^{-1} (allylic and benzylic C–H), which makes a number of autoxidation processes feasible. The relative oxidizability is further increased by the presence of alkyl substituents on the benzylic carbon (see Table 4.6). The autoxidation of isopropylbenzene (Hock process, Fig. 4.49) accounts for the majority of the world production of phenol [131]:

In the Amoco/MC process terephthalic acid (TPA) is produced by aerobic oxidation of p-xylene. This bulk chemical ($> 10 \times 10^6$ t a^{-1}) is chiefly used for poly-

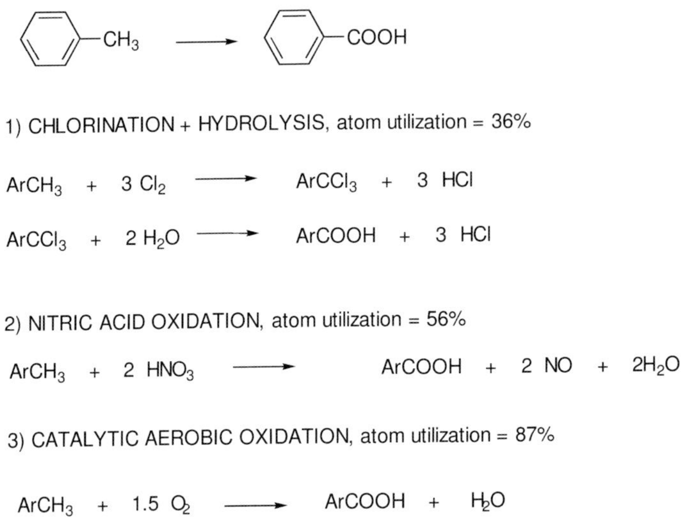

1) CHLORINATION + HYDROLYSIS, atom utilization = 36%

$$ArCH_3 \ + \ 3 \ Cl_2 \ \longrightarrow \ ArCCl_3 \ + \ 3 \ HCl$$

$$ArCCl_3 \ + \ 2 \ H_2O \ \longrightarrow \ ArCOOH \ + \ 3 \ HCl$$

2) NITRIC ACID OXIDATION, atom utilization = 56%

$$ArCH_3 \ + \ 2 \ HNO_3 \ \longrightarrow \ ArCOOH \ + \ 2 \ NO \ + \ 2H_2O$$

3) CATALYTIC AEROBIC OXIDATION, atom utilization = 87%

$$ArCH_3 \ + \ 1.5 \ O_2 \ \longrightarrow \ ArCOOH \ + \ H_2O$$

Fig. 4.48 Classical versus 'green' aromatic side-chain oxidation.

Table 4.6 Relative oxidizability of aromatic side-chains.

Substrate	Rel. Oxidizability
$PhCH_3$	1.0
$PhCH_2CH_3$	15
$PhCH(CH_3)_2$	107

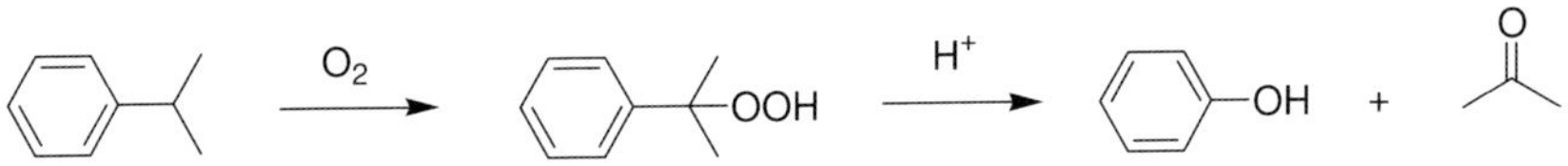

Fig. 4.49 Hock process for production of phenol.

etheneterephthalate (PET, a polymer used in e.g. plastic bottles). The solvent of choice is acetic acid, which shows a high solubility for substrates and low for products, is safe, recyclable (up to 40 times), has a wide temperature range and is relatively inert [132], although some solvent is "burnt" into CO_2 and H_2O (Fig. 4.50).

The rather drastic conditions are required because in this particular case the COOH group deactivates the intermediate *p*-toluic acid towards further oxidation, and some *p*-carboxybenzaldehyde is found as a side-product, which is hydrogenated back to *p*-toluic acid. Other than that, a large number of functional groups are tolerated (see Table 4.7) [129]. The combination of cobalt, manganese and bromide ions is essential for optimum performance. The benzylic radicals are best generated with bromine atoms (see above) which in turn are more easily produced

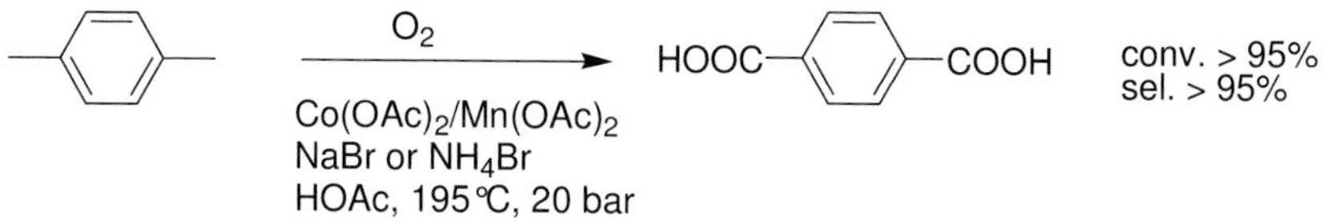

O2
Co(OAc)2/Mn(OAc)2
NaBr or NH4Br
HOAc, 195°C, 20 bar
HOOC
COOH
conv. > 95%
sel. > 95%

Fig. 4.50 Amoco/MC process for the production of terephthalic acid.

Table 4.7 Tolerance of functional groups in Co/Mn/Br-catalyzed aerobic oxidation of substituted toluenes.[a)]

Complete conversion 90–98% yield	Side reactions 30–70% yield	Total radical consumption no yield
F, Cl, Br (o,m,p)	OAc (m,p)	I (o,m,p)
OCH_3 (p)	OCH_3 (o,m)	OH (o,m,p)
OPh (m,p)	NHAc (m,p)	NH_2 (o,m,p)
COOH (m,p)	COOH (o)	OPh (o)
NO_2 (m,p)	Ph (o)	NO_2 (o)
Ph (m,p)	SO_2NH_2 (o)	
COPh (o,m,p)		
SO_2NH_2 (m,p)		
SO_2R (m,p)		
POR_2 (o,m,p)		

a) Taken from Ref. [129].

by reaction of bromide with Mn^{III} rather than Co^{III} (see Fig. 4.51). The resulting peracid, however, is more easily reduced by Co^{II} than by Mn^{II}. In the Eastman Kodak and related processes fairly large amounts of e.g. acetaldehyde (0.21 t/t) are used as additives instead of bromide, leading to less corrosion.

The use of manganese catalysts maintains a low concentration of Co(III), which is important since the latter promotes extensive decarboxylation of acetic acid at $T > 130\,°C$. A somewhat similar process which operates under biphasic conditions rather than in acetic acid also gives high conversions and selectivities [133].

O2, (CoII / CoIII)
CH2 ·
CO3H
Br-
MnIII
CoII
CH3
CO2H
Br ·
MnII
CoIII

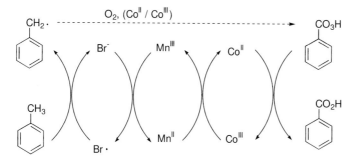

Fig. 4.51 Mechanism for the aerobic oxidation of aromatic side-chains.

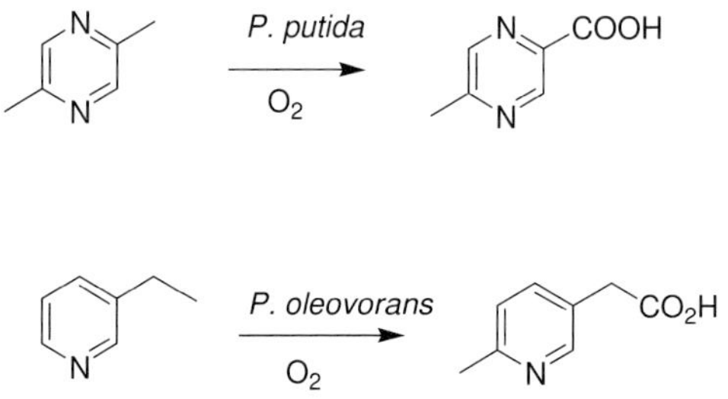

Fig. 4.52 Biocatalytic oxidation of alkyl groups on aromatic heterocycles.

Lonza has succesfully developed a biocatalytic method for the oxidation of alkyl groups on aromatic heterocycles [5]. The oxidation is carried out in a fermenter using *Pseudomonas putida* grown on xylenes as the microorganism. Product concentrations of up to 24 g L^{-1} could be reached. Another microorganism *Pseudomenas oleoverans* grown on n-octane could be used for the terminal oxidation of ethyl groups (see Fig. 4.52). These are important examples of industrial biocatalystic applications where the high selectivity and activity demonstrated by enzymes for these difficult alkane transformations has a clear advantage. The reactions need to be carried out with living cells, because the monoxygenase enzymes, which carry out these transformations are cofactor-dependent (see Chapter 1, Section 1.2). The success of this methodology is related to the relatively high water-solubility of these heterocycles, which makes them more amenable to oxidation by enzymes.

Another example of industrially viable biooxidation of *N*-heterocycles side chain oxidation is the oxidation of 2-methylquinoxaline to 2-quinoxalinecarboxylic acid by Pfizer [134].

4.4.3
Aromatic Ring Oxidation

Oxidation of aromatic rings is quite complicated for a number of reasons. Firstly, radical intermediates preferentially abstract hydrogen atoms from the aromatic side chain, rather than the nucleus (the dissociation energies of the Ar–H bond and the ArCR$_2$–H bond are ~ 110 kcal mol^{-1} and ~ 83 kcal mol^{-1}, respectively). Secondly, the phenol products are much more reactive towards oxidation than the hydrocarbon substrates. Recently a solution was found to the problem of over-oxidation by using aqueous-organic solvent mixtures. The optimum conditions are represented in Fig. 4.53. As catalyst, FeSO$_4$, which is a conventional Fenton catalyst, was applied in the presence of 5-carboxy-2-methylpyrazine N-oxide as ligand and trifluoroacetic acid as cocatalyst [135]. The choice of the N,O ligand turned out to be crucial. Using 10-fold excess of benzene to hydrogen peroxide, 97% selectivity of phenol (relative to benzene) could be achieved at almost full conversion of H$_2$O$_2$. Similar results could be obtained by using aqueous-ionic liquid biphasic mixtures and iron-dodecanesulfonate salts as catalyst [136].

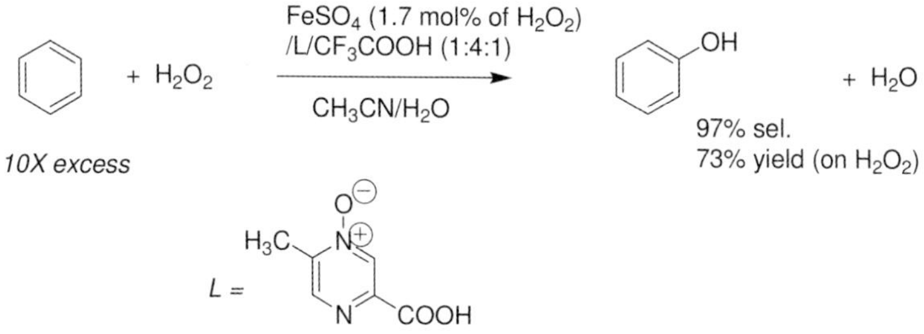

Fig. 4.53 Direct oxidation of benzene to phenol using H_2O_2.

This technology could also be applied for the oxidation of methane (70 °C, 5 Mpa) to formic acid (46% selectivity based on hydrogen peroxide). Another catalyst which can be applied for the direct oxidation of benzene with H_2O_2 is the TS-1 catalyst, as described above for propylene epoxidation [41]. Conversion is generally kept low, because introduction of a hydroxy group activates the aromatic nucleus to further oxidation to hydroquinone, catechol and eventually to tarry products [137].

Arguable the best technology – due to its simplicity – for directly converting benzene to phenol is represented by the Fe-ZSM-5/N_2O gas phase technology as developed by Panov and coworkers [138]. Nitrous oxide is the terminal oxidant, and MFI-type zeolite-containing low amounts (<1.0 wt%) of iron acts as the catalyst. The use of N_2O instead of oxygen circumvents the occurrence of free radical reactions. In this way highly selective conversion to phenol (>95%) could be reached at 350 °C (Fig. 4.54). The technology was adopted by Monsanto, but no chemical plant was built. Oxidation of other aromatics and alkenes using this technology generally leads to low selectivities.

For the direct hydroxylation of N-heterocycles a microbial oxidation is most suitable. Lonza developed a process to produce 6-hydroxynicotinic acid on a 10 tonnes scale [5]. The enzymatic production is carried out in two-steps. Firstly, *Achromobacter xylosoxydans* LK1 biomass is produced, which possesses a highly

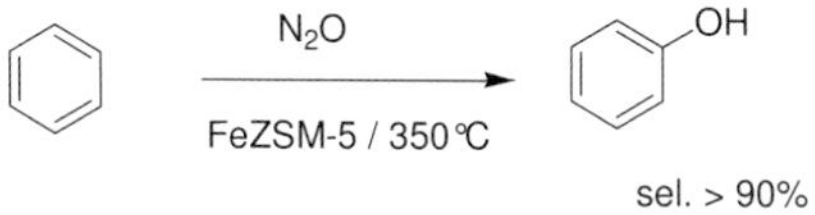

Fig. 4.54 Direct oxidation of benzene to phenol using N_2O as the terminal oxidant.

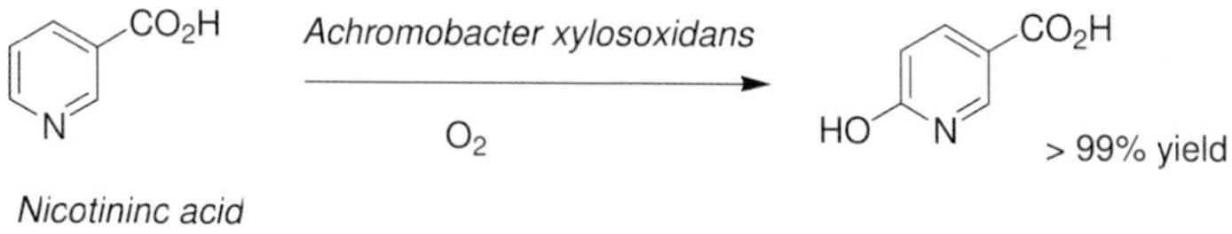

Fig. 4.55 Hydroxylation of nicotinic acid.

active nicotinic acid hydroxylase. In the second step the biomass is used for the hydroxylation (Fig. 4.55). The impressive yield of >99% is due to the selectivity of the enzyme involved.

4.5
Oxygen-containing Compounds

4.5.1
Oxidation of Alcohols

The oxidation of primary and secondary alcohols into the corresponding carbonyl compounds plays a central role in organic synthesis [1, 139, 140]. Traditional methods for performing such transformations generally involve the use of stoichiometric quantities of inorganic oxidants, notably chromium(VI) reagents [141]. However, from both an economic and environmental viewpoint, atom efficient, catalytic methods that employ clean oxidants such as O_2 and H_2O_2 are more desirable.

The catalytic oxidation of alcohols is a heavily studied field, where many metals can be applied. New developments in the 21st century can be discerned, such as the use of nanocatalysts (notably Pd and Au) which combine high stability with a high activity. Furthermore catalysis in water, and the use of non-noble metals such as copper are important green developments. Thus both homogeneous and heterogeneous catalysts are employed. In addition, biocatalysis is on the rise. In combination with a mediator, the copper-dependent oxidase, laccase is very promising for alcohol oxidation. Furthermore, alcohol dehydrogenases can also be employed for (asymmetric) alcohol oxidation, by using acetone as a cosubstrate. It should be noted that hydrogen peroxide is not really needed for alcohol conversion, compared to the use of oxygen it has little advantage. The focus in this section will therefore be on molecular oxygen. For a complete overview in this field the reader is referred to reviews on this topic [142–144]. We will start with some mechanistic aspects and a separate section highlighting the green aspects of using industrially important oxoammonium ions as catalyst. The latter methodology is widely used in industrial batch processes.

As discussed above, metal-catalyzed oxidations with hydrogen peroxide or alkyl hydroperoxide can be conveniently divided into two categories, involving peroxometal and oxometal species, respectively, as the active oxidant [13]. This is illustrated for alcohol oxidations in Fig. 4.56 [4]. In the peroxometal pathway the metal ion does not undergo any change in oxidation state during the catalytic cycle and no stoichiometric oxidation is observed in the absence of H_2O_2. In contrast, oxometal pathways involve a two-electron change in oxidation state of the metal ion and a stoichiometric oxidation is observed, with the oxidized state of the catalyst, in the absence of H_2O_2. Indeed, this is a test for distinguishing between the two pathways.

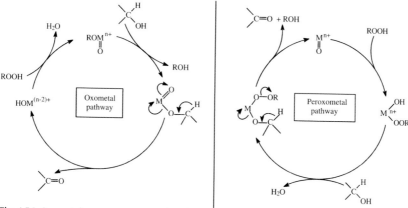

Fig. 4.56 Oxometal versus peroxometal pathways for alcohol oxidation.

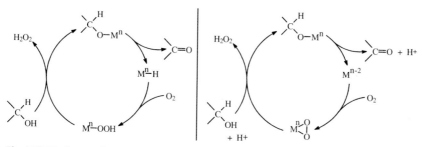

Fig. 4.57 Mechanism for metal catalyzed aerobic oxidation of alcohols.

In aerobic oxidations of alcohols a third pathway is possible with late transition metal ions, particularly those of Group VIII elements. The key step involves dehydrogenation of the alcohol, via β-hydride elimination from the metal alkoxide to form a metal hydride (see Fig. 4.57). This constitutes a commonly employed method for the synthesis of such metal hydrides. The reaction is often base-catalyzed which explains the use of bases as cocatalysts in these systems. In the catalytic cycle the hydridometal species is reoxidized by O_2, possibly via insertion into the M–H bond and formation of H_2O_2. Alternatively, an alkoxymetal species can afford a proton and the reduced form of the catalyst, either directly or via the intermediacy of a hydridometal species (see Fig. 4.57). Examples of metal ions that operate via this pathway are Pd(II), Ru(III) and Rh(III). We note the close similarity of the β-hydride elimination step in this pathway to the analogous step in the oxometal pathway (see Fig. 4.56). Some metals, e.g. ruthenium, can operate via both pathways and it is often difficult to distinguish between the two.

4.5.1.1 Ruthenium Catalysts

Ruthenium compounds have been extensively studied as catalysts for the aerobic oxidation of alcohols [142]. They operate under mild conditions and offer possibilities for both homogeneous and heterogeneous catalysts. The activity of common ruthenium precursors such as $RuCl_2PPh_3$, can be increased by the use of ionic liquids as solvents (Fig. 4.58). Tetramethylammoniumhydroxide and aliquat® 336 (tricaprylylmethylammonium chloride) were used as solvent and rapid conversion of benzyl alcohol was observed [145]. Moreover the tetramethylammonium hydroxide/$RuCl_2(PPh_3)_3$ could be reused after extraction of the product.

Ruthenium compounds are widely used as catalysts for hydrogen transfer reactions. These systems can be readily adapted to the aerobic oxidation of alcohols by employing dioxygen, in combination with a hydrogen acceptor as a cocatalyst, in a multistep process. These systems demonstrate high activity. For example, Bäckvall and coworkers [146] used low-valent ruthenium complexes in combination with a benzoquinone and a cobalt-Schiff's base complex. Optimization of the electron-rich quinone, combined with the so-called "Shvo" Ru-catalyst, led to one of the fastest catalytic systems reported for the oxidation of secondary alcohols (Fig. 4.59).

The regeneration of the benzoquinone can also be achieved with dioxygen in the absence of the cobalt cocatalyst. Thus, Ishii and coworkers [147] showed that a combination of $RuCl_2(Ph_3P)_3$, hydroquinone and dioxygen, in $PhCF_3$ as solvent, oxidized primary aliphatic, allylic and benzylic alcohols to the corresponding aldehydes in quantitative yields.

Another example of a low-valent ruthenium complex, is the combination of $RuCl_2(Ph_3P)_3$ and the stable nitroxyl radical, 2,2′,6,6′-tetramethylpiperidine-N-oxyl (TEMPO). This system is a remarkably effective catalyst for the aerobic oxidation of a variety of primary and secondary alcohols, giving the corresponding aldehydes and ketones, respectively, in >99% selectivity [148]. The best results were obtained using 1 mol% of $RuCl_2(Ph_3P)_3$ and 3 mol% of TEMPO (Fig. 4.60). Primary alcohols give the corresponding aldehydes in high selectivity, e.g. 1-octanol affords 1-octanal in >99% selectivity. Over-oxidation to the corresponding carboxylic acid, normally a rather facile process, is completely suppressed in the presence of a catalytic amount of TEMPO. TEMPO suppresses the autoxidation of aldehydes by efficiently scavenging free radical intermediates

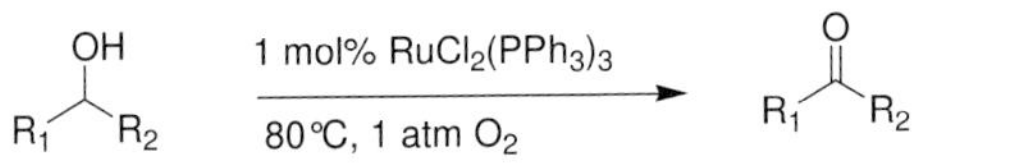

solvent:	tetramethylammonium hydroxide	aliquat
R_1=PhCH$_2$, R_2=H	91% conv. (5h)	58% conv. (5h)
R_1, R_2, = c-C$_7$H$_{14}$	61% conv. (11h)	92% conv. (11h)
R_1=C$_6$H$_{13}$, R_2=CH$_3$	43% conv. (25h)	81% conv. (25h)

Fig. 4.58 Ruthenium salts in ionic liquids as catalysts for oxidation of alcohols.

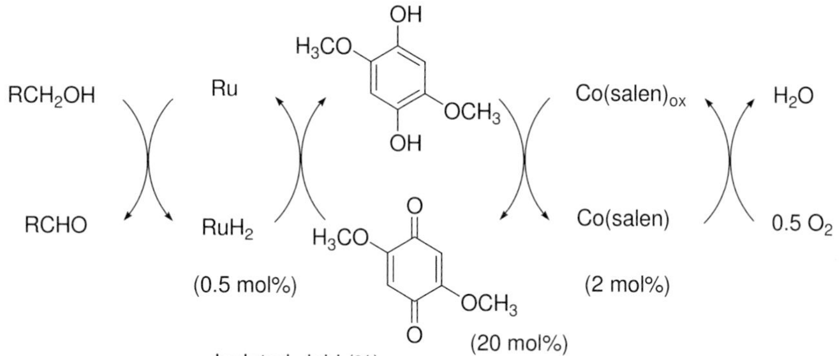

Fig. 4.59 Low-valent ruthenium complexes in combination with benzoquinone.

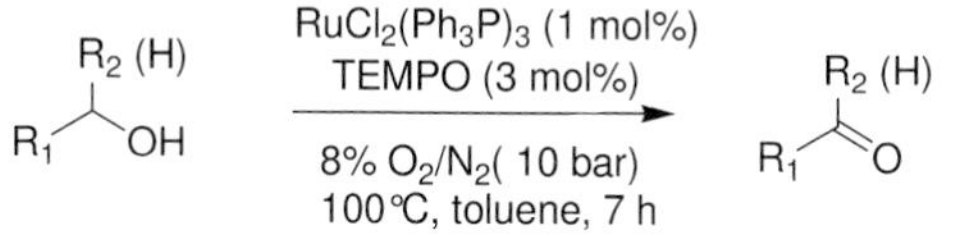

Isolated yield (%)

2-octanol	92 (1 h)
cyclohexanol	92 (1 h)
2-phenylethanol	89 (1 h)
L-menthol	80 (2 h)

	S/C ratio	yield (%)
1-octanol	50	85
2-octanol	100	98
geraniol	67	91
benzyl alcohol	200	>99 (2.5 h)
2-phenyl ethanol	100	>99 (4 h)

Fig. 4.60 Ruthenium/TEMPO catalyzed oxidation of alcohols.

resulting in the termination of free radical chains, i.e. it acts as an antioxidant. Allylic alcohols were selectively converted to the corresponding unsaturated aldehydes in high yields.

Perruthenate catalysts, i.e. TPAP, are superior Ru-catalysts, which are air-stable, non-volatile and soluble in a wide range of organic solvents. It was shown that TPAP is an excellent catalyst for the selective oxidation of a wide variety of alcohols using N-methylmorpholine-N-oxide (NMO) or oxygen as the stoichiometric oxidant [139, 149–151]. In particular, polymer-supported perruthenate (PSP), prepared by anion exchange of $KRuO_4$ with a basic anion exchange resin (Amberlyst A-26), has emerged as a versatile catalyst for the aerobic oxidation (Fig. 4.61) of alcohols [152]. However the activity was ca. 4 times lower than homogeneous TPAP, and this catalyst could not be recycled, which was attrib-

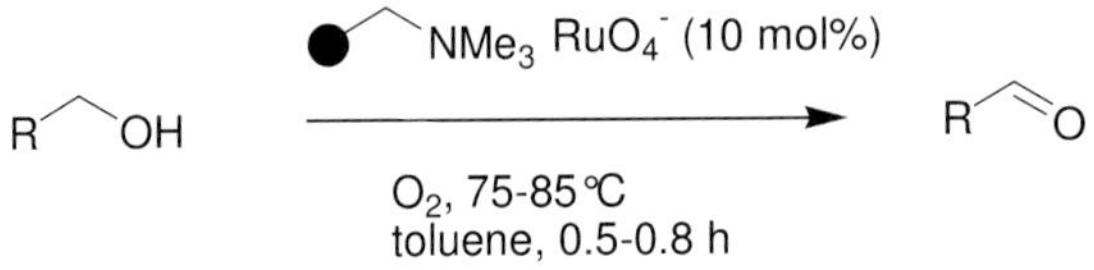

NMe$_3$ RuO$_4^-$ (10 mol%)

R⌒OH ⟶ R⌒O

O$_2$, 75-85 °C
toluene, 0.5-0.8 h

yield

1-octanol   91% (8h)

benzyl alcohol   >95% (0.5 h)

cinnamyl alcohol >95% (1 h)

Fig. 4.61 TPAP catalyzed oxidation of alcohols.

uted to oxidative degradation of the polystyrene support. PSP displays a marked preference for primary versus secondary alcohol functionalities [152]. The problem of deactivation was also prominent for the homogeneous TPAP oxidation, which explains the high (10 mol%) loading of catalyst required.

Recently two heterogeneous TPAP-catalysts were developed, which could be recycled successfully and displayed no leaching: In the first example the tetra-alkylammonium perruthenate was tethered to the internal surface of mesoporous silica (MCM-41) and was shown [153] to catalyze the selective aerobic oxidation of primary and secondary allylic and benzylic alcohols. Surprisingly, both cyclohexanol and cyclohexenol were unreactive although these substrates can easily be accommodated in the pores of MCM-41. The second example involves straightforward doping of methyl modified silica, denoted as ormosil, with tetra-propylammonium perruthenate via the sol–gel process [154]. A serious disadvantage of this system is the low-turnover frequency (1.0 and 1.8 h^{-1}) observed for primary aliphatic alcohol and allylic alcohol respectively.

Many examples of heterogeneous ruthenium systems are available. One of the more promising seems to be ruthenium on alumina, which is an active and recyclable catalyst [155]. This system displayed a large substrate scope (see Fig. 4.62) and tolerates the presence of sulfur and nitrogen groups. Only primary aliphatic alcohols required the addition of hydroquinone. Turnover frequencies in the range of 4 h^{-1} (for secondary allylic alcohols) to 18 h^{-1} (for 2-octanol) were obtained in trifluorotoluene, while in the solvent-free oxidation at 150 °C a TOF of 300 h^{-1} was observed for 2-octanol.

Ruthenium-exchanged hydrotalcites were shown by Kaneda and coworkers [156], to be heterogeneous catalysts for the aerobic oxidation of reactive allylic and benzylic alcohols. Ruthenium could be introduced in the Brucite layer by ion exchange [156]. The activity of the ruthenium-hydrotalcite was significantly enhanced by the introduction of cobalt(II) in addition to ruthenium(III), in the Brucite layer [157]. For example, cinnamyl alcohol underwent complete conversion in 40 min in toluene at 60 °C, in the presence of Ru/Co-HT, compared with 31% conversion under the same conditions with Ru-HT. A secondary aliphatic

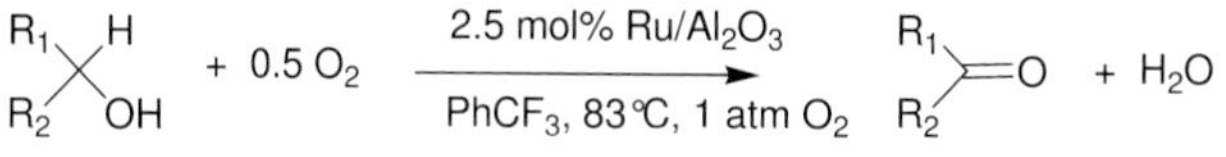

$$\underset{R_2}{\overset{R_1}{\diagdown}}\underset{OH}{\overset{H}{\diagup}} + 0.5\,O_2 \xrightarrow[\text{PhCF}_3,\,83\,°C,\,1\,atm\,O_2]{\text{2.5 mol\% Ru/Al}_2\text{O}_3} \underset{R_2}{\overset{R_1}{\diagdown}}{=}O + H_2O$$

	yield
1-octanol*	91% (2 h)
2-octanol	85% (4 h)
benzyl alcohol	> 99% (1 h)
2-phenyl ethanol	> 99% (1 h)
geraniol	86% (6 h)

* 5 mol% Ru/Al$_2$O$_3$ and 5 mol% hydroquinone needed

Fig. 4.62 Ru on alumina for aerobic alcohol oxidation.

alcohol, 2-octanol, was smoothly converted into the corresponding ketone but primary aliphatic alcohols, e.g. 1-octanol, exhibited extremely low activity. The results obtained in the oxidation of representative alcohols with Ru-HT and Ru/Co-HT are compared in Table 4.8.

Other examples of heterogeneous ruthenium catalysts are a ruthenium-based hydroxyapatite catalyst [158], a ferrite spinel-based catalyst MnFe$_{1.5}$Ru$_{0.35}$Cu$_{0.15}$O$_4$ [159], and Ru supported on CeO$_2$ [160], which all gave lower activities.

Another class of ruthenium catalysts, which has attracted considerable interest due to their inherent stability under oxidative conditions, are the polyoxometalates [161]. Recently, Mizuno et al. [162] reported that a mono-ruthenium-substituted silicotungstate, synthesized by the reaction of the lacunary polyoxometalate [SiW$_{11}$O$_{39}$]$^{8-}$ with Ru^{3+} in an organic solvent, acts as an efficient heterogeneous catalyst with high turnover frequencies for the aerobic oxidation of alcohols (see Fig. 4.63). Among the solvents used 2-butyl acetate was the most

Table 4.8 Oxidation of various alcohols to their corresponding aldehydes or ketones with Ru-hydrotalcites using molecular oxygen. [a]

Substrate	Ru-Mg-Al-CO$_3$-HT [b]		Ru-Co-Al-CO$_3$-HT [c]	
	Time/h	Yield (%)	Time/h	Yield (%)
PhCH=CHCH$_2$OH	8	95	0.67	94
PhCH$_2$OH	8	95	1	96
PhCH(CH$_3$)OH	18	100	1.5	100
n-C$_6$H$_{13}$CH(CH$_3$)OH			2	97

a) 2 mmol substrate, 0.3 g hydrotalcite (∼14 mol%), in toluene, 60 °C, 1 bar O$_2$.
b) Ref. [156].
c) Ref. [157].

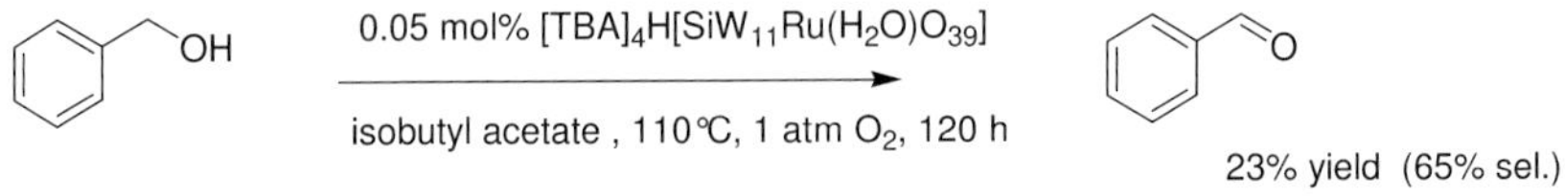

Fig. 4.63 Ru-HPA catalysed oxidation of alcohols.

effective and this Ru-heteropolyanion could be recycled. The low loading used resulted in very long reaction times of >2 days and low selectivities.

4.5.1.2 Palladium-catalyzed Oxidations with O_2

Much effort has been devoted to finding synthetically useful methods for the palladium-catalyzed aerobic oxidation of alcohols. For a detailed overview the reader is referred to several excellent reviews [163]. The first synthetically useful system was reported in 1998, when Peterson and Larock showed that simple $Pd(OAc)_2$ in combination with $NaHCO_3$ as a base in DMSO as solvent catalyzed the aerobic oxidation of primary and secondary allylic and benzylic alcohols to the corresponding aldehydes and ketones, respectively, in fairly good yields [164, 165]. Recently, it was shown that replacing the non-green DMSO by an ionic liquid (imidazole-type) resulted in a three times higher activity of the Pd-catalyst [166].

Uemura and coworkers [167, 168] reported an improved procedure involving the use of $Pd(OAc)_2$ (5 mol%) in combination with pyridine (20 mol%) and 3 Å molecular sieves in toluene at 80 °C. This system smoothly catalyzed the aerobic oxidation of primary and secondary aliphatic alcohols to the corresponding aldehydes and ketones, respectively, in addition to benzylic and allylic alcohols. Although this methodology constitutes an improvement on those previously reported, turnover frequencies were still generally $<10\,h^{-1}$ and, hence, there is considerable room for further improvement. Recent attempts to replace either pyridine by triethylamine [169], or $Pd(OAc)_2$ by palladacycles [170] all resulted in lower activities. Notably, the replacement of pyridine by pyridine derivatives having a 2,3,4,5 tetraphenyl substitutent completely suppressed the Pd black formation using only 1 atm air [171]. However the reaction times in this case were rather long (3–4 days).

A much more active catalyst is constituted by a water-soluble palladium(II) complex of sulfonated bathophenanthroline as a stable, recyclable catalyst for the aerobic oxidation of alcohols in a two-phase aqueous-organic medium, e.g. in Fig. 4.64 [16, 172, 173]. Reactions were generally complete in 5 h at 100 °C/30 bar air with as little as 0.25 mol% catalyst. No organic solvent is required (unless the substrate is a solid) and the product ketone is easily recovered by phase separation. The catalyst is stable and remains in the aqueous phase which can be recycled to the next batch.

A wide range of alcohols were oxidized with TOFs ranging from $10\,h^{-1}$ to $100\,h^{-1}$, depending on the solubility of the alcohol in water (since the reaction occurs in the aqueous phase the alcohol must be at least sparingly soluble in water). Representative examples of primary and secondary alcohols that were

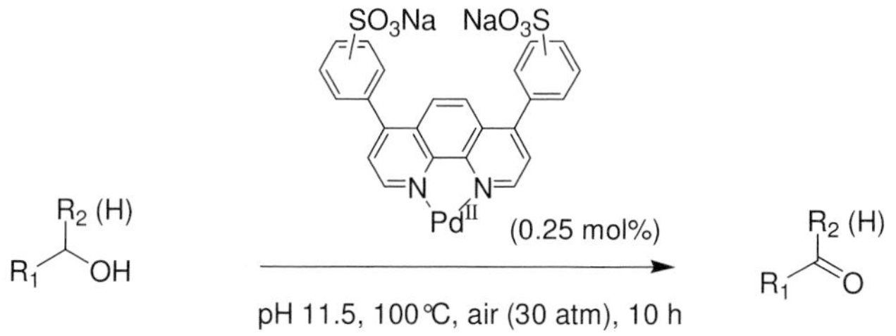

Fig. 4.64 Pd-bathophenanthroline as catalyst for alcohol oxidation in water.

Substrate	Time	Isolated Yield (%)
2-pentanol	5 h	90
2-phenyl ethanol	10 h	85
3-penten-2-ol	10 h	79
1-pentanol[a,b]	15 h	90
Benzyl alcohol[a]	10 h	93

a) 0.5 mol% Pd^{II}-catalyst; b) 2 mol% TEMPO added.

smoothly oxidized using this system are collected in Fig. 4.64. The corresponding ketones were obtained in >99% selectivity in virtually all cases.

Primary alcohols afforded the corresponding carboxylic acids via further oxidation of the aldehyde intermediate, e.g. 1-hexanol afforded 1-hexanoic acid in 95% yield. It is important to note, however, that this was achieved without the requirement of one equivalent of base to neutralize the carboxylic acid product (which is the case with supported noble metal catalysts). In contrast, when 1 mol% TEMPO (4 equivalents per Pd) was added, the aldehyde was obtained in high yield, e.g. 1-hexanol afforded 1-hexanal in 97% yield. Under cosolvent conditions using water/ethylene carbonate, Pd-neocuproine was found to be even more active (Fig. 4.65) [174]. This system is exceptional because of its activity (TOF $>> 500$ h^{-1} could be reached for 2-octanol) and functional group tolerance, such as C=C bonds, C≡C bonds, halides, α-carbonyls, ethers, amines etc. Thereby this system is expected to have a broad synthetic utility.

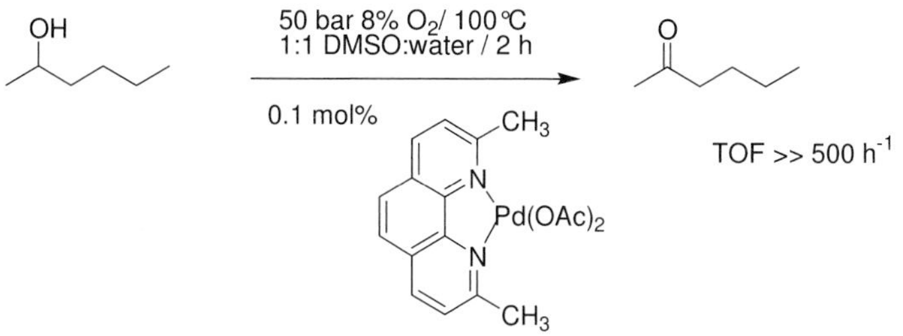

Fig. 4.65 Pd-neocuproine as catalyst for alcohol oxidation in water/cosolvent mixtures.

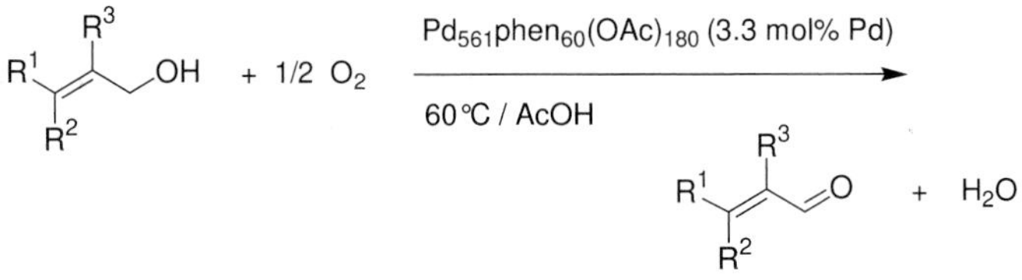

$$R^1\text{C}(R^2)\text{=C}(R^3)\text{-CH}_2\text{OH} + 1/2\ O_2 \xrightarrow[\text{60 °C / AcOH}]{\text{Pd}_{561}\text{phen}_{60}(\text{OAc})_{180}\ (3.3\ \text{mol}\%\ \text{Pd})} R^1\text{C}(R^2)\text{=C}(R^3)\text{-CHO} + H_2O$$

Fig. 4.66 Alcohol oxidation by giant palladium clusters.

In the context of heterogeneous palladium catalysts, Pd/C catalysts are commonly used for water-soluble substrates, i.e. carbohydrates [175]. Palladium can also be introduced in the brucite-layer of the hydrotalcite [176]. As with Ru/Co-hydrotalcite (see above), besides benzylic and allylic also aliphatic and cyclic alcohols are smoothly oxidized using this palladium-hydrotalcite. However a major shortcoming is the necessity of at least 5 mol% catalyst and the co-addition of 20–100 mol% pyridine. A seemingly very active heterogeneous catalyst is $PdCl_2(PhCN)_2$ on hydroxyapatite [177]. Using trifluorotoluene as the solvent, very high TONs (>20 000) could be obtained for benzyl alcohol. However for aliphatic alcohols long reaction times were needed (24 h).

Major trends can be discerned for Pd-catalysts, aimed at increasing the stability and activity. First is the use of palladium-carbene complexes [178]. Although activities are still modest, much can be expected in this area. Second is the synthesis and use of palladium nanoparticles. For example, the giant palladium cluster, $Pd_{561}\text{phen}_{60}(\text{OAc})_{180}$ [179], was shown to catalyze the aerobic oxidation of primary allylic alcohols to the corresponding α,β-unsaturated aldehydes (Fig. 4.66) [180].

Some other recent examples are the use of palladium nanoparticles entrapped in aluminum hydroxide [181], resin-dispersed Pd nanoparticles [182], and poly(ethylene glycol)-stabilized palladium nanoparticles in $scCO_2$ [183]. Although in some cases the activities for activated alcohols obtained with these Pd-nanoparticles are impressive, the conversion of aliphatic alcohols is still rather slow.

4.5.1.3 Gold Catalysts

Recently, gold has emerged as one of the most active catalysts for alcohol oxidation and is especially selective for poly alcohols. In 2005, Corma [184] and Tsukuda [185], independently demonstrated the potential of gold nanoparticles for the oxidation of aliphatic alcohols. For example, in the case of gold nanoparticles deposited on nanocrystalline cerium oxide [184], a TOF of $12\,500\ \text{h}^{-1}$ was obtained for the conversion of 1-phenylethanol into acetophenone at 160 °C (Fig. 4.67). Moreover this catalyst is fully recyclable. Another example of a gold catalyst with exceptional activity is a 2.5% Au–2.5% Pd/TiO_2 as catalyst [186]. In this case for 1-octanol a TOF of $2000\ \text{h}^{-1}$ was observed at 160 °C (reaction without solvent, Fig. 4.67).

As reported below, Au is now considered as the catalyst of choice for carbohydrate oxidation. Similarly, glycerol can be oxidized to glyceric acid with 100% se-

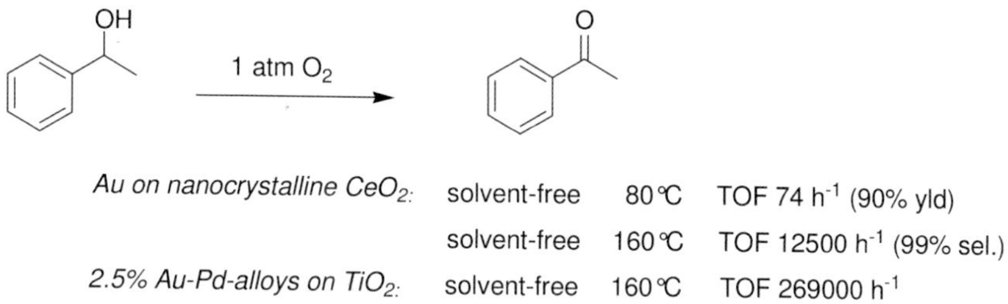
OH
1 atm O2
O
Au on nanocrystalline CeO2: solvent-free 80°C TOF 74 h-1 (90% yld)
solvent-free 160°C TOF 12500 h-1 (99% sel.)
2.5% Au-Pd-alloys on TiO2: solvent-free 160°C TOF 269000 h-1

Fig. 4.67 Au nanoparticles for alcohol oxidation.

lectivity using either 1% Au/charcoal or 1% Au/graphite catalyst under mild reaction conditions (60 °C, 3 h, water as solvent) [187].

4.5.1.4 **Copper Catalysts**

Copper would seem to be an appropriate choice of metal for the catalytic oxidation of alcohols with dioxygen since it comprises the catalytic centre in a variety of enzymes, e.g. galactose oxidase, which catalyze this conversion *in vivo* [188, 189]. Several catalytically active biomimetic models for these enzymes have been designed which are seminal examples in this area [190–193]. A complete overview of this field can be found in a review [194].

Marko and coworkers [195, 196] reported that a combination of Cu_2Cl_2 (5 mol%), phenanthroline (5 mol%) and di-*tert*-butylazodicarboxylate, DBAD (5 mol%), in the presence of 2 equivalents of K_2CO_3, catalyzes the aerobic oxidation of allylic and benzylic alcohols (Fig. 4.68). Primary aliphatic alcohols, e.g. 1-decanol, could be oxidized but required 10 mol% catalyst for smooth conversion.

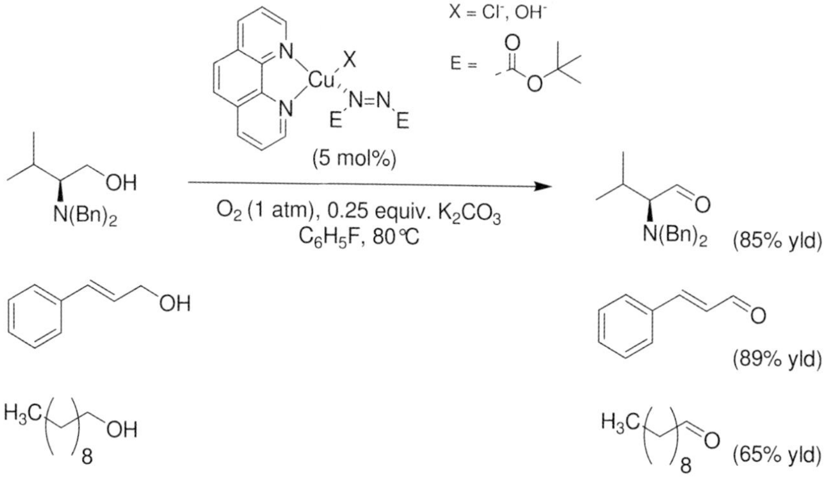
X = Cl-, OH-
E =
N
Cu
X
N=N
E
E
(5 mol%)
OH
N(Bn)2
O2 (1 atm), 0.25 equiv. K2CO3
C6H5F, 80°C
O
N(Bn)2 (85% yld)
OH
O
(89% yld)
H3C
OH
8
H3C
O
8
(65% yld)

Fig. 4.68 Cu-phenanthroline-DBAD catalyzed aerobic oxidation of alcohols.

An advantage of the system is that it tolerates a variety of functional groups. Serious drawbacks of the system are the low activity, the need for two equivalents of K_2CO_3 (relative to substrate) and the expensive DBAD as a cocatalyst. According to a later report [197] the amount of K_2CO_3 can be reduced to 0.25 equivalents by changing the solvent to fluorobenzene. The active catalyst is heterogeneous, being adsorbed on the insoluble K_2CO_3 (filtration gave a filtrate devoid of activity). Besides fulfilling a role as a catalyst support the K_2CO_3 acts as a base and as a water scavenger. In 2001, Marko et al. reported a neutral variant of their $Cu^ICl(phen)$–$DBADH_2$–base system [198]. Furthermore, it was found that 1-methylimidazole as additive in the basic system dramatically enhanced the activity [199].

The use of Cu in combination with TEMPO also affords an attractive catalyst [200, 201]. The original system however operates in DMF as solvent and is only active for activated alcohols. Knochel et al. [202] showed that $CuBr.Me_2S$ with perfluoroalkyl substituted bipyridine as the ligand and TEMPO as cocatalyst was capable of oxidizing a large variety of primary and secondary alcohols in a fluorous biphasic system of chlorobenzene and perfluorooctane (see Fig. 4.69). In the second example Ansari and Gree [203] showed that the combination of CuCl and TEMPO can be used as a catalyst in 1-butyl-3-methylimidazolium hexafluorophosphate, an ionic liquid, as the solvent. However in this case turnover frequencies were still rather low even for benzylic alcohol (around $1.3\ h^{-1}$).

Recently, an alternative to the catalytic system described above was reported [204]. The new catalytic procedure for the selective aerobic oxidation of primary alcohols to aldehydes was based on a $Cu^{II}Br_2(Bpy)$–TEMPO system (Bpy = 2,2'-bipyridine). The reactions were carried out under air at room temperature and were catalyzed by a [copper^{II}(bipyridine ligand)] complex and TEMPO and base (KOtBu) as co-catalysts (Fig. 4.70).

Several primary benzylic, allylic and aliphatic alcohols were successfully oxidized with excellent conversions (61–100%) and high selectivities. The system displays a remarkable selectivity towards primary alcohols. This selectivity for

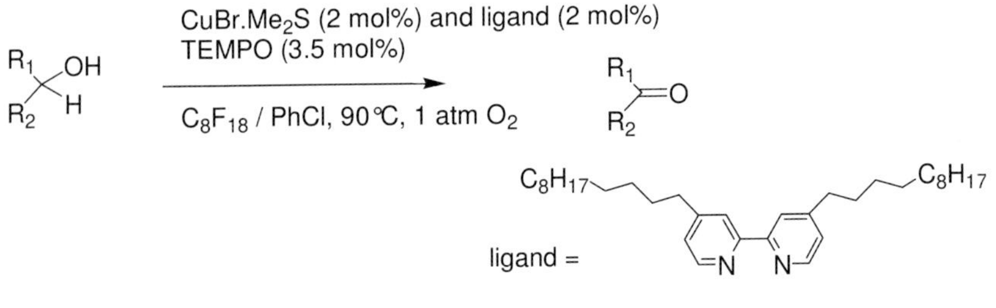

1-decanol	7-13 h	73% yld.
2-decanol	7-13 h	71% yld.
cinnamyl alcohol	2-7 h	79% yld.

Fig. 4.69 Fluorous $CuBr_2$-bipy-TEMPO system for alcohol oxidation.

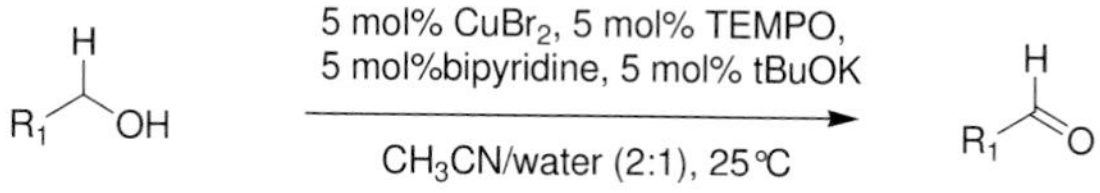

5 mol% CuBr$_2$, 5 mol% TEMPO,
5 mol%bipyridine, 5 mol% tBuOK

CH$_3$CN/water (2:1), 25 °C

benzyl alcohol	2.5 h	100% yld
2-phenyl ethanol	5 h	no reaction
1-octanol	24 h	61% yld
geraniol	5 h	100% yld

Fig. 4.70 CuBr$_2$(bipy)-TEMPO catalyzed oxidation of alcohols

the oxidation of primary alcohols resembles that encountered for certain copper-dependent enzymes. In the mechanism proposed, TEMPO coordinates side-on to the copper during the catalytic cycle.

4.5.1.5 Other Metals as Catalysts for Oxidation with O$_2$

Co(acac)$_3$ in combination with N-hydroxyphthalimide (NHPI) as cocatalyst mediates the aerobic oxidation of primary and secondary alcohols, to the corresponding carboxylic acids and ketones, respectively, e.g. Fig. 4.71 [205]. By analogy with other oxidations mediated by the Co/NHPI catalyst studied by Ishii and coworkers [206, 207], Fig. 4.71 probably involves a free radical mechanism. We attribute the promoting effect of NHPI to its ability to efficiently scavenge alkylperoxy radicals, suppressing the rate of termination by combination of alkylperoxy radicals (see above for alkane oxidation).

After their leading publication on the osmium-catalyzed dihydroxylation of olefins in the presence of dioxygen [208], Beller et al. [209] recently reported that alcohol oxidations could also be performed using the same conditions. The reactions were carried out in a buffered two-phase system with a constant pH of 10.4. Under these conditions a remarkable catalyst productivity (TON up to 16 600 for acetophenone) was observed. The pH value is critical in order to ensure the reoxidation of Os(VI) to Os(VIII). The scope of this system seems to be limited to benzylic and secondary alcohols.

As heterogeneous oxidation catalyst, 5% Pt, 1% Bi/C has been identified as an efficient catalyst for the conversion of 2-octanol to 2-octanone and 1-octanol to octanoic acid (see Fig. 4.72) [210]. Also manganese-substituted octahedral mo-

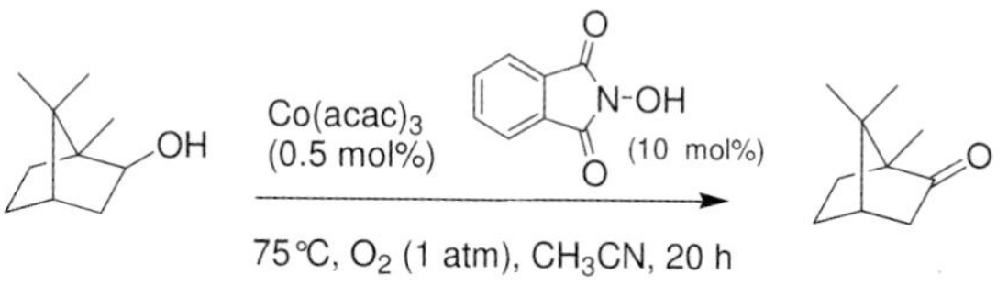

Co(acac)$_3$
(0.5 mol%)

N-OH
(10 mol%)

75 °C, O$_2$ (1 atm), CH$_3$CN, 20 h

Fig. 4.71 Co/NHPI catalyzed oxidation of alcohols.

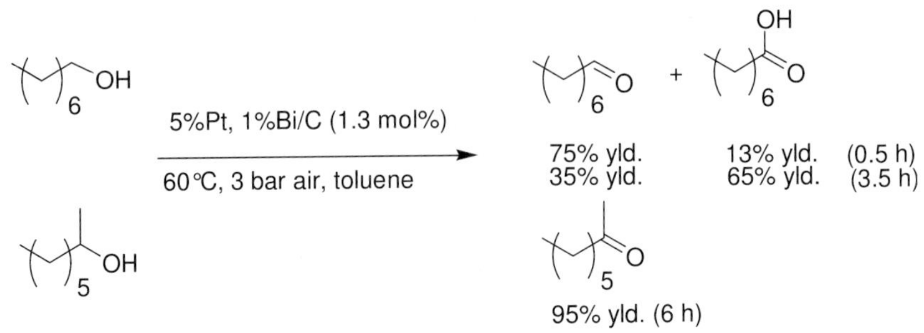

Fig. 4.72 Pt/Bi on carbon as effective catalyst for alcohol oxidation.

lecular sieves have been reported [211]. In this case benzylic and allylic alcohols could be converted within 4 h. However 50 mol% of catalyst was needed to achieve this.

Scant attention has been paid to vanadium-catalyzed oxidation of alcohols, despite its ability to act according to the oxometal mechanism. Punniyamurthy recently reported that indeed vanadium turns out to be a remarkably simple and selective catalyst with a wide substrate scope, which requires few additives [212], albeit, the activities are still rather low.

4.5.1.6 Catalytic Oxidation of Alcohols with Hydrogen Peroxide

The use of tungsten affords a highly active catalytic system for the oxidation of alcohols. Noyori and coworkers [213, 214] have achieved substantial improvements in the sodium tungstate-based, biphasic system by employing a phase transfer agent containing a lipophilic cation and bisulfate as the anion, e.g. $CH_3(n-C_8H_{17})_3NHSO_4$. This system requires 1.1 equivalents of 30% aq. H_2O_2 in a solvent-free system. For example, 1-phenylethanol was converted to acetophenone with turnover numbers up to 180 000. As with all Mo- and W-based systems, the Noyori system shows a marked preference for secondary alcohols, e.g. Fig. 4.73.

Titanium silicalite (TS-1), an isomorphously substituted molecular sieve [215], is a truly heterogeneous catalyst for oxidations with 30% aq. H_2O_2, including the oxidation of alcohols [216].

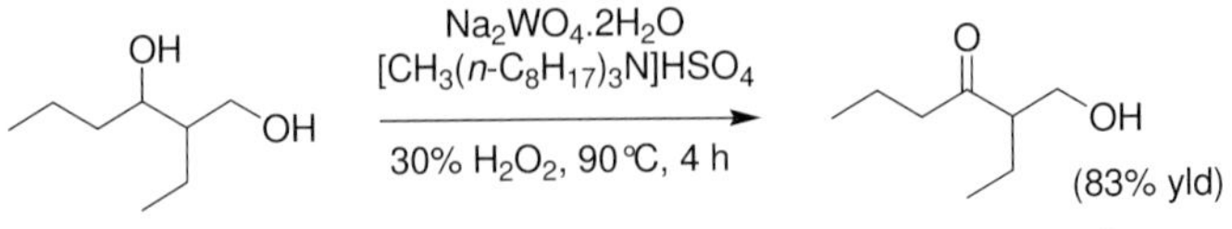

Fig. 4.73 Tungsten catalyzed alcohol oxidation with hydrogen peroxide.

4.5.1.7 Oxoammonium Ions in Alcohol Oxidation

In addition a very useful and frequently applied method in the fine chemical industry to convert alcohols into the corresponding carbonyl compounds is the use of oxoammonium salts as oxidants [15]. These are very selective oxidants for alcohols, which operate under mild conditions and tolerate a large variety of functional groups.

The oxoammonium is generated *in situ* from its precursor, TEMPO (or derivatives thereof), which is used in catalytic quantities, see Fig. 4.74. Various oxidants can be applied as the final oxidant. In particular, the TEMPO-bleach protocol using bromide as cocatalyst introduced by Anelli et al. is finding wide application in organic synthesis [217]. TEMPO is used in concentrations, as low as, 1 mol% relative to substrate and full conversion of substrates can commonly be achieved within 30 min. The major drawbacks of this method are the use of NaOCl as the oxidant, the need for addition of bromine ions and the necessity to use chlorinated solvents. Recently a great deal of effort has been devoted towards a greener oxoammonium-based method, by e.g. replacing TEMPO by heterogeneous variations or replacing NaOCl with a combination of metal as cocatalyst and molecular oxygen as oxidant. Examples of heterogeneous variants of TEMPO are anchoring TEMPO to solid supports such as silica [218, 219] and the mesoporous silica, MCM-41 [220] or by entrapping TEMPO in sol–gel [221]. Alternatively, an oligomeric TEMPO can be used [222].

Alternatively TEMPO can be reoxidized by metal salts or enzyme. In one approach a heteropolyacid, which is a known redox catalyst, was able to generate oxoammonium ions *in situ* with 2 atm of molecular oxygen at 100 °C [223]. In the other approach, a combination of manganese and cobalt (5 mol%) was able to generate oxoammonium ions under acidic conditions at 40 °C [224]. Results for both methods are compared in Table 4.9. Although these conditions are still open to improvement both processes use molecular oxygen as the ultimate oxidant, are chlorine free and therefore valuable examples of progress in this area. Alternative Ru and Cu/TEMPO systems, where the mechanism is me-

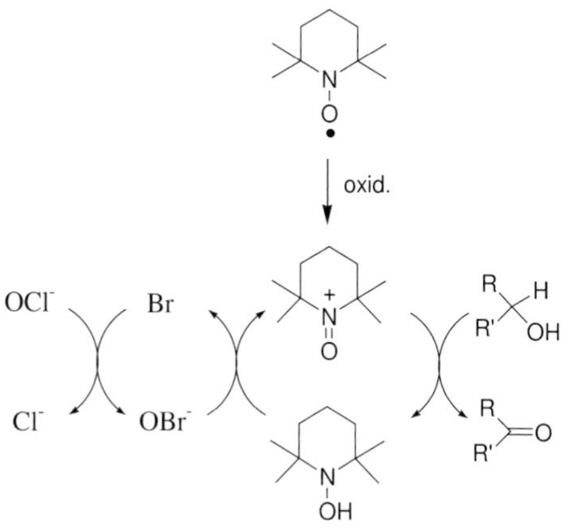

Fig. 4.74 Oxoammonium salts as active intermediates in alcohol oxidation.

Table 4.9 Aerobic oxoammonium-based oxidation of alcohols.

Substrate	Aldehyde or ketone yield [a]		
	2 mol% Mn(NO$_3$)$_2$ 2 mol% Co(NO$_3$)$_2$ 10 mol% TEMPO acetic acid, 40 °C 1 atm O$_2$ [b]	1 mol% H$_5$PMo$_{10}$V$_2$O$_{40}$ 3 mol% TEMPO acetone, 100 °C 2 atm O$_2$ [c]	Laccase (3 U/ml) Water pH 4.5 30 mol% TEMPO 25 °C, 1 atm O$_2$ [d]
n-C$_6$H$_{13}$-CH$_2$OH	97% (6 h)		
n-C$_7$H$_{15}$-CH$_2$OH		98% (18 h)	
n-C$_9$H$_{19}$-CH$_2$OH			15% (24 h)
n-C$_7$H$_{15}$-CH(CH$_3$)OH	100% (5 h)		
n-C$_6$H$_{13}$-CH(CH$_3$)OH		96% (18 h)	
PhCH$_2$OH	98% (10 h) [e]	100% (6 h)	92% (24 h)
PhCH(CH$_3$)OH	98% (6 h) [e]		
cis-C$_3$H$_7$-CH=CH–CH$_2$OH		100% (10 h)	
Ph–CH=CH–CH$_2$OH	99% (3 h)		94% (24 h)

a) GLC yields.
b) Minisci et al. Ref. [224].
c) Neumann et al. Ref. [223].
d) Fabbrini et al. Ref. [225].
e) Reaction performed at 20 °C with air.

tal-based rather than oxoammonium based, and which display higher activity, will be discussed below. Another approach to generate oxoammonium ions *in situ* is an enzymatic one. Laccase, that is an abundant highly potent redox enzyme, is capable of oxidizing TEMPO to the oxoammonium ion (Table 4.9) [225]. A recent report shows that 15 mol% TEMPO and 5 h reaction time can lead to similar results using laccase from *Trametes versicolor* [226].

4.5.1.8 Biocatalytic Oxidation of Alcohols

Enzymatic methods for the oxidation of alcohols are becoming more important. An excellent overview of biocatalytic alcohol oxidation is given in a review [227]. Besides the already mentioned oxidases (laccase, see above and e.g. glucose oxidase), the enzymes widespread in nature for (asymmetric) alcohol dehydrogenation are the alcohol dehydrogenases. However, their large scale application has predominantly been impeded by the requirement for cofactor-recycling. The vast majority of dehydrogenases which oxidize alcohols require NAD(P)$^+$ as cofactor, are relatively unstable and too expensive when used in molar amounts. Recently, a stable NAD$^+$-dependent alcohol dehydrogenase from *Rhodococcus rubber* was reported, which accepts acetone as co-substrate for NAD$^+$ regeneration and at the same time performs the desired alcohol oxidation (Fig. 4.75). Alcohol concentrations up to 50%v/v could be applied [228]. However it must be realized that this oxidation generally results in kinetic resolution: Only one of the enan-

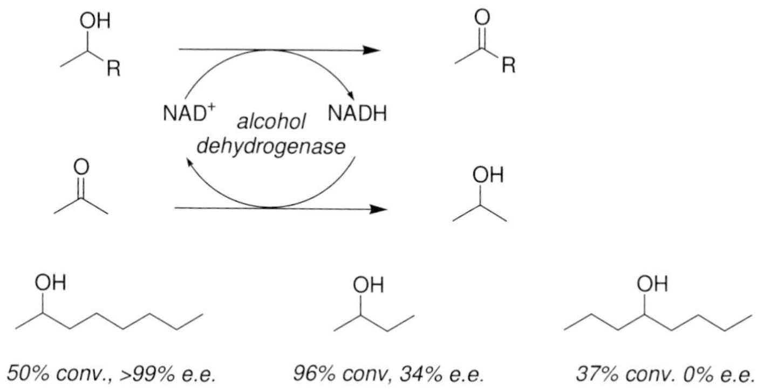

NAD$^+$ alcohol   NADH
dehydrogenase

50% conv., >99% e.e.   96% conv, 34% e.e.   37% conv. 0% e.e.

Fig. 4.75 Biocatalytic oxidation of alcohol using acetone for cofactor recycling.

tiomers is converted. For example, for 2-octanol, maximum 50% conversion is obtained and the residual alcohol has an *ee* of >99%. On the other hand for 2-butanol, no chiral discrimination occurs, and 96% conversion was observed. Also steric requirements exist: while cyclohexanol is not oxidized, cyclopentanol is a good substrate.

4.5.2
Oxidative Cleavage of 1,2-Diols

The oxidation of *vic*-diols is often accompanied by cleavage of the C–C bond to yield ketones and/or aldehydes. Especially the clean conversion of (cyclohexene to) 1,2-cyclohexanediol to adipic acid (Fig. 4.76) has received tremendous interest [229].

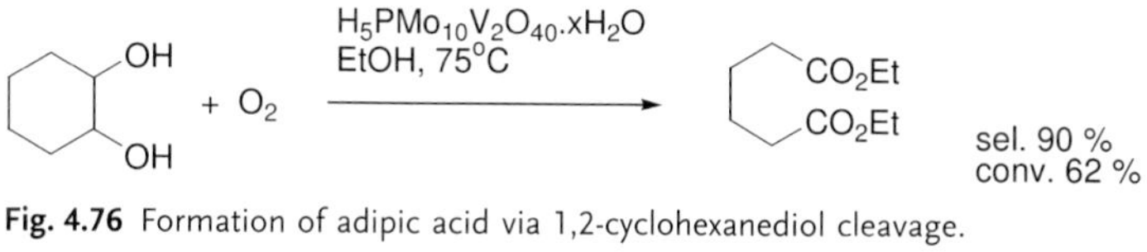

$H_5PMo_{10}V_2O_{40} \cdot xH_2O$
EtOH, 75°C

CO_2Et
CO_2Et

sel. 90 %
conv. 62 %

Fig. 4.76 Formation of adipic acid via 1,2-cyclohexanediol cleavage.

4.5.3
Carbohydrate Oxidation

Carbohydrate oxidations are generally performed with dioxygen in the presence of heterogeneous catalysts, such as Pd/C or Pt/C [230]. An example of homogeneous catalysis is the ruthenium-catalyzed oxidative cleavage of protected mannitol with hypochlorite (Fig. 4.77) [231].

Glucose oxidation to gluconate, is carried out using glucose oxidase from *Aspergillus niger* mould [232]. Very recently it was found that unsupported gold particles in aqueous solution with an average diameter of 3–5 nm, show a sur-

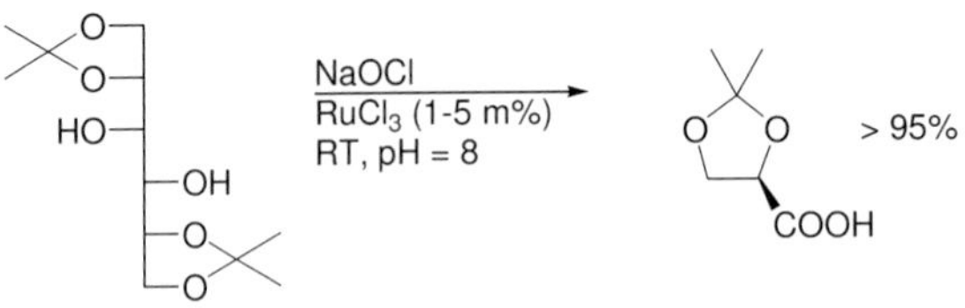

Fig. 4.77 Ruthenium/hypochlorite for oxidation of protected mannitol.

prisingly high activity in the aerobic oxidation of glucose, not far from that of an enzymatic system [233]. Both gold and glucose oxidase share the common stoichiometric reaction producing gluconate and hydrogen peroxide: $C_6H_{12}O_6 + O_2 + H_2O \rightarrow C_6H_{12}O_7 + H_2O_2$. In both cases the hydrogen peroxide is decomposed, either through alkali-promoted decomposition, or catalase-promoted decomposition.

4.5.4
Oxidation of Aldehydes and Ketones

Aldehydes undergo facile autoxidation (see Fig. 4.78), which is frequently used to form peracids in situ. The peracid itself will react with aldehydes in a Baeyer-Villiger (BV) reaction to form carboxylic acids [6].

The reaction is used commercially in the oxidation of acetaldehyde to peracetic acid [234], acetic anhydride [235] and acetic acid [236], respectively (Fig. 4.79). In the production of acetic anhydride, copper(II) salt competes with dioxygen for the intermediate acyl radical affording acetic anhydride via the acyl cation.

Also heterogeneous catalysts can be employed, such as Pd or Pt on carbon [237]. Recently, it was found that gold supported on a mesoporous CeO_2 matrix

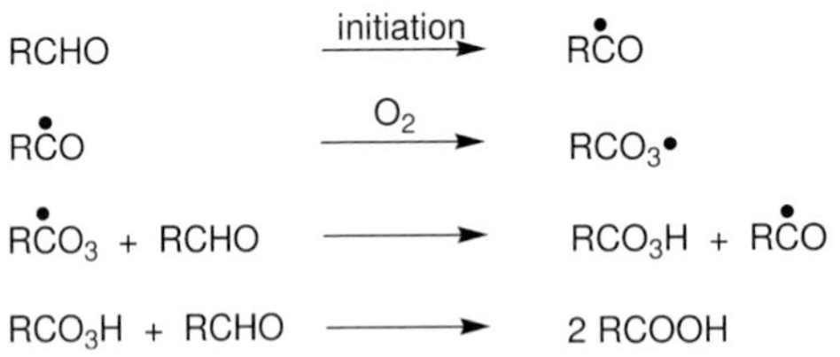

Fig. 4.78 Autoxidation of aldehydes.

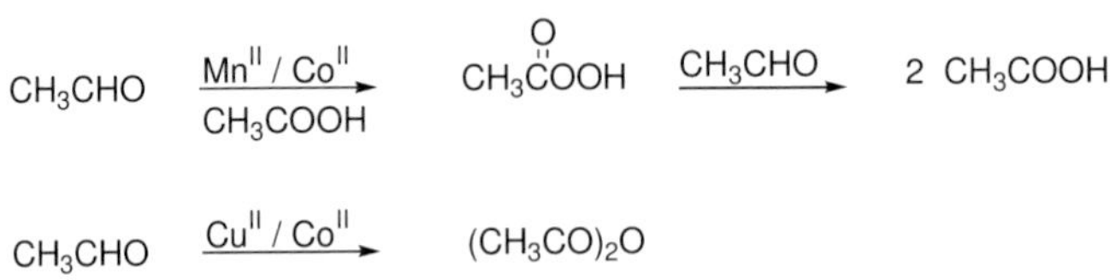

Fig. 4.79 Production of acetic acid and acetic anhydride starting from acetaldehyde.

catalyzes the selective aerobic oxidation of aliphatic and aerobic aldehydes better than other heterogeneous reported materials [238]. The activity is due to the nanomeric particle size of Au and CeO_2.

4.5.4.1 Baeyer-Villiger Oxidation

The Baeyer-Villiger reaction is the oxidation of a ketone or aldehyde to the corresponding ester. The mechanism of the reaction is relatively straightforward (see Fig. 4.80). In the first step a peracid undergoes a nucleophilic attack on the carbonyl group to give the so-called Criegee intermediate [239]. In the second step a concerted migration of one of the alkyl groups takes place with the release of the carboxylate anion.

The reaction has a broad scope and is widely used in organic synthesis. The oxidant is usually an organic peracid although, in principle, any electrophilic peroxide can be used. A large number of functional groups are tolerated and a variety of carbonyl compounds may react: ketones are converted into esters, cyclic ketones into lactones, benzaldehydes into phenols or carboxylic acids and α-diketones into anhydrides. The regiochemistry of the reaction is highly predictable and the migrating group generally retains its absolute configuration. These characteristic features are best illustrated by the reaction in Fig. 4.81, in which a number of functional groups are left intact [240].

Dioxygen can be used as the oxidant in combination with a "sacrificial" aldehyde [241], in which case the aldehyde first reacts with dioxygen and the resulting peracid is the oxidant. This is similar to the classical Baeyer-Villiger oxidation.

A large number of catalysts have been shown to be active in the oxidation of cycloalkanones to lactones using only hydrogen peroxide as the oxidant. Methyltrioxorhenium (MTO) is moderately active in the oxidation of linear ketones [242] or higher cycloalkanones [243] but it is particularly active in the oxidation of cyclobutanone derivatives (Fig. 4.82), which are oxidized faster with MTO than with other existing methods [244].

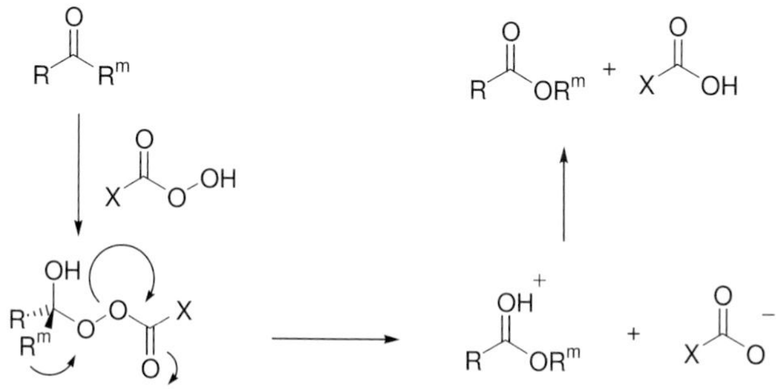

Fig. 4.80 Mechanism of the Baeyer-Villiger reaction.

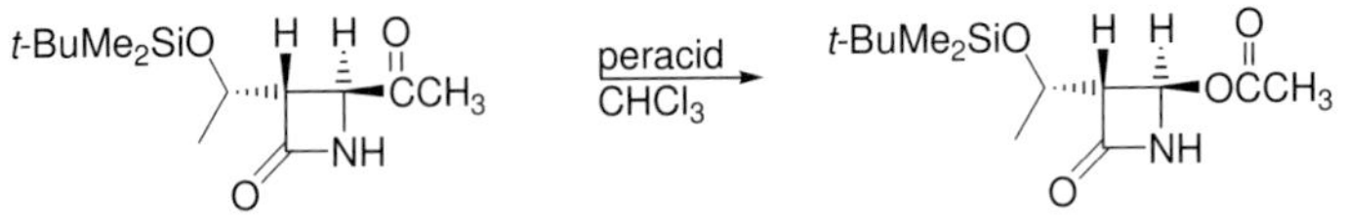

Fig. 4.81 Functional group tolerance in Baeyer-Villiger oxidation.

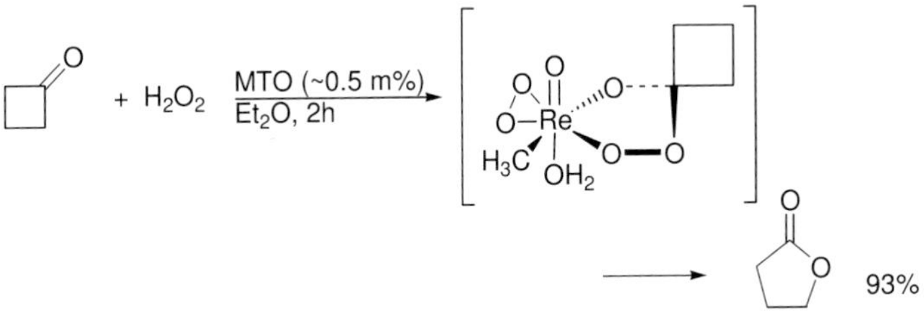

Fig. 4.82 MTO-catalyzed oxidation of cycloalkanones.

With <1 mol% MTO cyclobutanones are fully converted within one hour. Another approach consists of the use of a fluorous Sn-catalyst under biphasic conditions [245]. A perfluorinated tin(IV) compound, $Sn[NSO_2C_8F_{17}]_4$, was recently shown to be a highly effective catalyst for BV oxidations of cyclic ketones with 35% hydrogen peroxide in a fluorous biphasic system (Fig. 4.83). The catalyst, which resides in the fluorous phase, could be easily recycled without loss of activity.

Migration of hydrogen can yield carboxylic acids as the product of BV-oxidation of aldehydes. SeO_2 is especially good for oxidation of linear aldehydes to acids, but also of heteroaromatic aldehydes to acids [246]. To avoid contamination of the products with selenium catalysts, Knochel [247] immobilised 2,4-bis (perfluorooctyl)phenyl butyl selenide in a fluorous phase. The system was improved slightly by using the 3,5-bis(perfluorooctyl)phenyl butyl selenide isomer of the catalyst and by using the more polar dichloroethane as a co-solvent system instead of benzene (Fig. 4.84) [248]. The latter selenium isomer is slightly more difficult to synthesize, but both catalysts could be recycled without serious loss of activity.

A (lipophilic) quaternary ammonium salt has been shown to catalyze the Beayer-Villiger reaction of aldehydes under halide-and metal-free conditions with aqueous H_2O_2 (Fig. 4.85) [249].

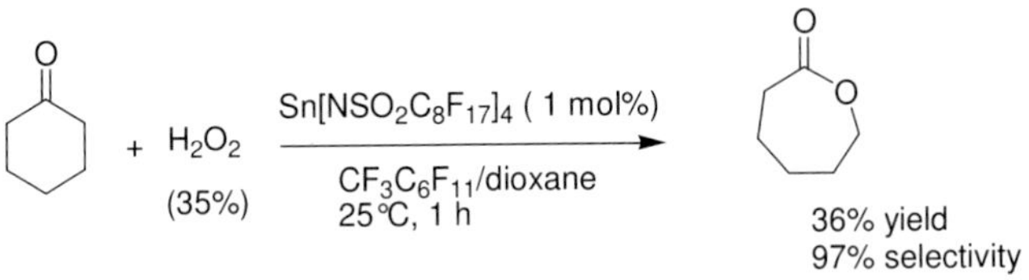

Fig. 4.83 Use of a fluorous Sn-catalyst for BV oxidations in a fluorous biphasic system.

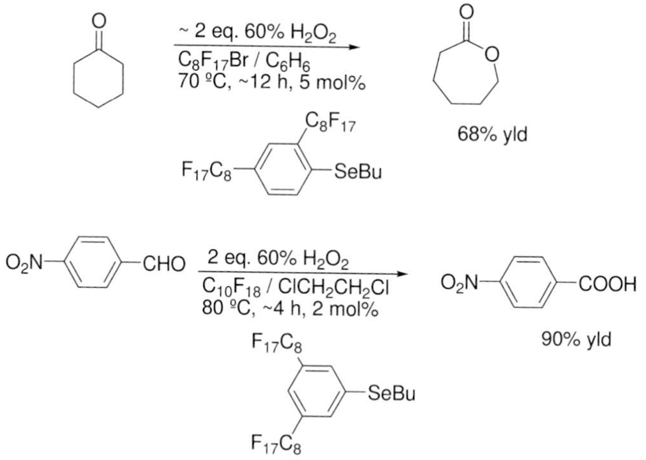

Fig. 4.84 Fluorous phase BV-oxidation using fluorous selenium catalysts.

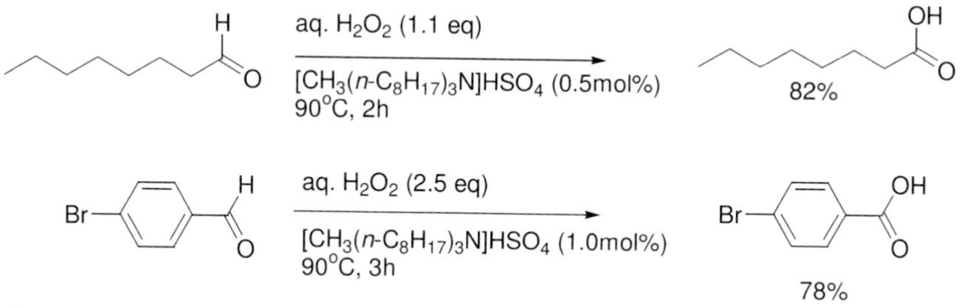

Fig. 4.85 Quaternary ammonium salts as catalyst for BV-oxidation with H_2O_2.

Heterogeneous catalysts are also active for the Baeyer-Villiger oxidation with H_2O_2. One of the best – based on 1.6 wt% tin in zeolite beta – was developed in the group of Corma [250, 251]. It provides a green method for BV oxidations in that it utilizes hydrogen peroxide as the oxidant, a recyclable catalyst, and avoids the use of chlorinated hydrocarbon solvents. It is believed that the Sn Lewis acid sites solely activate the ketone for BV-reaction, and leave hydrogen peroxide non-activated. Corma proposed that, in this way, side-reactions, such as epoxidation, which require electrophilic activation of hydrogen peroxide are largely avoided. Oxidation of bicyclo[3.2.0]hept-2-en-6-one with hydrogen peroxide in the presence of tin zeolite beta gave selective oxidation of the ketone, leaving the olefin intact. The tin catalyst may be particularly effective, because it can expand its coordination sphere to 6, thereby providing room for coordination of both the ketone and the hydroxide leaving group.

Surprisingly, both in the oxidation of bicyclo[3.2.0]hept-2-en-6-one and of di-hydrocarvone (not shown) isomeric lactones formed by migration of the second-ary carbon, as is usual in Baeyer-Villiger reactions, were not observed. Other

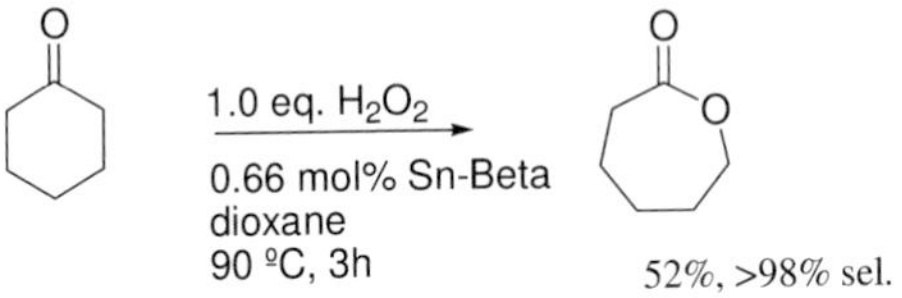

Fig. 4.86 Sn-Beta catalyzed BV-oxidation.

1.0 eq. H$_2$O$_2$

0.66 mol% Sn-Beta
dioxane
90 °C, 3h

52%, >98% sel.

ketones (e.g. cyclohexanone, Fig. 4.86) were oxidized with remarkable selectivity (>98%) considering the reaction conditions and the catalyst was recycled several times without loss of activity. Dioxane or the more attractive, methyl-*tert*-butyl ether, were used as solvents in these transformations.

Also very important is the asymmetric version of the BV-reaction. In this area biocatalysts are very promising, see Section 4.7.

4.5.5
Oxidation of Phenols

The aerobic oxidation of phenols in the presence of cobalt-Schiff's base complexes as catalysts is facilitated by (electron-donating) alkyl substituents in the ring and affords the corresponding *p*-quinones, e.g. the Vitamin E intermediate drawn in Fig. 4.87. When the *para*-position is occupied the reaction may be directed to the *ortho*-position [252, 253]. Copper compounds also mediate this type of oxidation, e.g. the Mitsubishi Gas process for the Vitamin E intermediate [254].

Phenol undergoes hydroxylation with H$_2$O$_2$ in the presence of Brønsted (HClO$_4$, H$_3$PO$_4$) or Lewis acid (TS-1) catalysts, affording a mixture of catechol and hydroquinone [255]. The reaction proceeds via a heterolytic mechanism involving an electrophilic Ti–OOH species or, in the case of Brønsted acids, perhaps even a hydroxy cation. On the other hand, catalysis by Fe^{II}/Co^{II} (Brichima process) involves a homolytic mechanism and hydroxyl radical intermediates [255] (see Table 4.10).

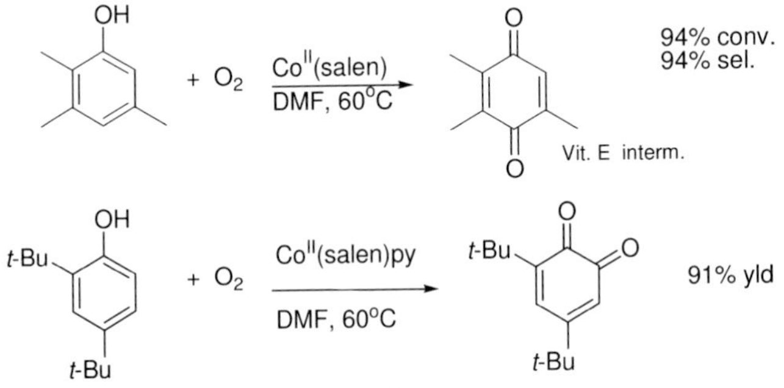

94% conv.
94% sel.

Vit. E interm.

91% yld

Fig. 4.87 Aerobic oxidation of phenols.

Table 4.10 Comparison of phenol conversion processes [255].

Process (catalyst)	Rhone-Poulenc $(HClO_4, H_3PO_4)$	Brichima (Fe^{II}/Co^{II})	Enichem (TS-1)
Phenol conversion %	5	10	25
Selectivity on phenol %	90	80	90
Selectivity on H_2O_2	70	50	70
Catechol/hydroquinone	1.4	2.3	1.0

4.5.6
Oxidation of Ethers

Ethers are readily autoxidized at the α-position to the corresponding α-hydroperoxyethers, which is of no synthetic utility, but important to realize when ethers are evaporated to leave an explosive residue of α-hydroperoxyethers behind. Ethers are oxidized with stoichiometric amounts of high valent metal oxides such as RuO_4 [256], CrO_3 [257] and $KMnO_4$ [258], but very few catalytic oxidations are known and these generally pertain to activated, e.g. benzylic, substrates. One example is the NHPI [259] catalyzed oxidation of methyl benzyl ether to the corresponding acetate (Fig. 4.88).

Recently a convenient and selective Ru/hypochlorite oxidation protocol for the selective transformation of a series of ethers and alcohols was reported [260]. The catalytic systems described in the literature, especially regarding ether oxidation, have limited reproducibility and/or require an excess of terminal oxidant, i.e. NaOCl solutions (household bleach). When the pH is maintained at 9–9.5 the reactions proceed smoothly, using the theoretical amount of hypochlorite, with fast, complete conversion of substrate and high selectivity to the target molecules at low concentration of Ru precursor. A variety of Ru precursors can be used, although it was found that $[Pr_4N][RuO_4]$ (TPAP) gave the best results. The catalyst can also be recycled with negligible loss of selectivity and minimum loss of activity, which can be further optimised by scale-up. In the oxidation of dibutyl ether, only 0.25 mol% of TPAP was used in ethyl acetate as the solvent, leading to 95% yield and 97% selectivity towards butyl butyrate (Fig. 4.89).

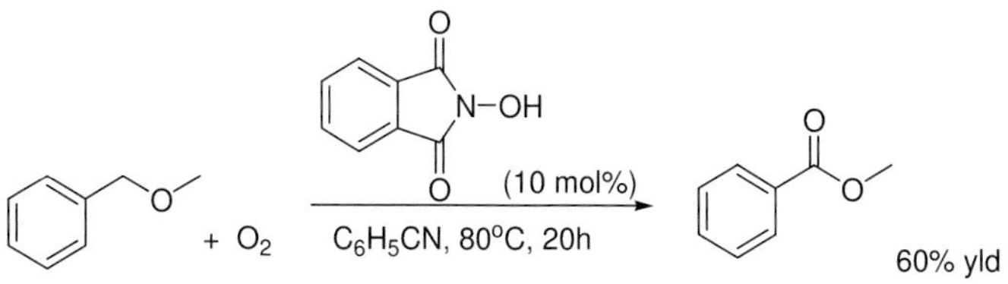

Fig. 4.88 NHPI catalyzed oxidation of ethers.

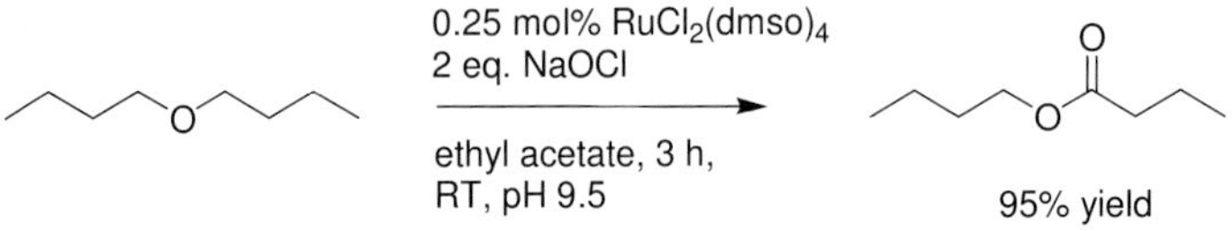

0.25 mol% RuCl$_2$(dmso)$_4$
2 eq. NaOCl

ethyl acetate, 3 h,
RT, pH 9.5

95% yield

Fig. 4.89 Green protocol for TPAP/NaOCl catalyzed oxidation of ethers.

4.6
Heteroatom Oxidation

Autoxidations of substrates containing heteroatoms tend to be complex and non-selective processes. Nonetheless, a number of selective catalytic reactions with hydrogen peroxide are known.

4.6.1
Oxidation of Amines

A large number of different nitrogen-containing compounds such as primary, secondary and tertiary amines, N,N-dimethylhydrazones, hydroxylamines, pyridines, etc. can be oxidized with e.g. hydrogen peroxide, leading to an equally vast number of different products. The presence of an a-hydrogen may have a major influence on the outcome of the reaction. Furthermore, a distinction can be made between oxidations involving high-valent peroxometal, or oxometal species and oxidative dehydrogenations mediated by low-valent transition metal catalysts. Nonetheless, with the right choice of solvent and equivalents of oxidants the reactions can be directed to a certain extent.

4.6.1.1 Primary Amines
Primary amines are dehydrogenated by high-valent oxometal species to give nitriles, or imines depending on the number of available a-hydrogens. Oxidation of the amine via peroxometal-intermediates (e.g. with MTO, Na$_2$MoO$_4$,

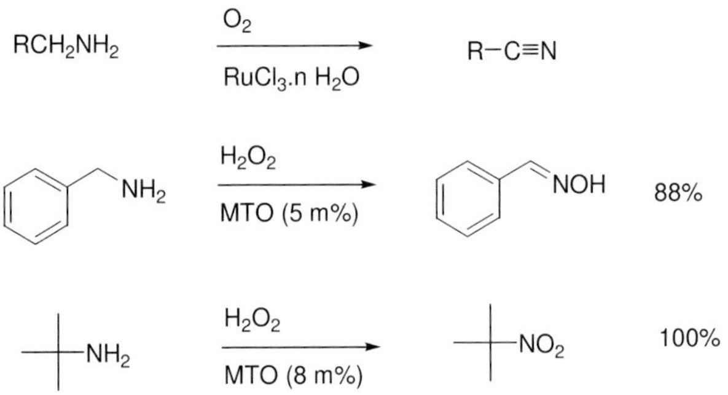

RCH$_2$NH$_2$ $\xrightarrow[\text{RuCl}_3.\text{n H}_2\text{O}]{\text{O}_2}$ R—C≡N

$\xrightarrow[\text{MTO (5 m\%)}]{\text{H}_2\text{O}_2}$ NOH   88%

$\xrightarrow[\text{MTO (8 m\%)}]{\text{H}_2\text{O}_2}$ —NO$_2$ 100%

Fig. 4.90 Oxidation of primary amines.

Na$_2$WO$_4$, NaVO$_3$ in combination with H$_2$O$_2$) leads to the formation of oximes [261, 262], or in the absence of α-hydrogens even to nitro-compounds (Fig. 4.90) [263].

4.6.1.2 Secondary Amines
Peroxometal-forming catalysts e.g. MTO, SeO$_2$ or Na$_2$WO$_4$, catalyze the oxidation of secondary amines to nitrones via the corresponding hydroxylamines, Fig. 4.91 a [262]. In the absence of α-hydrogens these substrates will also give nitroxides (R$_2$NO$^\bullet$), see Fig. 4.91 b [263]. Here again oxometal complexes react differently, giving the dehydrogenated imine rather than the nitrone [122, 264], as is illustrated in Fig. 4.92.

4.6.1.3 Tertiary Amines
Similarly, for tertiary amines a distinction can be made between oxometal and peroxometal pathways. Cytochrome P450 monooxygenases catalyze the oxidative N-demethylation of amines in which the active oxidant is a high-valent oxoiron species. This reaction can be mimicked with some oxometal complexes (Ru^V=O), while oxidation via peroxometal complexes results in oxidation of the N atom (Fig. 4.93 a and b) [261]. A combination of MTO/hydrogen peroxide can

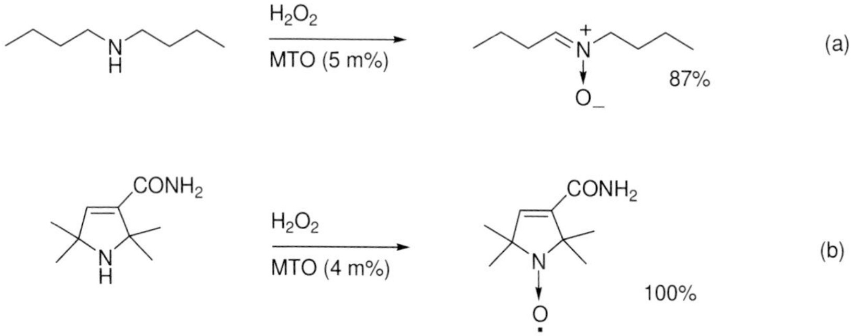

Fig. 4.91 MTO catalyzed oxidation of secondary amines.

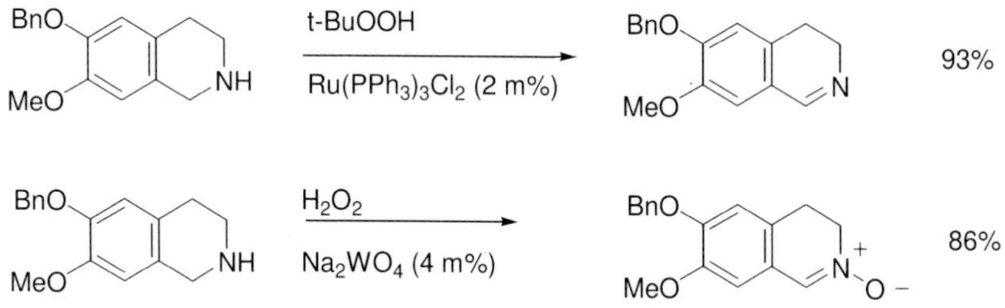

Fig. 4.92 Oxidation of secondary amines via Ru and W catalysts.

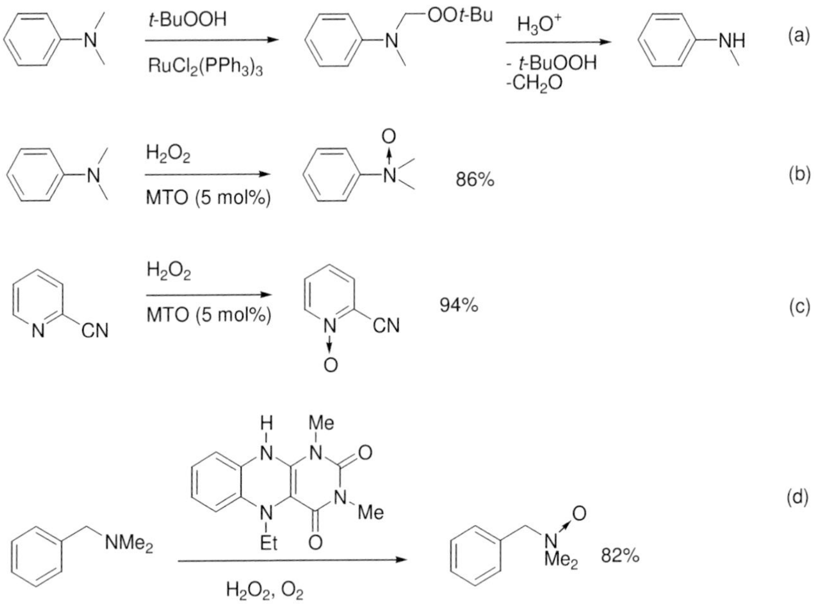

Fig. 4.93 Oxidation of tertiary amines.

also oxidize pyridines to pyridine *N*-oxides as was recently published by Sharpless et al. [265]. A rare example of clean oxidation of tertiary amines without metal compounds was reported by Bäckvall et al. [266] using flavin as the catalyst (Fig. 4.93 d).

4.6.1.4 Amides
Ruthenium oxo compounds catalyze the oxidation of amides, e.g. β-lactams, at the position α to the nitrogen, using peracetic acid as the oxidant, presumably via oxoruthenium intermediates [264].

4.6.2
Sulfoxidation

Dialkyl sulfides are readily oxidized by e.g. H_2O_2 in the presence of metal catalysts to give the corresponding sulfoxides. In turn the sulfoxides can be further oxidized to the corresponding sulfones $R-S(O)_2-R'$ with excess peroxide. Due to the polarity of the S–O bond the second oxidation step is more difficult for electrophilic oxidation catalysts. This is especially important in the oxidative destruction of mustard gas where the corresponding sulfoxide is relatively harmless, but the sulfone quite toxic again [267]. In order to see which catalyst is likely to give "overoxidation" of the sulfoxide the thianthrene 5-oxide (see Fig. 4.94) is an excellent probe to test catalysts active in oxygen transfer reactions for whether

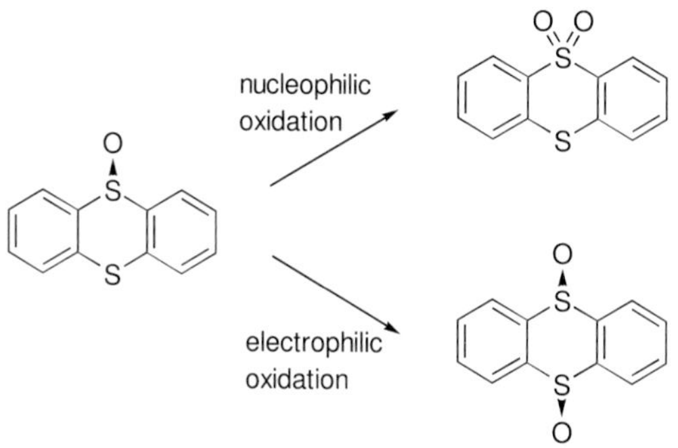

Fig. 4.94 Nucleophilic vs. electrophilic S oxidation pathways.

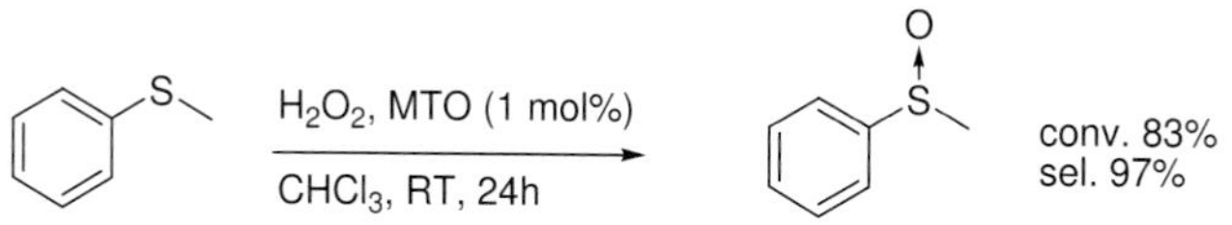

conv. 83%
sel. 97%

Fig. 4.95 MTO/H₂O₂ catalyzed sulfide oxidation.

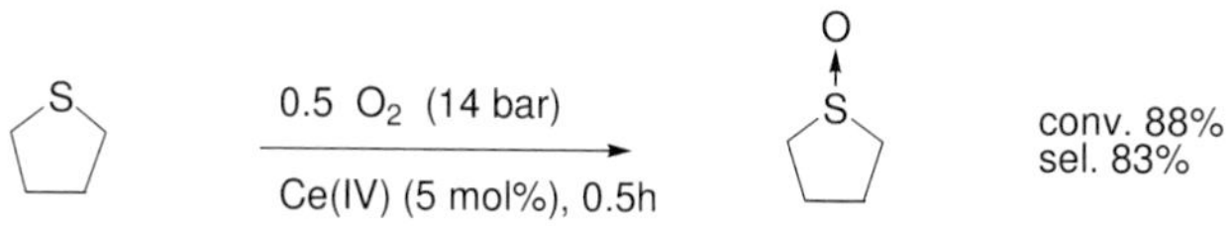

conv. 88%
sel. 83%

Fig. 4.96 Sulfide oxidation using air as the oxidant.

they are nucleophilic or electrophilic in character. Catalytic oxidations with hydrogen peroxide are generally performed with early transition elements, e.g. Mo(VI) [268], W(VI), Ti(IV), V(V), Re(V) [269] and Re(VII) (=MTO, Fig. 4.95) [270] or seleninic acids as catalysts [271] and involve peroxometal pathways. Sulfides can also be oxidized with dioxygen, using a cerium (IV)/cerium (III) redox couple [271], via a homolytic mechanism involving a sulfonium radical cation intermediate (Fig. 4.96).

4.7
Asymmetric Oxidation

Owing to the instability of many (chiral) ligands under oxidative reaction conditions, asymmetric oxidation is not an easy reaction to perform. However, much progress has been achieved over the last decades. Because of the relatively low volumes and high added value of the products asymmetric oxidation allows the use of more expensive and environmentally less attractive oxidants such as hypochlorite and N-methylmorpholine-N-oxide (NMO).

4.7.1
Asymmetric Epoxidation of Olefins

Chiral epoxides are important intermediates in organic synthesis. A benchmark classic in the area of asymmetric catalytic oxidation is the *Sharpless* epoxidation of allylic alcohols in which a complex of titanium and tartrate salt is the active catalyst [273]. Its success is due to its ease of execution and the ready availability of reagents. A wide variety of primary allylic alcohols are epoxidized in >90% optical yield and 70–90% chemical yield using *tert*-butyl hydroperoxide as the oxygen donor and titanium-isopropoxide-diethyltartrate (DET) as the catalyst (Fig. 4.97). In order for this reaction to be catalytic, the exclusion of water is absolutely essential. This is achieved by adding 3 Å or 4 Å molecular sieves. The catalytic cycle is identical to that for titanium epoxidations discussed above (see Fig. 4.20) and the actual catalytic species is believed to be a 2:2 titanium(IV) tartrate dimer (see Fig. 4.98). The key step is the preferential transfer of oxygen from a coordinated alkylperoxo moiety to one enantioface of a coordinated allylic alcohol. For further information the reader is referred to the many reviews that have been written on this reaction [274, 275].

The applicability of the "Sharpless" asymmetric epoxidation is however limited to functionalized alcohols, i.e. allylic alcohols (see Table 4.11). The best method for non-functionalized olefins is the Jacobsen-Kaksuki method. Only a few years after the key publication of Kochi and coworkers on salen-manganese complexes as catalysts for epoxidations, Jacobsen and Kaksuki independently described, in 1990, the use of chiral salen manganese(III) catalysts for the synthesis of optically active epoxides [276, 277] (Fig. 4.99). Epoxidations can be carried out using commercial bleach (NaOCl) or iodosylbenzene as terminal oxidants and as little as 0.5 mol% of catalyst. The active oxidant is an oxomanganese(V) species.

Over the years further tuning of the ligand by varying the substituents in the aromatic ring has led to >90% *ee* for a number of olefins [278]. Improvements were also achieved by adding amine-*N*-oxides such as 4-phenylpyridine-*N*-oxide

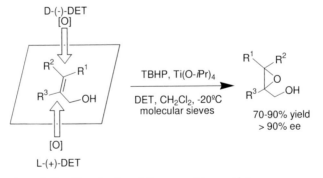

Fig. 4.97 The "Sharpless" catalytic asymmetric epoxidation

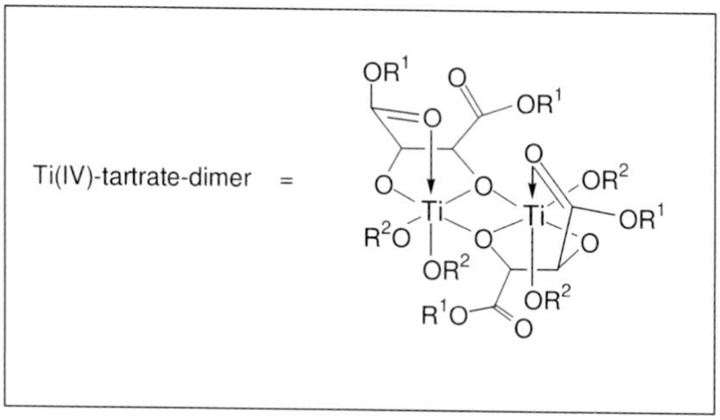

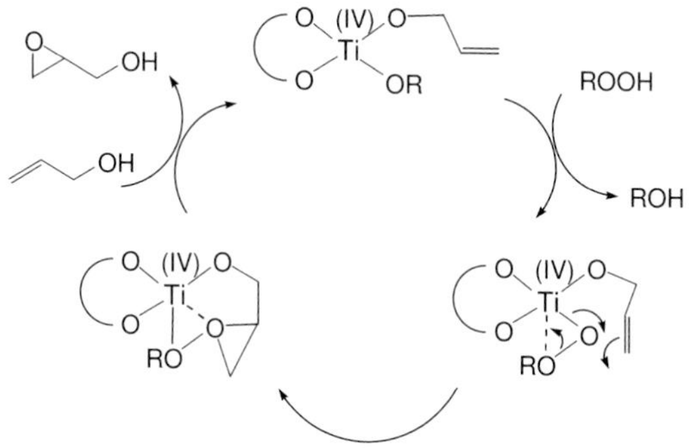

Ti(IV)-tartrate-dimer =

Fig. 4.98 The catalytic cycle of the "Sharpless" catalytic asymmetric epoxidation.

Table 4.11 Sharpless asymmetric epoxidation of allylic alcohols. [a]

Substrate	Yield (%)	ee (%)
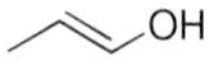 OH	65	90
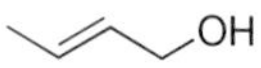 OH	70	96
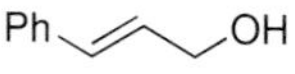 Ph OH	89	>98
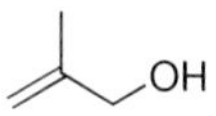 OH	50	>95
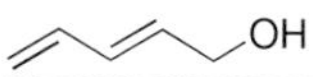 OH	40	95

a) Results with 5 mol% of catalyst, see Ref. [273].

which serve as axial ligands in the active oxomanganese(V) species. The method generally gives good results for cis-disubstituted olefins, whereas trans-disubstituted olefins are less suitable substrates (see Table 4.12). On the other hand, chromium-salen complexes catalyze the asymmetric epoxidation of trans-substituted olefins in reasonable ees (see Fig. 4.100) using iodosylarenes as oxidants [279].

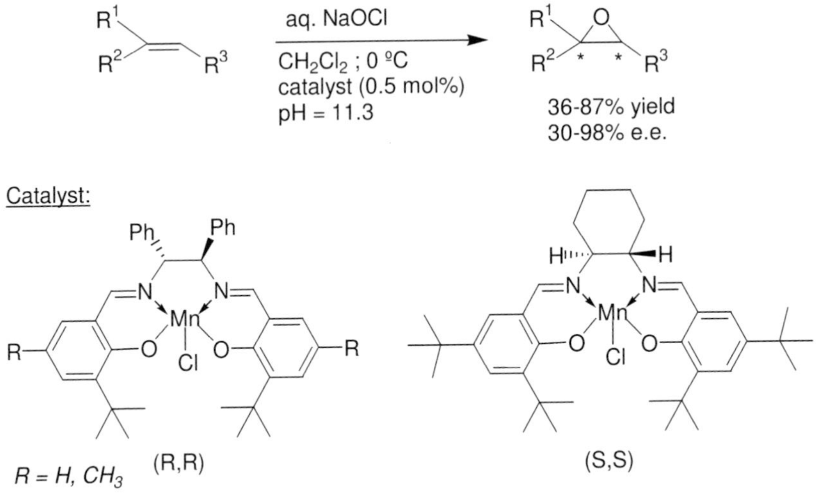

Catalyst:

$R = H, CH_3$ (R,R)   (S,S)

Fig. 4.99 The Jacobsen-Katsuki asymmetric epoxidation of unfunctionalized olefins.

Table 4.12 Jacobsen-Katsuki asymmetric epoxidation of unfunctionalized olefins.

Olefin	Method [a]	Equiv. catalyst	Isolated yield (%)	*ee* (%)
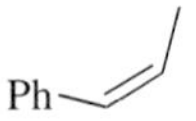 Ph—	A	0.04	84	92
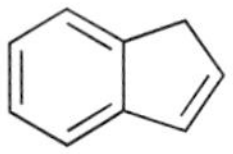 (indene)	A	0.01	80	88
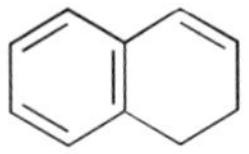 (dihydronaphthalene)	B	0.04	67	88
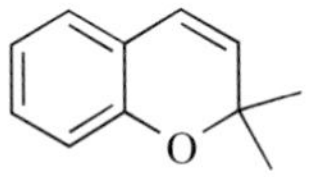 (chromene)	A	0.02	87	98
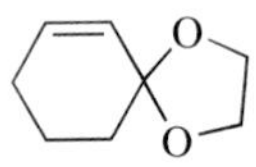 (dioxolane-cyclohexene)	B	0.15	63	94
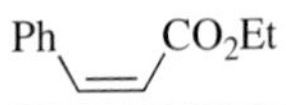 Ph CO₂Et	B	0.08	67	97

a)   Method A: NaOCl, pH 11.3, CH$_2$Cl$_2$, 0 °C. Method B: same
    as A + 0.2 eq. 4-phenylpyridine-*N*-oxide.

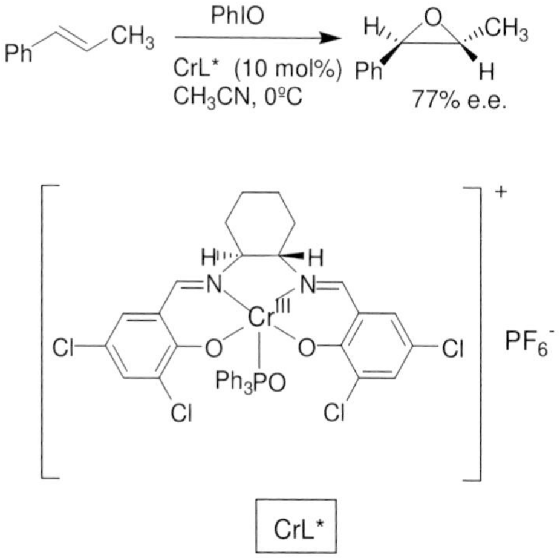

Fig. 4.100 Cr-salen catalyzed asymmetric epoxidation.

CrL*

The recent reports of Katsuki and Beller using Ti- [280, 281] and Ru-based [282] complexes for asymmetric epoxidation with H_2O_2, have however changed the state-of-the-art in the area of metal-catalyzed asymmetric epoxidations. Especially for titanium in combination with reduced salen-type ligands, excellent stereochemical control was achieved for terminal olefins using hydrogen peroxide as the oxidant (see Fig. 4.101). Also with ruthenium, notorious for its catalase activity, excellent yields and enantioselectivities were obtained for cis- and activated olefins. In Table 4.13 an overview is given of the performance of titanium and ruthenium complexes as catalysts with hydrogen peroxide.

For titanium, only 1 mol% of catalyst Ti(salalen) (Table 4.13) and 1.05 eq. H_2O_2 are required to obtain high yields and selectivities. 1,2-Dihydronaphthalene is an activated cis-olefin, which obviously gives the best results (yield and enantioselectivity of over 98%) using little or no excess of hydrogen peroxide. However also for simple styrene 93% *ee* can be achieved using this chiral titanium complex. What is most striking is the result obtained for 1-octene. For this simple non-activated olefin a reasonable 82% *ee* could be attained. Titanium-salan (see Table 4.13) as a catalyst requires higher loadings (5 mol%), but its ease of synthesis and *in situ* synthesis from Ti(O*i*Pr)$_4$ and salan ligand, makes this a very promising catalyst for practical applications.

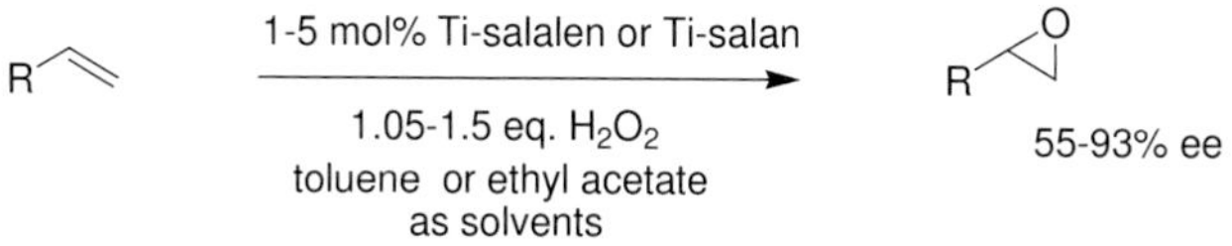

Fig. 4.101 Ti-catalyzed asymmetric epoxidation of olefins using aqueous hydrogen peroxide as the oxidant.

Table 4.13 Comparison of metal catalysts for the asymmetric epoxidation of alkenes using aqueous H_2O_2.

di-μ-oxo titanium(salalen) -**1**   di-μ-oxo titanium(salan) -**2**   Ru(pyboxazine)(pydic) -**3**

Substrate	Catalyst	Catalyst loading (mol%)	Eq. 30% aqueous H_2O_2	Yield epoxide (%)	*ee* epoxide (%)	Ref.
	Ti(salalen)-**1** [a)]	1	1.05	90	93	280
	Ti(salan)-**2** [b)]	5	1.5	47	82	281
	Ru(pyboxazine) (pydic)-**3** [c), d)]	5	3	85	59	282
	Ti(salalen)-**1** [a)]	1	1.05	70	82	280
	Ti(salan)-**2** [b)]	5	1.5	25	55	281
	Ti(salalen)-**1** [a)]	1	1.05	99	99	280
	Ti(salan)-**2** [b)]	5	1.5	87	96	281
	Ru(pyboxazine) (pydic)-**3** [c)]	5	3	95	72	282

a) RT, 12–48 h, CH_2Cl_2 as solvent.
b) 25 °C, 6–24 h, CH_2Cl_2 as solvent.
c) RT, 12 h, 2-methylbutan-2-ol as solvent.
d) 20 mol% acetic acid was added.

Mukaiyama has reported the use of manganese complexes for chiral epoxidations using a combination of molecular oxygen and an aldehyde as the oxidant [283]. With salen-manganese(III) complexes and pivaldehyde/oxygen the corresponding epoxides of several 1,2-dihydronaphthalenes were obtained in very good yields (77 to 92% *ee*) in the presence of *N*-alkyl imidazole as axial ligand. Altering the ligand structure from salen derivatives to optically active β-keto-imine-type ligands, gave the novel manganese catalysts in Fig. 4.102 which oxidize phenyl-conjugated olefins, such as dihydronaphthalenes and cis-β-methyl styrene in 53 to 84% *ee* [284]. The enantiofacial selection in these manganese-catalyzed epoxidations is opposite to that obtained with sodium hypochlorite or iodosylbenzene as the primary oxidant in the Jacobsen-Katsuki epoxidation. Apparently the catalytically active species differs from the putative oxomanganese(V) complex in the latter processes. An acylperoxomanganese complex was proposed as the active oxidant, i.e. a peroxometal rather than an oxometal mechanism.

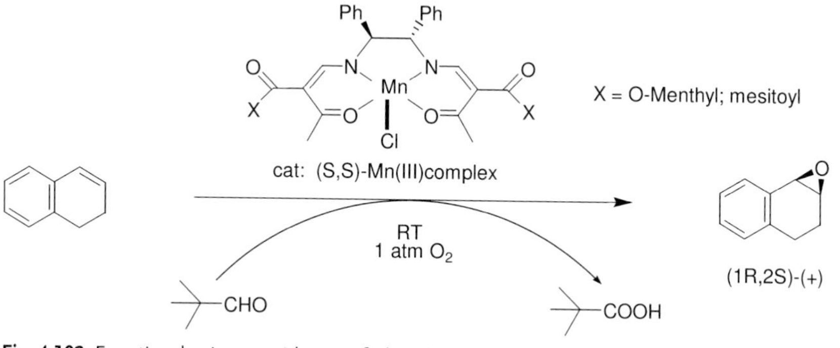

Fig. 4.102 Enantioselective epoxidation of phenyl-conjugated olefins employing aldehyde and molecular oxygen as the oxidant.

A very different, non-metal based, approach, for the synthesis of trans-disubstituted chiral epoxides is the use of the highly reactive chiral dioxiranes (Fig. 4.103). These oxidants can be generated *in situ* by using potassium persulfate as the primary oxidant. Although this method requires the use of 20–30 mol% of dioxirane precursor, the reaction proceeds with high chemical yields and *ees* over 90% for a wide variety of trans-disubstituted olefines [285]. The mechanism is given in Fig. 4.104, and involves a spiro transition state. More recently, it was shown that the persulfate can be replaced by H_2O_2 in CH_3CN, which generates the peroxyimidic acid (Payne reagent) *in situ* [286].

Another method is to use poly-L-amino acids as catalysts in alkaline media (Julia-Colonna epoxidation) for the asymmetric epoxidation of chalcones and other electron-poor olefins with H_2O_2 [287]. SmithKline Beecham workers used this method (see Fig. 4.105) as a key step in the synthesis of a leukotriene antagonist, although it required 20 equivalents of H_2O_2 and 12 equivalents of NaOH, based on substrate [288]. The mechanism probably involves the asym-

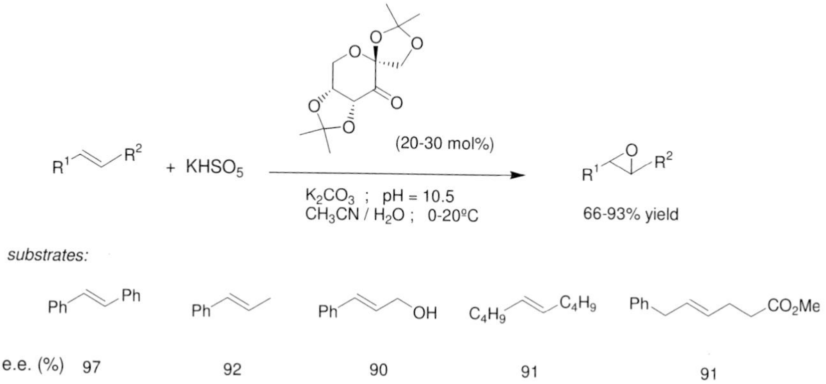

Fig. 4.103 Efficient asymmetric epoxidation of trans-alkenes.

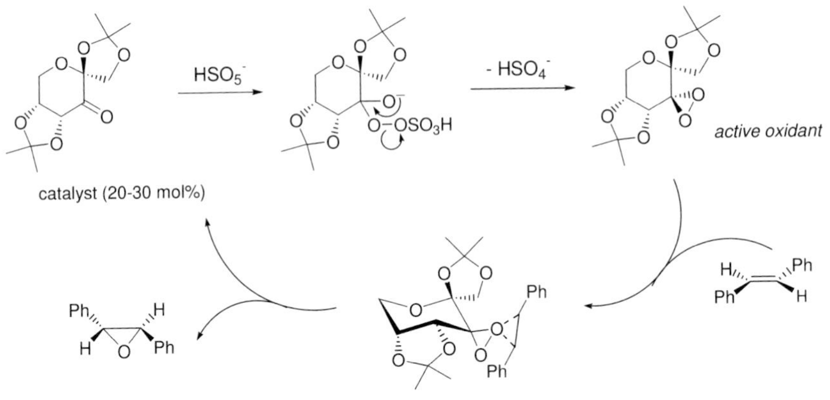

Fig. 4.104 Catalytic cycle of asymmetric epoxidation via chiral dioxiranes.

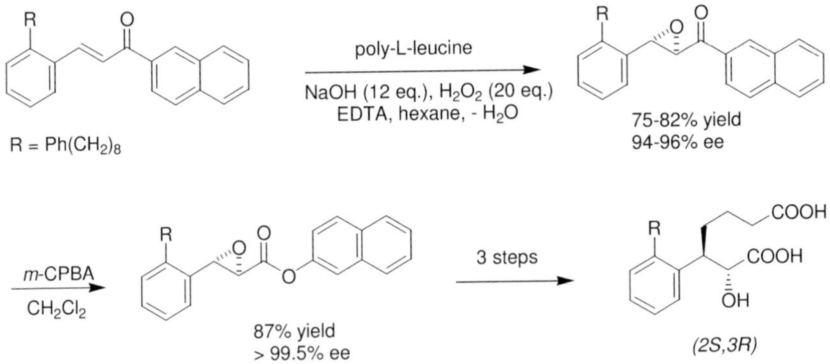

Fig. 4.105 Asymmetric epoxidation of chalcones.

metric addition of a hydroperoxy anion (HOO⁻) to the olefinic double bond, followed by epoxide ring closure. The poly-L-amino acid acts as a chiral phase transfer catalyst and may be considered as a synthetic enzyme mimic. Further improvements of the Julia-Colonna epoxidation have been achieved by using sodium percarbonate as the oxidant and recyclable silica-supported polyamino acids as catalysts in dimethoxyethane as solvent [289].

Finally, enzymes themselves can be used for the direct epoxidation of olefins. As already outlined in Section 4.2, peroxidases (which can directly apply hydrogen peroxide as the oxidant) as well as monooygenases are available. Because of its ease of application, chloroperoxidase (CPO) from *Caldariomyces fumago* is very promising. It is commercially available and displays a reasonable substrate range [290]. The main problem is its instability under reaction conditions and the sluggish reaction rates. Typical turnover rates are 0.1–2 s⁻¹. The epoxidation of styrene in the presence of 0.8% of surfactant is, at 5.5 s⁻¹ the fastest on record [291]. The advantage is that the *ees* in most cases are always good to excellent. For example 2-heptene could be oxidized with 1700 turnovers relative to enzyme in 30% t-BuOH. The *ee* of the resulting epoxide was 96%. The total turnover number could be further increased in this reaction (from 1700 to 11500) when the hydrogen peroxide was generated *in situ* by the glucose oxidase-mediated reaction of glucose and oxygen. A wide range of 2-methyl-1-alkenes could also be oxidized with *ees* ranging from 95 to 50% [292]. 1-Alkenes with the exception of styrene are suicide reactants that alkylate the heme in native CPO. The latter problem can be circumvented by using certain mutants [293]. Styrene resulted in only 49% *ee*, for this substrate the styrene monooxygenases seem to be better suited.

Recently, the first asymmetric cell-free application of styrene monooxygenase (StyAB) from *Pseudomonas* sp. VLB120 was reported [294]. StyAB catalyses the enantiospecific epoxidation of styrene-type substrates and requires the presence of flavin and NADH as cofactor. This two-component system enzyme consists of the actual oxygenase subunit (StyA) and a reductase (StyB). In this case, the reaction could be made catalytic with respect to NADH when formate together with oxygen were used as the actual oxidant and sacrificial reductant respectively. The whole sequence is shown in Fig. 4.106. The total turnover number on StyA enzyme was around 2000, whereas the turnover number relative to NADH ranged from 66 to 87. Results for individual substrates are also given in Fig. 4.106. Excellent enantioselectivities are obtained for *a*- and *β*-styrene derivatives.

Recently it was shown that the flavin (FAD in Fig. 4.106) could be directly regenerated by the organometallic complex $Cp^*Rh(bpy)(H_2O)]^{2+}$ and formate. In

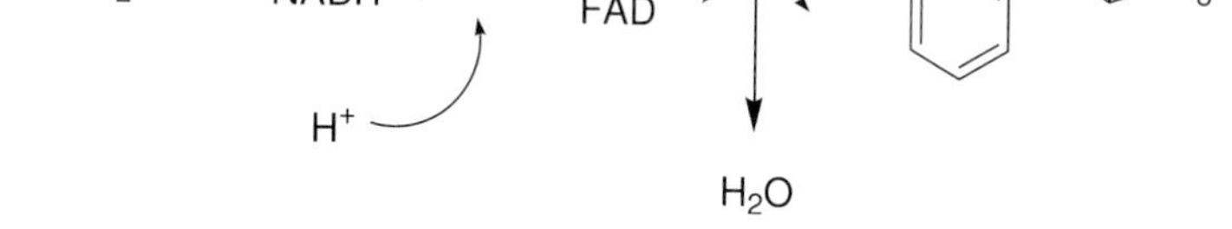

m-chlorostyrene (R_1=Cl, R_2,R_3=H)   73% yld.   90.5% conv.   >99.9% ee

α-methyl-styrene (R_2=CH$_3$, R_1,R_3=H)   75% yld.   87.9% conv.   98.1% ee

β-methyl-styrene (R_1,R_2=H, R_3=CH$_3$)   87% yld.   95.3% conv.   99.7% ee

Fig. 4.106 Biocatalytic epoxidation with styrene monooxygenase including cofactor regeneration.

this way the reaction becomes simpler, because only enzyme (StyA), FAD and Rh-complex are required. The resulting turnover number on Rh ranged from 9 to 18 depending on the substrate [295].

4.7.2
Asymmetric Dihydroxylation of Olefins

Following their success in asymmetric epoxidation, Sharpless and coworkers developed an efficient method for the catalytic asymmetric dihydroxylation of olefins. The method employs catalytic amounts of OsO$_4$ and derivatives of cinchona alkaloids as chiral ligands together with N-methylmorpholine-N-oxide (NMO) as the primary oxidant, see Fig. 4.107 [296]. The use of osmium(VIII) complexes as catalysts leads to stereospecific 1,2-cis addition of two OH groups to the olefin. The reaction greatly benefits from ligand accelerated catalysis, i.e. the ligation of osmium by the alkaloid enhances the rate of reaction with the alkene by one to two orders of magnitude relative to the reaction without ligand. The reaction is complicated by the fact that two catalytic cycles are possible, see Fig. 4.108 [297]. The primary cycle proceeds with high face selectivity and involves the chiral ligand in its selectivity determining step, the formation of the osmium(VI)glycolate. The latter is oxidized by the primary oxidant to the osmium(VIII)glycolate which is hydrolyzed to the product diol with concomitant

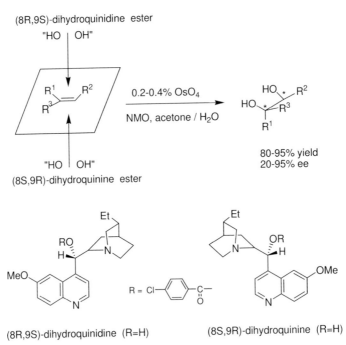

Fig. 4.107 Asymmetric dihydroxylation of alkenes.

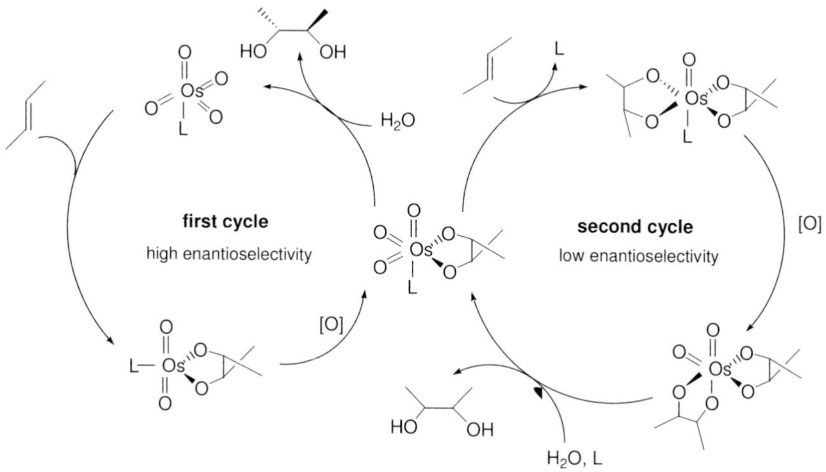

Fig. 4.108 Mechanism of asymmetric dihydroxylation.

generation of the catalyst. In the second catalytic cycle the osmium(VIII)glycolate reacts with a second molecule of olefin, displacing the chiral ligand and resulting in poor overall enantioselectivity. The desired pathway involves hydrolysis of the osmium(VI)glycolate in competition with coordination of another olefin molecule. Therefore, slow addition of olefin is essential to obtain high *ees*.

Alternatively, the process can be performed with $K_3Fe(CN)_6$ as the stoichiometric oxidant in *tert*-butanol/water mixtures [298]. In this case the olefin osmylation and osmium reoxidation steps are uncoupled, since they occur in different phases, resulting in improved enantioselectivities. However, from a practical point of view, the improved enantioselectivities (about 5–10% *ee*) are probably offset by the use of a less attractive oxidant.

When using NMO as the oxidant, on the other hand, the reduced form is readily recycled by oxidation with H_2O_2. The asymmetric dihydroxylation is successful with a broad range of substrates, in contrast to the Sharpless asymmetric epoxidation which is only suitable for allylic alcohols (Table 4.14). Further development of the system has led to the formulation of a reagent mixture, called AD-mix based on phthalazine-type ligands, which contains all the ingredients for the asymmetric dihydroxylation under heterogeneous conditions including OsO_4 [299], see Fig. 4.109. One equivalent of methanesulfonamide is added to accelerate the hydrolysis of the osmium(VI)glycolate and thus to achieve satisfactory turnover rates.

Beller et al. [85] recently described the aerobic dihydroxylation of olefins catalyzed by osmium at basic pH, as mentioned above. When using the hydroquinidine and hydroquinine bases, they were able to obtain reasonable enantioselectivities (54% *ee* to 96% *ee*) for a range of substrates. An alternative route towards enantiopure diols, is the kinetic resolution of racemic epoxides via enantioselective hydrolysis catalyzed by a Co(III)salen acetate complex, developed by Jacob-

Table 4.14 Comparison of asymmetric epoxidation (AE) and asymmetric dihydroxylation (AD) according to refs. [273] and [299].

Alkene	Substrate for	
	AD	AE
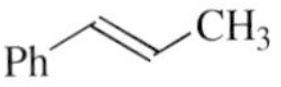 Ph⌒CH₃	>95% ee	NR
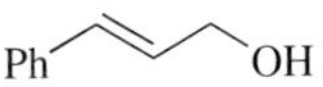 Ph⌒OH	80% ee	>95% ee
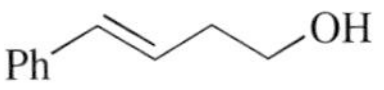 Ph⌒OH	>95% ee	30–50% ee
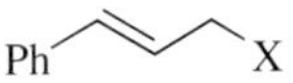 Ph⌒X X=OAc, OCH₂Ph, N₃, Cl	>95% ee	NR
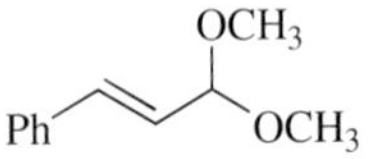 Ph⌒(OCH₃)OCH₃	>95% ee	NR
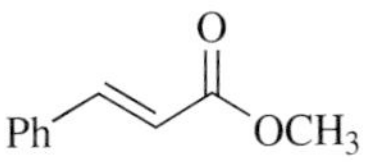 Ph⌒C(O)OCH₃	>95% ee	NR
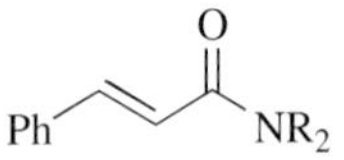 Ph⌒C(O)NR₂	>95% ee	NR

NR = non-relevant.

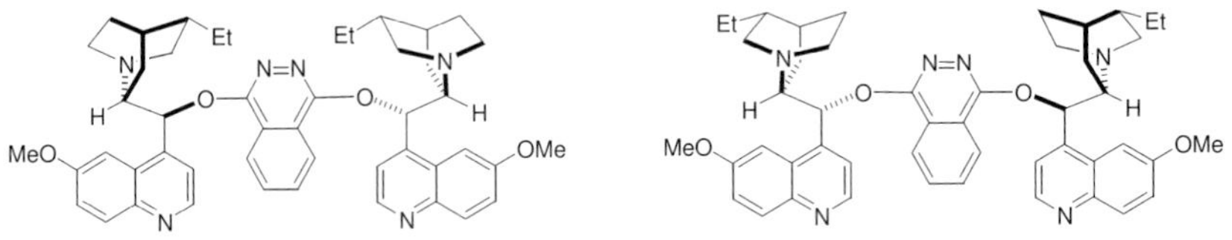

(DHQD)₂-PHAL,
ligand used in AD-mix-β

(DHQ)₂-PHAL,
ligand used in AD-mix-α

Fig. 4.109 Ligands used in commercial catalysts for asymmetric dihydroxylation.

sen [300]. In this case the maximum theoretical yield of the enantiomerically pure diol is 50%, compared with 100% in the asymmetric dihydroxylation method. Nonetheless, its simplicity makes it a synthetically useful methodology.

4.7.3
Asymmetric Sulfoxidation

In the area of metal catalyzed asymmetric sulfoxidation there is still much room for improvement. The most successful examples involve titanium tartrates, but at the same time often require near stoichiometric quantities of catalysts [301, 302]. Recently, this methodology has been successfully used for the production of (S)-Omeprazole by AstraZeneca [303] (see Fig. 4.110). A modified Kagan-procedure [302] was applied, using cumene hydroperoxide as the oxidant. Another example is the sulfoxidation of an aryl ethyl sulfide, which was in development by Astra Zeneca as a candidate drug for the treatment of schizophrenia. In this case the final *ee* could be improved from 60% to 80% by optimising the Ti:tartrate ratio [304].

From a practical viewpoint the recently discovered vanadium-based and iron-based asymmetric sulfoxidation with hydrogen peroxide is worth mentioning [305, 306]. For vanadium, in principle as little as 0.01 mol% of catalyst can be employed (Fig. 4.111). With tridentate Schiff-bases as ligands, formed from readily available salicylaldehydes and (S)-*tert*-leucinol, *ees* of 59–70% were obtained for thioanisole [305], 85% *ee* for 2-phenyl-1,3-dithiane [305] and 82–91% *ee* for *tert*-butyl disulfide [307]. For iron, similar results were obtained using 4 mol% of an iron catalyst, synthesized *in situ* from Fe(acac)$_3$ and the same type of Schiff base ligands as in Fig. 4.111 (see Ref. [306] for details).

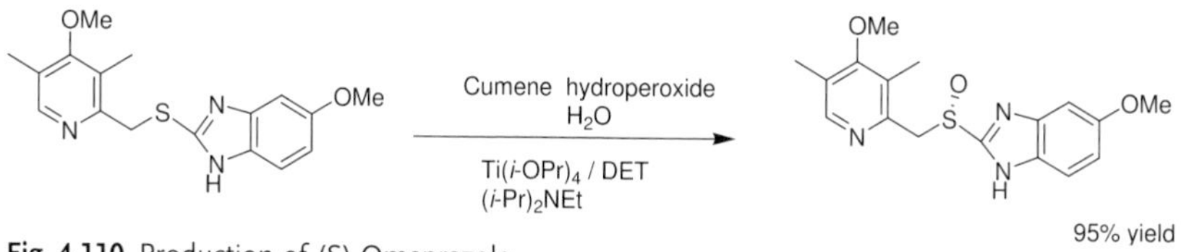

Fig. 4.110 Production of (S)-Omeprazole.

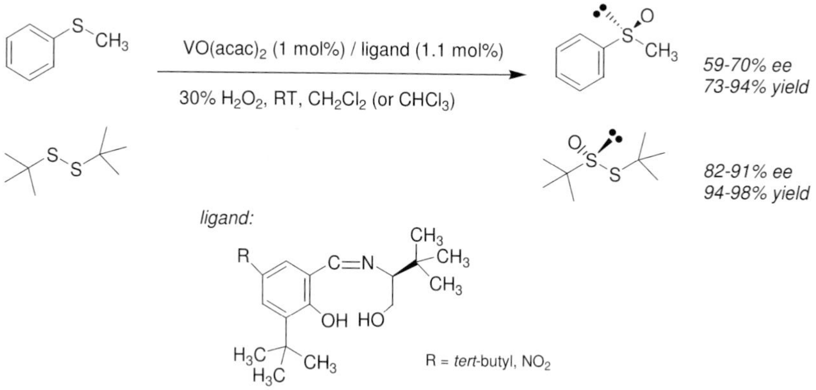

Fig. 4.111 Vanadium catalyzed asymmetric sulfoxidation.

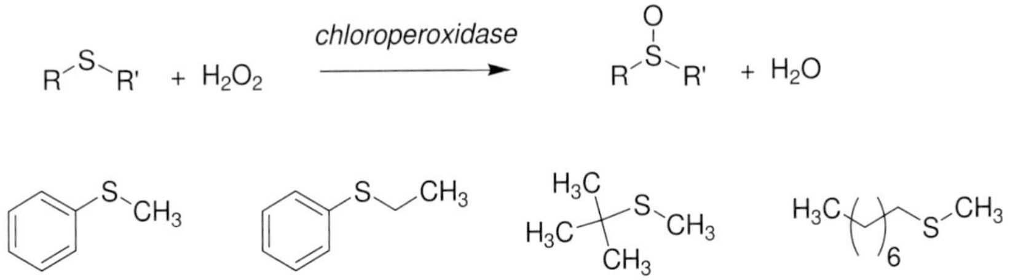

Fig. 4.112 Sulfoxidation reactions mediated by CPO.

Finally, but certainly not least, we note that the enzyme chloroperoxidase (CPO), catalyzes the highly enantioselective (>98% *ee*) sulfoxidation of a range of substituted thioanisoles [308]. In contrast to the epoxidation of alkenes, where turnover frequencies were low (see above), in the case of sulfoxidation of thioanisole a turnover frequency of around $16\ s^{-1}$ and a total turnover number of 125 000 could be observed. A selection of data is represented in Fig. 4.112. Besides aryl alkyl sulfides, also dialkylsulfides could be oxidized with reasonable enantioselectivities [27].

Another class of peroxidases which can perform asymmetric sulfoxidations, and which have the advantage of inherently higher stabilities because of their non-heme nature, are the vanadium peroxidases. It was shown that vanadium bromoperoxidase from *Ascophyllum nodosum* mediates the production of (*R*)-methyl phenyl sulfoxide with a high 91% enantiomeric excess from the corresponding sulfide with H_2O_2 [38]. The turnover frequency of the reaction was found to be around $1\ min^{-1}$. In addition this enzyme was found to catalyse the sulfoxidation of racemic, non-aromatic cyclic thioethers with high kinetic resolution [309].

4.7.4
Asymmetric Baeyer-Villiger Oxidation

The area of asymmetric catalytic Baeyer-Villiger oxidation was recently reviewed [310]. It was concluded that biocatalytic methods in this case would seem to have the edge with respect to enantioselectivity, regiochemistry, and functional group selectivity. Baeyer-Villiger monooxgenases (BVMOs) are versatile biocatalysts that have been widely used in synthetic biotransformations. The cyclohexanone monooxygenase (CHMO) from *Acinetobacter calcoaceticus* NCIMB is the best characterized and most studied of these enzymes [311]. They contain a flavin moiety as the prosthetic group and utilize NAD(P)H as a stoichiometric cofactor (see Fig. 4.113).

In view of the requirement for cofactor regeneration these reactions are generally performed with whole microbial cells in a fermentation mode. Degrada-

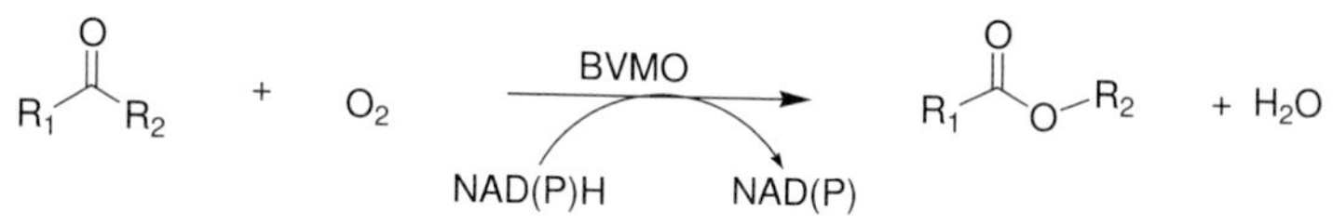

$$R_1 \overset{O}{\underset{}{\bigwedge}} R_2 \quad + \quad O_2 \quad \xrightarrow[\text{NAD(P)H} \quad \text{NAD(P)}]{\text{BVMO}} \quad R_1 \overset{O}{\underset{}{\bigwedge}} O{-}R_2 \quad + \quad H_2O$$

Fig. 4.113 BV reactions catalyzed by BVMOs.

tion of the lactone product can be circumvented by the addition of hydrolase in-hibitors and/or heterologous expression. Several CHMOs are now commercially available, e.g. from Fluka and they are easy to use, even for non-specialists. A disadvantage of the methodology is the low volume yield (concentrations are typically 10 mM compared to 1 M solutions in homogeneous catalysis). Further-more, many organic substrates have limited solubility in water. This problem could possibly be solved by performing the reactions in organic media [312]. The generally accepted mechanism for BV oxidations catalyzed by BVMOs was previously discussed in Section 4.2 (see Fig. 4.15). The state-of-the-art with re-gard to scale-up of BVMO catalyzed oxidations has been recently reviewed [313]. The reaction suffers from both substrate and product inhibition and the opti-mum ketone concentration was shown to be 0.2 to 0.4 g l^{-1} and at product con-centrations above 4.5 to 5 g l^{-1} the activity of the whole cell biocatalyst fell to zero. This suggests that a fed-batch operation with continuous product removal, by adsorption on a solid resin, is necessary in order to obtain reasonable volu-metric yields. In this way volumetric yields up to 20 g l^{-1} could be obtained.

In principle, the use of the isolated enzyme should allow the reaction to be performed at higher substrate concentrations, avoid side reactions and facilitate downstream processing. However, this requires an ancillary enzymatic cofactor regeneration system. Encouraging results have been obtained by coupling CHMO-catalyzed BV oxidation to cofactor regeneration mediated by an NADPH-dependent formate dehydrogenase (Fig. 4.114) [314, 315]. The overall reaction constitutes an enantioselective BV oxidation with stoichiometric con-sumption of O_2 and formate to give water and carbon dioxide as the coproducts. The reaction was carried to complete conversion with a 40 mM substrate con-centration and the enzymes separated by ultrafiltration and re-used.

The enzyme catalyzed hydrolysis of the lactone and subsequent oxidation are unwanted side-reactions, obviously. Several yeasts [316] and E. coli [317] have

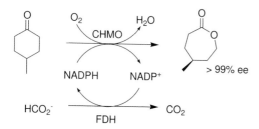

> 99% ee

Fig. 4.114 CHMO catalyzed BV reaction coupled to cofactor regeneration by formate dehydrogenase.

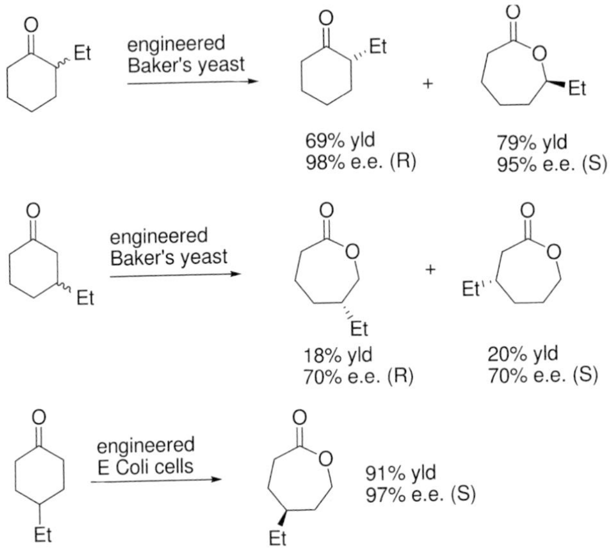

Fig. 4.115 Enzyme catalyzed asymmetric BV oxidation of 2-, 3- and 4-ethylcyclohexanones.

been engineered to convert the ketone to the lactone exclusively. Thus 2-substituted [316] and 4-substituted (or prochiral), [317, 318] cyclohexanones can be converted with high selectivity (Fig. 4.115) [319]. The (racemic) 3-substituted [320] cyclohexanones are synthetically less interesting as they can yield two regioisomeric lactones, each in two enantiomers. For small substituents there is little discrimination by the enzyme.

The first example of a dynamic kinetic resolution involving a CHMO-catalyzed BV oxidation was recently reported (Fig. 4.116) [321]. The reaction was performed with whole cells of CHMO-containing recombinant *E. coli* sp. at pH 9. Under these conditions the ketone substrate underwent facile racemization,

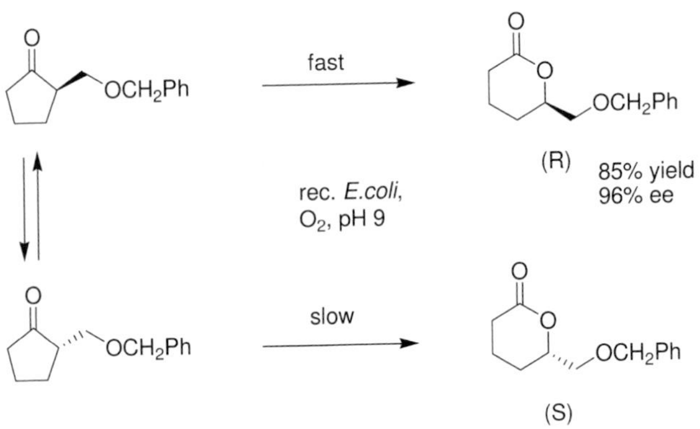

Fig. 4.116 Dynamic kinetic resolution of ketones using CHMO-containing recombinant *E. coli*.

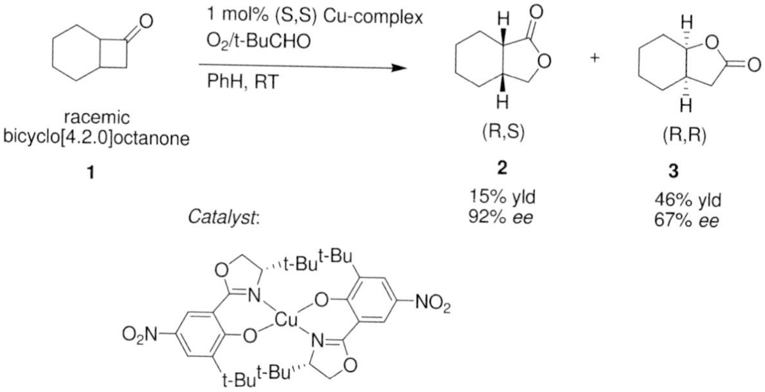

Fig. 4.117 Asymmetric copper-catalyzed BV oxidation of cyclobutanones.

via keto–enol tautomerism, and the lactone product was obtained in 85% yield and 96% *ee*.

Homogeneous catalysts can also be used for asymmetric BV oxidation [310]. One of the best catalysts appears to be 1 mol% of a chiral copper catalyst with oxazoline ligands (see Fig. 4.117), cyclobutanone derivatives are readily oxidized to give optically active lactones. The most promising results were obtained in re-actions using benzene solutions under an atmosphere of dioxygen (1 atm) with pivaldehyde as co-reductant at ambient temperature [322]. In general, 2-aryl sub-stituted cyclohexanones give good results. In the example of Fig. 4.117, metal catalyzed oxidation of rac-**1** gave isomeric γ-lactones **2** and **3** in a ratio of 3:1 (61% yield) with enantiomeric excesses of 67 and 92% respectively [323].

4.5
Conclusion

Summarizing, it is evident that catalytic oxidation is a mature technology which is still under strong development. Every day the performance of catalysts that use O_2 or H_2O_2 as green oxidants, increases. This is illustrated by two key areas, the oxidation of alcohols and alkenes. In the first case, the oxidation of al-cohols no longer requires stoichiometric use of heavily polluting metals. Many homogeneous and heterogeneous catalysts are available which use molecular oxygen as the oxidant. Especially promising seems to be the use of gold nano-particles as catalysts for alcohol and carbohydrate oxidation. The turnover fre-quencies obtained in this case overwhelm all results obtained so far. In the area of asymmetric epoxidation, the use of hydrogen peroxide now seems to give as good results as the more established methods using alkylhydroperoxide or hypo-chlorite as the oxidant.

When comparing homogeneous and heterogeneous catalysts, clearly homoge-neous systems take the lead. In addition, heterogeneous systems have the disadvantage that the leaching of metals in solution is always a possible complication in the reaction, and therefore they have to be subjected to a rigorous test of heterogeneity.

When comparing chemical and biocatalytic methods, one could say that, especially for asymmetric oxidations, enzymatic methods enter the scene. This is most evident in the area of asymmetric Baeyer-Villiger oxidation, where biocatalysts take the lead and homogeneous chiral catalysts lag far behind in terms of *ee* values. Significant progress can be expected in the area of biocatalysis due to the advancement in enzyme production technologies and the possibility of tailor-made enzymes.

Despite the huge effort put into research towards green and catalytic oxidation, the number of applications in the fine chemical industry is limited. For industrial batch processes simple, robust and reliable methods are required. For oxidations, this is in general less straightforward than for reductions. However, much is to be gained in this area, and therefore elegance should be a leading principle for all newly designed chemical oxidation processes.

References

1 R.A. Sheldon, J.K. Kochi, *Metal-Catalyzed Oxidations of Organic Compounds*, Academic Press, New York, 1981.

2 G. Franz, R.A. Sheldon, in *Ullmann's Encyclopedia of Industrial Chemistry*, 5th Edn, Vol. A18, VCH, Weinheim, 1991, p. 261.

3 R.A. Sheldon, R.A. van Santen (Eds.), *Catalytic Oxidation: Principles and Applications*, World Scientific, Singapore, 1995.

4 R.A. Sheldon, *CHEMTECH* 1991, *21*, 566; 1994, *24*, 38.

5 N.M. Shaw, K.T. Robins, A. Kiener, *Adv. Synth. Catal.* 2003, *345*, 425.

6 J.A. Howard, *Adv. Free-radical Chem.* 1972, *4*, 49.

7 P. Mars, D.W. van Krevelen, *Chem. Eng. Sci. Spec. Suppl.* 1954, *3*, 41.

8 T. Funabiki (Ed.), *Oxygenases and Model Systems*, Kluwer, Dordrecht, 1996.

9 R.A. Sheldon (Ed.), *Metalloporphyrins in Catalytic Oxidations*, Marcel Dekker, New York, 1994.

10 F. Montanari, L. Casella (Eds.), *Metalloporphyrin Catalyzed Oxidations*, Kluwer, Dordrecht, 1994.

11 M. Sono, M.P. Roach, E.D. Coulter, J.H. Dawson, *Chem. Rev.* 1996, *96*, 2841.

12 R.A. Sheldon, in *Catalytic Activation and Functionalisation of Light Alkanes*, E.G. Derouane et al. (Eds.), Kluwer, Dordrecht, 1998.

13 R.A. Sheldon, *Top. Curr. Chem.* 1993, *164*, 21; R.A. Sheldon, *Bull. Soc. Chim. Belg.* 1985, *94*, 651.

14 W. Adam, R. Curci, J.O. Edwards, *Acc. Chem. Res.* 1989, *22*, 205; W. Adam, A.K. Smerz, *Bull. Soc. Chim. Belg.* 1996, *105*, 581.

15 A.E.J. de Nooy, A.C. Besemer, H. van Bekkum, *Synthesis* 1996, *10*, 1153.

16 J.M. Bobbitt, M.C.L. Flores, *Heterocycles* 1988, *27*, 509.

17 T.J. Collins, *Acc. Chem. Res.* 1994, *27*, 279.

18 K. Drauz, H. Waldmann (Eds.), *Enzyme Catalysis in Organic Synthesis*, Vol. II, VCH, Weinheim, 1995.

19 M.A. McGuirl, D.M. Dooley, *Curr. Opin. Chem. Biol.*, 1999, *3*, 138.

20 D. Mansuy, P. Battioni, in *Bioinorganic Catalysis*, J. Reedijk (Ed.), Marcel Dekker, New York, 1993, p. 395.

21 P.R. Ortiz de Montellano (Ed.) *Cytochrome P450: Structure, mechanism & Biochemistry*, 2nd edn., Plenum, New York, 1995.

22 W.A. Duetz, J.B. van Beilen, B. Witholt, *Curr. Opin. Biotechnol.* **2001**, *12*, 419.

23 K. Faber, *Biotransformations in Organic Chemistry*, 5th edn. Springer, Berlin, 2004.

24 A. Schmid, K. Hofstetter, H.-J. Feiten, F. Hollmann, B. Witholt, *Adv. Synth. Catal.* **2001**, *343*, 732.

25 C.T. Walsh, Y.-C.J. Chen, *Angew. Chem. Int. Ed.* **1988**, *27*, 333.

26 S. Roberts, P.W.H. Wan, *J. Mol. Catal. B: Enzym.* **1998**, *4*, 111.

27 S. Colonna, N. Graggero, P. Pasta, G. Ottolina, *Chem. Commun.* **1996**, 2303.

28 G. Ottolina, S. Bianchi, B. Belloni, G. Carrea, B. Danieli, *Tetrahedron Lett.* **1999**, *40*, 8483.

29 A.L. Feig, S.J. Lippard, *Chem. Rev.* **1994**, *94*, 759.

30 M. Costas, M.P. Mehn, M.P. Jensen, L. Que Jr., *Chem. Rev.* **2004**, *104*, 939.

31 J. Green, H. Dalton, *J. Biol. Chem.* **1989**, *264*, 17698.

32 M. Merkx, D.A. Kopp, M.H. Sazinsky, J.L. Blazyk, J. Müller, S.J. Lippard, *Angew. Chem. Int. Ed.* **2001**, *40*, 4000; E.I. Solomon, T.C. Brunold, M.I. Davis, J.N. Kemsley, S.-K. Lee, N. Lehnert, F. Neese, A.J. Skulan, Y.-S. Yang, J. Zhou, *Chem. Rev.* **2000**, *100*, 235. For a more general overview of non-heme carboxylate-bridged diiron metalloproteins, see E.Y. Tshuva, S.J. Lippard, *Chem. Rev.* **2004**, *104*, 987.

33 M.D. Wolfe, J.D. Lipscomb, *J. Biol. Chem.* **2003**, *278*, 829.

34 K.D. Koehntop, J.P. Emerson, L. Que Jr. *J. Biol. Inorg. Chem.* **2005**, *10*, 87.

35 A. Decker, E.I. Solomon, *Curr Opin. Chem. Biol.* **2005**, *9*, 152.

36 J.P. Evans, K. Ahn, J.P. Klinman, *J. Biol. Chem.* **2003**, *278*, 29691.

37 F. van Rantwijk, R.A. Sheldon. *Curr. Opin. Biotechnol.* **2000**, *11*, 554.

38 H.B. ten Brink, A. Tuynman, H.L. Dekker, H.E. Schoenmaker, R. Wever, *Eur. J. Biochem.* **1998**, *258*, 906.

39 R.A. Sheldon, in *Applied Homogeneous Catalysis by Organometallic Compounds*, Vol. 1, B. Cornils, W.A. Herrmann (eds.), VCH, Weinheim, 1996, p. 411.

40 R.A. Sheldon, in *Aspects of Homogeneous Catalysis*, Vol. 4, R. Ugo (ed.), Reidel, Dordrecht, 1981, p. 3.

41 B. Notari, *Catal. Today* **1993**, *18*, 163; M.G. Clerici, P. Ingallina, *J. Catal.* **1993**, *140*, 71.

42 R.A. Sheldon, M. Wallau, I.W.C.E. Arends, U. Schuchardt, *Acc. Chem. Res.* **1998**, *31*, 485; R.A. Sheldon, M.C.A. van Vliet, in *Fine Chemicals through Heterogeneous Catalysis*, R.A. Sheldon, H. van Bekkum (eds.), Wiley-VCH, Weinheim, 2001, p. 473.

43 G. Venturello, E. Alneri, M. Ricci, *J. Org. Chem.* **1983**, *48*, 3831.

44 Y. Ishii, K. Yamawaki, T. Ura, H. Yamada, T. Yoshida, M. Ogawa, *J. Org. Chem.* **1988**, *53*, 3587.

45 For reviews see: I.V. Kozhevnikov, *Catal. Rev.* **1995**, *37*, 311; I.V. Kozhevnikov, *Chem. Rev.* **1998**, *98*, 171.

46 L. Salles, C. Aubry, R. Thouvenot, F. Robert, C. Doremieux-Morin, C. Chottard, H. Ledon, Y. Jeannin, J.M. Bregeault, *Inorg. Chem.* **1994**, *33*, 871.

47 K. Sato, M. Aoki, M. Ogawa, T. Hashimoto, R. Noyori, *J. Org. Chem.* **1996**, *61*, 8310; K. Sato, M. Aoki, M. Ogawa, T. Hashimoto, D. Panyella, R. Noyori, *Bull. Chem. Soc. Jpn.* **1997**, *70*, 905.

48 K. Sato, M. Aoki, R. Noyori, *Science* **1998**, *281*, 1646; see also: E. Antonelli, R. D'Aloisio, M. Gambaro, T. Fiorani, C. Venturello, *J. Org. Chem.* **1998**, *63*, 7190.

49 K. Kamata, K. Yonehara, Y. Sumida, K. Yamaguchi, S. Hikichi, N. Mizuno, *Science*, **2003**, *300*, 964.

50 X. Zuwei, Z. Ning, S. Yu, L. Kunlan, *Science*, **2001**, *292*, 1139.

51 D.E. De Vos, B.F. Sels, P.A. Jacobs, *Adv. Synth. Catal.* **2003**, *345*, 45, and references cited therein.

52 A.L. Villa de P., B.F. Sels, D.E. De Vos, P.A. Jacobs, *J. Org. Chem.* **1999**, *64*, 7267.

53 J. Ichihara, A. Kambara, K. Iteya, E. Sugimoto, T. Shinkawa, A. Takaoka, S. Yamaguchi, Y. Sasaki, *Green. Chem.*, **2003**, *5*, 491.

54 K. Yamaguchi, C. Yoshida, S. Uchida, N. Mizuno, *J. Am. Chem. Soc.* **2005**, *127*, 530.

55 B. Sels, D. De Vos, M. Buntinx, F. Pierard, A. Kirsch De Mesmaeker, P. Jacobs, *Nature* **1999**, *400*, 855; B. Sels, D. De Vos, P. Jacobs, *J. Am. Chem. Soc.* **2001**, *123*, 8350.

56 W. A. Herrmann, R. W. Fischer, D. W. Marz, *Angew. Chem. Int. Ed. Engl.* **1991**, *30*, 1638.

57 W. A. Herrmann, R. M. Kratzer, H. Ding, W. R. Thiel, H. Gras, *J. Organometal. Chem.* **1998**, *555*, 293.

58 C. Coperet, H. Adolfsson, K. B. Sharpless, *Chem. Commun.* **1997**, 1565; J. Rudolph, K. L. Reddy, J. P. Chiang, K. B. Sharpless, *J. Am. Chem. Soc.* **1997**, *119*, 6189; see also A. L. Villa de P., D. E. De Vos, C. Montes, P. A. Jacobs, *Tetrahedron Lett.* **1998**, *39*, 8521.

59 M. C. A. van Vliet, I. W. C. E. Arends, R. A. Sheldon, *Chem. Commun.* **1999**, 821.

60 M. C. A. van Vliet, I. W. C. E. Arends, R. A. Sheldon, *Synlett.* **2001**, 248.

61 K. Neimann, R. Neumann, *Org. Lett.* **2000**, *2*, 2861.

62 M. M. Abu-Omar, P. J. Hansen, J. H. Espenson, *J. Am. Chem. Soc.* **1996**, *118*, 4966.

63 The catalyst was introduced with other oxidants: H. Nishiyama, T. Shimada, H. Itoh, H. Sugiyama, Y. Motoyama, *Chem. Commun.* **1997**, 1863.

64 M. K. Tse, M. Klawonn, S. Bhor, C. Dobler, G. Anilkumar, H. Hugl, W. Magerlein, M. Beller, *Org. Lett.* **2005**, *7*, 987.

65 P. L. Anelli, S. Banfi, F. Montanari, S. Quici, *Chem. Commun.* **1989**, 779.

66 R. Hage, J. E. Iburg, J. Kerschner, J. H. Koek, E. L. M. Lempers, R. J. Martens, U. S. Racherla, S. W. Russell, T. Swarthoff, M. R. P. van Vliet, J. B. Warnaar, L. van der Wolf, B. Krijnen, *Nature* **1994**, *369*, 637.

67 D. E. De Vos, B. F. Sels, M. Reynaers, Y. V. Subba Rao, P. A. Jacobs, *Tetrahedron Lett.* **1998**, *39*, 3221; see also D. E. De Vos, T. Bein, *Chem. Commun.* **1996**, 917.

68 B. C. Gilbert, N. W. J. Kamp, J. R. Lindsay Smith, J. Oakes, *J. Chem. Soc., Perkin Trans. 2*, **1998**, 1841.

69 B. S. Lane, K. Burgess, *J. Am. Chem. Soc.* **2001**, *123*, 2933; B. S. Lane, M. Vogt, V. J. DeRose, K. Burgess, *J. Am. Chem. Soc.* **2002**, *124*, 11946.

70 K.-H. Tong, K.-Y. Wong, T. H. Chan, *Org. Lett.* **2003**, *5*, 3423.

71 A. Murphy, A. Pace, T. D. P. Stack, *Org. Lett.* **2004**, *6*, 3119.

72 K. Chen, L. Que, Jr., *Chem. Commun.* **1999**, 1375.

73 M. C. White, A. G. Doyle, E. N. Jacobsen, *J. Am. Chem. Soc.* **2001**, *123*, 7194.

74 M. Costas, A. K. Tipton, K. Chen, D. H. Jo, L. Que, Jr., *J. Am. Chem. Soc.* **2001**, *123*, 6722.

75 G. Dubois, A. Murphy, T. D. Stack, *Org. Lett.* **2003**, *5*, 2469.

76 M. C. A. van Vliet, I. W. C. E. Arends, R. A. Sheldon, *Chem. Commun.* **1999**, 263.

77 L. Syper, *Synthesis* **1989**, 167; B. Betzemeier, F. Lhermitte, P. Knochel, *Synlett.* **1999**, 489.

78 G. J. ten Brink, B. C. M. Fernandez, M. C. A. van Vliet, I. W. C. E. Arends, R. A. Sheldon, *J. Chem. Soc., Perkin Trans 1* **2001**, 224.

79 W. T. Reichle, S. Y. Kang, D. S. Everhardt, *J. Catal.* **1986**, *101*, 352.

80 K. Yamaguchi, K. Ebitani, K. Kaneda, *J. Org. Chem.* **1999**, *64*, 2966; S. Ueno, K. Yamaguchi, K. Yoshida, K. Ebitani, K. Kaneda, *Chem. Commun.* **1998**, 295.

81 M. C. A. van Vliet, D. Mandelli, I. W. C. E. Arends, U. Schuchardt, R. A. Sheldon,- *Green. Chem.* **2001**, *3*, 243; D. Mandelli, M. C. A. van Vliet, R. A. Sheldon U. Schuchardt, *Appl. Catal. A* **2001**, *219*, 209.

82 M. Schroder, *Chem. Rev.* **1980**, *80*, 187.

83 H. C. Kolb, K. B. Sharpless in *Transition Metals for Organic Synthesis*, Vol. 2, M. Beller, C. Bolm (eds.), Wiley-VCH, Weinheim, 1998, p. 219; M. Beller, K. B. Sharpless, in *Applied Homogeneous Catalysis with Organometallic Compounds*, Vol. 2, B. Cornils, W. A. Herrmann (eds.), VCH, Weinheim, 1996, p. 1009.

84 P. Dupau, R. Epple, A. A. Thomas, V. V. Fokin, K. B. Sharpless, *Adv. Synth. Catal.* **2002**, 344.

85 C. Döbler, G. Mehltretter, M. Beller, *Angew. Chem. Int. Ed. Engl.* **1999**, *38*, 3026; C. Döbler, G. M. Mehltretter, U. Sundermeier, M. Beller, *J. Am. Chem. Soc.* **2000**, *122*, 10289.

86 Y. Usui, K. Sato, M. Tanaka, *Angew. Chem. Int. Ed.* **2003**, *42*, 5623.

87 L. M. Berkowicz, P. N. Rylander, *J. Am. Chem. Soc.* **1958**, *75*, 3838.

88 P. H. J. Carlsen, T. Katsuki, V. S. Martin, K. B. Sharpless, *J Org. Chem.* **1981**, *46*, 3936.

89 D. Yang, C. Zhang, *J Org Chem.* **2001**, *66*, 4814.

90 S. Wolf, S. K. Hasan, J. R. Campbell, *J. Chem. Soc., Chem. Commun.* **1970**, 1420.

91 C.-M. Ho, W.-Y. Yu, C.-M. Che, *Angew. Chem. Int. Ed.* **2004**, *43*, 3303.

92 W. P. Griffith, E. Kong, *Synth. Commun.* **2003**, *33*, 2945.

93 S. Warwel, M. Sojka, M. Rüsch gen. Klaas, *Top. Curr. Chem.* **1993**, *164*, 81.

94 A. Johnstone, P. J. Middleton, W. R. Sanderson, M. Service, P. R. Harrison, *Stud. Surf. Sci. Catal.* **1994**, *82*, 609.

95 W. K. Wong, X. P. Chen, W. X. Pan, J. P. Guo, W. Y. Wong, *Eur. J. Inorg. Chem.* **2002**, 231.

96 E. Antonelli, R. D'Aloisio, M. Gambaro, T. Fiorani, C. Venurello, *J. Org. Chem.* **1998**, *63*, 7190.

97 R. Jira, in *Applied Homogeneous Catalysis by Organometallic Compounds*, Vol. 1, B. Cornils, W. A. Herrmann (eds.), VCH, Weinheim, 1996, p. 374.

98 I. A. Ansari, S. Joyasawal, M. K. Gupta, J. S. Yadav, R. Gree, *Tetrahedron Lett.* **2005**, *46*, 7507.

99 Z. Hou, B. Han, L. Gao, T. Jiang, Z. Liu, Y. Chang, X. Zhang, J. He, *New. J. Chem.* **2002**, *26*, 1246.

100 T. Mitsudome, T. Umetani, K. Mori, T. Mizugaki, K. Ebitani, K. Kaneda, *Tetrahedron Lett.* **2006**, *47*, 1425.

101 G. J. ten Brink, I. W. C. E. Arends, G. Papadogianakis, R. A. Sheldon, *Chem. Commun.* **1998**, 2359.

102 T. Mitsudome, T. Umetani, N. Nosaka, K. Mori, T. Mizugaki, K. Ebitani, K. Kaneda, *Angew. Chem. Int. Ed.* **2006**, *45*, 481.

103 H. Grennberg, J.-E Bäckvall in *Transition metals for Organic Synthesis*, M. Beller, C. Bolm (eds.), Wiley-VCH, Weinheim, Germany, 1998, p. 200.

104 M. Otake, *CHEMTECH*, **1995**, *25*, 36.

105 M. de Fatima Teixera Gomes, O. A. C. Antunes, *J. Mol. Catal. A: Chemical* **1997**, *121*, 145.

106 M. A. Umbreit, K. B. Sharpless, *J. Am. Chem. Soc.* **1977**, *99*, 5526.

107 D. Crich, Y. Zhou, *Org. Lett.* **2004**, *6*, 775.

108 D. J. Rawlinson, G. Sosnovsky, *Synthesis* **1972**, 1.

109 J. Le Bras, J. Muzart, *Tetrahedron Lett.* **2002**, *43*, 431.

110 F. Fache, O. Piva, *Synlett.* **2002**, 2035.

111 M. S. Kharash, G. Sosnovsky, N. C. Yang, *J. Am. Chem. Soc.* 1959, *81*, 5819.

112 J. Eames, M. Watkinson, *Angew. Chem. Int. Ed.* **2001**, *40*, 3567.

113 S. K. Ginotra, V. K. Singh, Tetrahedron **2006**, *62*, 3573.

114 I. I. Moiseev, M. N. Vargaftik, *Coord. Chem. Rev.* **2004**, *248*, 2381.

115 A. K. El-Qisiari, H. A. Qaseer, P. M. Henry, *Tetrahedron Lett.* **2002**, *43*, 4229.

116 A. E. Shilov, G. B. Shul'pin, *Chem. Rev.* **1997**, *97*, 2879.

117 K. Tanaka, *CHEMTECH*, **1974**, *4*, 555.

118 B. Meunier et al., in *Biomimetic Oxidations by Transition Metal Complexes*, B. Meunier (ed.), Imperial College Press, London, 1999.

119 D. H. R. Barton, D. Doller, *Acc. Chem. Res.* **1992**, *25*, 504.

120 F. Minisci, F. Fontana, *Chim. Ind. (Milan)* **1998**, *80*, 1309.

121 A. N. Bashkirov et al., in *The Oxidation of Hydrocarbons in the Liquid Phase*, N. M. Emanuel (ed.), Pergamon, Oxford, 1965, p. 183.

122 T. Naota, H. Takaya, S.-I. Murahashi, *Chem. Rev.* **1998**, *98*, 2599.

123 A. S. Goldstein, R. S. Drago, *J. Chem. Soc., Chem. Commun.* **1991**, 21.

124 R. A. Periana et al., in *Catalytic Activation and Functionalization of Light Alkanes*, E. G. Derouane, J. Haber, F. Lemos, F. R. Ribeiro, M. Guisnet (eds.), Kluwer Academic Publishers, Dordrecht, The Netherlands, 1998, p. 297.

125 Y. Yoshino, Y. Hayashi, T. Iwahama, S. Sakaguchi, Y. Ishii, *J. Org. Chem.* **1997**, *62*, 6810.

126 T. Iwahama, K. Syojyo, S. Sakaguchi, Y. Ishii, *Org. Proc. Res. Dev.* **1998**, *2*, 255.

127 X. Baucherel, I. W. C. E. Arends, S. Ell-wood, R. A. Sheldon, *Org. Proc. Res. Dev.* **2003**, *7*, 426.

128 A. S. Hay, H. S. Blanchard, *Can. J. Chem.* **1965**, *43*, 1306; Y. Kamiya, *J. Catal.* **1974**, *33*, 480.

129 W. Partenheimer, *Catal. Today* **1995**, *23*, 69.

130 W. Partenheimer, in *Catalysis of Organic Reactions*, D. W. Blackburn (ed.), Marcel Dekker, New York, 1990, p. 321.

131 J. Yamashita, S. Ishikawa, H. Hashimo-to. ACS/CSJ Chem. Congr., 1979; *Org. Chem. Div.* paper no. 76 (1979).

132 W. Partenheimer, R. K. Gipe, in *Catalytic Selective Oxidation*, S. T. Oyama, J. W. Hightower (eds.), *ACS Symp. Ser.* **1993**, *523*, 81.

133 Eur. Pat. Appl. 0300921 & 0300922 (1988) J. Dakka, A. Zoran, Y. Sasson, (to Gadot Petrochemical Industries).

134 J. W. Wong et al. *Org. Proc. Res. Dev.* **2002**, *6*, 477.

135 D. Bianchi, R. Bortoli, R. Tassinari, M. Ricci, R. Vignola, *Angew. Chem. Int. Ed.* **2000**, *39*, 4321; D. Bianchi, M. Bertoli, R. Tassinari, M. Ricci, R. Vignola, *J. Mol. Catal. A: Chemical* **2003**, *200*, 111.

136 J. Peng, F. Shi, Y. Gu, Y. Deng, *Green Chem.* **2003**, *5*, 224.

137 M. G. Clerici, in *Fine Chemicals through Hetrogeneous Catalysis*, R. A. Sheldon, H. van Bekkum (eds.), Wiley-VCH, Weinheim, 2001, p. 541.

138 G. I. Panov, CATTECH, **2000**, *4*, 18; A. A. Ivanov, V. S. Chernyavsky, M. J. Gross, A. S. Kharitonov, A. K. Uriarte, G. I. Panov, *Appl. Catal. A-General* **2003**, *249*, 327.

139 S. V. Ley, J. Norman, W. P. Griffith, S. P. Marsden, *Synthesis* **1994**, 639.

140 M. Hudlicky, *Oxidations in Organic Chemistry*, ACS, Washington DC, 1990.

141 G. Cainelli, G. Cardillo, *Chromium Oxidations in Organic Chemistry*, Springer, Berlin, 1984.

142 R. A. Sheldon, I. W. C. E. Arends, A. Dijksman, *Catal. Today* **2000**, *57*, 158.

143 I. W. C. E. Arends, R. A. Sheldon, in *Modern Oxidation Methods*, J. Bäckvall (ed.), Wiley-VCH, Weinheim, 2004, p. 83.

144 T. Mallat, A. Baiker, *Chem. Rev.* **2004**, *104*, 3037.

145 A. Wolfson, S. Wuyts, D. E. de Vos, I. F. J. Vancelecom, P. A. Jacobs, *Tetrahedron Lett.* **2002**, *43*, 8107.

146 J. E. Bäckvall, R. L. Chowdhury, U. Karlsson, *J. Chem. Soc., Chem. Commun.* **1991**, 473; G.-Z. Wang, U. Andreasson, J. E. Bäckvall, *J. Chem. Soc., Chem. Commun.* **1994**, 1037; G. Csjernyik, A. Ell, L. Fadini, B. Pugin, J. E. Backvall, *J. Org. Chem.* **2002**, *67*, 1657.

147 A. Hanyu, E. Takezawa, S. Sakaguchi, Y. Ishii, *Tetrahedron Lett.* **1998**, *39*, 5557.

148 A. Dijksman, I. W. C. E. Arends, R. A. Sheldon, *Chem. Commun.* **1999**, 1591; A. Dijksman, A. Marino-González, A. Mairata i Payeras, I. W. C. E. Arends, R. A. Sheldon, *J. Am. Chem. Soc.* **2001**, *123*, 6826. for a related study see T. Inokuchi, K. Nakagawa, S. Torii, *Tetrahedron Lett.* **1995**, *36*, 3223.

149 W. P. Griffith, S. V. Ley, *Aldrichim. Acta* **1990**, *23*, 13.

150 R. Lenz, S. V. Ley, *J. Chem. Soc., Perkin Trans. 1* **1997**, 3291.

151 I. E. Marko, P. R. Giles, M. Tsukazaki, I. Chelle-Regnaut, C. J. Urch, S. M. Brown, *J. Am. Chem. Soc.* **1997**, *119*, 12661.

152 B. Hinzen, R. Lenz, S. V. Ley, *Synthesis* **1998**, 977.

153 A. Bleloch, B. F. G. Johnson, S. V. Ley, A. J. Price, D. S. Shepard, A. N. Thomas, *Chem. Commun.* **1999**, 1907.

154 M. Pagliaro, R. Ciriminna, *Tetrahedron Lett.* **2001**, *42*, 4511.

155 K. Yamaguchi, N. Mizuno, *Angew. Chem. Int. Ed.* **2002**, *41*, 4538.

156 K. Kaneda, T. Yamashita, T. Matsushita, K. Ebitani, *J. Org. Chem.* **1998**, *63*, 1750.

157 T. Matsushita, K. Ebitani, K. Kaneda, *Chem. Commun.* **1999**, 265.

158 K. Yamaguchi, K. Mori, T. Mizugaki, K. Ebitani, K. Kaneda, *J. Am. Chem. Soc.* **2000**, *122*, 7144.

159 H. B. Ji, K. Ebitani, T. Mizugaki, K. Kaneda, *Catal. Commun.* **2002**, *3*, 511.

160 F. Vocanson, Y. P. Guo, J. L. Namy, H. B. Kagan, *Synth. Commun.* **1998**, *28*, 2577.

161 R. Neumann, *Prog. Inorg. Chem.* **1998**, *47*, 317; C. L. Hill, C. M. Prosser-McCartha, *Coord. Chem. Rev.* **1995**, *143*, 407; M. T. Pope, A. Müller, *Angew. Chem. Int. Ed. Engl.* **1991**, *30*, 34.

162 K. Yamaguchi, N. Mizuno, *New. J. Chem.* **2002**, *26*, 972.

163 S. S. Stahl, *Science* **2005**, *309*, 1824; S. S. Stahl, *Angew. Chem. Int. Ed.* **2004**, *43*, 3400; J. Muzart, Tetrahedron **2003**, *59*, 5789;

164 K. P. Peterson, R. C. J. Larock, *Org. Chem.* **1998**, *63*, 3185.

165 R. A. T. M. van Benthem, H. Hiemstra, P. W. N. M. van Leeuwen, J. W. Geus, W. N. Speckamp, *Angew. Chem. Int. Ed.* **1995**, *34*, 457.

166 K. R. Seddon, A. Stark, *Green Chem.* **2002**, *4*, 119.

167 T. Nishimura, T. Onoue, K. Ohe, S. Uemura, *Tetrahedron Lett.* **1998**, *39*, 6011.

168 T. Nishimura, T. Onoue, K. Ohe, S. J. Uemura, *J. Org. Chem.* **1999**, *64*, 6750; T. Nishimura, K. Ohe, S. J. Uemura, *J. Am. Chem. Soc.* **1999**, *121*, 2645.

169 M. J. Schultz, C. C. Park, M. S. Sigman, *Chem. Commun.* **2002**, 3034.

170 K. Hallman, C. Moberg, *Adv. Synth. Catal.* **2001**, *343*, 260.

171 T. Iwasawa, M. Tokunaga, Y. Obora, Y. Tsuji, *J. Am. Chem. Soc.* **2004**, *126*, 6554.

172 G.-J. ten Brink, I. W. C. E. Arends, R. A. Sheldon, *Science* **2000**, *287*, 1636.

173 G. J. ten Brink, I. W.C. E. Arends, R. A. Sheldon, *Adv. Synth. Catal.* **2002**, *344*, 355.

174 G. J. ten Brink I. W. C. E. Arends, M. Hoogenraad, G. Verspui, R. A. Sheldon, *Adv. Synth. Catal.* **2003**, *345*, 1341.

175 For an example in toluene see C. Keresszegi, T. Burgi, T. Mallat, A. Baiker, *J. Catal.* **2002**, *211*, 244.

176 T. Nishimura, N. Kakiuchi, M. Inoue, S. Uemura, *Chem. Commun.* **2000**, 1245; see also N. Kakiuchi, T. Nishimura, M. Inoue, S. Uemura, *Bull. Chem. Soc. Jpn.* **2001**, *74*, 165.

177 K. Mori, K. Yamaguchi, T. Hara, T. Mizugaki, K. Ebitani, K. Kaneda, *J. Am. Chem. Soc.* **2002**, *124*, 11573.

178 D. R. Jensen, M. J. Schultz, J. A. Mueller, M. S. Sigman, *Angew. Chem. Int. Ed.* **2003**, *42*, 3810.

179 M. N. Vargaftik, V. P. Zagorodnikov, I. P. Storarov, I. I. Moiseev, *J. Mol. Catal.* **1989**, *53*, 315; see also I. I. Moiseev, M. N. Vargaftik, in *Catalysis by Di- and Polynuclear Metal Cluster Complexes*, R. D. Adams, F. A. Cotton (eds.), Wiley-VCH, Weinheim, 1998, p. 395.

180 K. Kaneda, M. Fujii, K. Morioka, *J. Org. Chem.* **1996**, *61*, 4502; K. Kaneda, Y. Fujie, K. Ebitani, *Tetrahedron Lett.* **1997**, *38*, 9023.

181 M. S. Kwon, N. Kim, C. M. Park, J. S. Lee, K. Y. Kang, J. Park, *Org. Lett.* **2005**, *7*, 1077.

182 Y. Uozumi, R. Nakao, *Angew. Chem. Int. Ed.* **2003**, *42*, 194.

183 Z. Hou, N. Theyssen, A. Brinkmann, W. Leitner, *Angew. Chem. Int. Ed.* **2005**, *44*, 1346.

184 A. Abad, P. Concepcion, A. Corma, H. Garcia, *Angew. Chem. Int. Ed.* **2005**, *44*, 4066.

185 H. Tsunoyama, H. Sakurai, Y. Negishi, T. Tsukuda, *J. Am. Chem. Soc.* **2005**, *127*, 9374.

186 D. I. Enache, J. K. Edwards, P. Landon, B. Solsona-Espriu, A. F. Carley, A. A. Herzing, M. Watanabe, C. J. Kiely, D. W. Knight, G. J. Hutchings, *Science* **2006**, *311*, 362.

187 S. Carrettin, P. McMorn, P. Johnston, K. Griffin, G. J. Hutchings, *Chem. Commun.* **2002**, 696.

188 N. Ito, S. E. V. Phillips, C. Stevens, Z. B. Ogel, M. J. McPherson, J. N. Keen, K. D. S. Yadav, P. F. Knowles, *Nature* **1991**, *350*, 87.

189 K. Drauz, H. Waldmann, *Enzyme Catalysis in Organic Synthesis*, VCH, Weinheim, 1995, Ch. 6.

190 Y. Wang, J. L. DuBois, B. Hedman, K. O. Hodgson, T. D. P. Stack, *Science*, **1998**, *279*, 537.

191 P. Chauhuri, M. Hess, U. Flörke, K. Wieghardt, *Angew. Chem. Int. Ed.*, **1998**, *37*, 2217.

192 P. Chauhuri, M. Hess, T. Weyhermüller, K. Wieghardt, *Angew. Chem. Int. Ed.*, **1998**, *38*, 1095.

193 V. Mahadevan, R. J. M. Klein Gebbink, T. D. P. Stack, *Curr. Opin. Chem. Biol.*, **2000**, *4*, 228.

194 I. W. C. E. Arends, P. Gamez, R. A. Sheldon in *Biomimetic Oxidation Catalysts*, Topics in Inorganic Chemistry, R. v. Eldik, J. Reedijk (eds.), Springer, Hamburg, **2005**.

195 I. E Marko, P. R. Giles, M. Tsukazaki, S. M. Brown, C. J. Urch, *Science*, **1996**, *274*, 2044; I. E. Marko, M. Tsukazaki, P. R. Giles, S. M. Brown, C. J. Urch, *Angew. Chem. Int. Ed. Engl.*, **1997**, *36*, 2208.

196 I. E. Marko, P. R. Giles, M. Tsukazaki, I. Chellé-Regnaut, A. Gautier, S. M. Brown, C. J. Urch, *J. Org. Chem.*, **1999**, *64*, 2433.

197 I. E. Marko, A. Gautier, I. Chellé-Regnaut, P. R. Giles, M. Tsukazaki, C. J. Urch, S. M. Brown, *J. Org. Chem.*, **1998**, *63*, 7576.

198 I. E. Marko, A. Gautier, J. L. Mutonkole, R. Dumeunier, A. Ates, C. J. Urch, S. M. Brown, *J. Organomet. Chem.* **2001**, *624*, 344.

199 I. E. Marko, A. Gautier, R. Dumeunier, K. Doda, F. Philippart, S. M. Brown, C. J. Urch, *Angew. Chem. Int. Ed.* **2004**, *43*, 1588.

200 M. F. Semmelhack, C. R. Schmid, D. A. Cortés, C. S. Chou, *J. Am. Chem. Soc.*, **1984**, *106*, 3374.

201 A. Dijksman, I. W. C. E. Arends, R. A. Sheldon, *Org. Biomol. Chem.* **2003**, *1*, 3232

202 B. Betzemeier, M. Cavazzine, S. Quici, P. Knochel, *Tetrahedron Lett.* **2000**, *41*, 4343.

203 I. A. Ansari, R. Gree, *Org. Lett.* **2002**, *4*, 1507.

204 P. Gamez, I. W. C. E. Arends, J. Reedijk, R. A. Sheldon, *Chem. Commun.* **2003**, 2414; P. Gamez, I. W. C. E. Arends, R. A. Sheldon, J. Reedijk, *Adv. Synth. Catal.* **2004**, *346*, 805.

205 T. Iwahama, S. Sakaguchi, Y. Nishiyama, Y. Ishii, *Tetrahedron Lett.* **1995**, *36*, 6923.

206 Y. Yoshino, Y. Hanyashi, T. Iwahama, S. Sakaguchi, Y. Ishii, *J. Org. Chem.*, **1997**, *62*, 6810; S. Kato, T. Iwahama, S. Sakaguchi, Y. Ishii, *J. Org. Chem.*, **1998**, *63*, 222; S. Sakaguchi, S. Kato, T. Iwahama, Y. Ishii, *Bull. Chem. Soc. Jpn.* **1988**, *71*, 1.

207 see also F. Minisci, C. Punta, F. Recupero, F. Fontana, G. F. Pedulli, *Chem. Commun.* **2002**, *7*, 688.

208 C. Döbler, G. Mehltretter, M. Beller, *Angew. Chem. Int. Ed.* **1999**, *38*, 3026; C. Döbler, G. Mehltretter, G. M. Sundermeier, M. J. Beller, *J. Am. Chem. Soc.* **2000**, *122*, 10289.

209 C. Döbler, G. M. Mehltretter, U. Sundermeier, M. Eckert, H-C. Militzer, M. Beller, *Tetrahedron Lett.* **2001**, *42*, 8447.

210 R. Anderson, K. Griffin, P. Johnston, P. L. Alsters, *Adv. Synth. Catal.* **2003**, *345*, 517.

211 Y.-C. Son, V. D. Makwana, A. R. Howell, S. L. Suib, *Angew. Chem. Int. Ed.* **2001**, *40*, 4280.

212 S. R. Reddy, S. Das, T. Punniyamurthy, *Tetrahedron Lett.* **2004**, *45*, 3561; S. Velusamy, T. Punniyamurthy, *Org. Lett.* **2004**, *6*, 217.

213 K. Sato, M. Aoki, J. Takagi, R. Noyori, *J. Am. Chem. Soc.* **1997**, *119*, 12386.

214 K. Sato, J. Takagi, M. Aoki, R. Noyori, *Tetrahedron Lett.* **1998**, *39*, 7549.

215 I. W. C. E. Arends, R. A. Sheldon, M. Wallau, U. Schuchardt, *Angew. Chem. Int. Ed. Engl.*, **1997**, *36*, 1144.

216 F. Maspero, U. Romano, *J. Catal.* **1994**, *146*, 476.

217 P. L. Anelli, C. Biffi, F. Montanari, S. Quici, *J. Org. Chem.* **1987**, *52*, 2559.

218 C. Bolm, T. Fey, *Chem. Commun.* **1999**, 1795; D. Brunel, F. Fajula, J. B. Nagy, B. Deroide, M. J. Verhoef, L. Veum, J. A. Peters, H. van Bekkum, *Appl. Catal. A: General* **2001**, *213*, 73.

219 D. Brunel, P. Lentz, P. Sutra, B. Deroide, F. Fajula, J. B. Nagy, *Stud. Surf. Sci. Catal.* **1999**, *125*, 237.

220 M. J. Verhoef, J. A. Peters, H. van Bekkum, *Stud. Surf. Sci. Catal.* **1999**, *125*, 465.

221 R. Ciriminna, J. Blum, D. Avnir, M. Pagliaro, *Chem. Commun.* **2000**, 1441.

222 A. Dijksman, I. W. C. E. Arends, R. A. Sheldon, *Chem. Commun.* **2000**, 271.

223 R. B. Daniel, P. Alsters, R. Neumann, *J. Org. Chem.* **2001**, *66*, 8650.

224 A. Cecchetto, F. Fontana, F. Minisci, F. Recupero, *Tetrahedron Lett.* **2001**, *42*, 6651.

225 M. Fabbrini, C. Galli, P. Gentilli, D. Macchitella, *Tetrahedron Lett.* **2001**, *42*, 7551; F. d'Acunzo, P. Baiocco, M. Fabbrini, C. Galli, P. Gentili, *Eur. J. Org. Chem.* **1995**, 4195.

226 I. W. C. E. Arends, Y.-X. Li, R. Ausan, R. A. Sheldon, *Tetrahedron*, **2006**, *62*, 6659.

227 W. Kroutil, H. Mang, K. Edegger, K. Faber, *Adv. Synth. Catal.* **2004**, *346*, 125.

228 W. Stampfer, B. Kosjek, C. Moitzi, W. Kroutil, K. Faber, *Angew. Chem. Int. Ed. Engl.* **2002**, *41*, 1014; W. Stampfer, B. Kosjek, K. Faber, W. Kroutil, *J. Org. Chem.* **2003**, *68*, 402; K. Edegger, H. Mang, K. Faber, J. Gross, W. Kroutil, *J. Mol. Catal. A: Chemical* **2006**, *251*, 66.

229 J. M. Brégeault, B. El Ali, J. Mercier, C. Martin, *C. R. Acad. Sci. Ser. 2*, **1989**, *309*, 459.

230 T. Mallat, A. Baiker, *Catal. Today* **1994**, *19*, 247; T. M. Besson, P. Gallezot in *Fine Chemicals through Heterogeneous Catalysis*, R. A. Sheldon, H. van Bekkum (eds.), Wiley-VCH, Weinheim, 2001, p. 507.

231 C. H. H. Emons, B. F. M. Kuster, J. A. J. M. Vekemans, R. A. Sheldon, *Tetrahedron Asymm.* **1991**, *2*, 359.

232 *Ullmann's Encyclopedia of Industrial Chemistry*, VI edn., Vol. 15, p. 645, Wiley-VCH, Weinheim, 2003.

233 M. Comotti, C. D. Pina, R. Matarrese, M. Rossi, *Angew. Chem. Int. Ed.* **2004**, *43*, 5812.

234 J. A. John, F. J. Weymouth, *Chem. Ind. (London)* **1962**, 62.

235 G. Benson, *Chem. Metall. Eng.* **1962**, *47*, 150.

236 G. C. Allen, A. Aguilo, *Adv. Chem. Series* **1968**, *76*, 363.

237 M. Besson, P. Gallezot, *Catal. Today*, **2000**, *57*, 127.

238 A. Corma, M. E. Domine, *Chem. Commun.* **2005**, 4042.

239 R. Criegee, *Justus Liebigs Ann. Chem.* **1948**, *560*, 127.

240 M. Shiozaki, N. Ishida, H. Maruyama, T. Hiraoka, *Tetrahedron* **1983**, *39*, 2399.

241 T. Mukaiyama in *The Activation of Dioxygen and Homogeneous Catalytic Oxidation*, D. H. R. Barton, A. E. Bartell, D. T. Sawyer (eds.), Plenum, New York, 1993, p. 133.

242 M. Abu-Omar, J. H. Espenson, *Organometallics* **1996**, *15*, 3543.

243 R. Bernini, E. Mincione, M. Cortese, G. Aliotta, A. Oliva, R. Saladino, *Tetrahedron Lett.* **2001**, *42*, 5401.

244 A. M. F. Phillips, C. Romao, *Eur. J. Org. Chem.* **1999**, 1767.

245 X. Hao, O. Yamazaki, A. Yoshida, J. Nishikido, *Green Chem.* **2003**, *5*, 525.

246 M. Brzaszcz, K. Kloc, M. Maposah, J. Mlochowski, *Synth. Commun.* **2000**, *30*, 4425; J. Mlochowski, S. B. Said, *Polish J. Chem.* **1997**, *71*, 149.

247 B. Betzemeier, F. Lhermitte, P. Knochel, *Synlett* **1999**, 489; B. Betzemeier, P. Knochel, *Deutsche Forschungs Gemeinschaft – Peroxide Chemistry (mechanistic and preparative aspects of oxygen transfer)*, Ch. 8, Wiley-VCH, Weinheim, 2000, Ch. 8, p. 454.

248 G.-J ten Brink, J.-M. Vis, I. W. C. E. Arends, R. A. Sheldon, *Tetrahedron*, **2002**, *58*, 3977.

249 K. Sato, M. Hyodo, J. Takagi, M. Aoki, R. Noyori, *Tetrahedron Lett.* **2000**, *41*, 1439.

250 A. Corma, L. T. Nemeth, M. Renz, S. Valencia, *Nature* **2001**, *412*, 423.

251 M. Renz, T. Blasco, A. Corma, V. Fornés, R. Jensen, L. Nemeth, *Chem. Eur. J.* **2002**, *8*, 4708.

252 O. A. Kholdeeva, A. V. Golovin, R. I. Maksimovskaya, I. V. Kozhevnikov, *J. Mol. Catal.* **1992**, *75*, 235.

253 R. A. Sheldon, J. K. Kochi, *Metal-Catalyzed Oxidations of Organic Compounds*, Ch. 12-VI: Phenols, Academic Press, New York, 1981, p. 368.

254 C. Mercier, P. Chabardes, *Pure Appl. Chem.* **1994**, *66*, 1509.

255 B. Notari, *Stud. Surf. Sci. Catal.* **1988**, *37*, 413.

256 A. B. Smith, R. M. Scarborough Jr., *Synth. Commun.* **1980**, 205.

257 I. T. Harrison, S. Harrison, *J. Chem. Soc., Chem. Commun.* **1966**, 752.

258 H. J. Schmidt, H. J. Schafer, *Angew. Chem. Int. Ed. Engl.* **1979**, *18*, 69.

259 Y. Ishii, K. Nakayama, M. Takeno, S. Sakaguchi, T. Iwahama, Y. Nishiyama, *J. Org. Chem.* **1995**, *60*, 3934.

260 L. Gonsalvi, I. W. C. E. Arends, P. K. Moilanen, R. A. Sheldon, *Adv. Synth. Catal.* **2003**, *345*, 1321.

261 Z. Zhu, J. H. Espenson, *J. Org. Chem.* **1995**, *60*, 1326.

262 S. Yamazaki, *Bull. Chem. Soc. Jpn.* **1997**, *70*, 877.

263 R. W. Murray, K. Iyanar, J. Chen, J. T. Wearing, *Tetrahedron Lett.* **1996**, *37*, 805.

264 S.-I. Murahashi, *Angew. Chem. Int. Ed. Engl.* **1995**, *34*, 2443.

265 C. Copéret, H. Adolfsson, T.-A. V. Khuong, A. K. Yudin, K. B. Sharpless, *J. Org. Chem.* **1998**, *63*, 1740.

266 K. Bergstad, J. E. Bäckvall, *J. Org. Chem.* **1998**, *63*, 6650.

267 Y.-U. Yang, J. A. Baker, J. R. Ward, *Chem. Rev.* **1992**, *92*, 1729.

268 H. S. Schultz, H. B. Freyermuth, S. R. Buc, *J. Org. Chem.* **1963**, *28*, 1140.

269 H. Q. N. Gunaratne, M. A. McKervey, S. Feutren, J. Finlay, J. Boyd, *Tetrahedron Lett.* **1998**, *39*, 5655.

270 W. Adam, C. M. Mitchell, C. R. Saha-Möller, *Tetrahedron* **1994**, *50*, 13121.

271 H. J. Reich, F. Chow, S. L. Peake, *Synthesis*, **1978**, 299.

272 D. P. Riley, M. R. Smith, P. E. Correa, *J. Am. Chem. Soc.* **1988**, *110*, 177.

273 T. Katsuki, K. B. Sharpless, *J. Am. Chem. Soc.* **1980**, *102*, 5974.

274 R. A. Johnson, K. B. Sharpless, in *Catalytic Asymmetric Synthesis*, I. Ojima (ed.), VCH, Berlin, **1993**, p. 103.

275 K. B. Sharpless, *Janssen Chim. Acta* **1988**, *6*(1), 3.

276 W. Zhang, J. L. Loebach, S. R. Wilson, E. N. Jacobsen, *J. Am. Chem. Soc.* **1990**, *112*, 2801.

277 R. Irie, K. Noda, Y. Ito, T. Katsuki, *Tetrahedron Lett.* **1991**, *32*, 1055.

278 E. N. Jacobsen, in *Catalytic Asymmetric Synthesis*, I. Ojima (ed.), VCH, Berlin, 1993, p. 159.

279 C. Bousquet, D. G. Gilheany, *Tetrahedron Lett.* **1995**, *36*, 7739.

280 K. Matsumoto, Y. Sawada, B. Saito, K. Sakai, T. Katsuki, *Angew. Chem. Int. Ed.* **2005**, *44*, 4935.

281 Y. Sawada, K. Matsumoto, S. Kondo, H. Watanabe, T. Ozawa, K. Suzuki, B. Saito, T. Katsuki, *Angew. Chem. Int. Ed.* **2006**, *45*, 3478.

282 M. K. Tse, C. Döbler, S. Bhor, M. Klawonn, W. Mägerlein, H. Hugl, M. Beller, *Angew. Chem. Int. Ed.* **2004**, *43*, 5255; M. K. Tse, S. Bhor, M. Klawonn, G. Anilkumar, H. Jiao, A. Spannenberg, C. Döbler, W. Mägerlein, H. Hugl, M. Beller, *Chem. Eur. J.* **2006**, *12*, 1875.

283 T. Yamada, K. Imagawa, T. Nagata, T. Mukaiyama, *Chem. Lett.* **1992**, 2231.

284 T. Mukaiyama, T. Yamada, T. Nagata, K. Imagawa, *Chem. Lett.* **1993**, 327.

285 Y. Tu, Z. X. Wang, Y. Shi, *J. Am. Chem. Soc.* **1996**, *118*, 9806; Z. X. Wang, Y. Tu, M. Frohn, J. R. Zhang, Y. Shi, *J. Am. Chem. Soc.* **1997**, *119*, 11224.

286 L. Shu, Y. Shi, *Tetrahedon Lett.* **1999**, *40*, 8721.

287 S. Julia, J. Guixer, J. Masana, J. Roca, S. Colonna, R. Annunziata, H. Molinari, *J. Chem. Soc., Perkin Trans. I* **1982**, 1314.

288 I. Lantos, V. Novack, in *Chirality in Drug Design and Synthesis*, C. Brown (ed.), Academic Press, New York, **1990**, pp. 167–180.

289 K. H. Dranz, personal communication.

290 F. van Rantwijk, R. A. Sheldon, *Curr. Opin. Biotechnol.* **2000**, *11*, 554.

291 J.-B. Park, D. S. Clark, *Biotechn. Bioeng.* **2006**, *94*, 189.

292 F. J. Lakner, K. P. Cain, L. P. Hager, *J. Am. Chem. Soc.* **1997**, *119*, 443; A. F. Dexter, F. J. Lakner, R. A. Campbell, L. P. Hager, *J. Am. Chem. Soc.* **1995**, *117*, 6412.

293 G. P. Rai, Q. Zong, L. P. Hager, *Isr. J. Chem.* **2000**, *40*, 63.

294 K. Hofstetter, J. Lutz, I. Lang, B. Witholt, A. Schmid, *Angew. Chem. Int. Ed.* **2004**, *43*, 2163.

295 F. Hollmann, P-C. Lin, B. Witholt, A. Schmid, *J. Am. Chem. Soc.* **2003**, *125*, 8209.

296 E. N. Jacobsen, I. Marko, W. S. Mungall, G. Schröder, K. B. Sharpless, *J. Am. Chem. Soc.* **1988**, *110*, 1968.

297 J.S.M. Wai, I. Marko, J.S. Svendsen, M.G. Finn, E.N. Jacobsen, K.B. Sharpless, *J. Am. Chem. Soc.* **1989**, *111*, 1123.

298 H.L. Kwong, C. Sorato, Y. Ogino, H. Chen, K.B. Sharpless, *Tetrahedron Lett.* **1990**, *31*, 2999.

299 K.B. Sharpless, W. Amberg, Y.L. Bennani, G.A. Crispino, J. Hartung, K.-S. Jeong, H.-L. Kwong, K. Morikawa, Z.-M. Wang, D. Xu, X.-L. Zhang, *J. Org. Chem.* **1992**, *57*, 2768; H.C. Kolb, K.B. Sharpless in *Transition metals for Organic Synthesis*, M. Beller, C. Bolm (eds.), Wiley-VCH, Weinheim, Germany, **1998**; p. 219.

300 M. Tokunaga, J.F. Larrow, E.N. Jacobsen, *Science* **1997**, *277*, 936.

301 V. Conte, F. Di Furia, G. Licini, G. Modena, in *Metal Promoted Selectivity in Organic Synthesis*, A.F. Noels, M. Graziani, A.J. Hubert (eds.), Kluwer, Amsterdam, **1991**, p. 91; F. Di Furia, G. Modena, R. Seraglia, *Synthesis*, **1984**, 325.

302 H.B. Kagan, in *Catalytic Asymmetric Synthesis*, I. Ojima (ed.), VCH, Berlin, **1993**, p. 203.

303 M. Larsson, lecture presented at Chiral Europe 2000, 26–27 Oct. 2000, Malta.

304 P.J. Hogan, P.A. Hopes, W.O. Moss, G.E. Robinson, I. Patel, *Org. Proc. Res. Dev.* **2002**, *6*, 225.

305 C. Bolm, F. Bienewald, *Angew. Chem., Int. Ed. Engl.* **1995**, *34*, 2640.

306 J. Legros, C. Bolm, *Angew. Chem. Int. Ed.* **2003**, *42*, 5487.

307 G. Liu, D.A. Cogan, J.A. Ellman, *J. Am. Chem. Soc.* **1997**, *119*, 9913.

308 M.P.J. van Deurzen, I.J. Remkes, F. van Rantwijk, R.A. Sheldon, *J. Mol. Catal. A: Chemical* **1997**, *117*, 329.

309 H.B. ten Brink, H.L. Holland, H.E. Schoenmaker, H. van Lingen, R. Wever, *Tetrahedron Asymm.* **1999**, *10*, 4563.

310 G.-J. ten Brink, I.W.C.E. Arends, R.A. Sheldon, *Chem. Rev.* **2004**, *104*, 4105.

311 J.D. Stewart, *Curr. Org. Chem.* **1998**, *2*, 195.

312 G. Carrea, S. Riva, *Angew. Chem. Int. Ed.* **2000**, *39*, 2226.

313 V. Alphand, G. Carrea, R. Wohlgemuth, R. Furstoss, J.M. Woodley, *Trends Biotechnol.* **2003**, *21*, 318.

314 S. Rissom, U. Schwarz-Linek, M. Vogel, V.I. Tishkov, U. Kragl, *Tetrahedron Asymm.* **1997**, *8*, 2523.

315 K. Seelbach, B. Riebel, W. Hummel, M.R. Kula, V.I. Tishkov, C. Wandrey, U. Kragl, *Tetrahedron Lett.* **1996**, *9*, 1377.

316 J.D. Stewart, K.W. Reed, J. Zhu, G. Chen, M.M. Kayser, *J. Org. Chem.* **1996**, *61*, 7652; K. Saigo, A. Kasahara, S. Ogawa, H. Nohiro, *Tetrahedron Lett.* **1983**, 511.

317 M.D. Mihovilovic, G. Chen, S. Wang, B. Kyte, F.Rochon, M. Kayser, J.D. Stewart, *J. Org. Chem.* **2001**, *66*, 733.

318 M.J. Taschner, D.J. Black, Q.-Z. Chen, *Tetrahedron Asymm.* **1993**, *4*, 1387; M.J. Taschner, L. Peddadda, *J. Chem. Soc., Chem. Commun.* **1992**, 1384; U. Schwartz-Linek, A. Krodel, F.-A. Ludwig, A. Schulze, S. Rissom, U. Kragl, V.I. Tishkov, M. Vogel, *Synthesis* **2001**, 947.

319 Usually, 2-or 3-substituted cyclopentanones are converted with considerably lower enantioselectivities: M.M. Kayser, G. Chen, J.D. Stewart, *J. Org. Chem.* **1998**, *63*, 7103.

320 J.D. Stewart, K.W. Reed, C.A. Martinez, J. Zhu, G. Chen, M.M. Kayser, *J. Am. Chem. Soc.* **1998**, *120*, 3541.

321 N. Berezina, V. Alphand, R. Furstoss, *Tetrahedron Asymm.* **2002**, *13*, 1953.

322 C. Bolm, G. Schlingloff, K. Weickhardt, *Angew. Chem., Int. Ed. Engl.* **1994**, *33*, 1848.

323 C. Bolm, G. Schlingloff, *J. Chem. Soc., Chem. Commun.* **1995**, 1247.

5
Catalytic Carbon–Carbon Bond Formation

5.1
Introduction

The formation of carbon–carbon bonds is central to organic chemistry, indeed to chemistry in general. The preparation of virtually every product, be it fine chemical or bulk chemical, will include a carbon–carbon bond formation at some stage in its synthesis. The importance of carbon–carbon bond forming reactions can therefore not be overemphasized. Consequently they have been a focus of interest ever since chemists made their first attempts at synthesis. In the course of the last 150 years many very selective and efficient reactions for the formation of carbon–carbon bonds have been introduced, too many for the scope of this book. At the same time carbon–carbon bond forming reactions are often textbook examples of wastefulness. The Nobel prize winning Wittig reaction being a particularly good illustration of why green chemistry is necessary [1]. At the same time it also becomes obvious how much there still remains to be done before chemistry is green, since not even the equally Nobel prize winning metathesis can completely replace the Wittig reaction.

This chapter focuses on the application of transition metal catalysts and enzymes for the formation of carbon–carbon bonds. Transition metal-catalyzed carbon–carbon bond formations are not always very green but they often replace even less favorable conventional approaches. The key to making them really green is that they have to be easily separable and reusable. Several of these reactions, such as the hydroformylation, oligomerisation, carbonylation of alcohols and the metathesis, are therefore also treated in Chapter 1, Section 1.8 and Chapter 7, since their greener variations are performed in novel reaction media.

5.2
Enzymes for Carbon–Carbon Bond Formation

The synthesis of carbon–carbon bonds in nature is performed by a vast variety of enzymes, however the bulk of the reactions are performed by a rather limited number of them. Indeed, for the synthesis of fatty acids just one carbon–carbon bond forming enzyme is necessary. Contrary to expectation, virtually none of

Green Chemistry and Catalysis. I. Arends, R. Sheldon, U. Hanefeld
Copyright © 2007 WILEY-VCH Verlag GmbH & Co. KGaA, Weinheim
ISBN: 978-3-527-30715-9

the enzymes that nature designed for building up molecules are used in organic synthesis. This is because they tend to be very specific for their substrate and can therefore not be broadly applied. Fortunately an ever increasing number of enzymes for the formation of carbon–carbon bonds are available [2–4]. Many of them are lyases and, in an indirect approach, hydrolases. These classes of enzymes were designed by nature not for making but for breaking down molecules. By reversing the equilibrium reactions that these enzymes catalyze in the natural substrates they can be applied to the synthesis of carbon–carbon bonds. As reagents that were designed by nature for degrading certain functional groups, they are robust but not very substrate specific. However, they are specific for the functional group they destroy in nature and generate in the laboratory or factory. Equally important, they are very stereoselective. Hence these enzymes are ideal for the application in organic synthesis.

5.2.1
Enzymatic Synthesis of Cyanohydrins

Cyanohydrins are versatile building blocks that are used in both the pharmaceutical and agrochemical industries [2–9]. Consequently their enantioselective synthesis has attracted considerable attention (Scheme 5.1). Their preparation by the addition of HCN to an aldehyde or a ketone is 100% atom efficient. It is, however, an equilibrium reaction. The racemic addition of HCN is base-catalyzed, thus the enantioselective, enzymatic cyanide addition should be performed under mildly acidic conditions to suppress the undesired background reaction. While the formation of cyanohydrins from aldehydes proceeds readily, the equilibrium for ketones lies on the side of the starting materials. The latter reaction can therefore only be performed successfully by either bio- or chemo-cat-

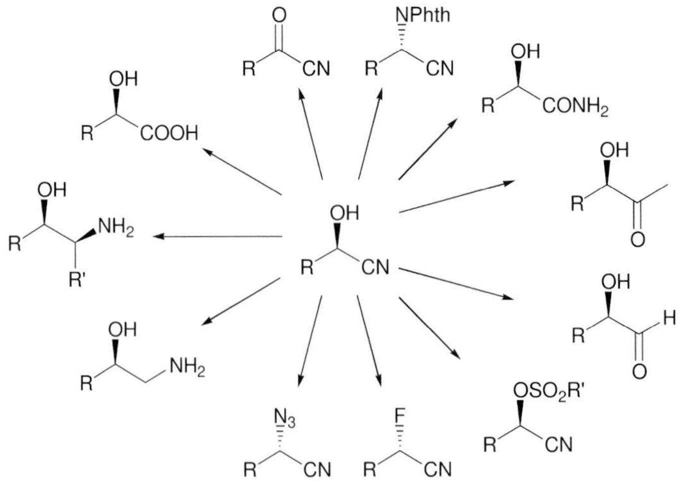

Scheme 5.1 Cyanohydrins are versatile building blocks.

alysis when an excess of HCN is employed, or when the product is constantly re-moved from the equilibrium. The only advantage of this unfavorable equilibrium is that the liquid acetone cyanohydrin can replace the volatile HCN in the labora-tory. It releases HCN during the synthesis of an aldehyde-based cyanohydrin, im-proving safety in the laboratory significantly. At the same time the atom efficiency of the reaction is, however, greatly reduced. Alternative methods for the safe han-dling of cyanides on a laboratory scale are for instance to use cyanide salts in solu-tion, again generating waste. In order to achieve high yields an excess of HCN, or of the other cyanide sources, is commonly used. This excess cyanide needs to be destroyed with iron (II) sulfate or bleach in the laboratory. It is much easier to han-dle HCN on an industrial scale and to regain any HCN. In general all work involv-ing cyanide (enzyme-catalyzed or not) must be performed in well-ventilated la-boratories and HCN detectors have to be used [6].

5.2.1.1 Hydroxynitrile Lyases

For the synthesis of cyanohydrins nature provides the chemist with *R*- and *S*-se-lective enzymes, the hydroxynitrile lyases (HNL) [4–7]. These HNLs are also known as oxynitrilases and their natural function is to catalyze the release of HCN from natural cyanohydrins like mandelonitrile and acetone cyanohydrin. This is a defense reaction of many plants. It occurs if a predator injures the plant cell. The reaction also takes place when we eat almonds. Ironically the benzaldhyde released together with the HCN from the almonds is actually the flavor that attracts us to eat them.

Since the release of HCN is a common defense mechanism for plants, the num-ber of available HNLs is large. Depending on the plant family they are isolated from, they can have very different structures; some resemble hydrolases or carbox-ypeptidase, while others evolved from oxidoreductases. Although many of the HNLs are not structurally related they all utilize acid–base catalysis. No co-factors need to be added to the reactions nor do any of the HNL metallo-enzymes require metal salts. A further advantage is that many different enzymes are available, *R*- or *S*-selective [10]. For virtually every application it is possible to find a stereoselective HNL (Table 5.1). In addition they tend to be stable and can be used in organic sol-vents or two-phase systems, in particular in emulsions.

The *R*-selective *Prunus amygdalus* HNL is readily available from almonds. Ap-proximately 5 g of pure enzyme can be isolated from 1 kg of almonds, alterna-tively crude defatted almond meal has also been used with great success. This enzyme has already been used for almost 100 years and it has successfully been employed for the synthesis of both aromatic and aliphatic *R*-cyanohydrins (Scheme 5.2) [11]. More recently it has been cloned into *Pichea pastoris*, guaran-teeing unlimited access to it and enabling genetic modifications of this versatile enzyme [12].

Prunus amygdalus HNL can be employed for the bulk production of (*R*)-*o*-chloromandelonitrile, however with a modest enantioselectivity (*ee*=83%) [9]. When replacing alanine 111 with glycine the mutant HNL showed a remarkably

Table 5.1 Commonly used hydroxynitrile lyases (HNL).

Name and origin	Natural substrate	Stereoselectivity
Prunus amygdalus HNL, Almonds	(R)-mandelonitrile	R
Linum usitatissimum HNL, Flax seedlings	(R)-butanone cyanohydrin and acetone canohydrin	R
Hevea brasiliensis HNL, Rubber-tree leaves	acetone cyanohydrin	S
Sorghum bicolor HNL, Millet seedlings	(S)-4-hydroxy-mandelonitrile	S
Manihot esculenta HNL, Manioc leaves	acetone cyanohydrin	S

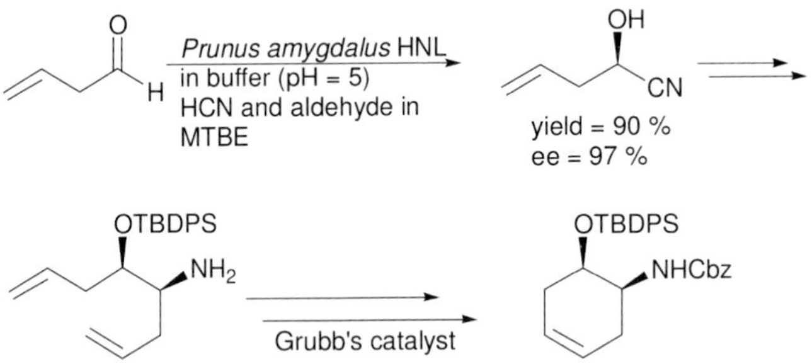

Scheme 5.2 Application of *Prunus amygdalus* HNL for the synthesis of R-cyanohydrins.

high enantioselectivity towards o-chlorobenzaldehyde and the corresponding cyanohydrin was obtained with an *ee* of 96.5% [12]. Hydrolysis with conc. HCl yields the enantiopure (S)-o-chloromandelic acid, an intermediate for the antithrombotic drug clopidogrel (Scheme 5.3).

Recently it was described that site directed mutagenesis has led to a *Prunus amygdalus* HNL that can be employed for the preparation of (R)-2-hydroxy-4-phenylbutyronitrile with excellent enantioselectivity (*ee* > 96%). This is a chiral building block for the enantioselective synthesis of ACE inhibitors such as enalapril (Scheme 5.4) [13].

The unmodified enzyme has also been used in ionic liquids, expanding the scope of its application even further. Significant rate enhancement, in particular for substrates that otherwise react only sluggishly, was observed [14].

The discovery of the S-selective HNLs is more recent. The application of the S-selective *Hevea brasiliensis* HNL was first described in 1993 [15]. The potential of this enzyme was immediately recognized and already four years after describing it for

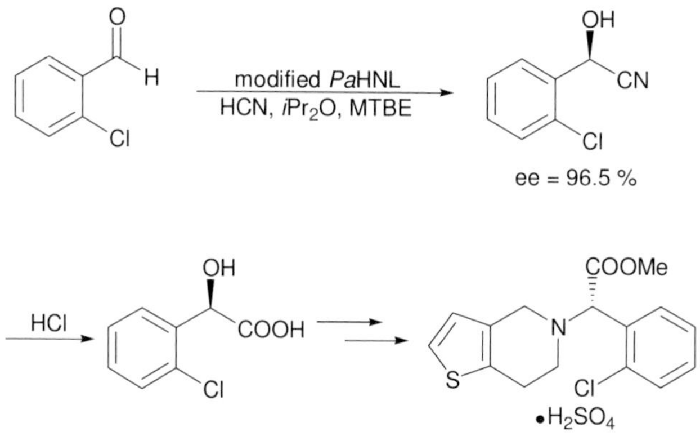

Scheme 5.3 Modified *Prunus amygdalus* HNL is applied for the synthesis of enantiopure R-o-chloromandelonitrile, a precursor for the synthesis of clopidogrel.

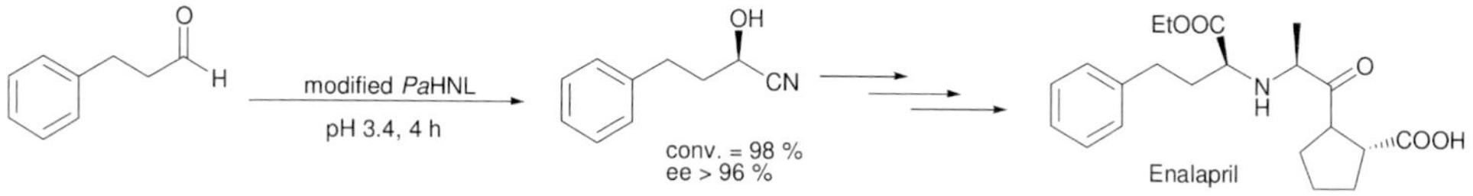

Scheme 5.4 Modified *Prunus amygdalus* HNL catalyzes the enantioselective formation of potential precursors for ACE inhibitors.

the first time it has been cloned and overexpressed, making it available for large-scale applications [16]. Indeed, this enzyme is not only a versatile catalyst in the laboratory, industrially it is used to prepare (*S*)-*m*-phenoxymandelonitrile with high enantiopurity (*ee* > 98%) [9]. This is a building block for the pyrethroid insecticides deltamethrin and cypermethrin and is used on a multi-ton scale (Scheme 5.5). The HNL-based process replaced another enzymatic process that was used earlier for the production of the enantiopure cyanohydrin, the kinetic resolution of *m*-phenoxymandelonitrile acetate with a hydrolase from *Arthrobacter globiformis* [17].

As mentioned above, the synthesis of cyanohydrins from ketones is difficult due to the unfavorable equilibrium. However, it can be achieved. When methyl ketones were treated with only 1.5 equivalents of HCN in an emulsion of citrate buffer (pH 4.0) and MTBE, the *S*-selective *Manihot esculenta* HNL catalyzed the synthesis of the corresponding cyanohydrins with good yields (85–97%) and enantioselectivities (69–98%). The corresponding esters were then used in a second carbon–carbon bond forming reaction. Since cyanohydrins from ketones cannot be deprotonated adjacent to the nitrile group, the esters could be selectively deprotonated and then a ring closing attack on the nitrile function yielded the unsaturated lactones (Scheme 5.6). When the nitrile function was first con-

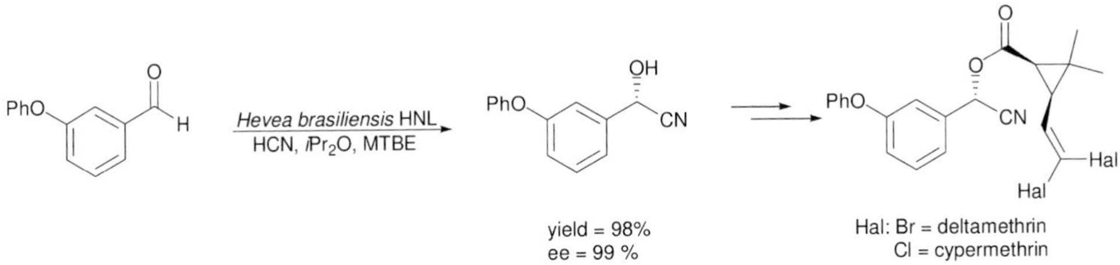

yield = 98%
ee = 99 %

Hal: Br = deltamethrin
Cl = cypermethrin

Scheme 5.5 *Hevea brasiliensis* HNL catalyzes a key step in the enantioselective synthesis of pyrethroid insecticides.

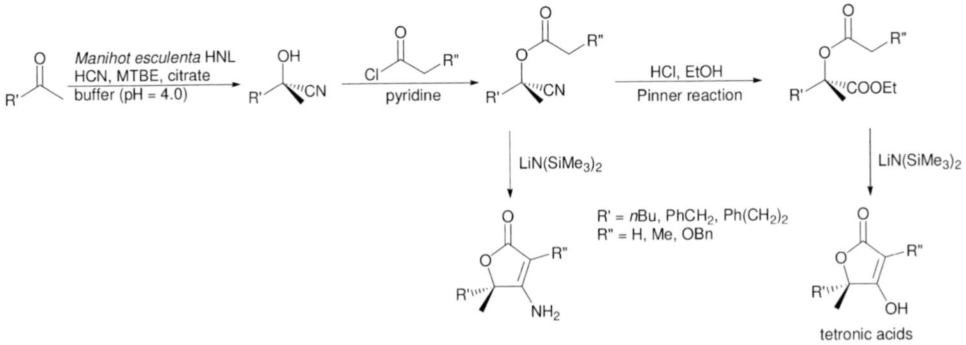

R' = *n*Bu, PhCH₂, Ph(CH₂)₂
R" = H, Me, OBn

tetronic acids

Scheme 5.6 HNL-catalyzed formation of cyanohydrins from ketones and their application in synthesis.

verted into an ester via a Pinner reaction, the intramolecular Claisen reaction gave chiral tetronic acids [18].

5.2.1.2 Lipase-based Dynamic Kinetic Resolution

A completely different enzyme-catalyzed synthesis of cyanohydrins is the lipase-catalyzed dynamic kinetic resolution (see also Chapter 6). The normally unde-sired, racemic base-catalyzed cyanohydrin formation is used to establish a dy-namic equilibrium. This is combined with an irreversible enantioselective ki-netic resolution via acylation. For the acylation, lipases are the catalysts of choice. The overall combination of a dynamic carbon–carbon bond forming equilibrium and a kinetic resolution in one pot gives the desired cyanohydrins protected as esters with 100% yield [19–22].

This methodology was employed to prepare many heterocyclic cyanohydrin acetates in high yields and with excellent enantioselectivities, *Candida antarctica* lipase A (CAL-A) being the lipase of choice (Scheme 5.7) [23]. A recent detailed study of the reaction conditions revealed that the carrier on which the lipase is immobilised is important; generally Celite should be used for aromatic sub-strates. With Celite R-633 as support for *Candida antarctica* lipase B (CAL-B)

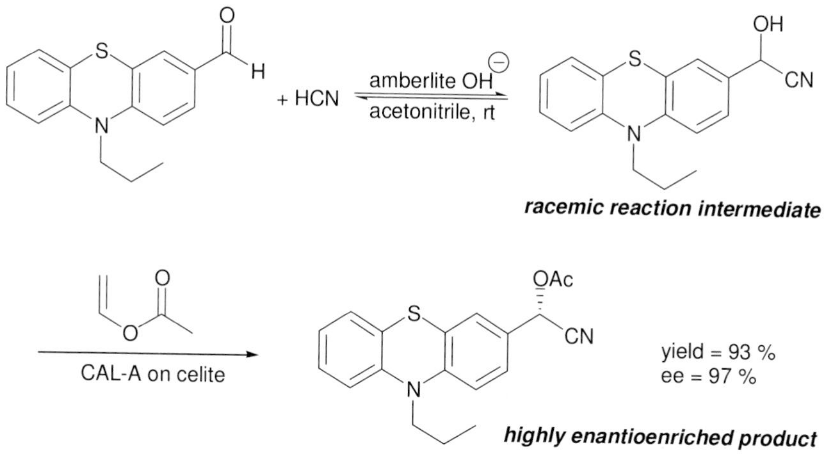

Scheme 5.7 CAL-A-catalyzed formation of chiral cyanohydrins.

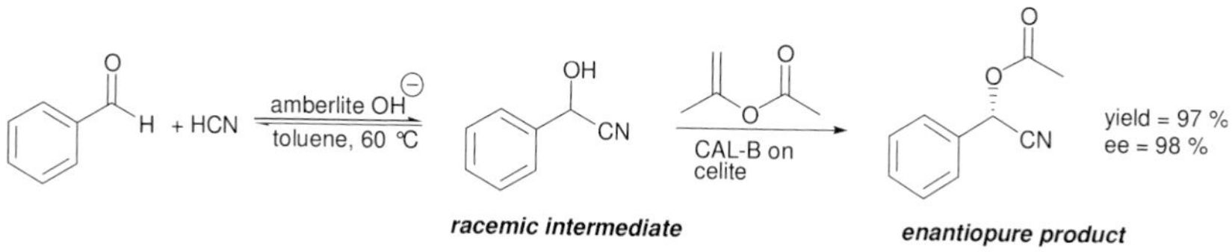

Scheme 5.8 Synthesis of enantiopure mandelonitrile acetate via a dynamic kinetic resolution.

mandelonitrile acetate was synthesised in 97% yield and ee=98% (Scheme 5.8) [24, 25]. Drawbacks of this approach are that an R-selective variant has so far not been developed and most lipases do not accept sterically demanding acids. Consequently, attempts at preparing the pyrethroid insecticides in one step via this approach were not successful.

5.2.2
Enzymatic Synthesis of α-Hydroxyketones (Acyloins)

The bifunctional nature and the presence of a stereocenter make α-hydroxyketones (acyloins) amenable to further synthetic transformations. There are two classical chemical syntheses for these α-hydroxyketones: the acyloin condensation and the benzoin condensation. In the acyloin condensation a new carbon–carbon bond is formed by a reduction, for instance with sodium. In the benzoin condensation the new carbon–carbon bond is formed with the help of an umpolung, induced by the formation of a cyanohydrin. A number of enzymes catalyze this type of reaction, and as might be expected, the reaction conditions are considerably milder [2– 4, 26, 27]. In addition the enzymes such as benzaldehyde lyase (BAL) catalyze the formation of a new carbon–carbon bond enantioselectively. Transketolases (TK)

and decarboxylases, such as benzoylformate decarboxylase (BFD), catalyze the acyloin formation efficiently, too. Like the chemical benzoin reaction the latter two
break a carbon–carbon bond while forming a new one. This making and breaking
of bonds involves a decarboxylation in both the TK and the decarboxylase-catalyzed
reactions. When TK is applied it might also involve the discarding of a ketone and
not only CO_2 [27, 28]. The atom efficiency of these enzymes is therefore limited. In
contrast to TK, BFD can, however, also catalyze the formation of new carbon–carbon bonds without decarboxylation (Scheme 5.9).

All of the enzymes that catalyze the acyloin formation from two carbonyl groups
rely on a cofactor: thiamin diphosphate. The unphosphorylated thiamin is also
known as vitamin B1, deficiency thereof is the cause of the disease Beriberi. Thiamin diphosphate induces an umpolung of the carbonyl group, similar to the cyanide in the benzoin condensation. After deprotonation its ylide attacks the carbonyl group of the aldehyde. Shifting of the negative charge from the alcoholate to
the carbanion initiates this umpolung. This enables the nucleophilic attack on the
second aldehyde, stereoselectively forming the carbon–carbon bond for the α-hydroxyketones. Another shift of the charge and elimination releases the desired acyloin and the ylide is available for further action (Scheme 5.10).

In the decarboxylase- and TK-catalyzed reactions the ylide attacks the keto
function of α-keto acids to form an intermediate carboxylate ion. Decarboxylation, i.e. carbon–carbon bond fragmentation leads to the same reactive carbanion as in the BAL-catalyzed reaction (Scheme 5.10). Once again, it enantioselectively forms the new carbon–carbon bond of the desired α-hydroxyketone. TK
requires not only thiamin diphosphate but also a second co-factor, Mg^{2+} [28].

The enantioselective synthesis of α-hydroxyketones via a carbon–carbon bond
forming reaction has received a significant impulse during recent years. Four
different enzymes are commonly used for this reaction: BAL, BFD, pyruvate
decarboxylase (PDC) and TK. Many different compounds can be prepared with

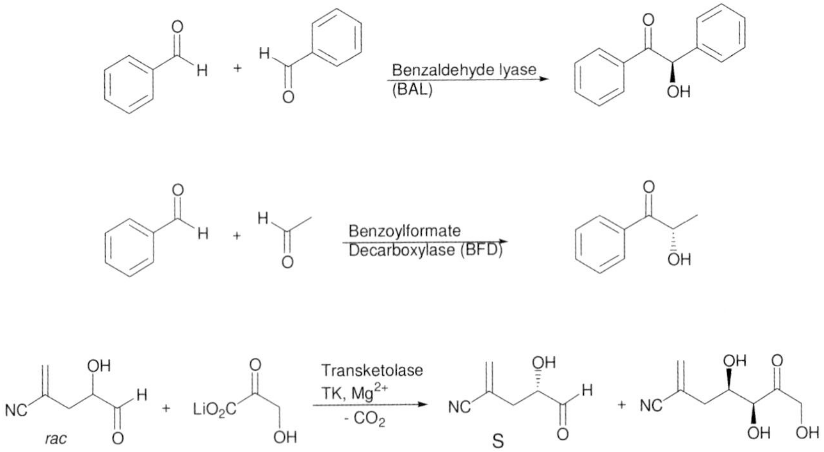

Scheme 5.9 BAL, BFD and TK catalyze the formation of α-hydroxyketones (acyloins).

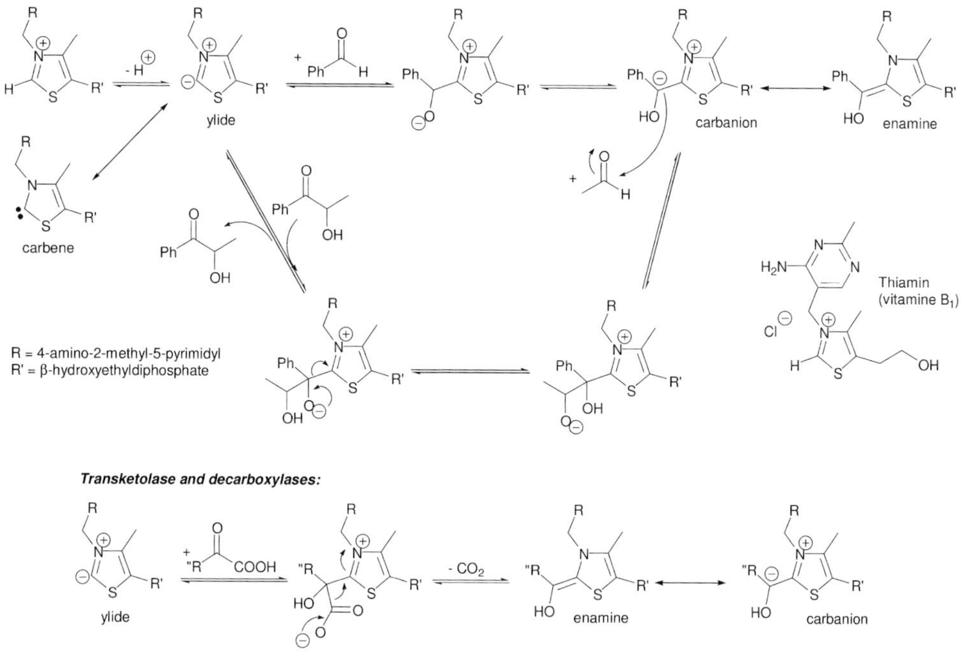

Scheme 5.10 Catalytic cycle in thiamin diphosphate based carbon–carbon bond forming enzymes.

their aid, however, not every stereoisomer is accessible yet. Moreover, it is still difficult to perform the reaction of two different aldehydes to obtain the mixed acyloins in a predictable manner.

The R-selective BAL accepts either two aromatic aldehydes or an aliphatic and an aromatic aldehyde as substrates. This enables enantioselective access to a variety of biologically active compounds and natural products such as Cytoxazone, a metabolite from *Streptomyces sp.* (Scheme 5.11) [29]. If formaldehyde is used instead of prochiral acetaldehyde or its derivatives achiral hydroxyacetophenones are obtained [30].

Both BAL and BFD can be applied for R-selective cross coupling reactions between two aromatic aldehydes (Scheme 5.12). This opens a direct access to a huge variety of compounds. Their stereoselectivity in these reactions is identical, their substrate specificity however, is not. Of particular interest is that BFD is only R-selective for the coupling of two aromatic aldehydes. For the coupling of

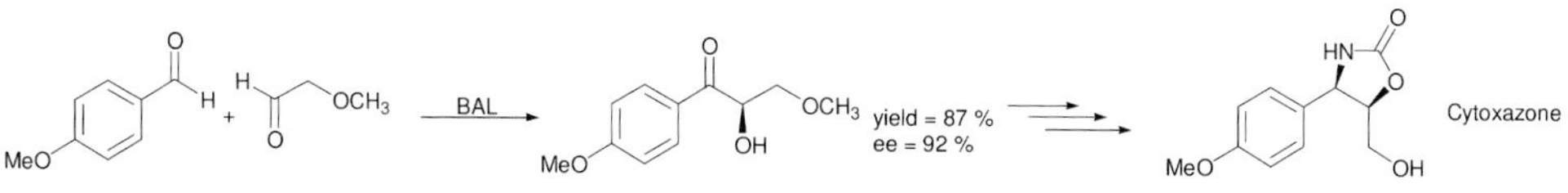

Scheme 5.11 BAL-catalyzed synthesis of carbon–carbon bonds.

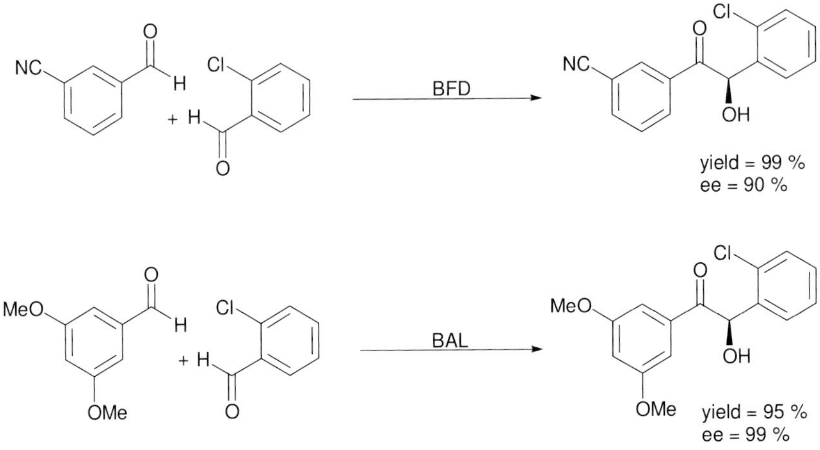

Scheme 5.12 BAL and BFD-catalyzed cross coupling reactions
between two aromatic aldehydes.

two aliphatic aldehydes or of aliphatic aldehydes with aromatic aldehydes BFD
is *S*-selective (Schemes 5.9 and 5.12). Consequently BAL and BFD complement
each other's stereoselectivity in coupling reactions between an aromatic and aliphatic aldehyde, widening the scope of the cross coupling approach [31, 32].

In the above-described cross coupling reactions between an aromatic and an aliphatic aldehyde the keto group of the product is always adjacent to the benzene ring
(Schemes 5.9 and 5.12). With pyruvate decarboxylase (PDC) this selectivity is turned
around. It catalyzes the reaction of pyruvate with benzaldehyde. After decarboxylation of the pyruvate a new carbon–carbon bond is formed. Fortunately acetaldehyde
can be used instead of pyruvate, improving the atom efficiency of the reaction. With
either pyruvate or acetaldehyde the cross coupling with benzaldehyde yields (*R*)-
phenylacetylcarbinol in good optical purity. This process has been performed on
an industrial scale since the 1930s. Whole yeast cells are used and pyruvate is employed as a starting material, together with benzaldehyde. This acyloin is converted
via a platinum-catalyzed reductive amination on a commercial scale into (–)-ephe-
drine (Scheme 5.13) [26, 33]. The newest development for the industrial application
is the introduction of the PDC from *Zymomonas mobilis*, a bacterial PDC [33, 34].
This PDC accepts acetaldehyde instead of the significantly more expensive pyruvate. While the industrial synthesis of (*R*)-phenylacetylcarbinol with PDC was very
successful, PDC has not been used for many other acyloin syntheses.

The sugar metabolism is a source of many enzymes, the transketolase (TK)
being one of them. TK transfers an α-hydroxy carbonyl fragment from D-xylu-
lose-5-phosphate onto D-ribose-5-phosphate, forming D-sedoheptulose-7-phos-
phate and D-glyceraldehyde-3-phosphate (Scheme 5.14). Since this reaction is an
equilibrium reaction and starting materials and products are of similar stability,
it is not very versatile for organic synthesis. Fortunately TK also accepts pyruvate instead of xylulose. Under these modified circumstances carbon dioxide

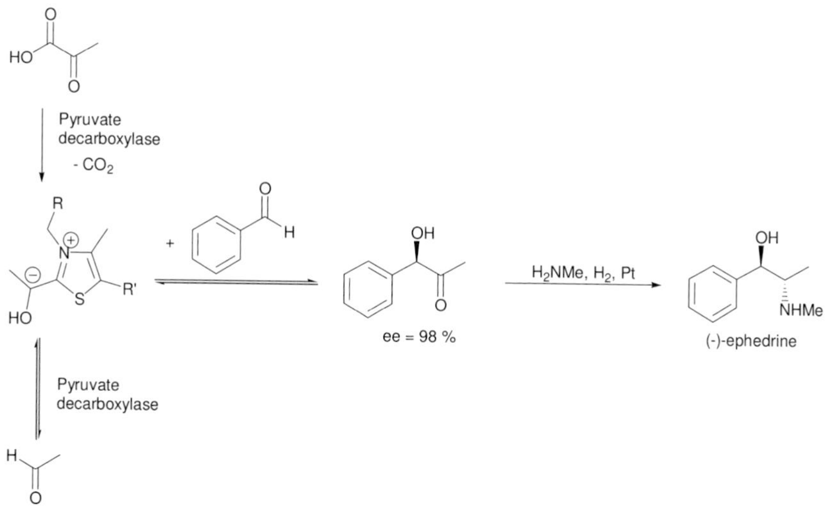

Scheme 5.13 The PDC-catalyzed synthesis of (R)-phenylacetyl-carbinol induces the stereochemistry in the industrial synthesis of (–)-ephedrine.

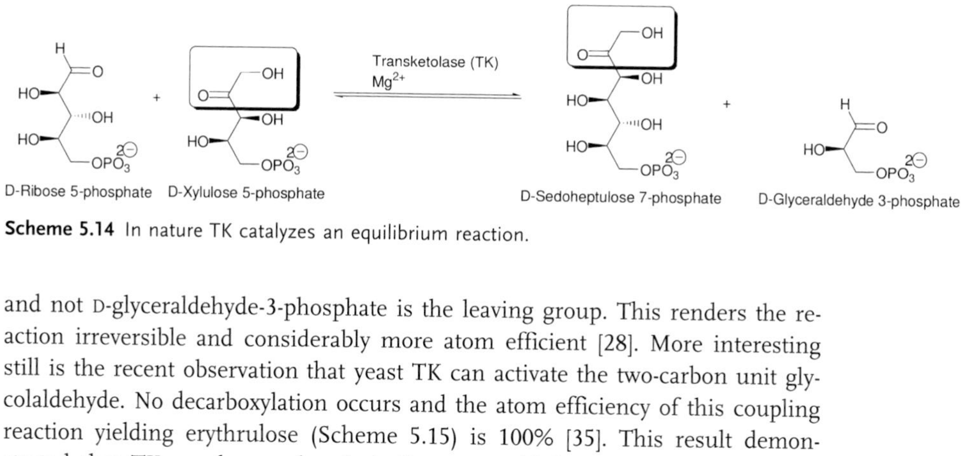

D-Ribose 5-phosphate D-Xylulose 5-phosphate   D-Sedoheptulose 7-phosphate   D-Glyceraldehyde 3-phosphate

Scheme 5.14 In nature TK catalyzes an equilibrium reaction.

and not D-glyceraldehyde-3-phosphate is the leaving group. This renders the reaction irreversible and considerably more atom efficient [28]. More interesting still is the recent observation that yeast TK can activate the two-carbon unit glycolaldehyde. No decarboxylation occurs and the atom efficiency of this coupling reaction yielding erythrulose (Scheme 5.15) is 100% [35]. This result demonstrated that TK can be employed similar to an aldolase and the path is now open for efficient cross coupling reactions.

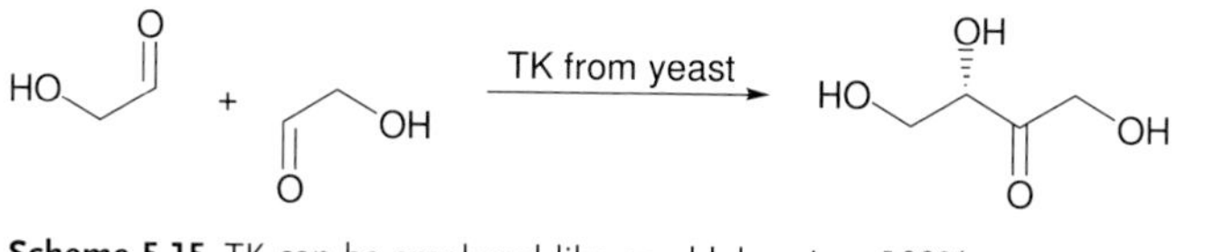

Scheme 5.15 TK can be employed like an aldolase in a 100% atom efficient reaction.

5.2.3
Enzymatic Synthesis of α-Hydroxy Acids

α-Hydroxy acids are valuable building blocks and can be prepared by the chemical hydrolysis of enantiopure cyanohydrins, as described above. Utilising a dynamic kinetic resolution, it is, however, possible to prepare them from aldehydes in a single step. This is achieved by combining the base-catalyzed equilibrium between the aldehyde and HCN with a racemic cyanohydrin with an irreversible, enantioselective follow up reaction (see Section 5.2.1.2). The *in situ* hydrolysis of just one enantiomer of the cyanohydrin, catalyzed by an enantioselective nitrilase [36, 37], yields the desired α-hydroxy acids in excellent enantiopurity. Thus the actual carbon–carbon bond formation is not catalyzed by the enzyme, the nitrilase does, however, impart the enantioselectivity.

Nature provides a vast variety of R-selective nitrilases that accept aliphatic and aromatic substrates. This disconnection can therefore be applied for the synthesis of many structurally different α-hydroxy acids. Industrially the process is applied on a multi-ton scale, to prepare (R)-mandelic acid and its analogs (Scheme 5.16) [9]. What makes the process particularly interesting from a green point of view, is that no organic solvents are used and the reaction is performed at ambient temperatures.

A drawback of this reaction has recently been addressed. Only very few S-selective nitrilases were known; this problem has been solved: a systematic screening program yielded a number of S-selective nitrilases that have successfully been employed in this dynamic kinetic resolution (Scheme 5.17) [38]. In an alternative approach, combining the enantioselectivity of an HNL with the hydrolytic power of a not very selective nitrilase that did accept cyanohydrins as substrates, the synthesis of optically enriched α-hydroxy acids starting from alde-

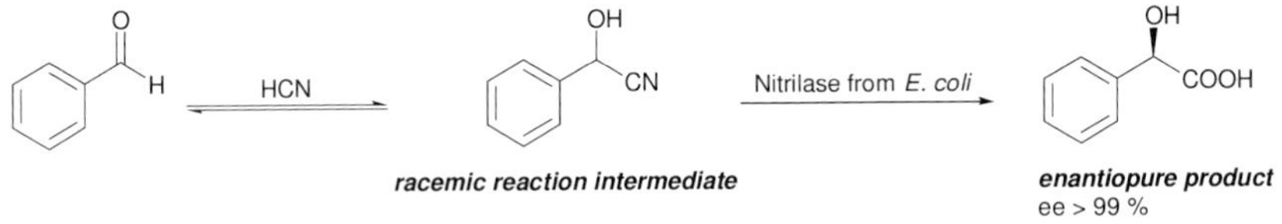

Scheme 5.16 The industrial synthesis of R-mandelic acid proceeds via a base and nitrilase-catalyzed dynamic kinetic resolution.

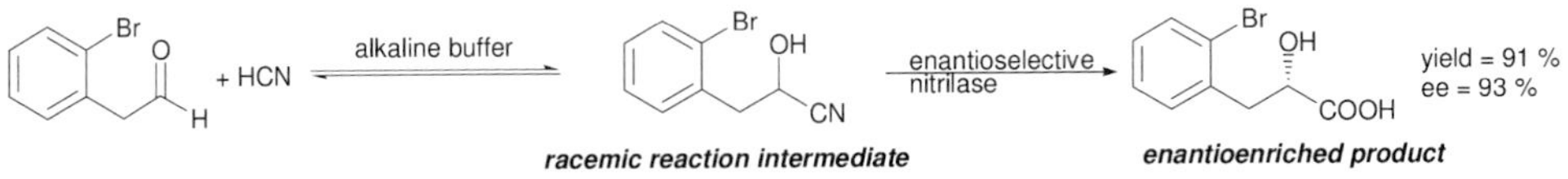

Scheme 5.17 S-α-hydroxy acids are prepared in high yield and optical purity via the dynamic kinetic resolution.

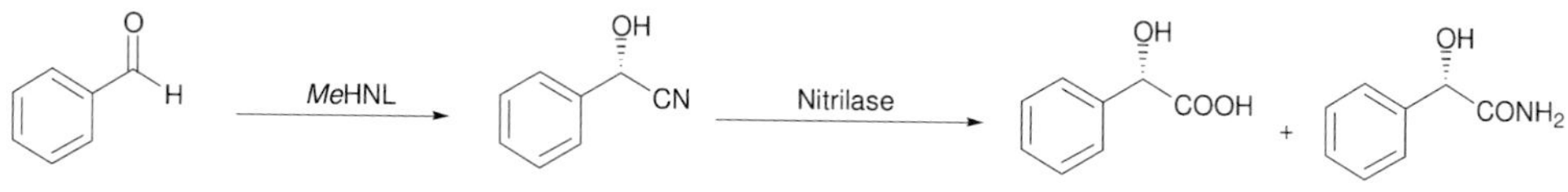

Scheme 5.18 Combining a HNL with an unselective nitrilase allows the synthesis of α-hydroxy acids.

hydes succeeded (Scheme 5.18). Surprisingly not only the desired acid was formed but the amide was also obtained [39].

To date, this reaction cascade has no direct enantioselective chemo-catalytic equivalent. It is therefore a welcome addition to synthetic organic chemistry. Moreover, nitrile hydratases can potentially be used in this reaction, too, expanding its scope even further.

5.2.4
Enzymatic Synthesis of Aldols (β-Hydroxy Carbonyl Compounds)

The aldol reaction is one of the most versatile reactions for the formation of a new carbon–carbon bond and it is also highly atom efficient. Since its first description more than 120 years ago it has been the basis of the 1,3-disconnection in retrosynthesis. In contrast to the cyanohydrin synthesis there is a significant difference between the chemically catalyzed and the enzyme-catalyzed aldol reaction. The acid or base catalyzed aldol reaction is performed with starting materials that tend to have no functional groups in the α-position. The product is a β-hydroxy carbonyl compound, with the classical 1,3-functionality (Scheme 5.19). In the case of a cross coupling, care has to be taken to direct the reaction in such a manner that only one product is obtained, i.e. that only one molecule acts as a donor and the other as an acceptor. Moreover, there is always the risk that elimination will take place. The aldolases that catalyze the aldol reaction in nature are, similar to the TK, often obtained from the sugar metabolism. Consequently they convert highly substituted substrates and the products often have functional groups in the 1-, 2-, 3-, and 4-positions. Unlike the chemical reaction, the enzyme controls the cross coupling reactions very well and they almost al-

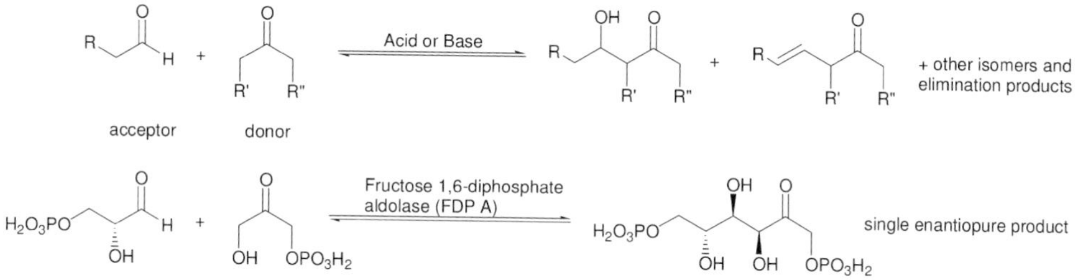

Scheme 5.19 Comparison of the chemically catalyzed aldol reaction with an aldolase-catalyzed reaction.

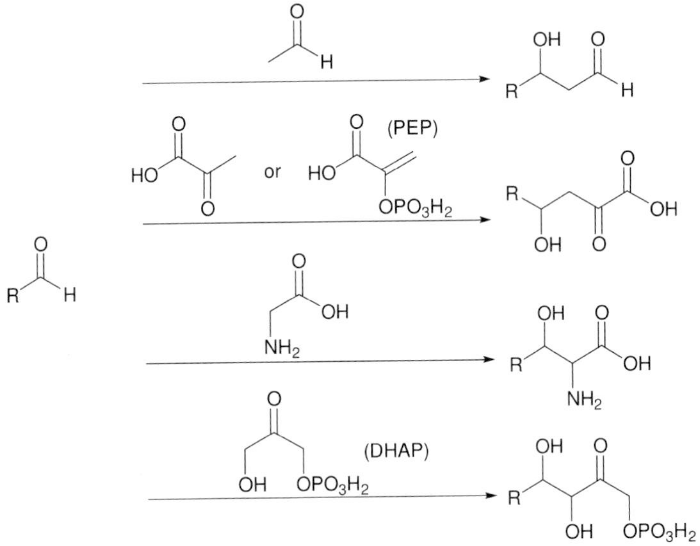

Scheme 5.20 Aldolases are classified according to the donor molecule they activate: DHAP, pyruvate, glycine and acetaldehyde.

ways yield just one product (Scheme 5.19), i.e. they display an excellent control over which molecule is the acceptor and which the donor. This has, however, one drawback: only a very limited number of carbonyl compounds can be used as donors [40]. Of course, the chemical reaction has successfully been converted into a selective reaction; proline and its derivates act as catalysts, which can be used directly with aldehydes and ketones [41]. In many cases, however, chiral auxiliaries in combination with highly reactive and toxic Lewis acids are employed, introducing extra steps and generating much waste [40, 42].

From a synthetic point of view, the chemically catalyzed and the aldolase-catalyzed reactions complement each other. Both can catalyze the synthesis of compounds that are difficult to obtain with the other type of catalyst. Aldolases have an excellent control over the regiochemistry and accept a wide variety of acceptor molecules. As mentioned above they allow only a few donor molecules. The aldolases that are commonly used activate four different donor molecules and are classified according to them (Scheme 5.20) [2–4, 40, 43]. Other aldolases are known, too, but their application for synthesis has so far been very limited and they will therefore not be discussed here.

5.2.4.1 DHAP-dependent Aldolases

Dihydroxyacetone phosphate (DHAP) is the donor ketone that is utilized by the DHAP-dependent aldolases. These aldolases come under the class of lyases, just like the hydroxynitrile lyases (see Section 5.2.1.1). As for the HNLs, no cofactor needs to be added to the reaction mixture and the DHAP-dependent aldolases are straightforward to use. There are two types of these aldolases: Type I aldolases, which are primarily found in higher plants and animals. They work by an enamine mechanism: the amino group of the active site lysine residue forms a Schiff's base with the carbonyl group of the donor DHAP and this activates the DHAP. The imine thus formed tautomerises to the enamine and this adds stereoselectively to the acceptor molecule (Scheme 5.21). Type II aldolases are found in fungi and bacteria; they contain a Zn^{2+} ion in the active site of the enzyme. This is thought to induce Lewis acidity in the enzyme. The zinc polarises the carbonyl group of the donor and forms the reactive enediolate. This nucleophile then adds to the acceptor aldehyde stereoselectively. A glutamate residue (Enz-COO-) and a tyrosine residue (Enz-OH) present in the active site of the enzyme assist in the removal and donation of protons respectively (Scheme 5.22). Both types of aldolases control the stereochemistry of the reaction, it is in most cases independent of the structure of the acceptor and many different acceptors can be employed with a high predictability to the stereochemical outcome of the synthesis. It is important to mention that both, Type I and Type II aldolases catalyze the same reactions, the formation of an aldol from DHAP and an acceptor aldehyde. Although they have completely different modes of action and are obtained from different organisms [40, 43]. This is similar to the HNLs: they too can be structurally very different while they all catalyze the formation of cyanohydrins [10].

Two new stereocenters are established in the DHAP-dependent aldolases-catalyzed carbon–carbon bond formation. Consequently four different stereoisomers can be formed (Scheme 5.23). Enantioselective aldolases that catalyze the formation of just one of each of the stereoisomers are available: fructose 1,6-diphosphate aldolase (FDP A), rhamnulose 1-phosphate aldolase (Rha 1-PA), L-fuculose 1-phosphate aldolase (Fuc 1-PA) and tagatose 1,6-diphosphate aldolase (TDP A). In particular the FDP A, that catalyzes the formation of the D-threo stereochemistry, has been employed in many syntheses. One such FDP A that

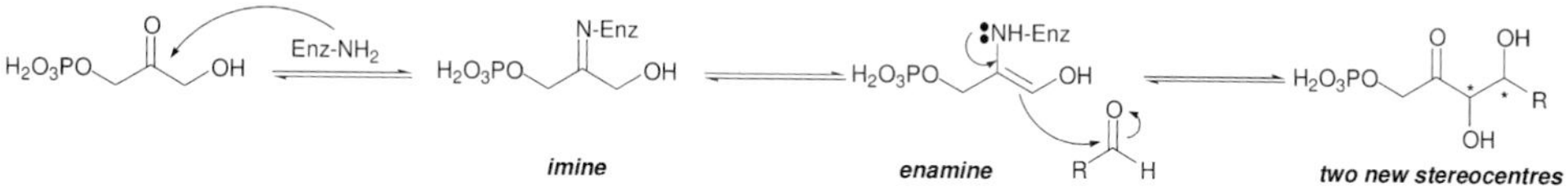

Scheme 5.21 Reaction mechanism of a Type I aldolase.

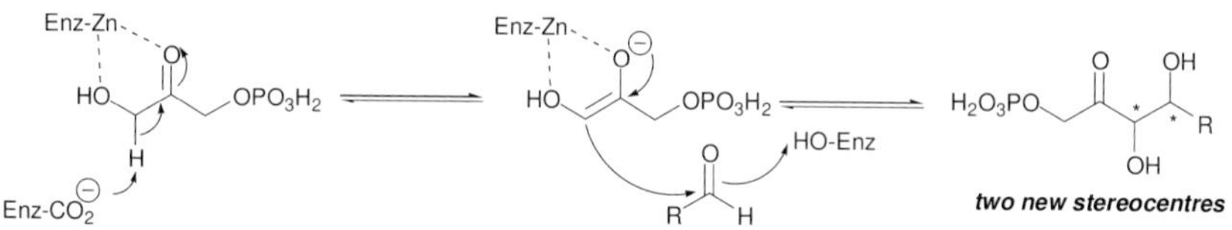

Scheme 5.22 Reaction mechanism of a Type II aldolase.

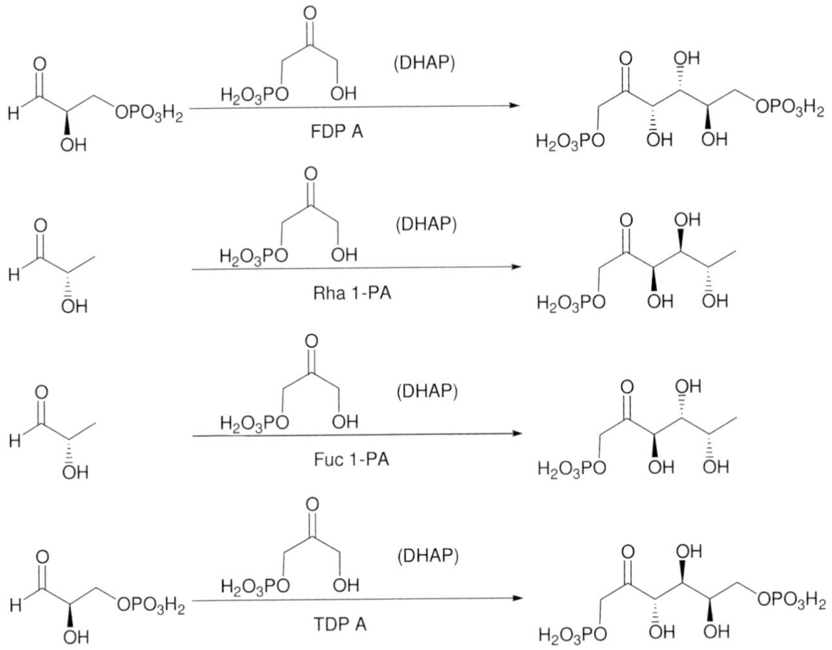

Scheme 5.23 The four different stereoisomers that can be formed are synthesized selectively by four different DHAP-dependent aldolases.

can be isolated from rabbit muscles is better known as RAMA (Rabbit Muscle Aldolase) [43].

FDP A was employed in a study of pancratistatin analogs to catalyze the formation of the D-threo stereochemistry (Scheme 5.24). When rhamnulose 1-phosphate aldolase (Rha 1-PA) was used the L-threo stereoisomer was obtained with excellent selectivity. Thus these two enzymes allow the stereoselective synthesis of the two threo-stereoisomers [44]. They were also utilised successfully for the synthesis of different diastereoisomers of sialyl Lewis X mimetics as selectin inhibitors. Not only the two threo-selective aldolases RAMA and Rha 1-PA, but also the D-erythro-selective Fuc 1-PA was employed. In this way it was possible to synthesise three of the four diastereoisomers enantioselectively (Scheme 5.25). The L-erythro stereochemistry as the only remaining diastereoisomer was not prepared [45]. This is because the aldolase that might catalyze its formation, TDP A, is not very stereoselective and therefore often yields mixtures of diastereoisomers.

A significant drawback of the DHAP-dependent aldolases is that DHAP cannot be replaced by dihydroxyacetone. DHAP is expensive and it is labile at neutral and basic pH values. It can be synthesized chemically, however the procedures are not very atom efficient. Therefore alternative enzyme-catalyzed approaches have been developed. A variety of them can be performed in the

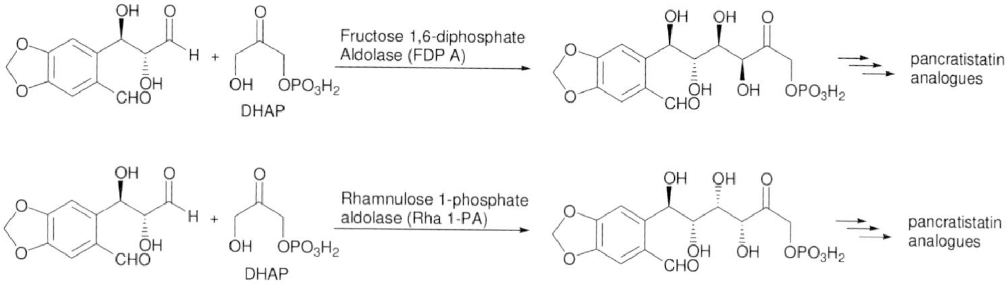

Scheme 5.24 Aldolase-catalyzed synthesis of different pancrastatin analogues.

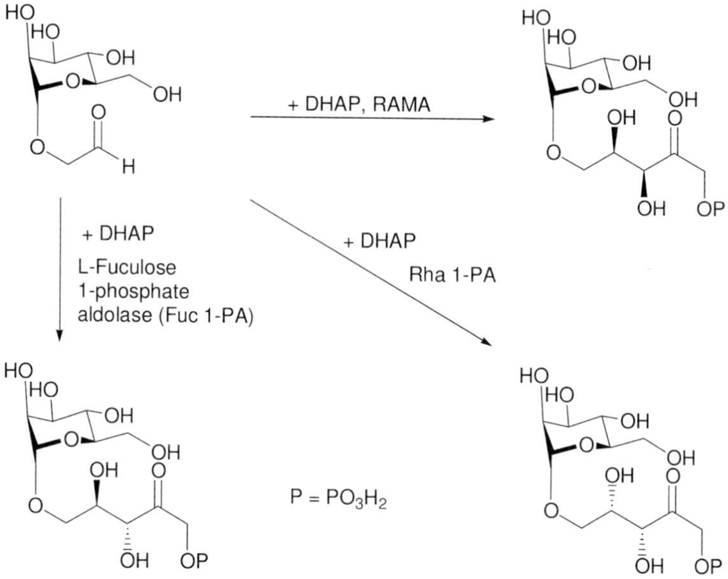

Scheme 5.25 Aldolase-catalyzed synthesis of three stereoisomers of sialyl Lewis X mimetics.

presence of the aldolase, allowing one-pot procedures. When starting from glycerol at lower pH values phytase can be employed for its selective phosphorylation, utilising cheap pyrophosphate as the phosphate source. Upon increase in the pH, glycerolphosphate oxidase (GPO) comes into action, catalyzing the formation of the desired DHAP and hydrogen peroxide. While FDP A catalyzes the enantioselective aldol reaction to yield the 5-deoxy-5-ethyl-D-xylulose monophosphate a catalase rapidly destroys the side product, thus preventing any oxidative damage. When the pH is lowered further, the phytase becomes active once more and catalyzes the dephosphorylation of the aldolproduct, thus the desired 5-deoxy-5-ethyl-D-xylulose is prepared in one pot from glycerol, pyrophosphate and butanal, catalyzed by four different enzymes [46]. Their efficient cooperation in one pot is controlled by the variation in pH levels (Scheme 5.26).

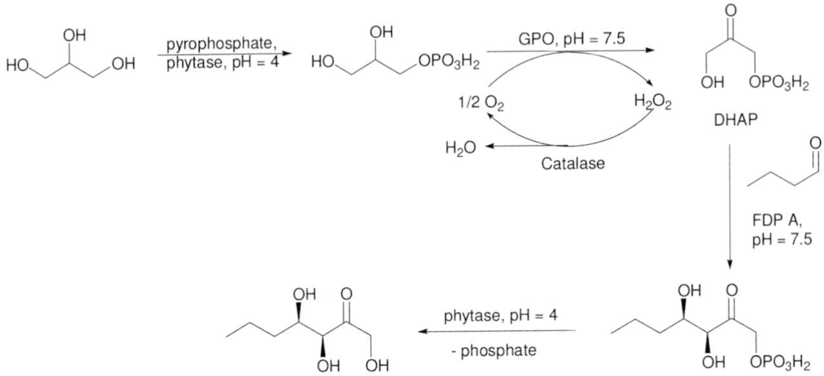

Scheme 5.26 The careful control of the pH value allows a one-pot four enzyme cascade; DHAP is prepared *in situ* and the product is deprotected by phytase.

The same authors then introduced a second system, in which arsenate esters were used as phosphate ester mimics. Dihydroxyacetone readily reacts with arsenate and the resulting ester is accepted by FDP A as a substrate. Since the arsenate ester formation is an equilibrium reaction the desired product is released *in situ* and the arsenate is available again for the next catalytic cycle (Scheme 5.27) [47]. However, although only catalytic amounts of arsenate are necessary it remains a toxic metal.

Recently a variation of the above described approach was introduced. *Rac*-glycidol was treated with disodiumhydrogenphosphate and then oxidized by catalase and L-glycerophosphate oxidase to DHAP. These two steps were integrated with a RAMA-catalyzed step, yielding the phosphorylated aldol product. After adjusting the pH, phosphatase was added and the aldol could be isolated in good yield (Scheme 5.28) [48]. It can be assumed that all the above-mentioned

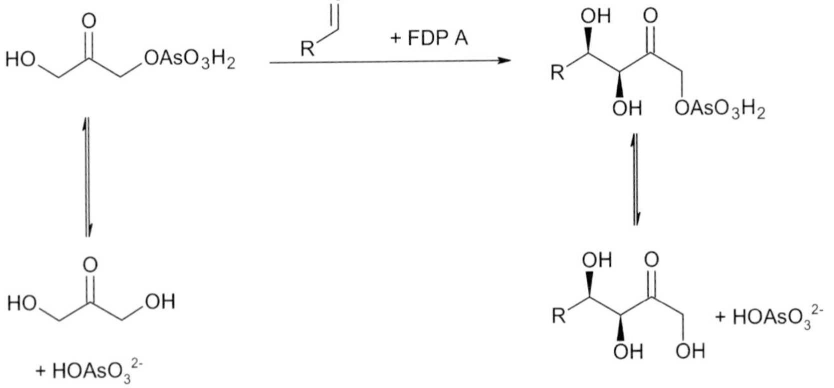

Scheme 5.27 Arsenate esters can replace phosphate in DHAP.

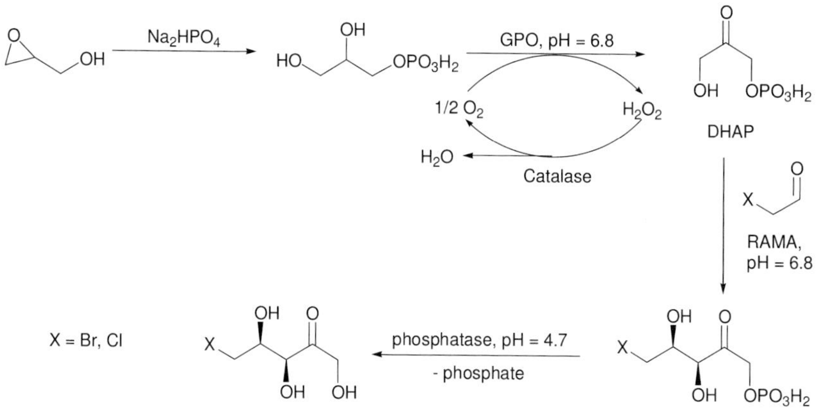

Scheme 5.28 Four enzyme-catalyzed synthesis of aldol products.

phosphorylation procedures can be coupled with other aldolases, giving access to the other stereoisomers.

Another problem of the DHAP-dependent aldolases is that the product is phosphorylated. The aldol reaction has therefore to be followed by a dephosphorylation step. In the sequences in Schemes 5.26–5.28 this deprotection is performed *in situ*. Otherwise several enzymes, such as phytase, are available for this purpose and the reaction normally proceeds under mild conditions [40, 43, 46].

5.2.4.2 PEP- and Pyruvate-dependent Aldolases

Pyruvate-dependent aldolases catalyze the breaking of a carbon–carbon bond in nature. This reaction can, however, be reversed if an excess of pyruvate is used, establishing one new stereocenter in the course of it. The natural function of phosphoenolpyruvate (PEP)-dependent aldolases on the other hand is to catalyze the synthesis of α-keto acids. Since PEP is a very reactive, unstable and difficult to prepare substrate, they are not commonly used in synthesis.

N-Acetylneuraminic acid aldolase (NeuAc aldolase) is commercially available and has been the subject of much attention [49]. NeuAc aldolase catalyzes the aldol reaction between pyruvate and mannose or mannose derivatives. The enzyme activates the donor as its enamine, similar to the Type I aldolase described above (Scheme 5.21). The enzyme has been used for the synthesis of aza sugars and var-

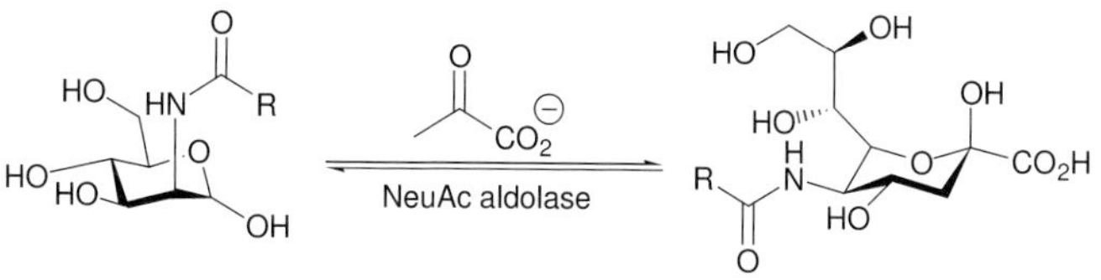

Scheme 5.29 Industrial synthesis of N-acetylneuraminic acid.

ious sialic acid derivatives and it is therefore also known as sialic acid aldolase. A number of chemo-enzymatic syntheses of sialic acid derivatives have been developed over the years. Acylated mannosamine derivatives could be converted into sialic acids with a variety of substituents at C-5 (Scheme 5.29). The versatility of this aldolase is best demonstrated by the fact that it is used for the multi-ton conversion of N-acetylmannosamine (R=Me) into N-acetylneuraminic acid [49–51].

5.2.4.3 Glycine-dependent Aldolases

The glycine-dependent aldolases contain a cofactor: pyridoxal phosphate (PLP). Binding of glycine to it as an imine enables the deprotonation necessary for the carbon–carbon bond forming reaction, with pyridine acting as an electron sink. The subsequent 100% atom efficient reaction with an aldehyde establishes the new bond and two new stereocenters (Scheme 5.30). Of all the glycine-dependent aldolases only L-threonine aldolase (LTA) is commonly used [40, 43, 52].

LTA has been successfully employed for the synthesis of peptide mimetics (Scheme 5.31). This example also reveals the drawback of this enzyme [53]. It only establishes the stereochemistry of the amino group, the second stereocenter, however, is not well defined. Thus diastereomeric mixtures of *threo* and *erythro* products are obtained, although the enzyme was evolved by nature to catalyze the formation/destruction of the *threo* product.

5.2.4.4 Acetaldehyde-dependent Aldolases

Acetaldehyde-dependent aldolases are the only aldolases that can catalyze the aldol formation between two aldehydes, i.e. with an aldehyde both as donor and acceptor [2–4, 40, 43]. More importantly, since they utilize, with high selectivity,

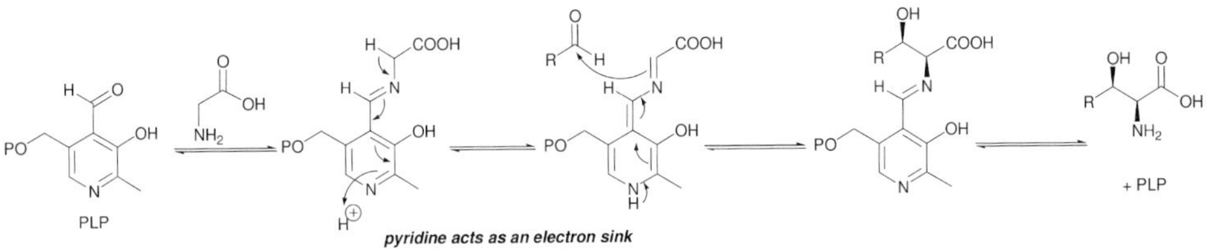

pyridine acts as an electron sink

Scheme 5.30 Mechanism of the glycine-dependent aldolases.

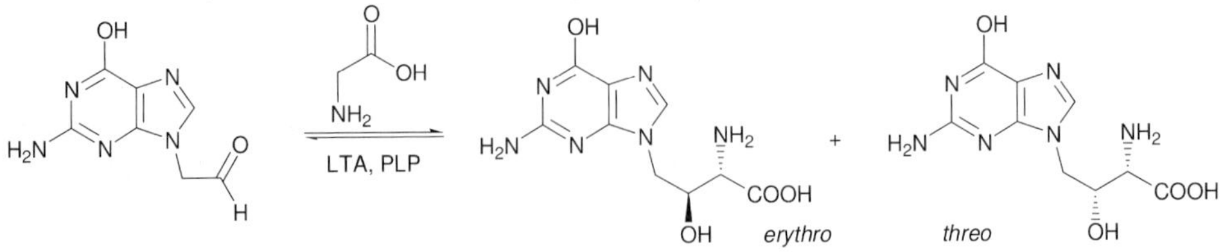

Scheme 5.31 Application of LTA in synthesis.

acetaldehyde as a donor, many cross coupling reactions are possible. To date only one of these aldolases has been applied extensively in organic synthesis: 2-deoxyribose-5-phosphate aldolase (DERA). DERA is a Type I aldolase: the active site lysine attacks acetaldehyde and activates the donor as an enamine (Scheme 5.21). It seems that a water molecule present in the active site plays an essential role in the catalysis and that the reaction mechanism is like that of DHAP-dependent Type I aldolases [54]. Also similar to the DHAP-dependent aldolases, DERA accepts a broad range of acceptors. The stereochemistry at the a-carbon of the acceptor as well as the polarity of the substituent determines whether the acceptor is a good substrate for DERA. If racemic aldehydes are used DERA often catalyzes the conversion of only one enantiomer, performing a kinetic resolution. However, if the "wrong" enantiomer of the acceptor is used as the sole substrate, DERA does occasionally convert it, too [55].

A particularly successful synthesis of Epothilone A is based on two DERA-catalyzed steps. In these two of the seven stereocentres of Epothilone A were established. When a racemic aldehyde was released in situ from its acetal, DERA converted only the R-enantiomer into the stable cyclic hemiacetal. This is a combined kinetic resolution and carbon–carbon bond formation yielding a building block with two chiral centers. Since the alcohol function was oxidized, the optical information obtained from the kinetic resolution was lost. Thus, for the overall yield it would have been better if DERA had displayed no stereoselectivity towards the acceptor (Scheme 5.32). In the DERA-catalyzed synthesis of another part of Epothilone A DERA is again highly stereoselective. Fortunately its preference is for the S-enantiomer of the acceptor aldehyde, the enantiomer that has to be submitted to the carbon–carbon bond formation in order to obtain the desired building block, again a stable hemiacetal (Scheme 5.32). Indeed, both DERA-catalyzed reactions yield open chain products that form stable cyclic hemiacetals. This ensures that the equilibria of these aldol reactions are shifted towards the desired products. Further synthetic manipulations converted these intermediates into Epothilone A [55].

A particularly elegant application of DERA is the sequential synthesis of thermodynamically stable cyclic hemiacetal. Two DERA-catalyzed aldol reactions convert one equivalent of acceptor and two equivalents of acetaldehyde into this stable compound. A mild subsequent oxidation yielded the corresponding lactone in ex-

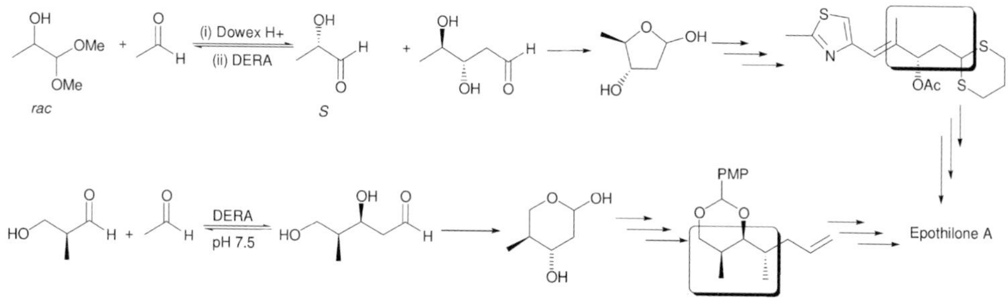

Scheme 5.32 DERA catalyzes key steps in the synthesis of Epothilone A.

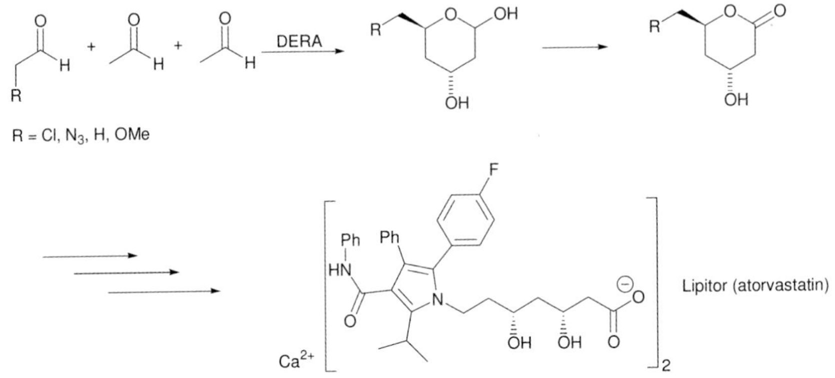

Scheme 5.33 DERA-catalyzed enantioselecitive synthesis of the atorvastatin side chain.

cellent optical purity, proving the great versatility of this class of enzymes. The lactones with R = N$_3$ or Cl are intermediates for the synthesis of Lipitor (atorvastatin), a cholesterol-lowering drug (Scheme 5.33) [56]. A mutant DERA has been employed in this synthesis and alternatively the screening of a very diverse library of environmental samples was performed and the DERA obtained from this library was used. The DERA derived from the screening was utilized for further optimisation and a process was developed in which the aldolase-catalyzed reaction cascade proceeds with an enantioselectivity of over 99% and the *de* of the product is above 96%. Equally important, the reaction proceeds at high substrate concentrations and up to 30 g L^{-1} h^{-1} of product are obtained [57]. Currently this is one of the most competitive approaches to the chiral side chain of the statins [58, 59].

5.2.5
Enzymatic Synthesis of β-Hydroxynitriles

Recently it was described that the halohydrin dehalogenases can be employed for the enantioselective formation of a new carbon–carbon bond. Their natural task is the dehalogenation of halohydrins, however this reaction can be reversed. When utilizing cyanide instead of a halide as the nucleophile in the reverse reaction the desired β-hydroxynitrile is formed with good to excellent enantioselectivity (Scheme 5.34) [60].

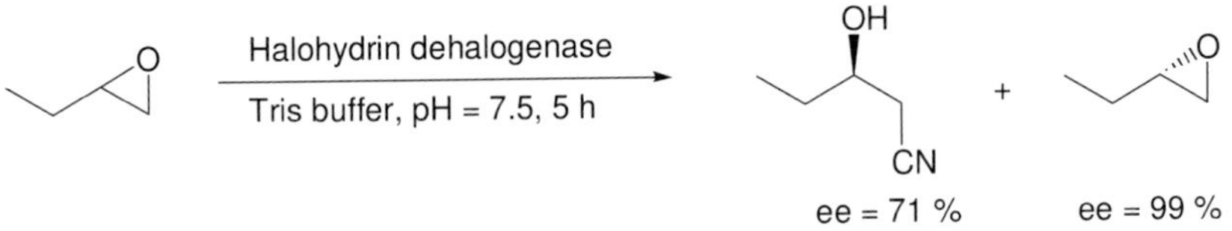

Scheme 5.34 Application of a halohydrin dehalogenase for the synthesis of a carbon–carbon bond.

5.3
Transition Metal Catalysis

While the application of enzymes and proline as catalysts for the (commercial) formation of carbon–carbon bonds is relatively new, transition metal catalysts are well established for the industrial synthesis of carbon–carbon bonds. Although in themselves not always perfectly green, transition metal catalysts often allow the replacement of multi-step and stoichiometric reaction sequences with one single catalytic step. Thus, the overall amount of waste generated and energy used is reduced drastically [61–64].

5.3.1
Carbon Monoxide as a Building Block

Carbon monoxide is a readily available C-1 building block that can be generated from many different sources [61]. It is, however, also a highly toxic gas and does not therefore fulfill the criteria of green chemistry as in the working definition in Chapter 1, Section 1.1. Similar to HCN, it fortunately can be safely handled on large and particularly industrial scales, making it a very attractive reagent for synthesis. Recently strategies for the replacement of carbon monoxide have been developed. In particular methyl formate can be used as a substitute and the current state of the art has been reviewed [65].

Carbon monoxide is an excellent ligand for transition metals. It coordinates via the free electron pair of the carbon atom as a σ-donor (and weakly as a π-donor) to the metal, back bonding shifts electrons from the transition metal into the π^* antibonding orbital of carbon monoxide, making it susceptible to the catalytic chemistry discussed here (Scheme 5.35).

5.3.1.1 Carbonylation of R–X (CO "Insertion/R-migration")
A long-standing success in transition metal catalysis is the carbonylation reaction [66], in particular the synthesis of acetic acid [67]. Formally this is the insertion of CO into another bond, in particular into a carbon–halogen bond. After the oxidative addition to the transition metal (the breaking of the carbon–halogen bond), a reaction with a CO ligand takes place. This reaction is often called an insertion. Mechanistic studies have, however, shown that the actual reaction

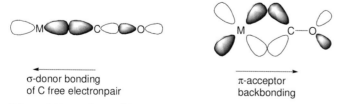

σ-donor bonding
of C free electronpair

π-acceptor
backbonding

Scheme 5.35 Bonding and backbonding in metal carbonyl complexes.

is a migration of the alkyl or aryl group to the CO ligand, not an insertion of the CO in the transition metal carbon bond [61, 63]. This generates a vacant site on the transition metal catalyst, which will be taken up by a new ligand. A subsequent reductive elimination releases the "insertion product" (Scheme 5.36).

Mankind has produced acetic acid for many thousand years but the traditional and green fermentation methods cannot provide the large amounts of acetic acid that are required by today's society. As early as 1960 a 100% atom efficient cobalt-catalyzed industrial synthesis of acetic acid was introduced by BASF, shortly afterwards followed by the Monsanto rhodium-catalyzed low-pressure acetic acid process (Scheme 5.36); the name explains one of the advantages of the rhodium-catalyzed process over the cobalt-catalyzed one [61, 67]. These processes are rather similar and consist of two catalytic cycles. An activation of methanol as methyl iodide, which is catalytic, since the HI is recaptured by hydrolysis of acetyl iodide to the final product after its release from the transition metal catalyst, starts the process. The transition metal catalyst reacts with methyl iodide in an oxidative addition, then catalyzes the carbonylation via a migration of the methyl group, the "insertion reaction". Subsequent reductive elimination releases the acetyl iodide. While both processes are, on paper, 100%

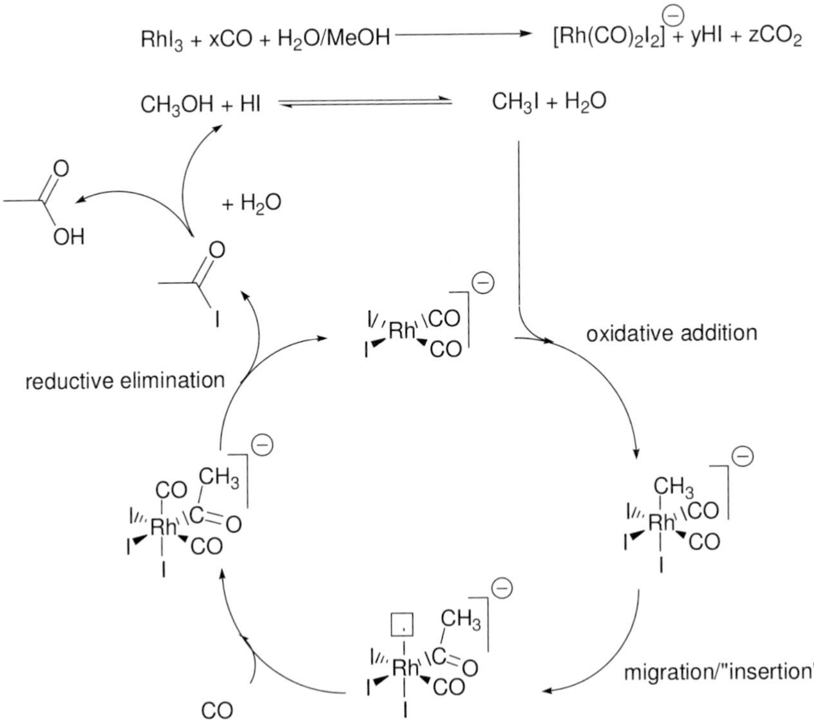

Scheme 5.36 The Monsanto acetic acid process.

atom efficient, methanol and CO are not used in a ratio 1:1 and there was room for improvement in particular for the CO conversion [68].

With the introduction of an iridium catalyst the carbonylation of methanol could be further improved [68, 69]. While the basic process remains the same, i.e. methanol is first converted into methyl iodide, the iridium-catalyzed carbonylation (Cativa process, Scheme 5.37) differs in several ways from the other processes. The selectivity of the reaction when correlated to methanol is greater than 99% but even for CO the yields are above 90%, indicating that the Cativa process is not only theoretically highly atom efficient. This improved selectivity improves the product purity and reduces the work-up costs. In addition, the water concentration in the reactor could be reduced, again easing the work up. Moreover, the iridium catalyst is more active that either rhodium or cobalt-catalyst and significantly more stable, ensuring a longer lifetime [69].

Next to the large-scale application for bulk chemical production, carbonylation reactions have successfully been applied for the synthesis of fine chemicals. In particular the synthesis of Ibuprofen, an analgesic and anti-rheumatic drug could be greatly improved. Instead of the wasteful multi-step Darzen's glycidic ester condensation reaction that was originally employed for the introduction of the acid group a formal insertion of carbon monoxide into the benzylic alcohol drastically shortened the synthesis [70–72].

The reaction proceeds in MEK and conc. HCl at 130 °C. First, the alcohol is converted into the benzylic chloride. In this case a palladium catalyst

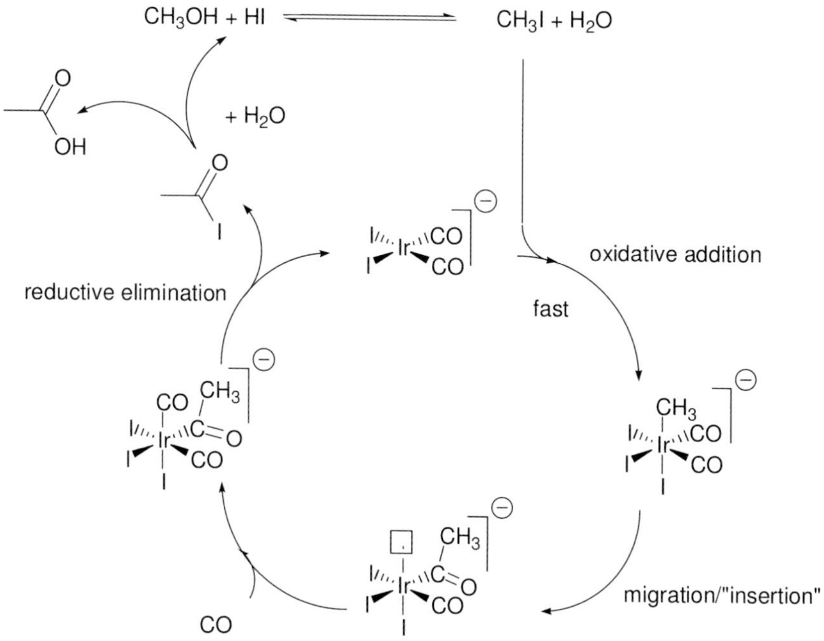

Scheme 5.37 The CATIVA process.

PdCl$_2$(PPh$_3$)$_2$ was applied. After initial reduction of the palladium to obtain the active species, an oxidative addition of the benzylic chloride takes place. Migration of the benzylic group leads to the formation of the new carbon–carbon bond; a reductive elimination then releases ibuprofen as its acid chloride which is hydrolysed, closing the second catalytic cycle, too. This process is used to produce 3500 t/year of ibuprofen and TON of 10 000 have been reached for the palladium catalyst (Scheme 5.38).

In a similar approach, naproxen is prepared from an olefin, the product of a Heck reaction (see Section 5.3.2.1). As described above, the reaction proceeds in the presence of water and HCl, additionally copper(II)chloride is added, possibly to prevent the formation of palladium black (Scheme 5.39). Addition of HCl to the double bond and subsequent oxidative addition reaction of the benzylic chloride with the active Pd^0 species initiates the catalytic cycle, which proceeds similarly to the ibuprofen synthesis [70–73].

In an attempt to replace the organohalides as starting materials, diazonium salts have been investigated as starting material. They, too, can react in an oxidative addition with the Pd^0 to form the first intermediate of the catalytic cycle. In a recent bench scale process for a herbicide in development (CGA 308 956) a diazonium salt proved to be suitable. The product of the oxidative addition will

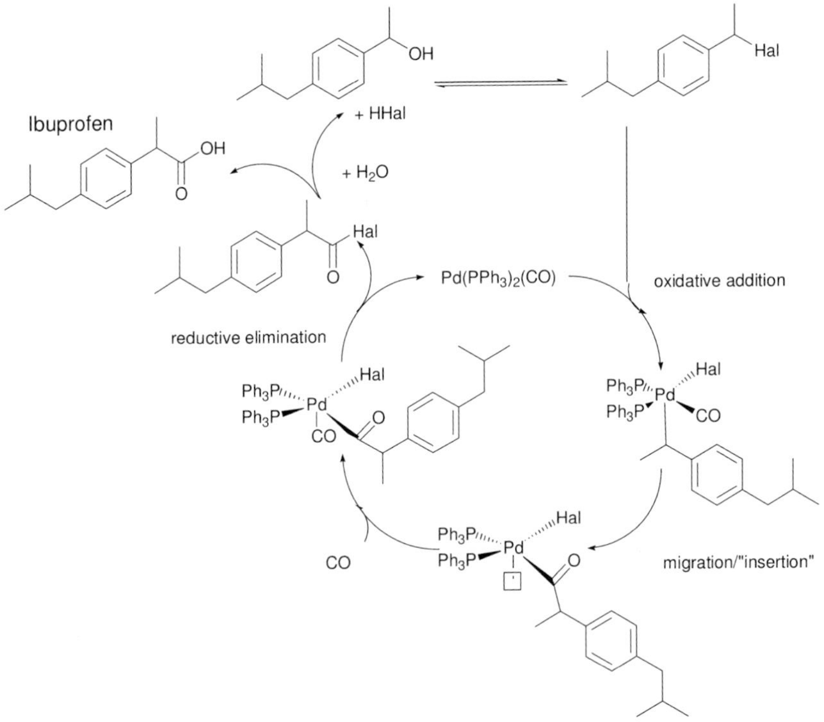

Scheme 5.38 Industrial carbonylation in the Ibuprofen synthesis of Celanese.

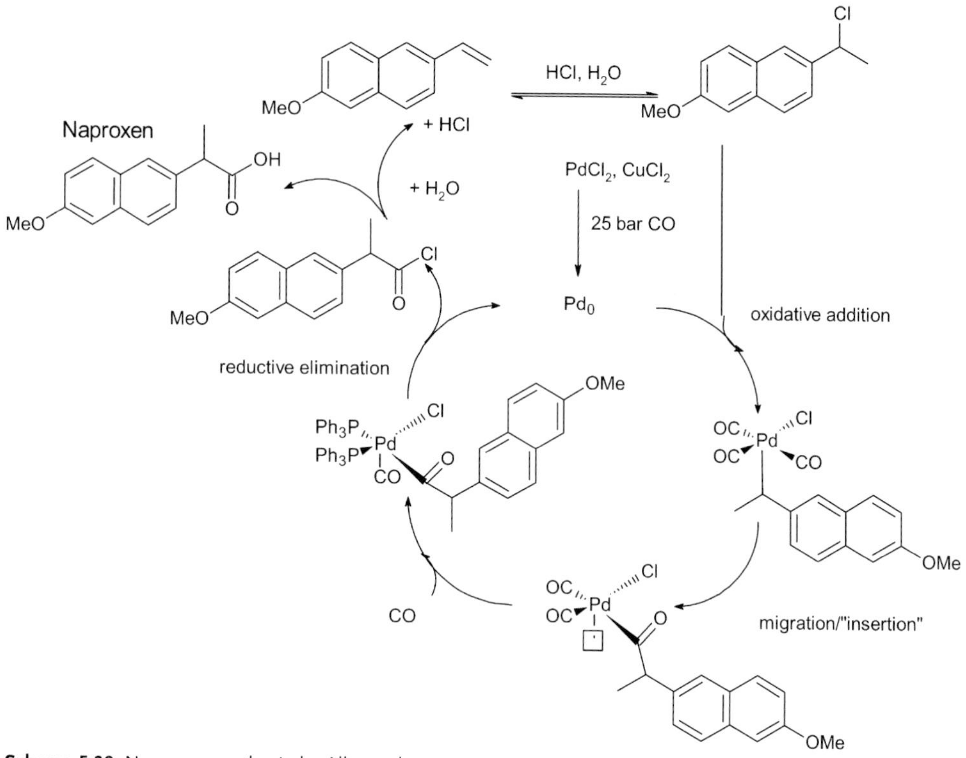

Scheme 5.39 Naproxen synthesis by Albemarle.

most likely be a complex with a bidentate ligand, the sulfonic acid group being in close proximity to the Pd. This oxidative addition also releases nitrogen, which causes an increase in pressure during the reaction, however this can readily be controlled Scheme 5.40). 2-Sulfo-4-methoxybenzoic acid (SMBA) was synthesized in excellent yields with 1 mol% palladium chloride (TON 100, TOF 30 h^{-1}), acetonitrile was employed as solvent and ligand for the catalyst [70, 74].

5.3.1.2 Aminocarbonylation

In the above described carbonylation reactions the nucleophile that reacts with the species released from the catalytic cycle is water or possibly an alcohol. This can be replaced by a more nucleophilic amine, yielding an amide as the product [66]. With this minor variation a different group of products becomes accessible. A striking application of this reaction is the synthesis of the monoamine oxidase B inhibitor Lazabemide [70, 75]. The first laboratory synthesis could be shortened from 8 steps to just one catalytic reaction with a TON of 3000 (Scheme 5.41). The only drawback to the greenness of this reaction is that the metal is removed via an extraction with aqueous NaCN.

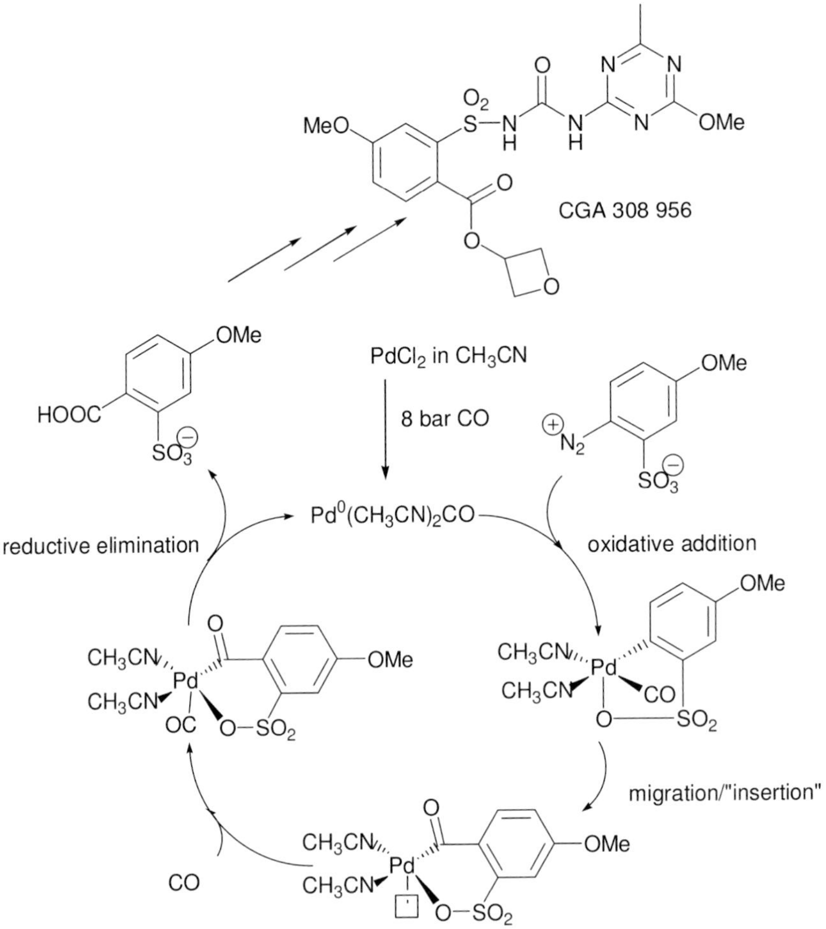

Scheme 5.40 Organohalides can be replaced by diazonium salts.

5.3.1.3 Hydroformylation or "Oxo" Reaction

Hydroformylation, also known by its old name as "oxo" reaction, is the carbonylation of olefins in the presence of hydrogen, i.e. treatment of olefins with synthesis gas (syn gas). Formally it is the addition of a formyl group and one hydrogen atom to a carbon–carbon double bond [61, 76, 77]. This 100% atom efficient reaction was described as early as 1938. At 100–400 bar and 90–250 °C, olefins, in particular ethylene (Scheme 5.42), were converted into the corresponding aldehydes. HCo (CO)$_4$ generated from Co$_2$(CO)$_8$ was used as catalyst, but these reactions had several disadvantages, among them a relatively low selectivity for the n-aldehyde rather than the isoaldehyde, the volatility of the cobalt catalyst and loss of alkene due to hydrogenation. Nowadays this carbon–carbon bond formation is normally performed with a rhodium catalyst. These tend to be 10^4 times faster than the comparable cobalt catalysts. A detailed description of a particularly green hydrofor-

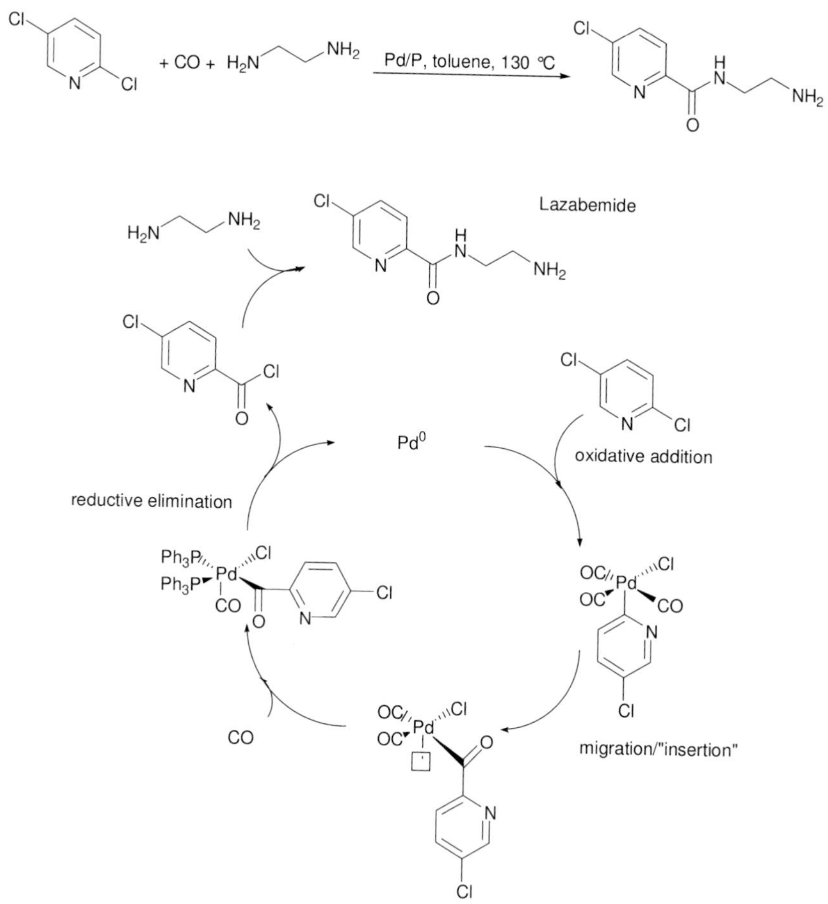

Scheme 5.41 Hoffmann-La Roche production of Lazabemide

mylation, the Rh-catalyzed Ruhrchemie/Rhone-Poulenc process (performed in water), is given in Chapter 7, Section 7.3.1.

5.3.1.4 Hydroaminomethylation

An interesting variation of hydroformylation with a great potential for the industrial preparation of primary amines is hydroaminomethylation. In this process two catalytic reactions are combined, a hydroformylation and a reductive amination of the resulting aldehyde. Although first described more than 60 years ago a really successful procedure was only published recently [78]. To ensure the success of this sequence a rhodium catalyst for the hydroformylation was combined with an iridium catalyst for the imine reduction in a two-phase system, similar to the Ruhrchemie/Rhone-Poulenc process for the hydroformylation. It was demonstrated that less polar solvents such as toluene in combina-

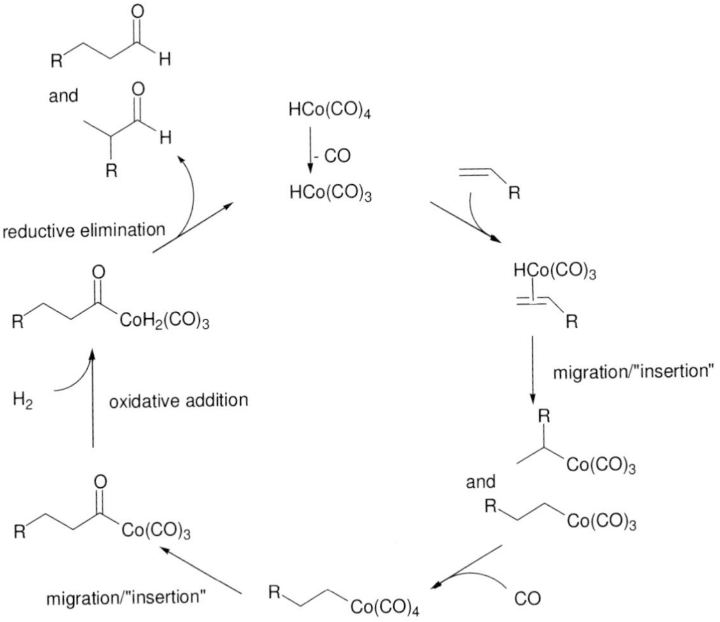

Scheme 5.42 Cobalt-catalyzed hydroformylation.

tion with water and high ammonia concentrations lead to very good selectivities for primary amines. When an ammonia/alkene ratio of 0.5 was employed in combination with BINAS as water-soluble ligand not only were almost pure secondary amines obtained but they had n:iso ratios of 99:1. Overall this approach is opening new opportunities for the green industrial preparation of low molecular mass primary or secondary amines (Scheme 5.43).

Indeed hydroaminomethylation is part of a new route to ε-caprolactam. Starting with readily available butadiene it enables the first carbon–carbon bond formation, extending the carbon chain and introducing the aminogroup. In a sub-

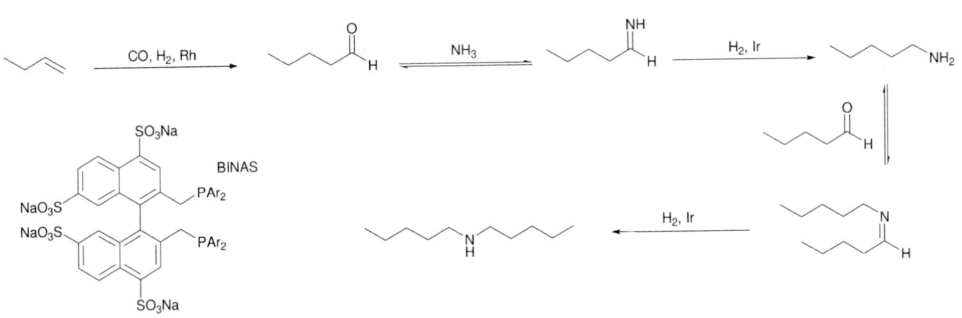

Scheme 5.43 A two-phase approach enabled the first efficient hydroaminomethylation.

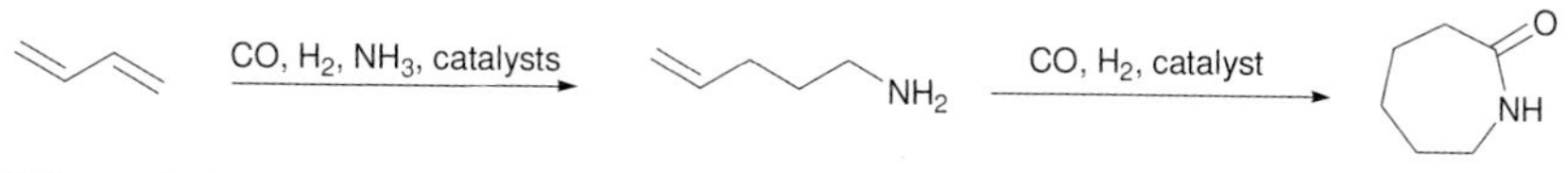

Scheme 5.44 A new route to nylon?

sequent cobalt-catalyzed intramolecular aminocarbonylation the lactam is formed. It was demonstrated that the addition of trialkylphosphines improved the formation of the lactam while suppressing the formation of the polymer [79]. This process is, however, still in the development stage (Scheme 5.44).

5.3.1.5 Methyl Methacrylate via Carbonylation Reactions

Methyl methacrylate (MMA) is one of the most important monomers [80–82]. It forms the basis of acrylic plastics and of polymer dispersion paints. The traditional production is by the formation of acetone cyanohydrin, elimination of water and hydrolysis of the nitrile group, followed by the ester formation. In the carbon–carbon bond forming reaction large amounts of excess HCN and ammonium bisulfate are left as waste. Although these problems have been addressed there is still much room for improvement. In particular the number of reaction steps should be reduced and, in order to achieve this, cyanide should be avoided. The building block to replace it is CO.

Indeed, this has been realised and BASF started a plant on a 5000 ton/year scale in 1989 [81, 82]. The process is based on the hydroformylation of ethylene. Subsequently the propanal is converted via a Mannich reaction into methacrolein, oxidation and esterification then leads to MMA (Scheme 5.45).

A much shorter route is the Reppe carbonylation [83] of propyne. Propyne is, together with propadiene (allene), part of the C3 stream of the cracking process. The order of metal substrate binding strength is allenes > alkynes > alkenes. Thus the desired reaction can only proceed if the propadiene has been removed from the feed, since it is an inhibitor of the Pd catalyst. Equally important, the alkyne complex reacts much faster than the alkene complex. Thus the product is neither a substrate nor an inhibitor for the catalyst (Scheme 5.46).

Pd complexed with 2-pyridyldiphenylphosphine (or its 6-methylpyridyl derivative) and weakly coordinating counter ions, such as methylsulfonate, are employed as catalyst [84, 85]. After the coordination of methanol and CO the migration of the methoxy group takes place to form the palladium carbomethoxy species. Then the coordination of propyne takes place. The steric bulk of the pyridyl group, and in particular of the 6-methylpyridyl group, induces the con-

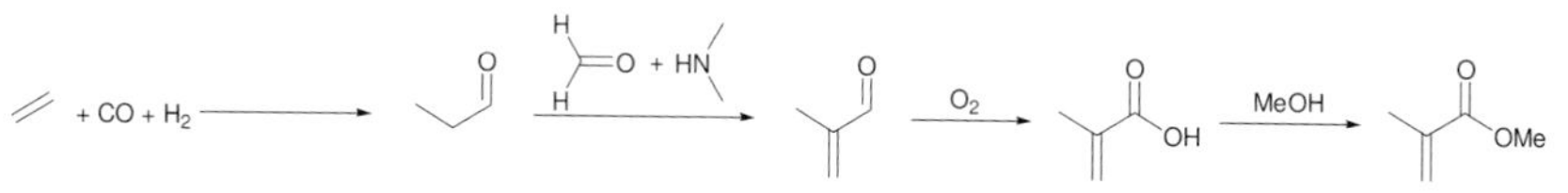

Scheme 5.45 BASF route to MMA.

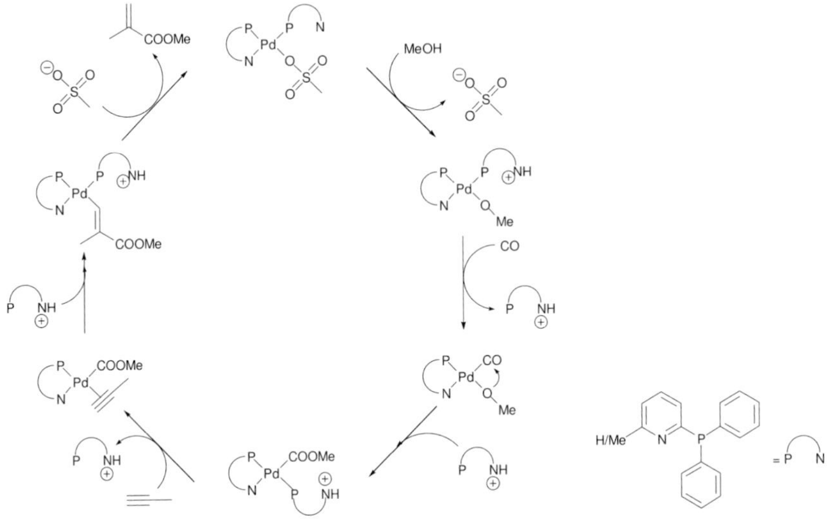

Scheme 5.46 Reppe carbonylation leads directly to MMA.

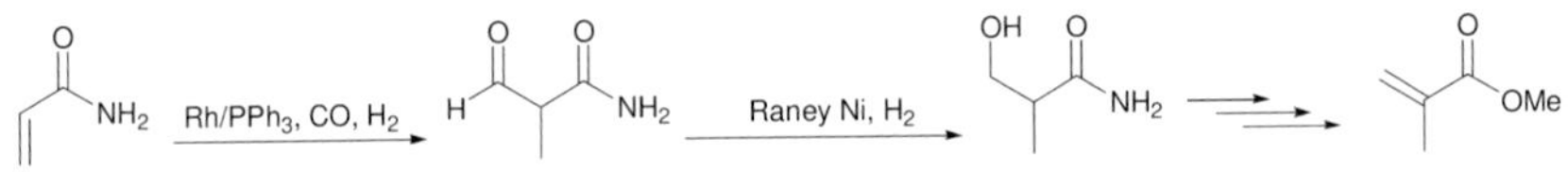

Scheme 5.47 MMA from acrylamide.

figuration necessary to achieve the regiochemistry in the migratory insertion reaction. Thus the carbomethoxy group adds to the triple bond forming 99.95%
pure MMA and only traces of methyl crotonate. Although this process has been
heralded as a green way to MMA it still awaits industrial application.

Recently a new approach towards MMA was proposed. Starting from acrylamide, a Rh-catalyzed hydroformylation yielded, with high selectivity, the desired
branched aldehyde. Given the many possible side reaction this is an achievement in its own right. Addition of Raney Ni allowed a sequential reduction of
the aldehyde [86]. The alcohol obtained in this manner can then be converted
via standard reactions into MMA (Scheme 5.47). Whether this approach does
not involve, once again, too many steps remains to be seen.

5.3.2
Heck-type Reactions

The coupling of aromatic rings with side chains is traditionally performed with
Friedel-Crafts reactions (Chapter 2). With the introduction of the Heck reaction
it is now possible to perform this type of carbon–carbon bond formation under
mild conditions, in the presence of many functional groups and without the
waste that is associated with the use of Lewis acids such as AlCl$_3$. Similarly, Pd-

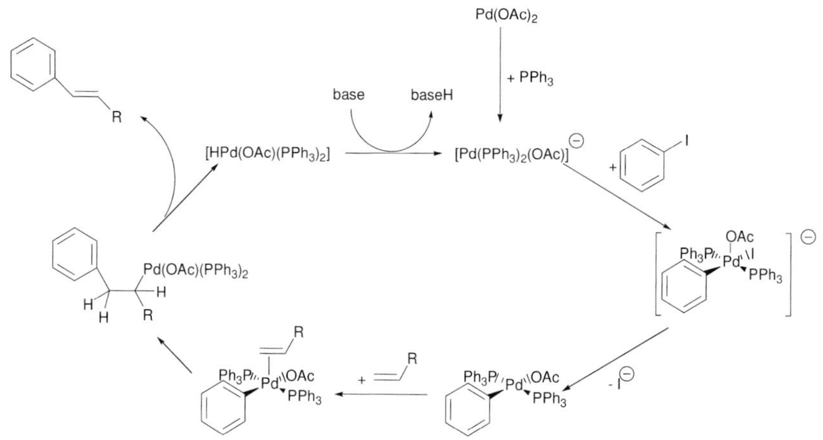

Scheme 5.48 Catalytic cycle of the Heck reaction.

catalyzed carbon–carbon bond forming reactions such as the Suzuki and the Sonogashira coupling can replace the Ullman, Grignard and Glaser coupling.

The Heck reaction is catalyzed by Pd⁰ and a base [72, 73, 87–93]. An oxidative addition of an arylhalide to Pd⁰ constitutes the first step, followed by the coordination of the olefin. The migratory insertion forms the new carbon–carbon bond and a β-elimination releases the desired product. Subsequent deprotonation reduces the Pd^{2+} to Pd^0 again, regenerating the catalyst (Scheme 5.48). In the Suzuki [94] and Sonogashira, as well as in the Negishi and Stille coupling the first step is identical [62]. An organohalide reacts in an oxidative addition with Pd⁰. It is essential that the organohalide has no β-hydrogen atoms since this would allow a rapid β-elimination, preventing the desired coupling reaction. Instead of employing an olefin, like in the Heck reaction, an organometallic compound is used. In the Suzuki reaction this is an organoboronic acid, in the Sonogashira reaction a Cu⁺ organyl is employed, while tin organyls and zinc or-

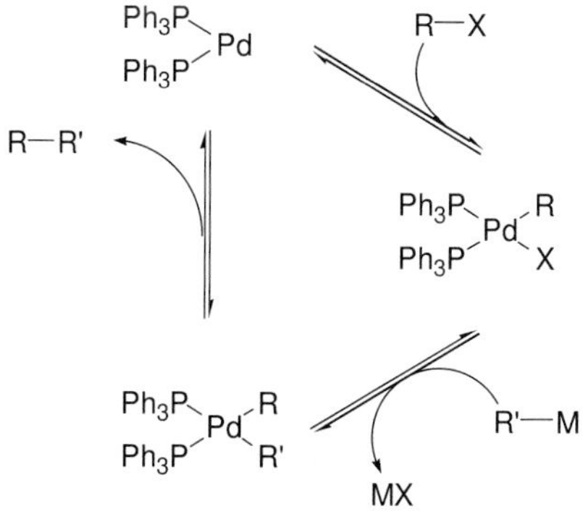

Scheme 5.49 Catalytic cycle of the Suzuki, Stille, Negishi and Sonogashira reactions.

ganyls are used in the Stille and the Negishi reactions, respectively. Transmetallation then leads to a Pd species carrying the two organic groups that are coupled in a reductive elimination (Scheme 5.49). All four coupling reactions have proven their value in organic synthesis, but given the high toxicity of tin the Stille reaction will not be discussed here. Due to the limited industrial application of the Negishi reaction it will also not be included in further discussion.

5.3.2.1 **Heck Reaction**

The great advantage of the Heck reaction is that it enables syntheses of fine chemicals that would otherwise only be accessible via long and wasteful routes. The first reported industrial Heck reaction is the Matsuda-Heck coupling of a diazonium salt, part of the Ciba-Geigy synthesis of the herbicide Prosulfuron [70, 73]. The application of a diazonium salt instead of an arylhalide reduces the amount of salts generated, allowing a particularly green reaction. Subsequent to the carbon–carbon bond formation, charcoal is added to remove the Pd. This at the same time generates the catalyst for the next step, a catalytic hydrogenation. Thus the expensive Pd is used twice and can then be recycled, significantly reducing costs (Scheme 5.50). Indeed this was and is one of the main problems in Pd-catalyzed reactions, to find catalysts that are so efficient that the application of this expensive metal is viable on an industrial scale.

A particularly straightforward way to circumvent the problems with catalyst recycling or the formation of palladium black from the homogeneous Pd catalysts is to use Pd on charcoal. This has been pioneered by IMI/Bromine Co Ltd for the industrial production of the UV-B sunscreen 2-ethylhexyl-*p*-methoxy-cinnamate (Scheme 5.51). At high temperatures *p*-bromoanisole is treated with 2-ethylhexylacrylate to yield the desired product and some side products – most likely due to the 180–190 °C reaction temperature. It is not quite clear what the actual catalytic species is [73, 95]. It can be assumed that separate Pd atoms that are released from the carrier at the elevated temperature are the catalytically active species.

Indeed, the observation that Pd black is often formed during Heck reactions led to detailed studies of the active species in many Heck reactions [96]. It was demonstrated that colloidal Pd, Pd that had not yet formed Pd black clusters

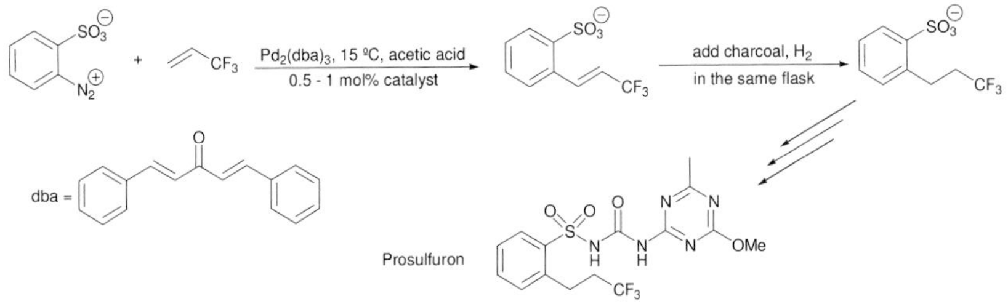

Scheme 5.50 First industrial application of the Heck reaction.

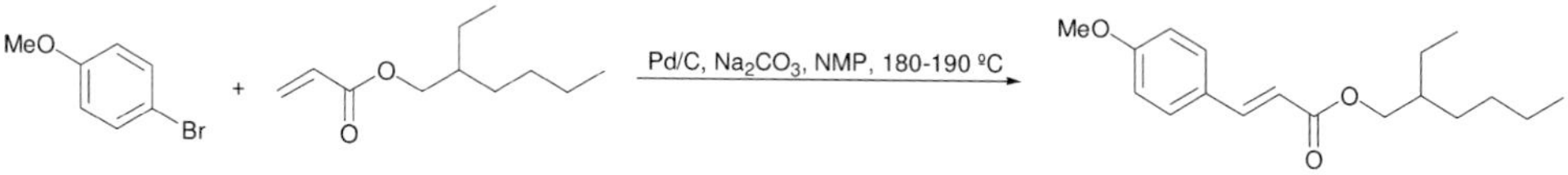

Scheme 5.51 Sunscreen production with immobilized Pd.

but was still in solution, did actually catalyze the reaction. This paved the way for the introduction of "homeopathic concentrations" of ligand-free Pd as catalyst of the Heck reaction [97, 98]. Very recently it was demonstrated that it is indeed the leached atoms from the Pd-clusters that catalyze the Heck reaction [99]. With this result in hand new variations of the Heck reaction are possible.

5.3.2.2 Suzuki and Sonogashira Reaction

The Suzuki coupling enables the high yield and clean linkage of two differently substituted aromatic rings, thus opening new opportunities for synthesis [94]. Based on this reaction many new entities could be synthesized and the Hoechst (now Clariant) process for 2-cyano-4′-methylbiphenyl [70, 72] forms the basis of the production of Valsartan and similar angiotensin II receptor antagonists (Scheme 5.52 a). The reaction is catalyzed by a straightforward catalyst, Pd/sulfonated triphenylphosphine in water/DMSO/glycol at 120 °C. At the end of the reaction two phases form, the polar catalyst can thus be easily recycled. In the Merck synthesis of Losartan the Suzuki coupling is performed at a much later stage [100], linking two highly functionalized benzene rings with excellent selectivity (Scheme 5.52 b), demonstrating the great flexibility of this carbon–carbon bond forming reaction.

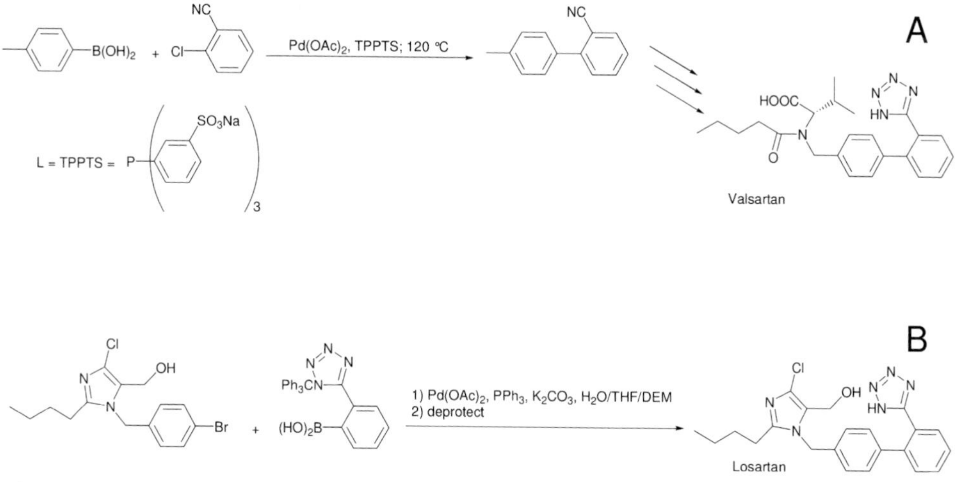

Scheme 5.52 Industrial application of the Suzuki reaction.

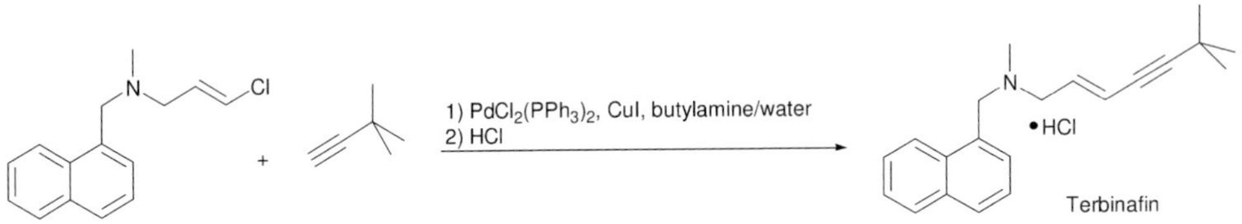

Scheme 5.53 Synthesis of the antifungal agent Terbinafin via Sonogashira coupling.

The Sonogashira coupling has enabled the efficient synthesis of Terbinafin, an antifungal drug [70, 72, 101]. The Sandoz/Novartis process is performed on a multi-ton scale, directly forming the drug. Less than 0.05 mol% of precatalyst $PdCl_2(PPh_3)_2$ in butylamine/water are used and $Cu^I I$ is the co-catalyst to activate the acetylene (Scheme 5.53).

5.3.3
Metathesis

More than half a century ago it was observed that Re_2O_7 and Mo or W carbonyls immobilized on alumina or silica could catalyze the metathesis of propylene into ethylene and 2-butene, an equilibrium reaction. The reaction can be driven either way and it is 100% atom efficient. The introduction of metathesis-based industrial processes was considerably faster than the elucidation of the mechanistic fundamentals [103, 104]. Indeed the first process, the Phillips tri-olefin process (Scheme 5.55) that was used to convert excess propylene into ethylene and 2-butene, was shut down in 1972, one year after Chauvin proposed the mechanism (Scheme 5.54) that earned him the Nobel prize [105]. Starting with a metal carbene species as active catalyst a metallocyclobutane has to be formed. The Fischer-type metal carbenes known at the time did not catalyze the metathesis reaction but further evidence supporting the Chauvin mechanism was published. Once the Schrock-type metal carbenes became known this changed. In 1980 Schrock and coworkers reported tungsten carbene complexes

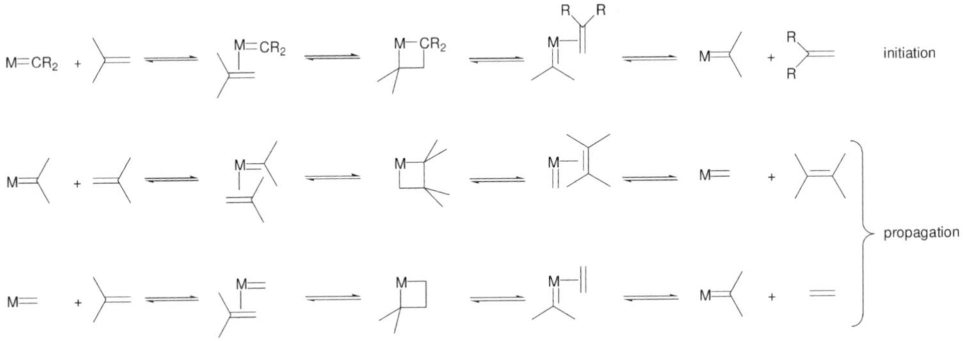

Scheme 5.54 The Chauvin mechanism.

that catalyzed the metathesis reaction [106–108]. The introduction of the first generation Grubbs catalyst then made the metathesis reaction attractive to synthetic organic chemists [109, 110] and since then the metathesis reaction is used in bulk chemicals production and polymerization and also in fine-chemical synthesis, thus spanning the entire range of chemistry.

5.3.3.1 Metathesis Involving Propylene

As mentioned above, the first metathesis reaction studied was the equilibrium between propylene and an ethylene 2-butene mixture. In the initial Phillips process this was used to convert excess propylene into ethylene and 2-butene (Scheme 5.55). When propylene demands surged, the process was reversed and is now known as olefins conversion technology (OCT). The OCT process is operated with a fixed-bed reactor, WO_3 on silica serves as a catalyst. In order to allow 1-butene in the feed MgO is added as an isomerisation catalyst [104]. The process is a gas phase reaction, proceeding at $>260\,°C$ and 30–35 bar and is used globally on a several million ton per year scale. An alternative process has been developed by the Institut Français du Pétrole and the Chinese Petroleum Corporation of Taiwan. In a liquid phase reactor at $35\,°C$ and 60 bar, ethylene and 2-butene are converted in the presence of Re_2O_7 and alumina.

5.3.3.2 Ring-opening Metathesis Polymerization (ROMP)

When cyclic alkenes are utilized as starting materials the metathesis reaction will lead to long chain polymers and/or cyclic oligomers [103, 104, 107, 108]. If a strained cyclic alkene is employed the reaction is effectively irreversible. Industrially cyclooctene (polymer: Vestenamer), 2-norbornene (polymer: Norsorex), and dicyclopentadiene (polymer: Telene, Metton, Pentam) are used as monomers. Upon polymerization cyclooctene and 2-norbornene yield straight chain polymers while dicyclopentadiene also allows cross-linking (Scheme 5.56).

During the last years ROMP has been developed to generate self-healing polymers. In these polymers droplets of dicyclopentadiene and of Grubbs-catalyst are incorporated. When the polymer cracks the droplets burst open, the catalyst comes into contact with the monomer and the plastic ideally heals itself [111]. This methodology is still far from application but it does indicate the power of ROMP.

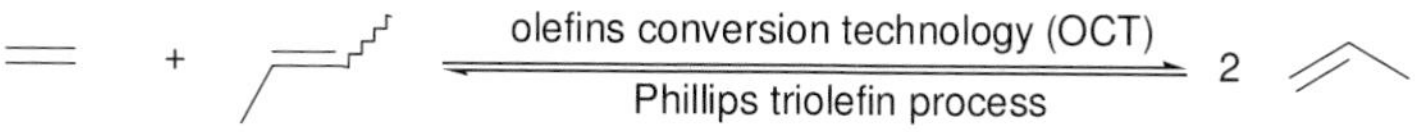

Scheme 5.55 First industrial application of the metathesis reaction.

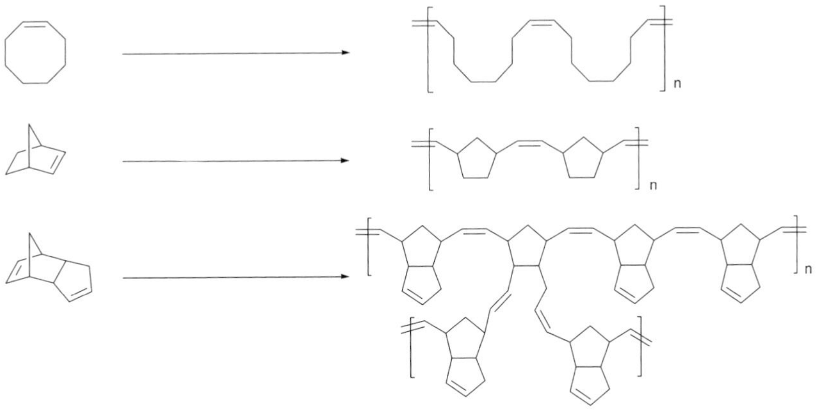

Scheme 5.56 ROMP as an approach to polymers.

5.3.3.3 Ring-closing Metathesis (RCM)

The metathesis reaction was introduced as a method to interconvert small olefins, such as ethylene, propylene and 2-butene. With a growing understanding of the mechanism and the introduction of defined catalysts, ROMP became established. Based on all the knowledge acquired, the opposite reaction could be developed, the ring-closing metathesis (RCM). One of the great challenges in natural product synthesis is the closure of large rings, often lactones or lactams. The classical approaches are often esterifications after the formation of amides, all this occurring in the presence of many delicate functional groups [108–110, 112]. RCM is a novel approach: an intramolecular metathesis reaction closes the ring, while releasing ethane, thus the reaction is virtually irreversible (Scheme 5.57). Given the highly improved metathesis catalysts, such as Grubbs first and second-generation catalyst, the reaction can now be applied in the presence of countless other functional groups. Thus an entirely new synthetic approach was introduced, opening up many new opportunities and establishing a new retrosynthetic disconnection.

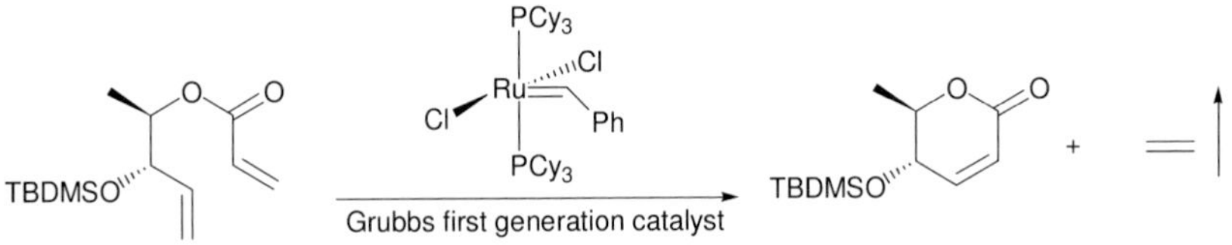

Scheme 5.57 Ring-closing metathesis in the presence of functional groups.

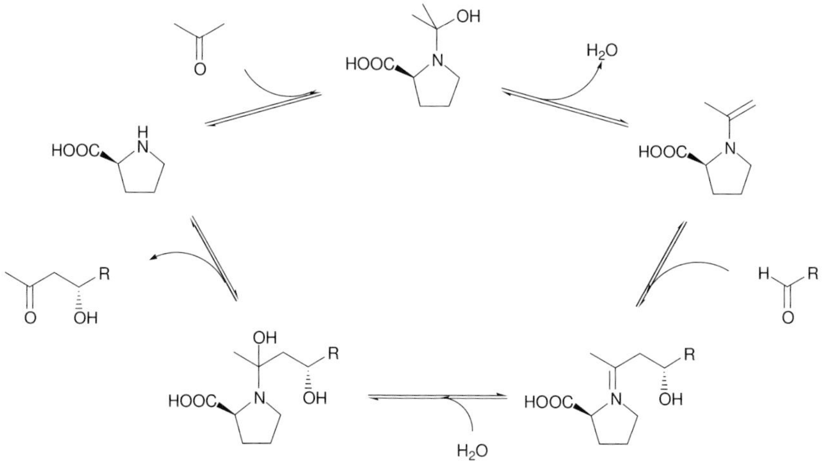

Scheme 5.58 Proline-catalyzed aldol reaction.

5.4
Conclusion and Outlook

The formation of carbon–carbon bonds is at the heart of organic synthesis and essential to the production of bulk and fine chemicals as well as pharmaceuticals. During the first half of the twentieth century, transition metal catalyzed reactions were introduced for the production of small molecules and the first enantioselective enzymatic processes were established. More recently a new type of catalysis has entered the stage, organocatalysis. Central to its success is the extreme versatility of the amino acid proline, which almost seems to be able to enantioselectively catalyze every reaction of the aldehyde/keto group, and with high selectivity too. This has introduced a new approach to the aldol chemistry and many reports on this and similar reactions have been published and reviewed [41, 113–116]. The proposed mechanism has recently been confirmed with an electrospray ionization MS study (Scheme 5.58) [117]. It can be expected that organocatalysis will evolve into a mature field of catalysis within the next few years, adding many green approaches towards essential building blocks in organic chemistry.

References

1 B.M. Trost, *Science* **1991**, *254*, 1471–1477.
2 G. Seoane, *Curr. Org. Chem.* **2000**, *4*, 283–304.
3 K. Faber, *Biotransformations in Organic Chemistry*, 5th edn., Springer, Berlin, Heidelberg, New York, 2004.
4 J. Sukumaran, U. Hanefeld, *Chem. Soc. Rev.* **2005**, *34*, 530–542.
5 R.J.H. Gregory, *Chem. Rev.* **1999**, *99*, 3649–3682.
6 J. Brussee, A. van der Gen, in *Stereoselective Biocatalysis*, P. N. Ramesh (Ed.), Mar-

cel Dekker, New York, 2000, pp. 289–320.

7 M. North, *Tetrahedron: Asym.* **2003**, *14*, 147–176.

8 J.-M. Brunel, I. P. Holmes, *Angew. Chem., Int. Ed.* **2004**, *43*, 2752–2778.

9 M. Breuer, K. Ditrich, T. Habicher, B. Hauer, M. Kesseler, R. Stürmer, T. Zelinski, *Angew. Chem. Int. Ed.* **2004**, *43*, 788–824.

10 K. Gruber, C. Kratky, *J. Polym. Sci. Part A: Polym. Chem.* **2004**, *42*, 479–486.

11 A. M. C. H. van den Nieuwendijk, A. B. T. Ghisaidoobe, H. S. Overkleeft, J. Brussee, A. van der Gen, *Tetrahedron* **2004**, *60*, 10385–10396.

12 A. Glieder, R. Weis, W. Skranc, P. Poechlauer, I. Dreveny, S. Majer, M. Wubbolts, H. Schwab, K. Gruber, *Angew. Chem. Int. Ed.* **2003**, *42*, 4815–4818.

13 R. Weis, R. Gaisberger, W. Skranc, K. Gruber, A. Glieder, *Angew. Chem. Int. Ed.* **2005**, *44*, 4700–4704.

14 R. P. Gaisberger, M. H. Fechter, H. Griengl, *Tetrahedron: Asym.* **2004**, *15*, 2959–2963.

15 N. Klempier, H. Griengl, *Tetrahedron Lett.* **1993**, *34*, 4169–4172.

16 H. Griengl, A. Hickel, D. V. Johnson, C. Kratky, M. Schmidt, H. Schwab, *Chem. Commun.* **1997**, 1933–1949.

17 H. Hirohara, M. Nishizawa, *Biosci. Biotechnol. Biochem.* **1998**, *62*, 1–9.

18 H. Bühler, A. Bayer, F. Effenberger, *Chem. Eur. J.* **2000**, *6*, 2564–2571.

19 M. Inagaki, J. Hiratake, T. Nishioka, J. Oda, *J. Am. Chem. Soc.* **1991**, *113*, 9360–9361.

20 M. Inagaki, J. Hiratake, T. Nishioka, J. Oda, *J. Org. Chem.* **1992**, *57*, 5643–5649.

21 M. Inagaki, A. Hatanaka, M. Mimura, J. Hiratake, T. Nishioka, J. Oda, *Bull. Chem. Soc. Jpn* **1992**, *65*, 111–120.

22 U. Hanefeld, *Org. Biomol. Chem.* **2003**, *1*, 2405–2415.

23 C. Paizs, P. Tähtinen, M. Tosa, C. Majdik, F.-D. Irimie, L. T. Kanerva, *Tetrahedron* **2004**, *60*, 10533–10540.

24 L. Veum, U. Hanefeld, *Tetrahedron: Asym.* **2004**, *15*, 3707–3709.

25 L. Veum, L. T. Kanerva, P. J. Halling, T. Maschmeyer, U. Hanefeld, *Adv. Synth. Catal.* **2005**, *347*, 1015–1021.

26 M. Pohl, B. Lingen, M. Müller, *Chem. Eur. J.* **2002**, *8*, 5289–5295.

27 F. Jordan, *Nat. Prod. Rep.* **2003**, *20*, 184–201.

28 N. J. Turner, *Curr. Opin. Biotechnol.* **2000**, *11*, 527–531.

29 A. S. Demir, Ö. Sesenoglu, P. Dünkelmann, M. Müller, *Org. Lett.* **2003**, *5*, 2047–2050.

30 A. S. Demir, P. Ayhan, A. C. Igdir, A. N. Duygu, *Tetrahedron* **2004**, *60*, 6509–6512.

31 P. Dünkelmann, D. Kolter-Jung, A. Nitsche, A. S. Demir, P. Siegert, B. Lingen, M. Baumann, M. Pohl, M. Müller, *J. Am. Chem. Soc.* **2002**, *124*, 12084–12085.

32 H. Iding, T. Dünnwald, L. Greiner, A. Liese, M. Müller, P. Siegert, J. Grötzinger, A. S. Demir, M. Pohl, *Chem. Eur. J.* **2000**, *6*, 1483–1495.

33 B. Rosche, M. Breuer, B. Hauer, P. L. Rogers, *Biotechnol. Bioeng.* **2004**, *86*, 788–794.

34 B. Rosche, V. Sandford, M. Breuer, B. Hauer, P. L. Rogers, *J. Mol. Catal. B: Enzym.* **2002**, *19/20*, 109–115.

35 I. A. Sevostyanova, O. N. Solovjeva, G. A. Kochetov, *Biochem. Biophys. Res. Commun.* **2004**, *313*, 771–774.

36 A. Banerjee, R. Sharma, U. C. Banerjee, *Appl. Microbiol. Biotechnol.* **2002**, *60*, 33–44.

37 C. O'Reilly, P. D. Turner, *J. Appl. Microbiol.* **2003**, *95*, 1161–1174.

38 G. DeSantis, Z. Zhu, W. A. Greenberg, K. Wong, J. Chaplin, S. R. Hanson, B. Farwell, L. W. Nicholson, C. L. Rand, D. P. Weiner, D. E. Robertson, M. J. Burk, *J. Am. Chem. Soc.* **2002**, *124*, 9024–9025.

39 C. Mateo, A. Chmura, S. Rustler, F. van Rantwijk, A. Stolz, R. A. Sheldon, *Tetrahedron: Asym.* **2006**, *17*, 320–323.

40 T. D. Machajewski, C.-H. Wong, *Angew. Chem. Int. Ed.* **2000**, *39*, 1352–1374.

41 Z. Tang, F. Jiang, L.-T. Yu, X. Cui, L.-Z. Gong, A.-Q. Mi, Y.-Z. Jiang, Y.-D. Wu, *J. Am. Chem. Soc.* **2003**, *125*, 5262–5263.

42 R. Mestres, *Green Chem.* **2004**, *6*, 583–603.

43 M. G. Silvestri, G. DeSantis, M. Mitchell, C.-H. Wong, in *Topics in Stereochemistry*, S. E. Denmark (Ed.), John Wiley & Sons, Hoboken, NJ, 2003, Vol. 23, pp. 267–342.

44 A. N. Phung, M. T. Zannetti, G. Whited, W.-D. Fessner, *Angew. Chem. Int. Ed.* **2003**, *42*, 4821–4824.

45 C.-C. Lin, F. Moris-Varas, G. Weitz-Schmidt, C.-H. Wong, *Bioorg. Med. Chem.* **1999**, *7*, 425–433.

46 R. Schoevaart, F. van Rantwijk, R. A. Sheldon, *J. Org. Chem.* **2000**, *65*, 6940–6943.

47 R. Schoevaart, F. van Rantwijk, R. A. Sheldon, *J. Org. Chem.* **2001**, *66*, 4559–4562.

48 F. Charmantray, P. Dellis, S. Samreth, L. Hecquet, *Tetrahedron Lett.* **2006**, *47*, 3261–3263.

49 C.-C. Lin, C.-H. Lin, C.-H. Wong, *Tetrahedron Lett.* **1997**, *38*, 2649–2652.

50 Z. Liu, R. Weis, A. Glieder, *Food Technol. Biotechnol.* **2004**, *42*, 237–249.

51 C. Wandrey, A. Liese, D. Kihumbu, *Org. Proc. Res. Devel.* **2000**, *4*, 286–290.

52 S.-J. Lee, H.-Y. Kang, Y. Lee, *J. Mol. Cat. B Enzym.* **2003**, *26*, 265–272.

53 T. Tanaka, C. Tsuda, T. Miura, T. Inazu, S. Tsuji, S. Nishihara, M. Hisamatsu, T. Kajimoto, *Synlett* **2004**, 243–246.

54 A. Heine, J. G. Luz, C.-H. Wong, I. A. Wilson, *J. Mol. Biol.* **2004**, *343*, 1019–1034.

55 J. Liu, C.-H. Wong, *Angew. Chem. Int. Ed.* **2002**, *41*, 1404–1407.

56 J. Liu, C.-C. Hsu, C.-H. Wong, *Tetrahedron Lett.* **2004**, *45*, 2439–2441.

57 W. A. Greenberg, A. Varvak, S. R. Hanson, K. Wong, H. Huang, P. Chen, M. J. Burk, *Proc. Natl. Acad. Sci.* **2004**, *101*, 5788–5793.

58 S. Panke, M. Wubbolts, *Curr. Opin. Chem. Biol.* **2005**, *9*, 188–194.

59 M. Müller, *Angew. Chem. Int. Ed.* **2005**, *44*, 362–365.

60 M. M. Elenkov, B. Hauer, D. B. Janssen, *Adv. Synth. Catal.* **2006**, *348*, 579–585.

61 P. W. N. M. van Leeuwen, *Homogeneous Catalysis*, Kluwer Academic Publishers, Dordrecht, 2004.

62 R. Bates, *Organic Synthesis using Transition Metals*, Sheffield Academic Press, Sheffield, 2000.

63 R. B. Jordan, *Reaction Mechanisms of Inorganic and Organometallic Systems*, 2nd edn, Oxford University Press, Oxford, 1998.

64 C.-J. Li, *Chem. Rev.* **2005**, *105*, 3095–3165.

65 T. Morimoto, K. Kakiuchi, *Angew. Chem. Int. Ed.* **2004**, *43*, 5580–5588.

66 R. Skoda-Földes, L. Kollar, *Curr. Org. Chem.* **2002**, *6*, 1097–1119.

67 H. Cheung, R. S. Tanke, G. P. Torrence, Acetic Acid, in *Ullmann's Encyclopedia of Industrial Chemistry*, electronic edition, Wiley-VCH, Weinheim, 2005.

68 M. J. Howard, M. D. Jones, M. S. Roberts, S. A. Taylor, *Catal. Today* **1993**, *18*, 325–354.

69 G. J. Sunley, D. J. Watson, *Catal. Today* **2000**, *58*, 293–307.

70 H.-U. Blaser, A. Indolese, F. Naud, U. Nettekoven, A. Schnyder, *Adv. Synth. Catal.* **2004**, *346*, 1583–1598.

71 J. G. de Vries, Homogeneous catalysis for fine chemicals, in *Encyclopedia of Catalysis*, I. T. Horvath (Ed.), Wiley, New York, 2003, Vol. 3, p. 295.

72 A. Zapf, M. Beller, *Top. Catal.* **2002**, *19*, 101–109.

73 J. G. de Vries, *Can. J. Chem.* **2001**, *79*, 1086–1092.

74 U. Siegrist, T. Rapold, H. U. Blaser, *Org. Process Res. Dev.* **2003**, *7*, 429–431.

75 R. Schmidt, *Chimia* **1996**, *50*, 110–113.

76 B. Breit, W. Seiche, *Synthesis* **2001**, 1–36.

77 H. W. Bohnen, B. Cornils, *Adv. Catal.* **2002**, *47*, 1–64.

78 B. Zimmermann, J. Herwig, M. Beller, *Angew. Chem. Int. Ed.* **1999**, *38*, 2372–2375.

79 S. Liu, A. Sen, R. Parton, *J. Mol. Catal. A Chem.* **2004**, *210*, 69–77.

80 B. Cornils, W. A. Herrmann, *J. Catal.* **2003**, *216*, 23–31.

81 K. Nagai, *Appl. Catal. A Gen.* **2001**, *221*, 367–377.

82 W. Bauer, Jr., Methacrylic Acid and Derivatives, in *Ullmann's Encyclopedia of Industrial Chemistry*, electronic edition, Wiley-VCH, Weinheim, 2005.

83 G. Kiss, *Chem. Rev.* **2001**, *101*, 3435–3456.

84 E. Drent, P. Arnoldy, P. H. M. Budzelaar, *J. Organomet Chem.* **1993**, *455*, 247–253.

85 E. Drent, P. Arnoldy, P. H. M. Budzelaar, *J. Organomet Chem.* **1994**, *475*, 57–63.

86 L. Garcia, C. Claver, M. Dieguez, A. M. Masdeu-Bulto, *Chem. Commun.* **2006**, 191–193.

87 I. P. Beletskaya, A. V. Cheprakov, *Chem. Rev.* **2000**, *100*, 3009–3066.

88 V. Farina, *Adv. Synth. Catal.* **2004**, *346*, 1553–1582.

89 A. C. Frisch, M. Beller, *Angew. Chem. Int. Ed.* **2005**, *44*, 674–688.

90 A. Zapf, M. Beller, *Chem. Commun.* **2005**, 431–440.

91 C. E. Tucker, J. G. de Vries, *Top. Catal.* **2002**, *19*, 111–118.

92 F. Alonso, I. P. Beletskaya, M. Yus, *Tetrahedron* **2005**, *61*, 11771–11835.

93 A. B. Dounay, L. E. Overman, *Chem. Rev.* **2003**, *103*, 2945–2963.

94 A. Suzuki, *Chem. Commun.* **2005**, 4759–4763.

95 A. Eisenstadt, D. J. Ager, in *Fine Chemicals through Heterogeneous Catalysis*, R. A. Sheldon, H. van Bekkum (Eds.), Wiley-VCH, Weinheim, 2001, pp. 576–587.

96 M. Nowotny, U. Hanefeld, H. van Koningsveld, T. Maschmeyer, *Chem. Commun.* **2000**, 1877–1878.

97 A. H. M. de Vries, J. M. C. A. Mulders, J. H. M. Mommers, H. J. W. Henderickx, J. G. de Vries, *Org. Lett.* **2003**, *5*, 3285–3288.

98 M. T. Reetz, J. G. de Vries, *Chem. Commun.* **2004**, 1559–1563.

99 M. B. Thathagar, J. E. ten Elshof, G. Rothenberg, *Angew. Chem.* **2006**, *118*, 2952–2956.

100 R. D. Larsen, A. O. King, C. Y. Chen, E. G. Corley, B. S. Foster, F. E. Roberts, C. Yang, D. R. Lieberman, R. A. Reamer, D. M. Tschaen, T. R. Verhoeven, P. J. Reider, Y. S. Lo, L. T. Romano, A. S. Brookes, D. Meloni, J. R. Moore, J. F. Arnett, *J. Org. Chem.* **1994**, *59*, 6391–6394.

101 U. Beutler, J. Macazek, G. Penn, B. Schenkel, D. Wasmuth, *Chimia* **1996**, *50*, 154–156.

103 D. Astruc, *New J. Chem.* **2005**, *29*, 42–56.

104 J. C. Mol, *J. Mol. Catal. A Chem.* **2004**, *213*, 39–45.

105 Y. Chauvin, *Angew. Chem. Int. Ed.* **2006**, *45*, 3741–3747.

106 J. H. Wengrovius, R. R. Schrock, M. R. Churchill, J. R. Missert, W. J. Youngs, *J. Am. Chem. Soc.* **1980**, *102*, 4515–4516.

107 R. R. Schrock, *Acc. Chem. Res.* **1990**, *23*, 158–165.

108 R. R. Schrock, *Angew. Chem. Int. Ed.* **2006**, *45*, 3748–3759.

109 R. H. Grubbs, *Angew. Chem. Int. Ed.* **2006**, *45*, 3760–3765.

110 R. H. Grubbs, *Tetrahedron* **2004**, *60*, 7117–7140.

111 B. C. Bernstein, *IEEE Elect. Insul. Mag.*, **2006**, *22*, 15–20.

112 K. C. Nicolaou, P. G. Bulger, D. Sarlah, *Angew. Chem. Int. Ed.* **2005**, *44*, 4490–4527.

113 J. Seayad, B. List, *Org. Biomol. Chem.* **2005**, *3*, 719–724.

114 P. I. Dalko, L. Moisan, *Angew. Chem. Int. Ed.* **2004**, *43*, 5138–5175.

115 B. List, *Tetrahedron* **2002**, *58*, 5573–5590.

116 B. List, *Chem. Commun.* **2006**, 819–824.

117 C. Marquez, J. O. Metzger, *Chem. Commun.* **2006**, 1539–1541.

6
Hydrolysis

6.1
Introduction

Hydrolysis reactions are straightforward to carry out and they have a favorable equilibrium, since the solvent, i.e. water, is one of the reagents. This also has the advantage of being particularly green, since water is an environmentally benign solvent. However, alongside these advantages traditional hydrolysis reactions have many disadvantages. They are commonly performed with concentrated and thus corrosive acids or bases. These need to be neutralized at the end of the reaction, generating salts as waste. In addition high temperatures are often required, wasting much energy. If the substrate submitted to the hydrolysis reaction has more than one functional group lack of selectivity becomes a major problem; delicate structures might even be completely degraded. At the end of the reaction the product needs to be separated from the water, which is often difficult and energy consuming, especially if it involves a distillation. Furthermore hydrolysis reactions are never 100% atom efficient, depending on the group that is cleaved off they can even be quite atom inefficient. From a green point of view much can still be improved in traditional hydrolysis reactions [1]. Different approaches, such as solid acids and bases (Chapter 2), have been developed to address some of these problems. However, only one type of catalysis offers a solution to almost all of the aspects, namely: Biocatalysis.

Enzymes can hydrolyse esters, amides and nitriles under very mild, non-corrosive conditions at neutral pH values. Temperatures can be kept low, however, given the great thermostability of many enzymes, they can be raised if necessary. Due to the mild reaction conditions even fragile substrates can be converted selectively. Indeed, this is one of the major advantages of hydrolases (hydrolyzing enzymes). They are highly selective, allowing chemo-, regio-, and enantioselective hydrolyses of different functional groups in the substrates.

This chapter consequently focuses on the application of enzymes for the selective cleavage of esters, amides and nitriles [2]. Out of all the reported industrial applications of enzymes these type of hydrolyses constitute more than 40% [3]. Enzymatic hydrolyses are often performed because of the enantioselectivity of enzymes, and in particular of the lipases that are used for the production of enantiopure fine chemicals.

Green Chemistry and Catalysis. I. Arends, R. Sheldon, U. Hanefeld
Copyright © 2007 WILEY-VCH Verlag GmbH & Co. KGaA, Weinheim
ISBN: 978-3-527-30715-9

6.1.1
Stereoselectivity of Hydrolases

Hydrolases in general are enantioselective and in some cases even almost enantiospecific. Indeed, the great value of lipases, esterases, proteases, amidases and nitrilases for synthesis is, in part, due to their high enantioselectivity. While there are no general rules for all hydrolases, a brief description of the enantiopreferences of lipases and proteases (subtilisin) is given. Since lipases are widely used in hydrolytic kinetic resolutions of secondary alcohols their enantioselectivity is of particular interest. A model describing this has been established; it is known as Kazlauskas rule [2, 4–7]. This rule is based on the empirical observation that many lipases preferentially catalyse the conversion of one of the enantiomers of secondary alcohols. This holds true for both the synthesis and the hydrolysis reaction (Scheme 6.1). This stereoselectivity has been explained by the spatial arrangement of the catalytic residues on the basis of X-ray studies [8]. Interestingly, different lipases seem to have different methods for inducing chirality, even though they all work with a catalytic triad and the net result is the same [9, 10]. In the case of chiral primary amines Kazlauskas rule tends to give reliable results, too [11]. It is important to note that proteases (subtilisin) commonly show the opposite enantioselectivity (Scheme 6.2) [12]. Interestingly, the enantioselectivity of subtilisin can be solvent dependent and particularly in water the selectivities can be reversed [13]. Looking at all these types of hydrolases there is, in general, an enzyme available to hydrolyse either enantiomer of a secondary alcohol or chiral primary amine [2].

A general rule describing *Burkholderia cepacia* lipase (formerly called *Pseudomonas cepacia* lipase, PCL) catalysed conversions of primary alcohols with a chiral carbon in the β-position is depicted in Scheme 6.3. This rule, however, is only reliable if there are no oxygen substituents on the chiral carbon [14].

As well as the kinetic resolution of chiral alcohols, lipases can be employed to resolve chiral acids. Their enantioselectivity towards chiral acids or their esters is, however, less predictable. *Candida rugosa* lipase (CRL) has a general stereochemical preference for one enantiomer of acids with a chiral α-carbon (Scheme 6.4) [15].

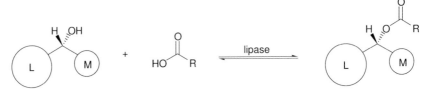

Scheme 6.1 Enantioselectivity of lipases for secondary alcohols and their esters according to the rule of Kazlauskas: The faster reacting enantiomer of a secondary alcohol in most acylations or the faster reacting enantiomer of an ester in most hydrolyses is the enantiomer depicted. Most primary amines also follow this rule. L: largest substituent, M: medium-sized substituent.

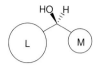

reacts in subtilisin catalysed reactions   does not react

Scheme 6.2 Subtilisin commonly displays the opposite enantioselectivity to lipases. L: largest substituent, M: medium-sized substituent.

reacts   does not react

Scheme 6.3 PCL preferentially catalyses the hydrolysis/acylation of only one of the depicted enantiomers of the chiral primary alcohol (ester). The selectivity is low if oxygen is bound to the chiral carbon. L: largest substituent, M: medium-sized substituent.

The enantioselectivity that the enzymes display is not always very large. Indeed, there are even examples known for the opposite selectivity. Nonetheless, with these general rules, it is possible to address a large number of synthetic problems. For the detailed choice of the right biocatalyst for each particular substrate, several excellent reviews are available [2, 16–23].

For the other hydrolases general rules for their enantioselectivity are not available. This does not mean that their enantioselectivities are unknown; indeed for many enzymes much knowledge is available, however, less is known about the entire class of enzymes of which these individual, well-researched enzymes are an element.

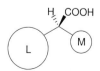

Scheme 6.4 Preferred enantiomer in CRL-catalysed hydrolysis and esterification reactions of chiral acids. L: largest substituent, M: medium-sized substituent.

6.1.2
Hydrolase-based Preparation of Enantiopure Compounds

Hydrolases catalyse the hydrolysis of esters, amides and nitriles under mild conditions. As mentioned above, they also display an often-remarkable enantioselectivity. This opens the opportunity to employ these enzymes for the preparation of enantiopure compounds. This desymmetrization has been reviewed [16–24] and different approaches [25–27] are possible.

6.1.2.1 Kinetic Resolutions

The most straightforward hydrolase-catalysed preparation of an enantiopure product from a racemic starting material is a kinetic resolution. In a kinetic resolution a racemic ester or amide is hydrolysed enantioselectively (Scheme 6.5 A). At the end of the reaction an enantiopure alcohol or amine is obtained. The unreactive enantiomer of the starting material should ideally also be enantiopure. Consequently the maximum yield for either compound in a kinetic resolution is only 50%. The enantiopurity of the products is dependent on the enantioselectivity of the enzyme; this is expressed as the enantiomeric ratio, E, of the enzyme [28]. If the E value for the enzyme is low (<25) neither the unreacted ester nor the obtained product are really pure at 50% conversion. Consequently kinetic re-

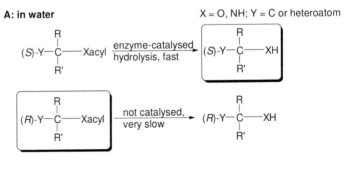

Scheme 6.5 A: Kinetic resolution of chiral esters or amides in water; B: kinetic resolution of chiral alcohols and amines in organic solvents.

solutions are only viable if the E of the enzyme is high (>50) and even then 50% of the starting material ends up as waste and needs to be recycled.

Already in 1900 it was demonstrated that hydrolase-catalysed reactions are reversible [29]. By switching from water as a solvent to organic solvents and by employing activated acids as acyl donors the reactions can even be converted into irreversible acylations [2, 22, 27]. An essential step towards utilizing this potential of the hydrolases was their application in organic solvents [30, 31]. The solvents of choice are hydrophobic, so that the enzyme does not dissolve and deactivate, nor lose the often-essential water that is bound to it. Common organic solvents such as *tert*-butyl methyl ether (MTBE), toluene and hexane (which should be replaced by the less toxic heptane) have proven their value in this type of reaction [22, 27]. The replacement of water as a solvent and reactant with the alcohol commonly leads to enzyme deactivation and is not normally used. Based on these findings kinetic resolutions of chiral alcohols or amines can be performed in organic solvents (Scheme 6.5 B). Again a maximum yield of only 50% can be obtained. Additional disadvantages are that an excess of acyl donor needs to be added (2–4 equivalents) and that a potentially toxic solvent is used.

Of course the stereocenter can also be in the acid moiety, again kinetic resolutions can be performed either in water or in organic solvents. If the starting material is a chiral nitrile kinetic resolutions in water are equally possible, in that case a mixture of enantiopure nitrile and acid/amide are formed.

6.1.2.2 Dynamic Kinetic Resolutions

An elegant way to avoid the low yields and the need for recycling half of the material in the case of kinetic resolutions is a dynamic kinetic resolution (DKR). The dynamic stands for the dynamic equilibrium between the two enantiomers that are kinetically resolved (Scheme 6.6 A). This fast racemisation ensures that the enzyme is constantly confronted with an (almost) racemic substrate. At the end of the reaction an enantiopure compound is obtained in 100% yield from racemic starting material. Mathematical models describing this type of reaction have been published and applied to improve this important reaction [32, 33]. There are several examples, in which the reaction was performed in water (see below). In most cases the reaction is performed in organic solvents and the hydrolase-catalysed reaction is the irreversible formation of an ester (for example see Figs. 9.3, 9.4, 9.6, 9.12) or amide (for example see Figs. 9.13, 9.14, 9.16).

When looking at the above described dynamic kinetic resolution from a green point of view, then one thing can immediately be noticed: This reaction would be unnecessary if the starting material had been synthesized enantioselectively. A much more efficient way of performing a dynamic kinetic resolution is thus to start with a prochiral material. The reversible addition of another building block to this prochiral starting material is not only the formation of a new bond but at the same time a pathway for the rapid racemisation of the intermediate

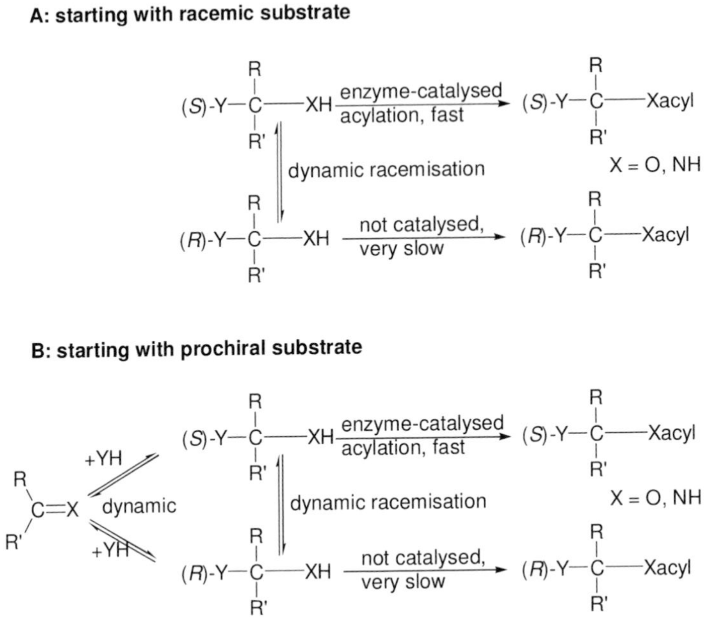

A: starting with racemic substrate

B: starting with prochiral substrate

Scheme 6.6 A: Dynamic kinetic resolution of a racemic starting material yields 100% enantiopure product; B: in a synthetic dynamic kinetic resolution a new bond is formed enantioselectively with 100% yield.

racemic alcohol or amine. The irreversible hydrolase-catalysed reaction then induces the stereochemistry of the final product. Overall a new bond is formed enantioselectively (Scheme 6.6 B). This dynamic kinetic resolution is thus a truly synthetic reaction, indeed it is applied for the enantioselective synthesis of chiral α-hydroxy acids on an industrial scale (Schemes 5.16, 5.17) [34]. This type of dynamic kinetic resolution is also mainly performed in organic solvents (for examples see: Schemes 5.7 and 5.8 as well as Fig. 9.15).

6.1.2.3 Kinetic Resolutions Combined with Inversions

Another possibility to obtain 100% yield of the enantiopure product is to combine the kinetic resolution with an inversion reaction [25, 35, 36]. In this case an enzymatic hydrolysis is followed by a Mitsunobu inversion. It is, however, in fact a three-step reaction with solvent changes between the reactions. Similarly the sulfatase-catalysed enantioselective inversion of a racemic sulfate yields a homo-chiral mixture of alcohol and sulfate. This yields 100% enantiopure product after a second, acid-catalysed hydrolysis step, which is performed in organic solvent/water mixtures [26].

6.1.2.4 Hydrolysis of Symmetric Molecules and the "*meso*-trick"

Instead of starting with racemic starting material it is also possible to use symmetric substrates [25]. The hydrolase selectively catalyses the hydrolysis of just one of the two esters, amides or nitriles, generating an enantiopure product in 100% yield (Scheme 6.7). No recycling is necessary, nor need catalysts be combined, as in the dynamic kinetic resolutions, and no follow-up steps are required, as in the kinetic resolutions plus inversion sequences. Consequently this approach is popular in organic synthesis. Moreover, symmetric diols, diamines and (activated) diacids can be converted selectively into chiral mono-esters and mono-amides if the reaction is performed in dry organic solvents. This application of the reversed hydrolysis reaction expands the scope of this approach even further [22, 24, 27].

In the "*meso*-trick" the same principle is applied [25]. A symmetric compound, in this case a *meso* compound, is submitted to selective hydrolysis. The asymmetric compound that is generated in this manner is obtained in 100% yield and ideally high optical purity (Scheme 6.8). *Meso* compounds with diamino, diol or diacid functions can be converted to chiral mono-esters or mono-amides, too, if the reaction is performed in organic solvents [22, 24, 27].

6.2
Hydrolysis of Esters

The lipase or esterase-catalysed hydrolyses are straightforward and mild reactions. Consequently they can readily be performed on a large scale and they are suitable for first year undergraduate teaching [37]. Indeed, these enzymes are

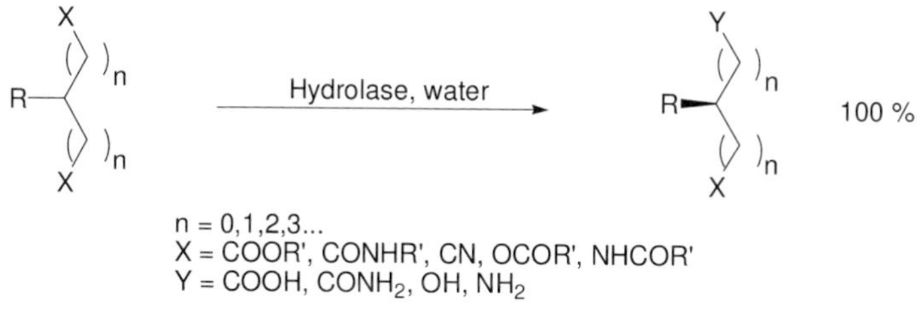

X
n
R
Hydrolase, water
Y
R
100 %
X
n = 0,1,2,3...
X = COOR', CONHR', CN, OCOR', NHCOR'
Y = COOH, CONH2, OH, NH2

Scheme 6.7 Hydrolases desymmetrise symmetrical compounds.

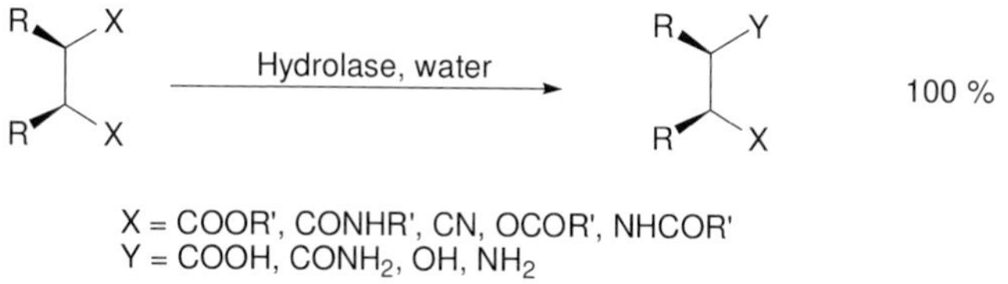

R
X
Hydrolase, water
Y
100 %
X = COOR', CONHR', CN, OCOR', NHCOR'
Y = COOH, CONH2, OH, NH2

Scheme 6.8 The "*meso*-trick".

applied on an industrial scale, for the production of both fine and bulk chemicals [3, 34, 38–41].

Kinetic Resolutions of Esters

As early as 1984 the porcine pancreas lipase-catalysed enantioselective synthesis of (R)-glycidol was described. At pH 7.8 and ambient temperatures the reaction was allowed to proceed to 60% conversion (Scheme 6.9). This means that the enzyme was not extremely enantioselective, otherwise it would have stopped at 50% conversion. Nonetheless, after workup the (R)-glycidol was obtained in a yield of 45% with an ee of 92% [42]. This was a remarkable achievement and the process was developed into an industrial multi-ton synthesis by Andeno-DSM [34, 43]. While on the one hand a success story, it also demonstrated the shortcomings of a kinetic resolution. Most enzymes are not enantiospecific but enantioselective and thus conversions do not always stop at 50%, reactions need to be fine-tuned to get optimal ees for the desired product [28]. As mentioned above kinetic resolutions only yield 50% of the product, the other enantiomer needs to be recycled. As a result of all these considerations this reaction is a big step forward but many steps remain to be done.

Scientists from Bristol-Myers Squibb developed a new side chain for Taxol, making it water-soluble. A kinetic resolution with *Pseudomonas cepacia* lipase (lipase PS-30 from Amano) was applied to obtain the desired material enantiopure (Scheme 6.10). After the lipase-catalysed hydrolysis of the wrong enantiomer (49% conversion) the ester was obtained with an ee of >99%. Separation and subsequent chemical cleavage of the ester yielded the desired enantiomer of the lactame, which could then be coupled to baccatin III [44].

The potential analgesic and replacements of the opiate-agonist tramadol, ε-hydroxyl-tramadol, was resolved enzymatically [43, 45]. Screening of several enzymes revealed that selective enzymes for both enantiomers were available. While pig liver esterase (PLE) hydrolysed the ester of the active enantiomer,

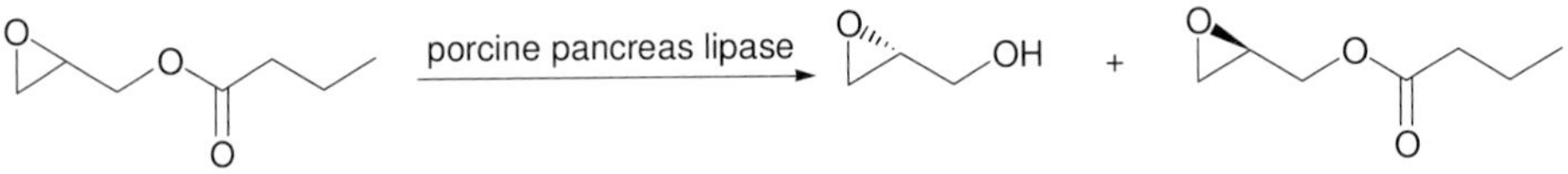

Scheme 6.9 The Andeno-DSM approach towards (R)-glycidol.

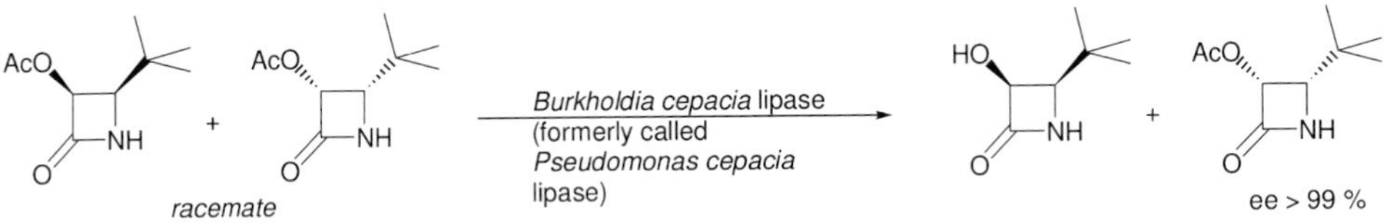

Scheme 6.10 Synthesis of a side-chain for an orally active taxane.

Candida rugosa lipase (CRL) hydrolysed the other enantiomer selectively (Scheme 6.11). This proves once again that in most cases enzymes with the desired stereoselectivity are available.

Utilising *Candida cylindracea* lipase (CCL) a chiral propionic acid was resolved by DSM [34, 46]. Only the undesired enantiomer of the ester was hydrolysed and at a conversion of 64% the remaining desired ester had an *ee* of 98% (Scheme 6.12). Although this means that the yield of the enantiopure ester is below 40% it did enable a new access to enantiopure captopril.

The atom efficiency of a kinetic resolution is increased if the starting material is not an ester but a lactone. Indeed, kinetic resolutions of lactones are used on an industrial scale. Fuji/Daiichi Chemicals produces D-pantothenic acid on a multi-ton scale based on such a resolution. D-Pantolactone is hydrolysed at pH 7 by a hydrolase from *Fusarium oxysporum* yielding D-pantoic acid with an *ee* of 96% while L-pantoic acid was barely detectable. The immobilized *Fusarium oxysporum* cells were recycled 180 times and retained 60% of their activity, demonstrating the great stability of this catalytic system [47–50].

D-Pantoic acid is again lactonised and then converted into D-pantothenic acid, better known as vitamin B5 (Scheme 6.13). The remaining L-pantolactone can be racemised and recycled. Similar approaches based on L-specific lactonohydro-

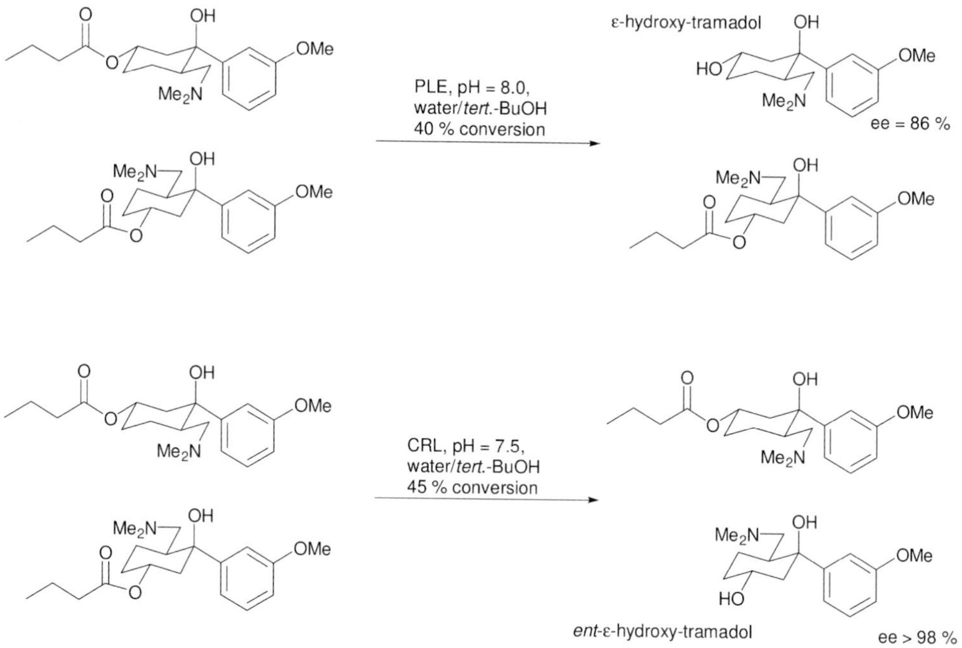

Scheme 6.11 Depending on the enzyme chosen either enantiomer of ε-hydroxyl-tramadol can be hydrolysed enantioselectively.

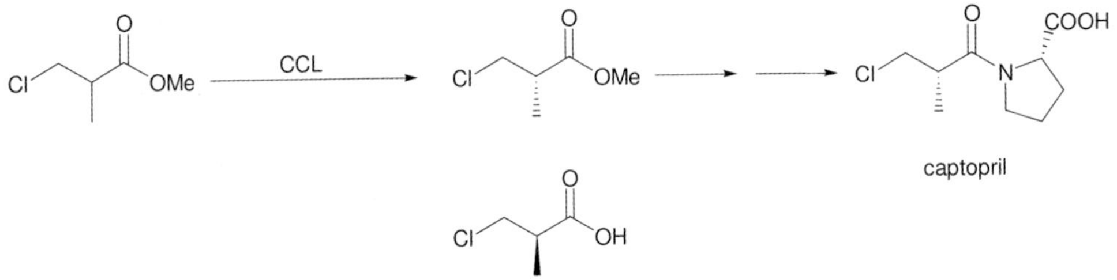

Scheme 6.12 Enantioselective synthesis of captopril.

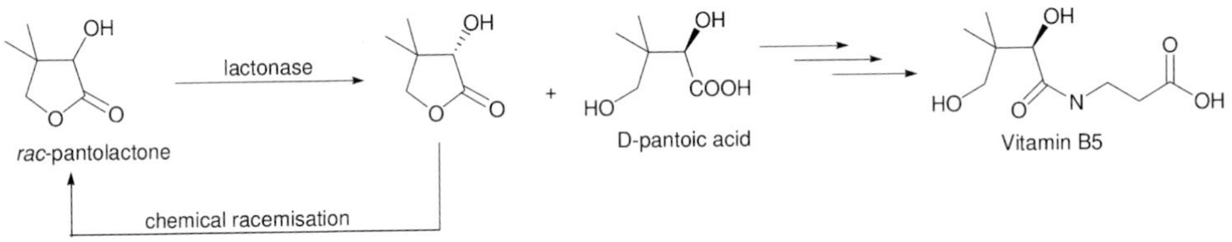

Scheme 6.13 Enantioselective synthesis of vitamin B5 based
on a kinetic resolution with a lactonase.

lases have shown first success but are not yet commercialized [34, 51]. For alternative approaches toward vitamin B5 see also Chapter 8.

For the enantiopure production of human rhinovirus protease inhibitors, scientists from Pfizer developed a kinetic resolution and recycling sequence (Scheme 6.14 A). The undesired enantiomer of the ester is hydrolysed and can be racemised under mild conditions with DBU. This enzymatic kinetic resolution plus racemisation replaced a significantly more expensive chemical approach [52]. An enzymatic kinetic resolution, in combination with an efficient chemically catalysed racemisation, is the basis for a chiral building block for the synthesis of Talsaclidine and Revatropate, neuromodulators acting on cholinergic muscarinic receptors (Scheme 6.14 B). In this case a protease was the key to success [53]. Recently a kinetic resolution based on a *Burkholderia cepacia* lipase-catalysed reaction leading to the fungicide Mefenoxam was described [54]. Immobilisation of the enzyme ensured >20 cycles of use without loss of activity (Scheme 6.14 C).

6.2.2
Dynamic Kinetic Resolutions of Esters

When the esters of chiral acids are submitted to a kinetic resolution, the wrong enantiomer cannot be racemised easily. However, if the chiral center of the acid can be racemised via the enol of the acid, this can be utilized. Esters do not enolize easily, but thioester and other activated esters [27] do so much more readily. This was exploited to convert kinetic resolutions into dynamic kinetic resolutions, thus increasing the yield from a maximum of 50% to a maximum of

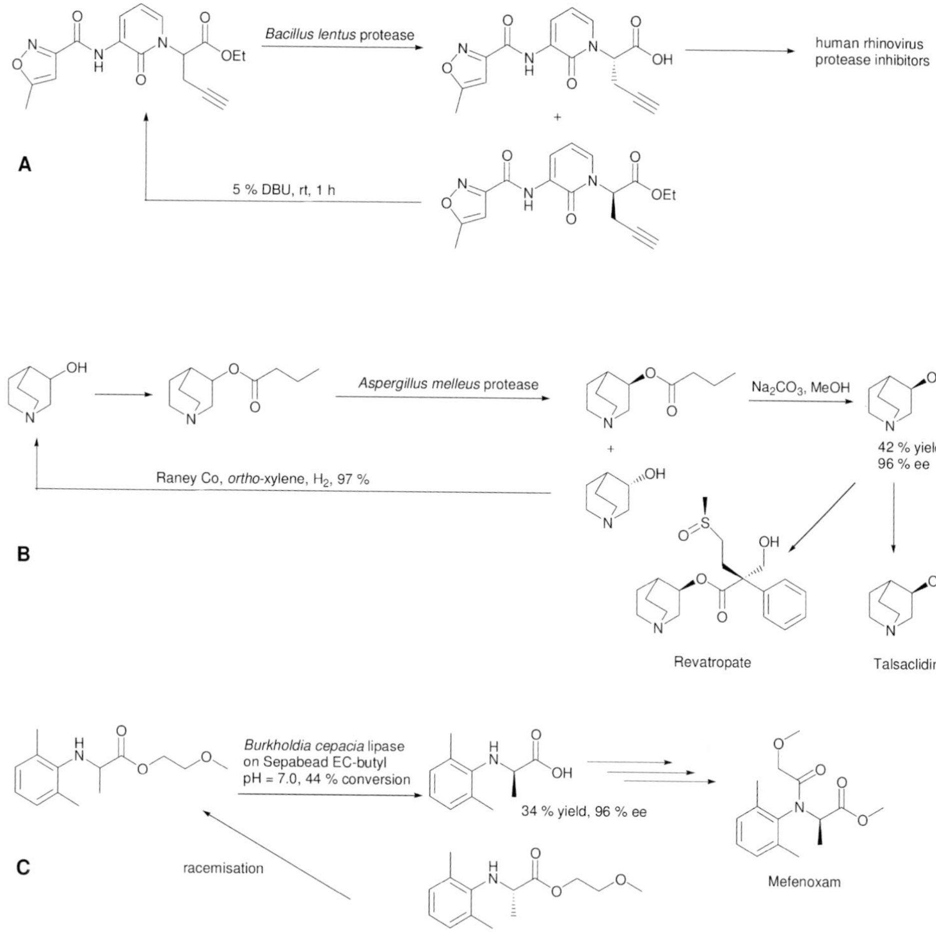

Scheme 6.14 A: Kinetic resolution and recycling of the wrong enantiomer for an intermediate of an antiviral drug; B: kinetic resolution and recycling of the wrong enantiomer for an intermediate of a neuromodulator; C: kinetic resolution and recycling of the wrong enantiomer for an intermediate of a fungicide.

100%. This approach was studied for naproxen trifluoroethylthioester [55], fenoprofen trifluoroethylthioester [56], naproxen trifluoroethylester [57] and ibuprofen 2-ethoxyethyl ester [58] (Scheme 6.15). Some of these reactions were not performed in water only, but in biphasic mixtures, due to solubility problems. This is a drawback from a green point of view, but the much higher yield and the fact that no recycling step is needed is a clear indication of the high efficiency of dynamic kinetic resolutions.

For the enantioselective preparation of roxifiban an entirely new approach for the *in situ* racemisation of the racemic starting material of the dynamic kinetic

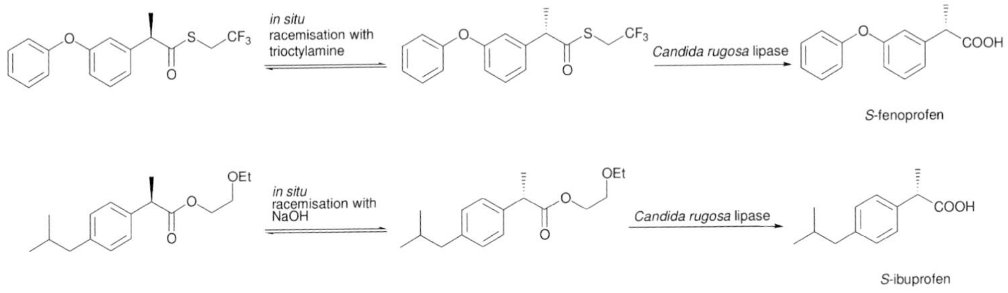

Scheme 6.15 Dynamic kinetic resolutions via activated esters.

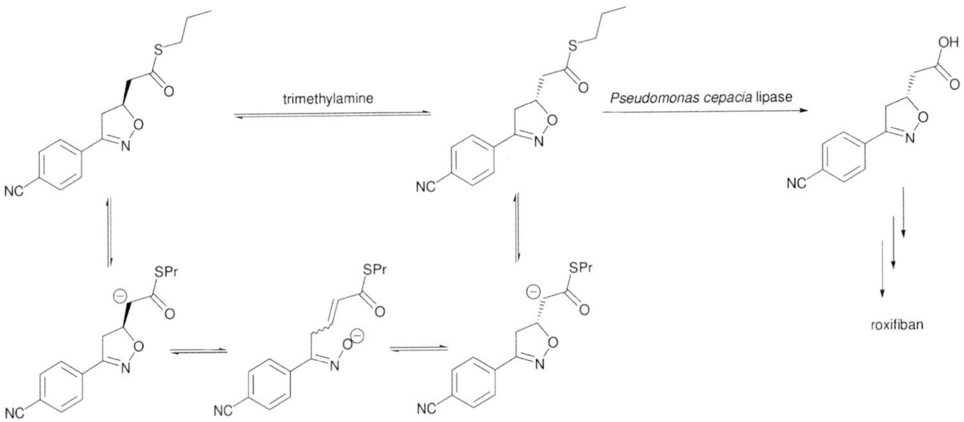

Scheme 6.16 Dynamic kinetic resolution of the key intermediate for the synthesis of roxifiban.

resolution was developed. Again the key to success was a thioester as activated ester. Instead of a racemisation of a stereocenter in the α-position, however, the base-catalysed racemisation occurred via a retro-Michael addition. Careful screening revealed that propylthiol was the thiol of choice, three equivalents of trimethylamine as a base should be used and that Amano PS-30 lipase catalyses the resolution step efficiently (Scheme 6.16). The overall procedure gave the acid in 80.4% yield with an *ee* of 94%. Recrystallisation improved this to >99.9% [59, 60].

6.2.3
Kinetic Resolutions of Esters Combined with Inversions

Scientists at Sumitomo developed this approach for the production of chiral insecticides [35, 61]. As discussed above, this approach involves several separate steps, an inversion replacing the racemisation that is normally necessary subsequent to a kinetic resolution. The advantage of this approach is that after the kinetic resolution no separation is necessary, since the reaction mixture is sub-

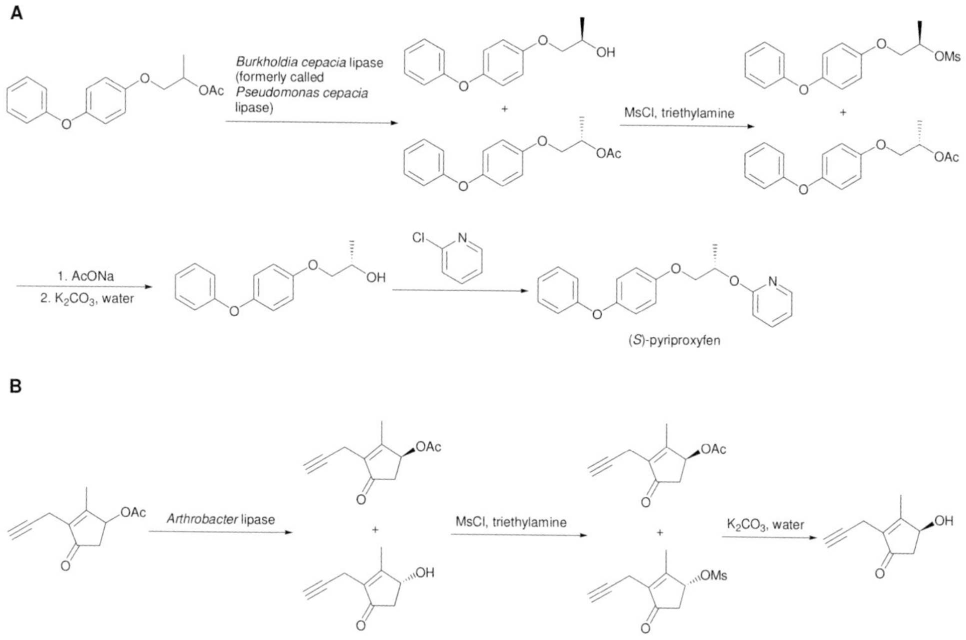

A

B

Scheme 6.17 A: Preparation of (S)-pyriproxyfen via a kinetic resolution–inversion sequence; B: chiral building blocks for the synthesis of pyrethroid insecticides are prepared via a kinetic resolution–inversion sequence.

mitted to the inversion directly. A PCL-catalysed kinetic resolution was followed by the formation of a methanesulfonyl ester. The mixture of (S)-acetate and (R)-sulfonate was then treated with sodium acetate. A S_N2 reaction yielded pure (S)-acetate in theoretically 100% yield. Subsequent hydrolysis and treatment with 2-chloropyridine produced the insecticide (S)-pyriproxyfen (Scheme 6.17 A).

For the production of various pyrethroid insecticides the (S)-4-hydroxy-3-methyl-2-(2-propynyl)-2-cyclopenten-1-one was required. *Arthrobacter* lipase hydrolysed only the R-enantiomer of the racemic acetate and, after extraction, the mixture of the alcohol and acetate were submitted to methanesulfonyl chloride and triethylamine. The thus-obtained acetate/mesylate mixture was hydrolysed/inverted yielding 82% of (S)-4-hydroxy-3-methyl-2-(2-propynyl)-2-cyclopenten-1-one with an *ee* of 90% (Scheme 6.17 B). The procedure was also proved to work with other secondary alcohols and instead of the mesylate a Mitsunobu inversion could be applied [35].

6.2.4
Hydrolysis of Symmetric Esters and the "*meso*-trick"

For a new potential *β*-3-receptor agonist a pig liver esterase-based enantioselective synthesis was devised (Scheme 6.18). The substituted malonic acid diester
was hydrolysed at pH 7.2 and yielded 86% of the (S)-monoester with an *ee* of
97% [62]. This reaction immediately demonstrates the great advantage of starting with a symmetric molecule. The enzyme very efficiently desymmetrizes the
diester and excellent yields with high optical purities are obtained. No extra
steps are necessary and no additional chemicals need to be added.

A similar strategy was chosen by the scientists from CIBA in their approach
towards the statin side chain. Starting with a symmetric glutaric acid ester derivative, they succeeded in desymmetrising it with excellent yields and *ee* [63].
This hydrolysis is also remarkable since it was performed while the hydroxyl
group was protected with an activated ester [27]. In the route towards atorvastatin this activated ester is selectively cleaved by PLE without hydrolyzing the remaining ethyl ester (Scheme 6.19). This once again demonstrates the great versatility of hydrolases in organic synthesis. A critical comparison of the different
enzymatic routes towards the atorvastatin side chain was recently published,
highlighting the power of enzymes for enantioselective synthesis [64].

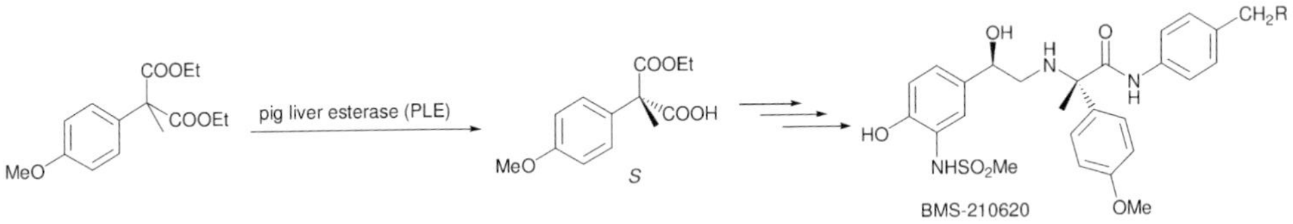

Scheme 6.18 High yields and excellent optical purities are
obtained in PLE-catalysed hydrolysis of an asymmetric starting
material.

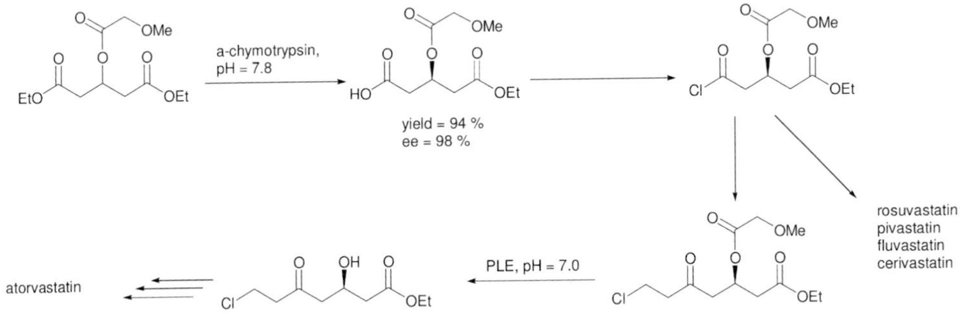

Scheme 6.19 Enantioselective CIBA route towards the statin
side chains, based on a symmetric starting material.

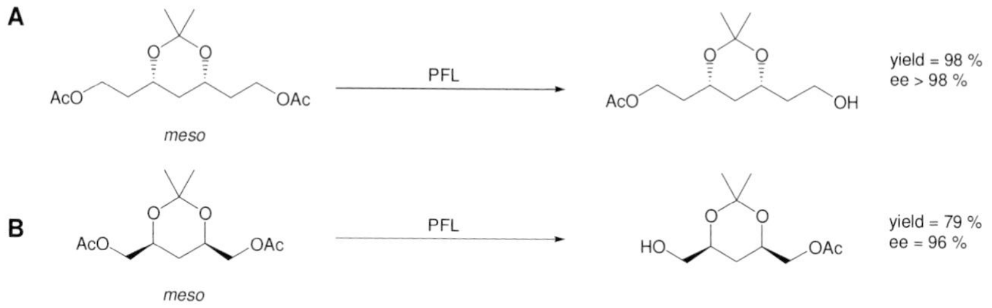

A

PFL

yield = 98 %
ee > 98 %

meso

B

PFL

yield = 79 %
ee = 96 %

meso

Scheme 6.20 A and B: PFL-catalysed enantioselective hydrolysis of *meso*-compounds.

Symmetric starting materials have often been applied, not only in the form of symmetric diacids. The strategy has often been used to prepare chiral diols [65–67]; this has been reviewed very recently [68].

Chiral diols have also been prepared starting from *meso*-compounds [68–71]. Since *meso*-compounds are, in essence, symmetric molecules, the same applies as for the other symmetric starting materials. Indeed, this is exactly what was found: Even though the stereocenters of the protected heptane tetrol are far away from the ester groups that are to be hydrolysed stereoselectively, this is what happens [69, 70]. The high selectivity is partly due to the fact that the secondary alcohol groups are protected as a cyclic acetal, giving additional structural information to the enzyme (Scheme 6.20A). A cyclic acetal also provides additional structural information in the enantioselective hydrolysis of a pentane tetrol derivative (Scheme 6.20B) [71]. In both cases *Pseudomonas fluorescens* lipase (PFL) proved to be the enzyme of choice.

6.3
Hydrolysis of Amides

Due to the delocalization of electrons the amide group is normally planar and is significantly more stable than esters [72]. Nonetheless amides can be hydrolysed enzymatically under very mild conditions. Initially it might be expected that only enzymes that were evolved for this function by nature could hydrolyse this stable bond, but by now many examples are known where lipases and esterases hydrolyse amides, too [2, 34, 73]. A recent review discusses the mechanisms, modes of action and enantioselectivities of all the important enzymes [74].

The most prominent green example, the regioselective hydrolysis of an amide on an industrial scale, is the production of penicillin. PenG acylase selectively hydrolyses the more stable amide bond, leaving the β-lactam ring intact [75, 76]. For a full discussion of this example see Chapter 1 (Fig. 1.37) and Chapter 8. Since the starting material is already enantiopure the enzyme induces no stereoinformation. In other industrial processes the enantioselectivity of the enzymes is used. This is, in particular, the case in the production of natural and unnatural amino acids.

6.3.1
Production of Amino Acids by (Dynamic) Kinetic Resolution

The industrial production of natural amino acids is mainly based on fermentation (Chapter 8), but not all 20 amino acids can be produced efficiently in this
way [77]. For these amino acids other chemo-enzymatic approaches have been
developed, several of them based on hydrolytic enzymes [78].

6.3.1.1 The Acylase Process
As early as 1969 an acylase was introduced for the industrial kinetic resolution of
racemic N-acyl amino acids by Tanabe Seiyaku (Japan) [34, 78]. Later Degussa followed and developed similar biocatalytic processes [77–79]. The success of these
industrial preparations of amino acids is based on the high selectivity and broad
substrate specificity of acylase I [80]. In particular L-methionine is prepared in this
way. The overall synthesis starts with a Michael addition of mercaptomethanol to
acrolein, followed by a Strecker synthesis and ammonium carbonate addition to
yield the hydantoin. Subsequent hydrolysis gives the racemic methionine in
95% yield calculated on the amount of acrolein used. Acylation results in the starting material for the enzyme reaction: N-acetylmethionine. Enantioselective hydrolysis with an acylase from *Aspergillus orycae* then releases the enantiopure L-
methionine (Scheme 6.21) [77]. Overall the synthesis is a remarkable feat, acrolein
is converted highly efficiently into racemic methionine and the final product is of
food grade. However, the process has two drawbacks: the formation of the hydantoin that is later destroyed is not very atom efficient and the kinetic resolution
wastes 50% of the product and causes a large recycle stream.

A straightforward approach to avoid low yields is to perform the reaction as a
dynamic kinetic resolution. Racemisation can be achieved chemically [33] or enzymatically, indeed a number of N-acyl amino acid racemases have been described and it has been demonstrated that they could be employed together with
the L-N-acyl amino acylase for the production of optically pure methionine [81].

The acylase process can also be applied for the production of D-amino acids.
These amino acids are valuable building blocks in pharmaceutical chemistry
and they can be prepared with high enantiopurity by the action of a D-N-acyl

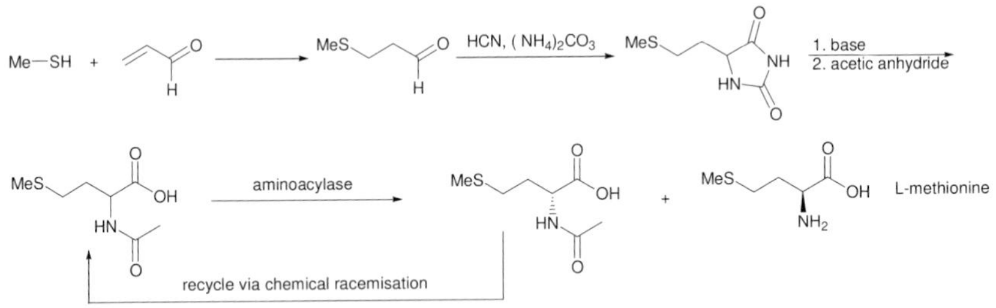

Scheme 6.21 The Degussa synthesis of L-methionine.

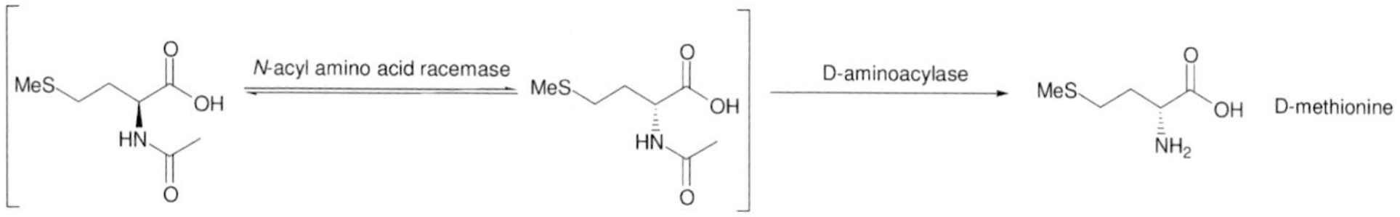

Scheme 6.22 Two-enzyme catalysed dynamic kinetic resolution for the preparation of D-amino acids.

amino acylase [82]. Again it is possible to combine the acylase reaction with a N-acyl amino acid racemase to obtain 100% product (Scheme 6.22) [81].

6.3.1.2 The Amidase Process

DSM developed a slightly different approach towards enantiopure amino acids. Instead of performing the Strecker synthesis with a complete hydrolysis of the nitrile to the acid it is stopped at the amide stage. Then a stereoselective amino acid amidase from *Pseudomonas putida* is employed for the enantioselective second hydrolysis step [83], yielding enantiopure amino acids [34, 77, 78]. Although the reaction is a kinetic resolution and thus the yields are never higher than 50% this approach is overall more efficient. No acylation step is necessary and the atom efficiency is thus much higher. A drawback is that the racemisation has to be performed via the Schiff's base of the D-amide (Scheme 6.23).

Recently it was reported that an α-amino-ε-caprolactam racemase from *Achromobacter obae* can racemise α-amino acid amides efficiently. In combination with a D-amino acid amidase from *Ochrobactrum anthropi* L-alanine amide could be converted into D-alanine. This tour de force demonstrates the power of the racemase [84]. If racemic amide is used as a starting material the application of this racemase in combination with a D- or L-amidase allows the preparation of 100% D- or L-amino acid, a dynamic kinetic resolution instead of DSM's kinetic resolution (Scheme 6.24).

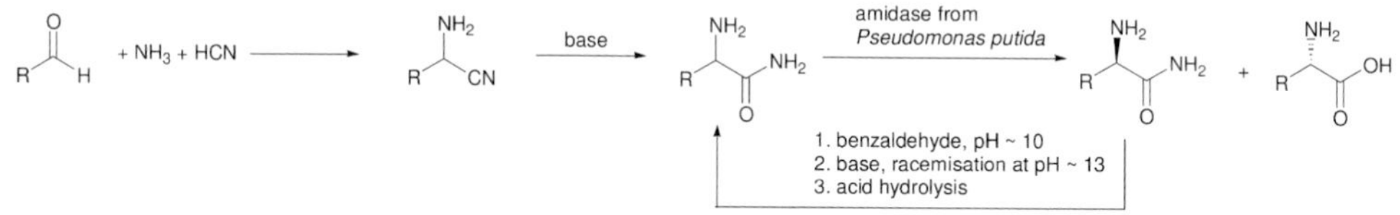

Scheme 6.23 The DSM amidase process.

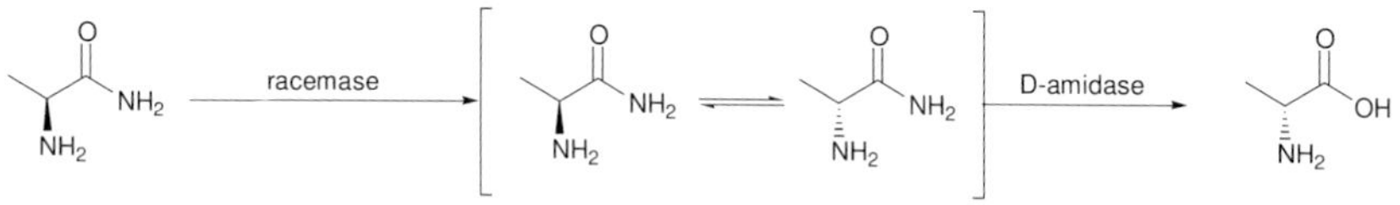

Scheme 6.24 α-Amino-ε-caprolactam racemase enables a dynamic kinetic resolution.

6.3.1.3 The Hydantoinase Process

The enzymes of the nucleic acid metabolism are used for several industrial processes. Related to the nucleobase metabolism is the breakdown of hydantoins. The application of these enzymes on a large scale has recently been reviewed [85]. The first step in the breakdown of hydantoins is the hydrolysis of the imide bond. Most of the hydantoinases that catalyse this step are D-selective and they accept many non-natural substrates [78, 86]. The removal of the carbamoyl group can also be catalysed by an enzyme: a carbamoylase. The D-selective carbamoylases show wide substrate specificity [85] and their stereoselectivity helps improving the overall enantioselectivity of the process [34, 78, 85]. Genetic modifications have made them industrially applicable [87]. Fortunately hydantoins racemise readily at pH >8 and additionally several racemases are known that can catalyze this process [85, 88]. This means that the hydrolysis of hydantoins is always a dynamic kinetic resolution with yields of up to 100% (Scheme 6.25). Since most hydantoinases are D-selective the industrial application has so far concentrated on D-amino acids. Since 1995 Kaneka Corporation has produced ~ 2000 tons/year of D-*p*-hydroxyphenylglycine with a D-hydantoinase, a D-carbamoylase [87] and a base-catalysed racemisation [85, 89].

While the production of D-amino acids is well established the preparation of L-amino acids is difficult due to the limited selectivity and narrow substrate spectrum of L-hydantoinases. This can be circumvented by employing rather unselective hydantoinases in combination with very enantioselective L-carbamoylases and carbamoyl racemases [90]. Furthermore, a D-hydantoinase has been genetically modified and converted into a L-hydantoinase. This enzyme can be used on a 100-kg scale for the production of L-*tert*-leucine [34]. Finally, the fact that the X-ray structure of an L-hydantoinase is known gives hope that side-directed mutagenesis will lead to improved L-hydantoinases [91].

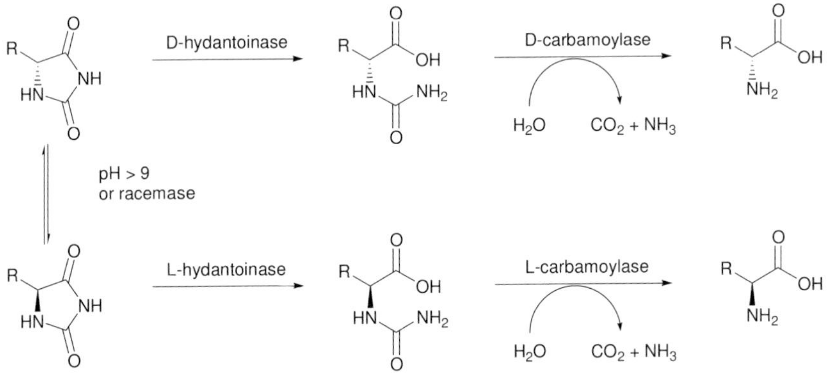

Scheme 6.25 The hydantoinase process for the production of amino acids.

6.3.1.4 Cysteine

In 1978 the conversion of the racemic 2-amino-Δ^2-thiazoline-4-carboxylic acid into L-cysteine, catalysed by *Pseudomonas thiazolinophilum* cells, was reported [92]. Due to a racemase the overall process is a dynamic kinetic resolution and yields of 95% are obtained in the industrial process (Scheme 6.26) [34].

6.3.2
Enzyme-catalysed Hydrolysis of Amides

The hydrolysis of amides is not limited to the industrial synthesis of enantiopure amino acids. Lonza has developed routes towards (S)-pipecolic acid and (R)- and (S)-piperazine-2-carboxylic acid that are based on amidases [93, 94]. The processes are based on whole bacterial cells. In the case of the pipecolic acid, an important building block for pharmaceutical chemistry, an S-selective amidase in *Pseudomonas fluorescens* cells, catalyses the reaction with high selectivity and the acid is obtained with an *ee* >99% (Scheme 6.27 A). For the preparation of piperazine-2-carboxylic acid from the racemic amide a R- and a S-selective amidase are available. Utilising *Klebsiella terrigena* cells the S-enantiomer is prepared with 42% isolated yield and *ee* >99%, while *Burkholderia sp.* cells catalyse the formation of the R-enantiomer (*ee*=99%, Scheme 6.27 B).

For the synthesis of the Glaxo anti-HIV drug abacavir (ziagen) a chiral cyclopentene derivative was needed. At Glaxo the application of savinase, a cheap en-

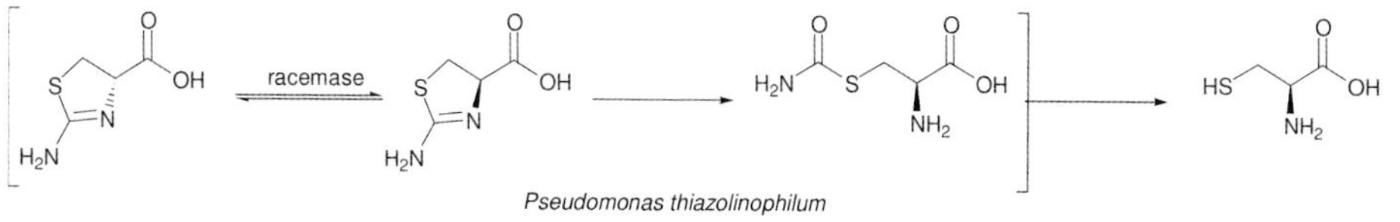

Scheme 6.26 Ajinomoto produces L-cysteine with 30 g l^{-1} via a *Pseudomonas thiazolinophilum*-catalysed process.

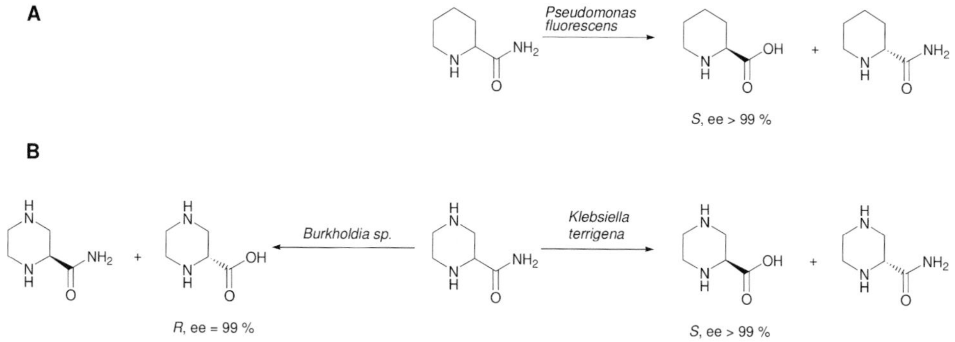

Scheme 6.27 Application of R- and S-selective amidases at Lonza for the production of pharmaceutical intermediates.

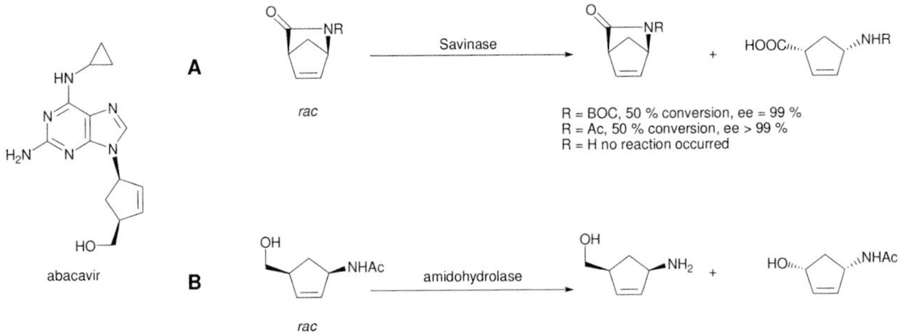

Scheme 6.28 A: the Glaxo route to enantiopure abacavir;
B: the Lonza route towards the chiral abacavir building block.

zyme that is used in bulk in the detergent industry, was chosen as the most suitable enzyme. Excellent conversions and enantioselectivities were obtained in the hydrolysis of the protected lactam (Scheme 6.28 A). Surprisingly the unprotected lactam was no substrate for the enzyme [95].

At Lonza [94] an alternative route was developed. Based on the acetylated amine a straightforward hydrolysis gave the desired amino alcohol (Scheme 6.28 B). The wrong enantiomer could not be recycled. Indeed, this is a problem with all the above-described kinetic resolutions: half of the starting material is waste and needs to be recycled.

At Bristol-Myers Squibb amidases were employed for the enantioselective preparation of new potential β-3-receptor agonists [62]. A kinetic resolution of the starting amides was achieved with very good to excellent enantioselectivities (Scheme 6.29). As already mentioned above, whole cells were used for these transformations.

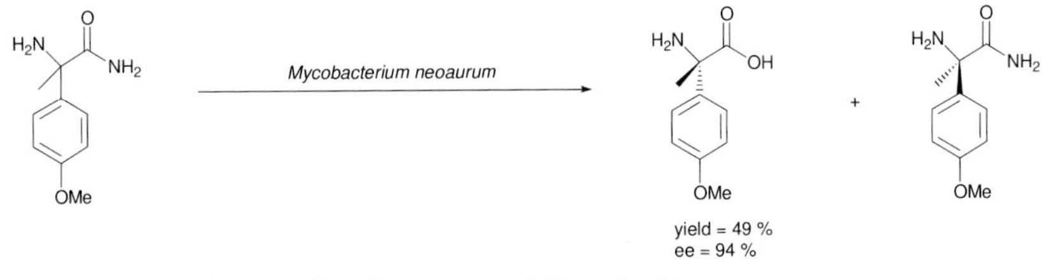

Scheme 6.29 Application of amidases at Bristol-Myers Squibb.

In general the hydrolysis of amides is performed as a kinetic resolution and not as a dynamic kinetic resolution. It is applied industrially [96] but in most cases the industrial kinetic resolution of amines, as performed for instance by BASF, is an acylation of racemic amines [38], rather than the hydrolysis of racemic amides. For the acylation of amines many different acyl donors [27] and enzymes can be used, including lipases (or a review see [97]).

6.3.3
Enzyme-catalysed Deprotection of Amines

The use of protecting groups is one of the reasons why synthetic organic chemistry is far from being green and sustainable. However, it is also obvious that protection group chemistry cannot and will not disappear within the next few years, since the methodology to prepare complex structures without the use of protecting groups does not yet exist. An alternative is therefore to develop protecting groups, which can be cleaved mildly and selectively with the help of enzymes. A particularly elegant example is the variation of the Cbz- or Z-group [98]. By attaching a hydrolysable group in the para-position the Z-group can be split off. The enzyme used selectively cleaves the acetate or phenylacetate group, generating a phenolate ion in the para-position. This causes an immediate *in situ* fragmentation, releasing the amino-group. Overall an amino-group can thus be deprotected without the use of an amidase but with a lipase (Scheme 6.30). Consequently this deprotection technique can readily be applied in peptide synthesis without the risk of the enzymes degrading the peptide.

A class of enzymes that was made accessible only recently is the peptide deformylases. It was demonstrated that peptide deformylase can be used for kinetic resolutions, but they can also be employed to cleave off formyl protection groups [99]. Due to the stereoselectivity of the enzyme the enantiopurity of the product is also improved during the deprotection (Scheme 6.31).

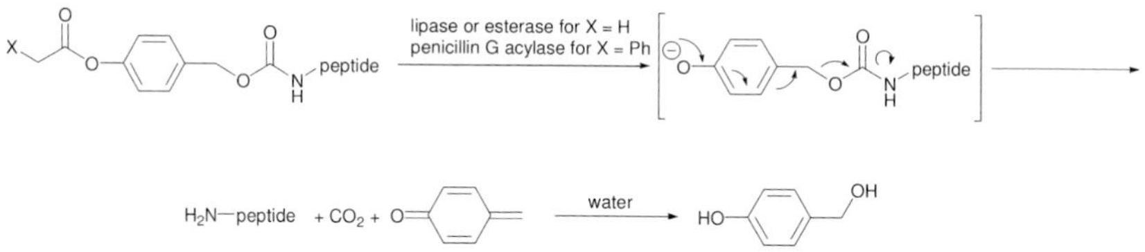

Scheme 6.30 Enzymatic cleavage of a modified Z-protection group.

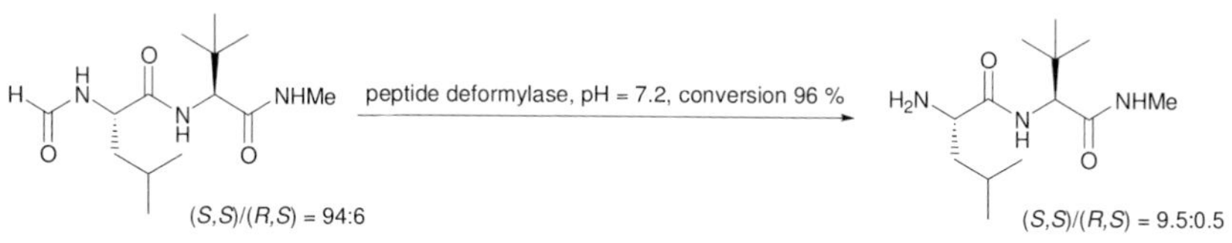

Scheme 6.31 Removal of a formyl protection group with a peptide deformylase.

6.4
Hydrolysis of Nitriles

The nitrile group is a versatile building block, in particular since it can be converted into acids or amides. It undergoes hydrolysis but requires relatively harsh reaction conditions. Nature provides two enzymatic pathways for the hydrolysis of nitriles. The nitrilases convert nitriles directly into acids, while the nitrile hydratases release amides. These amides can then be hydrolysed by amidases (see also above). Often nitrile hydratases are combined with amidases in one host and a nitrile hydratase plus amidase activity can therefore be mistaken as the activity of a nitrilase (Scheme 6.32). A large variety of different nitrilases and nitrile hydratases are available [100, 101] and both types of enzyme have been used in industry [34, 38, 94].

6.4.1
Nitrilases

The application of nitrilases is broad. A purified nitrilase from *Bacillus pallidus* was employed to hydrolyse a wide variety of aliphatic, aromatic and heteroaromatic nitriles and dinitriles (Scheme 6.33) [102]. Nitrilases have also been patented for the hydrolysis of α-substituted 4-methylthio-butyronitriles, however, no stereoselectivity was reported [103].

Although nitrilases do not always display high enantioselectivities [103] several examples of enantioselective nitrilases are known. Indeed they are used industrially for the synthesis of (R)-mandelic acid [34] and S-selective enzymes are also known [104]. In both cases the nitrilases were used for dynamic kinetic resolutions and they are discussed in Chapter 5 (Schemes 5.16 and 5.17).

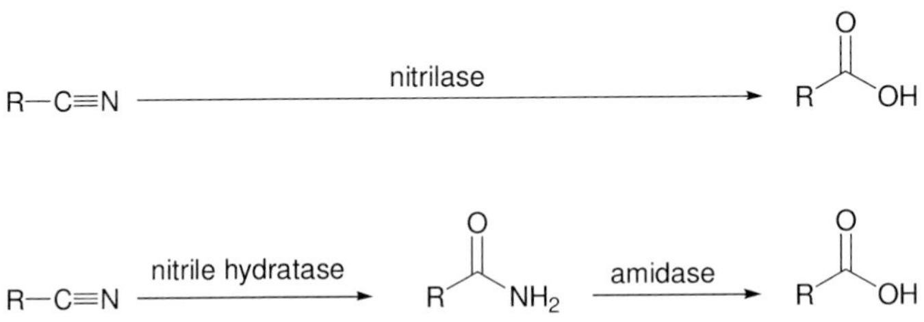

Scheme 6.32 Nitrile hydrolyzing enzymes.

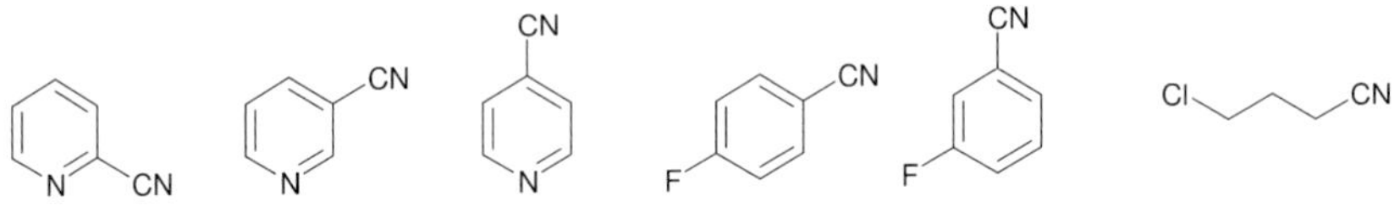

Scheme 6.33 Good substrates for nitrilase from *Bacillus pallidus*.

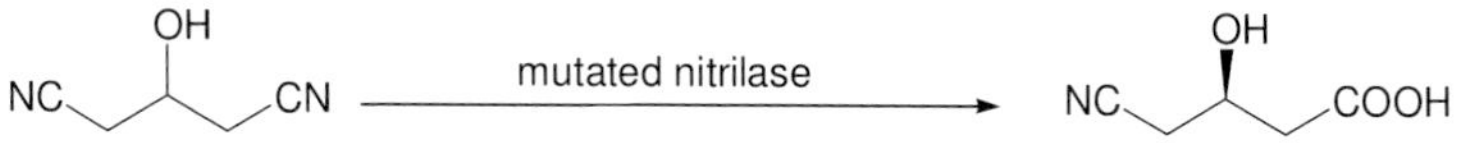

Scheme 6.34 Diversa route to optically pure atorvastatin side-chain.

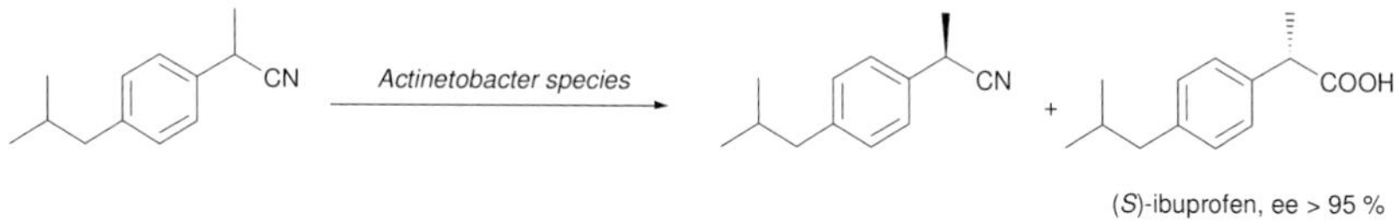

(S)-ibuprofen, ee > 95 %

Scheme 6.35 Preparation of (S)-ibuprofen via a kinetic resolution.

Nitrilases have been utilized successfully for the desymmetrisation of symmetric starting materials. At Diversa it was demonstrated that mutagenesis could create a highly selective nitrilase that was active at high substrate concentrations [105]. For their enzymatic route towards the atorvastatin (lipitor) side-chain this nitrilase now converts a symmetric precursor with ~ 600 g L^{-1} d^{-1} into the enantiopure (R)-4-cyano-3-hydroxybutyric acid (ee=98.5%, Scheme 6.34).

Selective nitrilases have also been developed for the enantiopure preparation of ibuprofen [106]. In a kinetic resolution with *Acinetobacter sp.* AK226 (S)-ibuprofen could be prepared in good optical purity (Scheme 6.35).

6.4.2
Nitrile Hydratases

Nitrile hydratases are, just like nitrilases, versatile enzymes and their structure and application has been reviewed extensively [100, 101, 103, 106–108]. This is due to the great success of these enzymes on an industrial scale. Nitrile hydratases are employed not only in fine chemistry but also in bulk chemistry. The enzymatic production of acrylamide is a large-scale process that is replacing the traditional processes (see also Chapter 1, Fig. 1.42). Indeed this product has a history starting with homogeneous catalysis, moving to heterogeneous catalysis and ending at biocatalysis, demonstrating why biocatalysis is the catalysis type of choice in hydrolysis reactions. Acrylamide was first produced on a large scale in 1954 by American Cyanamid. The homogeneous approach is based on the sulfuric acid process (acrylonitrile:sulphuric acid:water in the ratio 1:1:1 at 60–80 °C, followed by cooling and neutralisation) and the product is purified by recrystallisation from benzene [109]. Huge amounts of salt are generated and benzene is definitely an outdated solvent. This stoichiometric process was improved and replaced by the heterogeneous process based on a Raney copper catalyst [109]. The conversion of acrylonitrile is >50% and the selectivity for the amide is close to 100% but the process is performed at 120 °C. In addition some copper leaches and needs to be removed at the end of the reaction (Scheme 6.36).

In contrast, the nitrile hydratase-based catalytic hydrolysis of acrylonitrile is an example of how green and sustainable chemistry should be performed [107–

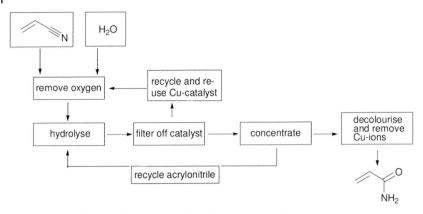

Scheme 6.36 Production of acrylamide with a Raney copper catalyst.

111]. Utilising an overexpressed nitrile hydratase from *Rhodococcus rhodochrous* in immobilised cells 30 000 tons per year are produced by Mitsubishi Rayon Co. Ltd. The selectivity of the nitrile hydratase is 99.99% for the amide, virtually no acrylic acid is detected and since all starting material is converted yields are excellent (>99%). The time space yield is 2 kg product per liter per day (Scheme 6.37). This clearly demonstrates the power of biocatalysis.

As mentioned in Chapter 1 the same *Rhodococcus rhodochrous* catalyses the last step in Lonza's >3500 tonnes/year nicotinamide synthesis [94, 111, 112]. Lonza has further developed this technology and currently synthesises a number of relevant fine chemical building blocks with nitrile hydratases [94, 113].

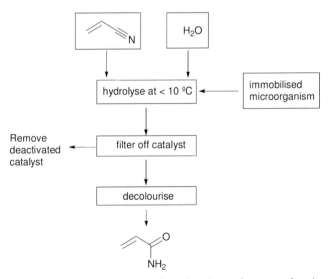

Scheme 6.37 Nitrile hydratase catalyses the selective formation of acrylamide.

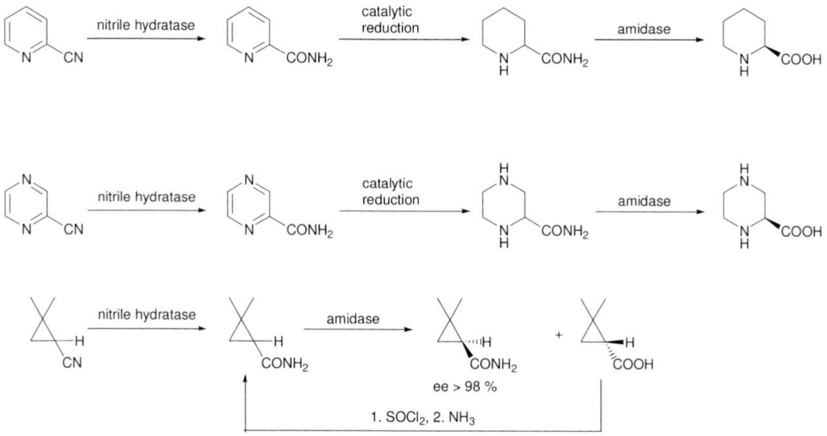

Scheme 6.38 Nitrile hydratases play an essential role in the production of fine chemicals.

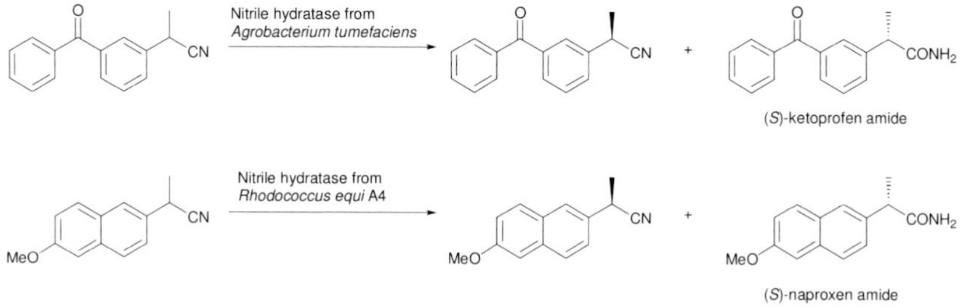

Scheme 6.39 Application of enantioselective nitrile hydratases for kinetic resolutions.

The nitrile hydrates employed are selective and stop at the amide stage. However, they display no relevant enantioselectivity. The enantioselectivity in the processes is always introduced by an amidase (see for instance Schemes 6.27 and 6.28) in a second hydrolysis step. Overall the syntheses are, remarkably, often purely catalytic and combine chemical catalysis for reductions with biocatalysis for hydrolyses and the introduction of stereoinformation (Scheme 6.38).

Although nitrile hydratases tend not to be stereoselective, examples of enantioselective enzymes are known [103, 106, 107, 114]. Of particular interest is the possibility to selectively hydrolyse 2-phenylproprionitriles, the core structure for ibuprofen and many other profens [103, 107, 114, 115]. This enables the enantioselective synthesis of the amides of ketoprofen and naproxen (Scheme 6.39).

6.5
Conclusion and Outlook

Biocatalysis is the answer to many problems in hydrolysis reactions. It enables the mild, selective and often enantioselective hydrolysis of many very different esters, amides and nitriles. Due to the very high selectivity of hydrolases a bulk chemical, acrylamide can now be produced in a clean and green manner, that is also more cost efficient than the chemical routes. The enantioselectivity of the hydrolases enables the production of commodity chemicals like enantiopure amino acids and also of many fine chemicals and drugs. A recent review summarizes the current state of the art for the biocatalytic synthesis of chiral pharmaceuticals [116], demonstrating the progress made. However, another recent review on the asymmetric synthesis of pharmaceuticals [117] reveals how much still needs to be done: less than five reactions in that review involve enzymes, most steps utilize stoichiometric reagents. Given the rapid development in biotechnology and the many new biocatalysts that have recently become available [118] and that will become available via the different genetic approaches [119, 120] it can be expected that the application of biocatalysis as described here is but a beginning.

References

1 P. T. Anastas, M. M. Kirchhoff, T. C. Williamson, *Appl. Catal. A: Gen.* **2001**, *221*, 3–13.
2 K. Faber, *Biotransformations in Organic Chemistry,* 5ht edn, Springer, Berlin, Heidelberg, New York, 2004.
3 A. J. J. Straathof, S. Panke, A. Schmid, *Curr. Opin. Biotechnol.* **2002**, *13*, 548–556.
4 R. J. Kazlauskas, A. N. E. Weissfloch, A. T. Rappaport, L. A. Cuccia, *J. Org. Chem.* **1991**, *56*, 2656–2665.
5 Z. F. Xie, *Tetrahedron Asymm.* **1991**, *2*, 773–750.
6 A. J. M. Janssen, A. J. H. Klunder, B. Zwanenburg, *Tetrahedron* **1991**, *47*, 7645–7662.
7 K. Burgess, L. D. Jennings, *J. Am. Chem. Soc.* **1991**, *113*, 6129–6139.
8 M. Cygler, P. Grochulski, R. J. Kazlauskas, J. D. Schrag, F. Buothillier, B. Rubin, A. N. Serreqi, A. K. Gupta, *J. Am. Chem. Soc.* **1994**, *116*, 3180–3186.
9 A. Mezzetti, J. D. Schrag, C. S. Cheong, R. J. Kazlauskas, *Chem. Biol.* **2005**, *12*, 427–437.

10 A. O. Magnusson, M. Takwa, A. Hamberg, K. Hult, *Angew. Chem. Int. Ed.* **2005**, *44*, 4582–4585.
11 L. E. Iglesias, V. M. Sanchez, F. Rebolledo, V. Gotor, *Tetrahedron Asymm.* **1997**, *8*, 2675–2677.
12 R. J. Kazlauskas, A. N. E. Weissfloch, *J. Mol. Catal. B Enzyme* **1997**, *3*, 65–72.
13 C. K. Savile, R. J. Kazlauskas, *J. Am. Chem. Soc.* **2005**, *127*, 12228–12229.
14 A. N. E. Weissfloch, R. J. Kazlauskas, *J. Org. Chem.* **1995**, *60*, 6959–6969.
15 M. C. R. Franssen, H. Jongejan, H. Kooijman, A. L. Spek, N. L. F. L. Camacho Mondril, P. M. A. C. Boavida dos Santos, A. de Groot, *Tetrahedron Asymm.* **1996**, *7*, 497–510.
16 F. Theil, *Chem. Rev.* **1995**, *95*, 2203–2227.
17 R. Azerad, *Bull. Soc. Chim. Fr.* **1995**, *132*, 17–51.
18 T. Sugai, *Curr. Org. Chem.* **1999**, *3*, 373–406.
19 R. D. Schmidt, R. Verger, *Angew. Chem. Int. Ed.* **1998**, *37*, 1608–1633.

20 E. Santaniello, S. Reza-Elahi, P. Ferra-boschi, in *Stereoselective Biocatalysis*, P. N. Ramesh (Ed.), Marcel Dekker, New York 2000, pp. 415–460.

21 D. Kadereit, H. Waldmann, *Chem. Rev.* 2001, *101*, 3367–3396.

22 R. Chênevert, N. Pelchat, F. Jacques, *Curr. Org. Chem.* 2006, *10*, 1067–1094.

23 A. Ghanem, H. Y. Aboul-Enein, *Chirality* 2005, *17*, 1–15.

24 E. Garcia-Urdiales, I. Alfonso, V. Gotor, *Chem. Rev.* 2005, *105*, 313–354.

25 K. Faber, *Chem. Eur. J.* 2001, *7*, 5005–5010.

26 P. Gadler, S. M. Glueck, W. Kroutil, B. M. Nestl, B. Larissegger-Schnell, B. T. Ueber-bacher, S. R. Wallner, K. Faber, *Biochem. Soc. Trans.* 2006, *34*, 296–300.

27 U. Hanefeld, *Org. Biomol. Chem.* 2003, *1*, 2405–2415.

28 A. J. J. Straathof, J. A. Jongejan, *Enzyme Microb. Technol.* 1997, *21*, 559–571.

29 J. H. Kastle, A. S. Loevenhart, *Am. Chem. J.* 1900, *24*, 491–525.

30 M. H. Coleman, A. R. Macrae, DE 2705 608, 1977.

31 A. M. Klibanov, *Acc. Chem. Res.* 1990, *23*, 114–120.

32 M. Kitamura, M. Tokunaga, R. Noyori, *Tetrahedron* 1993, *49*, 1853–1860.

33 V. Zimmermann, M. Beller, U. Kragl, *Org. Proc. Res. Dev.* 2006, *10*, 622–627.

34 M. Breuer, K. Ditrich, T. Habicher, B. Hauer, M. Kesseler, R. Stürmer, T. Ze-linski, *Angew. Chem. Int. Ed.* 2004, *43*, 788–824.

35 H. Danda, T. Nagatomi, A. Maehara, T. Umemura, *Tetrahedron* 1991, *47*, 8701–8716.

36 N. Bouzemi, L. Aribi-Zouioueche, J.-C. Fiaud, *Tetrahedron: Asymm.* 2006, *17*, 797–800.

37 D. Stetca, I. W. C. E. Arends, U. Hanefeld, *J. Chem. Educ.* 2002, *79*, 1351–1352.

38 A. Schmid, J. S. Dordick, B. Hauer, A. Kiener, M. Wubbolts, B. Witholt, *Nature* 2001, *409*, 258–268.

39 F. Hasan, A. Ali Shah, A. Hameed, *Enzyme Microb. Technol.* 2006, *39*, 235–251.

40 D. R. Yazbeck, C. A. Martinez, S. Hu, J. Tao, *Tetrahedron: Asymm.* 2004, *15*, 2757–2763.

41 V. Gotor-Fernandez, R. Brieva, V. Gotor, *J. Mol. Catal. B: Enzyme* 2006, *40*, 111–120.

42 W. E. Ladner, G. M. Whitesides, *J. Am. Chem. Soc.* 1984, *106*, 7250–7251.

43 Z. Liu, R. Weis, A. Glieder, *Food Technol. Biotechnol.* 2004, *42*, 237–249.

44 R. N. Patel, J. Howell, R. Chidambaram, S. Benoit, J. Kant, *Tetrahedron: Asymm.* 2003, *14*, 3673–3677.

45 H.-J. Gais, C. Griebel, H. Buschmann, *Tetrahedron: Asymm.* 2000, *11*, 917–928.

46 V. H. M. Elferink, J. G. T. Kierkels, M. Kloosterman, J. H. Roskam (Stamicarbon B.V.), EP 369553, 1990.

47 M. Kataoka, K. Shimizu, K. Sakamoto, H. Yamada, S. Shimizu, *Appl. Microbiol. Biotechnol.* 1995, *43*, 974–977.

48 S. Shimizu, M. Kataoka, *Ann. N.Y. Acad. Sci.* 1996, *799*, 650–658.

49 S. Shimizu, M. Kataoka, K. Honda, K. Sakamoto, *J. Biotechnol.* 2001, *92*, 187–194.

50 J. Ogawa, S. Shimizu, *Curr. Opin. Biotechnol.* 2002, *13*, 367–375.

51 M. Kesseler, T. Friedrich, H. W. Höffken, B. Hauer, *Adv. Synth. Catal.* 2002, *344*, 1103–1110.

52 C. A. Martinez, D. R. Yazbeck, J. Tao, *Tetrahedron* 2004, *60*, 759–764.

53 F. Nomoto, Y. Hirayama, M. Ikunaka, T. Inoue, K. Otsuka, *Tetrahedron: Asymm.* 2003, *14*, 1871–1877.

54 O.-J. Park, S.-H. Lee, T.-Y. Park, W.-G. Chung, S.-W. Lee, *Org. Proc. Res. Dev.* 2006, *10*, 588–591.

55 C.-S. Chang, S.-W. Tsai, *Biochem. Eng. J.* 1999, *3*, 239–242.

56 C.-Y. Chen, Y.-C. Cheng, S.-W. Tsai, *J. Chem. Technol. Biotechnol.* 2002, *77*, 699–705.

57 H.-Y. Lin, S.-W. Tsai, *J. Mol. Catal. B Enzyme* 2003, *24/25*, 111–120.

58 H. Fazlena, A. H. Kamaruddin, M. M. D. Zulkali, *Bioprocess Biosyst. Eng.* 2006, *28*, 227–233.

59 J. A. Pesti, J. Yin, L.-H. Zhang, L. Anza-lone, *J. Am. Chem. Soc.* 2001, *123*, 11075–11076.

60 J. A. Pesti, J. Yin, L.-H. Zhang, L. Anza-lone, R. E. Waltermire, P. Ma, E. Gorko, P. N. Confalone, J. Fortunak, C. Silver-man, J. Blackwell, J. C. Chung, M. D.

Hrytsak, M. Cooke, L. Powell, C. Ray, *Org. Proc. Res. Dev.* **2004**, *8*, 22–27.

61 H. Hirohara, M. Nishizawa, *Biosci. Biotechnol. Biochem.* **1998**, *62*, 1–9.

62 R. N. Patel, A. Banerjee, L. Chu, D. Brozozowski, V. Nanduri, L. J. Szarka, *J. Am. Oil Chem. Soc.* **1998**, *75*, 1473–1482.

63 R. Öhrlein, G. Baisch, *Adv. Synth. Catal.* **2003**, *345*, 713–715.

64 M. Müller, *Angew. Chem. Int. Ed.* **2005**, *44*, 362–365.

65 B. Wirz, H. Iding, H. Hilpert, *Tetrahedron: Asymm.* **2000**, *11*, 4171–4178.

66 C. R. Johnson, *Acc. Chem. Res.* **1998**, *31*, 333–341.

67 U. R. Kalkote, S. R. Ghorpade, S. P. Chavan, T. Ravindranathan, *J. Org. Chem.* **2001**, *66*, 8277–8281.

68 S. E. Bode, M. Wolberg, M. Müller, *Synthesis* **2006**, 557–588.

69 C. Bonini, R. Racioppi, G. Righi, L. Viggiani, *J. Org. Chem.* **1993**, *58*, 802–803.

70 C. Bonini, R. Racioppi, L. Viggiani, G. Righi, L. Rossi, *Tetrahedron: Asymm.* **1993**, *4*, 793–805.

71 Z.-F. Xie, H. Suemune, K. Sakai, *Tetrahedron: Asymm.* **1993**, *4*, 973–980.

72 K. Tani, B. M. Stoltz, *Nature* **2006**, *441*, 731–734.

73 A. Torres-Gavilan, E. Castillo, A. Lopez-Munguia, *J. Mol. Catal. B Enzyme* **2006**, *41*, 136–140.

74 A. Liljeblad, L. T. Kanerva, *Tetrahedron* **2006**, *62*, 5831–5854.

75 R. P. Elander, *Appl. Microbiol. Biotechnol.* **2003**, *61*, 385–392.

76 M. A. Wegmann, M. H. A. Janssen, F. van Rantwijk, R. A. Sheldon, *Adv. Synth. Catal.* **2001**, *343*, 559–576.

77 K. Drauz, B. Hoppe, A. Kleemann, H.-P. Krimmer, W. Leuchtenberger, C. Weckbecker, Industrial Production of Amino Acids, in *Ullmann's Encyclopedia of Industrial Chemistry*, electronic version, Article Online Posting Date: March 15, 2001, Wiley-VCH, Weinheim.

78 A. S. Bommarius, M. Schwarm, K. Drauz, *Chimia* **2001**, *55*, 50–59.

79 A. S. Bommarius, K. Drauz, K. Günther, G. Knaup, M. Schwarm, *Tetrahedron: Asymm.* **1997**, *8*, 3197–3200.

80 H. K. Chenault, J. Dahmer, G. M. Whitesides, *J. Am. Chem. Soc.* **1989**, *111*, 6354–6364.

81 S. Tokuyama, K. Hatano, *Appl. Microbiol. Biotechnol.* **1996**, *44*, 774–777.

82 M. Wakayama, K. Yoshimune, Y. Hirose, M. Moriguchi, *J. Mol. Catal. B Enzyme* **2003**, *23*, 71–85.

83 R. W. Feenstra, E. H. M. Stokkingreef, A. M. Reichwein, W. B. H. Lousberg, H. C. J. Ottenheim, *Tetrahedron* **1990**, *46*, 1745–1756.

84 Y. Asano, S. Yamaguchi, *J. Am. Chem. Soc.* **2005**, *127*, 7696–7697.

85 J. Ogawa, C-L. Soong, M. Kishino, Q.-S. Li, N. Horinouchi, S. Shimizu, *Biosci. Biotechnol. Biochem.* **2006**, *70*, 574–582.

86 O. Keil, M. P. Schneider, J. P. Rasor, *Tetrahedron: Asymm.* **1995**, *6*, 1257–1260.

87 Y. Ikenaka, H. Nanba, K. Yajima, Y. Yamada, M. Takano, S. Takahashi, *Biosci. Biotechnol. Biochem.* **1999**, *63*, 91–95.

88 J. Altenbuchner, M. Siemann-Herzberg, C. Syldatk, *Curr. Opin. Biotechnol.* **2001**, *12*, 559–563.

89 J. Ogawa, S. Shimizu, *Curr. Opin. Biotechnol.* **2002**, *13*, 367–375.

90 S. G. Burton, R. A. Dorrington, *Tetrahedron: Asymm.* **2004**, *15*, 2737–2741.

91 J. Abendroth, K. Niefind, O. May, M. Siemann, C. Syldatk, D. Schomburg, *Biochemistry* **2002**, *41*, 8589–8597.

92 K. Sano, K. Mitsugi, *Agric. Biol. Chem.* **1978**, *42*, 2315–2321.

93 E. Eichhorn, J.-P. Roduit, N. Shaw, K. Heinzmann, A. Kiener, *Tetrahedron: Asymm.* **1997**, *8*, 2533–2536.

94 N. M. Shaw, K. T. Robins, A. Kiener, *Adv. Synth. Catal.* **2003**, *345*, 425–435.

95 M. Mahmoudian, A. Lowdon, M. Jones, M. Dawson, C. Wallis, *Tetrahedron: Asymm.* **1999**, *10*, 1201–1206.

96 H. Smidt, A. Fischer, P. Fischer, R. D. Schmidt, U. Stelzer, EP 812363, 1995.

97 F. van Rantwijk, R. A. Sheldon, *Tetrahedron* **2004**, *60*, 501–519.

98 D. Kadereit, H. Waldmann, *Chem. Rev.* **2001**, *101*, 3367–3396.

99 T. Sonke, B. Kaptein, A. F. V. Wagner, P. J. L. M. Quaedflieg, S. Schultz, S. Ernste, A. Schepers, J. H. M. Mommers, Q. B. Broxterman, *J. Mol. Catal. B Enzyme* **2004**, *29*, 265–277.

100 A. Banerjee, R. Sharma, U.C. Banerjee, *Appl. Microbiol. Biotechnol.* **2002**, *60*, 33–44.

101 C. O'Reilly, P.D. Turner, *J. Appl. Microbiol.* **2003**, *95*, 1161–1174.

102 Q.A. Almatawah, R. Cramp, D.A. Cowan, *Extremophiles* **1999**, *3*, 283–291.

103 L. Martinkova, V. Kren, *Biocatal. Biotrans.* **2002**, *20*, 73–93.

104 G. DeSantis, Z. Zhu, W.A. Greenberg, K. Wong, J. Chaplin, S.R. Hanson, B. Farwell, L.W. Nicholson, C.L. Rand, D.P. Weiner, D.E. Robertson, M.J. Burk, *J. Am. Chem. Soc.* **2002**, *124*, 9024–9025.

105 G. DeSantis, K. Wong, B. Farwell, K. Chatman, Z. Zhu, G. Tomlinson, H. Huang, X. Tan, L. Bibbs, P. Chen, K. Kretz, M.J. Burk, *J. Am. Chem. Soc.* **2003**, *125*, 11476–11477.

106 M.-X. Wang, *Top. Catal.* **2005**, *35*, 117–130.

107 L. Martinkova, V. Mylerova, *Curr. Org. Chem.* **2003**, *7*, 1279–1295.

108 M. Kobayashi, S. Shimizu, *Curr. Opin. Chem. Biol.* **2000**, *4*, 95–102.

109 T. Ohara, T. Sato, N. Shimizu, G. Prescher, H. Schwind, O. Weiberg, H. Greim, Acrylic Acid and Derivatives, in *Ullmann's Encyclopedia of Industrial Chemistry*, Electronic version, Article Online Posting Date: March 15, 2003, Wiley-VCH, Weinheim.

110 M.R. Kula, *Elements: Degussa Science Newsletter* **2003**, *5*, 20–27.

111 H. Yamada, M. Kobayashi, *Biosci. Biotechnol. Biochem.* **1996**, *60*, 1391–1400.

112 T. Nagasawa, C.D. Mathew, J. Mauger, H. Yamada, *Appl. Environ. Microbiol.* **1988**, 1766–1769.

113 M. Petersen, M. Sauer, *Chimia* **1999**, *53*, 608–612.

114 I. Prepechalová, L. Martínková, A. Stolz, M. Ovesná, K. Bezouska, J. Kopecky, V. Kren, *Appl. Microbiol. Biotechnol.* **2001**, *55*, 150–156.

115 R. Bauer, H.-J. Knackmuss, A. Stolz, *Appl. Microbiol. Biotechnol.* **1998**, *49*, 89–95.

116 R.N. Patel, *Curr. Org. Chem.* **2006**, *10*, 1289–1321.

117 V. Farina, J.T. Reeves, C.H. Senanayake, J.J. Song, *Chem. Rev.* **2006**, *106*, 2734–2793.

118 A.S. Bommarus, K.M. Polizzi, *Chem. Eng. Sci.* **2006**, *61*, 1004–1016.

119 T.W. Johannes, H. Zhao, *Curr. Opin. Microbiol.* **2006**, *9*, 261–267.

120 M.T. Reetz, J.D. Carballeira, J. Peyralans, H. Höbenreich, A. Maichele, A. Vogel, *Chem. Eur. J.* **2006**, *12*, 6031–6038.

7

Catalysis in Novel Reaction Media

7.1
Introduction

7.1.1
Why Use a Solvent?

Organic reactions are generally performed in a solvent and there are several good reasons for this:

- Reactions proceed faster and more smoothly when the reactants are dissolved, because of diffusion. Although reactions in the solid state are known [1] they are often condensations in which a molecule of water is formed and reaction takes place in a thin film of water at the boundary of the two solid surfaces. Other examples include the formation of a liquid product from two solids, e.g. dimethylimidazolium chloride reacts with aluminum chloride to produce the ionic liquid, dimethylimidazolium tetrachloroaluminate [2]. It is worth noting, however, that not *all* of the reactant(s) have to be dissolved and reactions can often be readily performed with suspensions. Indeed, so-called solid-to-solid conversions, whereby a reactant is suspended in a solvent and the product precipitates, replacing the reactant, have become popular in enzymatic transformations [3]. In some cases, the solvent may be an excess of one of the reactants. In this case the reaction is often referred to as a solvolysis, or, when the reactant is water, hydrolysis.

- The solvent may have a positive effect on the rate and or selectivity of the reaction. The general rule is: reactions that involve ionic intermediates will be faster in polar solvents. For example, S_N1 substitutions proceed best in polar solvents while S_N2 substitutions, which involve a covalent intermediate, fare better in apolar solvents. The solvent may effect the position of an equilibrium, e.g. in a keto–enol mixture, thereby influencing selectivity in a reaction which involves competition between the two forms.
  Similarly, the solvent can have a dramatic effect on the rate and selectivity of catalytic reactions e.g. this is often the case in catalytic hydrogenations.

- The solvent acts as a heat transfer medium, that is it removes heat liberated in an exothermic reaction. It reduces thermal gradients in a reaction vessel, allowing a smooth and safe reaction. This is perfectly illustrated by the con-

Green Chemistry and Catalysis. I. Arends, R. Sheldon, U. Hanefeld
Copyright © 2007 WILEY-VCH Verlag GmbH & Co. KGaA, Weinheim
ISBN: 978-3-527-30715-9

cept of reflux, in which the reaction temperature is kept constant by allowing the solvent to boil and condense on a cold surface, before being returned to the reactor. In this way, a highly exothermic reaction can be prevented from 'running away'. Although reactions between two liquids obviously do not require a solvent, it is often prudent, for safety reasons, to employ a solvent (diluent).

7.1.2
Choice of Solvent

The choice of solvent depends on several factors. Obviously it should be liquid at the reaction temperature and, generally speaking, it is liquid at room temperature and below. Preferably it should be sufficiently volatile to be readily removed by simple distillation. Another important issue, in an industrial setting, is cost. Economic viability of the solvent will very much depend on the value of the product.

In the context of Green Chemistry, which we are primarily concerned with in this book, there are other major issues which have an important bearing on the choice of solvent. The solvent should be relatively nontoxic and relatively nonhazardous, e.g. not inflammable or corrosive. The word 'relatively' was chosen with care here, as Paracelsus remarked "the poison is in the dosage".

The solvent should also be contained, that is it should not be released to the environment. In recent years these have become overriding issues in the use of solvents in chemicals manufacture and in other industries. The FDA has issued guidelines for solvent use which can be found on the web site (www.fda.gov/cder/guidance/index.htm). Solvents are divided into four classes.

- Class 1 solvents should not be employed in the manufacture of drug substances because of their unacceptable toxicity or deleterious environmental effect. They include benzene and a variety of chlorinated hydrocarbons.
- Class 2 solvents should be limited in pharmaceutical processes because of their inherent toxicity and include more chlorinated hydrocarbons, such as dichloromethane, acetonitrile, dimethyl formamide and methanol.
- Class 3 solvents may be regarded as less toxic and of lower risk to human health. They include many lower alcohols, esters, ethers and ketones.
- Class 4 solvents, for which no adequate data were found, include di-isopropyl ether, methyltetrahydrofuran and isooctane.

Solvent use is being subjected to close scrutiny and increasingly stringent environmental legislation. Removal of residual solvent from products is usually achieved by evaporation or distillation and most popular solvents are, therefore, highly volatile. Spillage and evaporation inevitably lead to atmospheric pollution, a major environmental issue of global proportions. Moreover, worker exposure to volatile organic compounds (VOCs) is a serious health issue. Environmental legislation embodied in the Montreal (1987), Geneva (1991) and Kyoto (1997) protocols is aimed at strict control of VOC emissions and the eventual phasing out of greenhouse gases and ozone depleting compounds. Many chlorinated hydrocarbon solvents have already been banned or are likely to be in the near future. Unfortunately, many of these solvents are exactly those that have otherwise desirable properties and are, therefore, widely popular for performing organic reactions. We all have experienced reactions that only seem to go well in a chlorinated hydrocarbon solvent, notably dichloromethane.

Another class of solvents which presents environmental problems comprises the polar aprotic solvents, such as dimethylformamide and dimethyl sulfoxide, that are the solvents of choice for, e.g. many nucleophilic substitutions. They are high boiling and not easily removed by distillation. They are also water-miscible which enables their separation by washing with water. Unfortunately, this leads inevitably to contaminated aqueous effluent.

These issues surrounding a wide range of volatile and nonvolatile, polar aprotic solvents have stimulated the fine chemical and pharmaceutical industries to seek more benign alternatives. There is a marked trend away from hydrocarbons and chlorinated hydrocarbons towards lower alcohols, esters and, in some cases, ethers. Diethyl ether is frowned upon because of its hazardous, inflammable nature, and tetrahydrofuran because of its water miscibility but higher boiling point, water immiscible ethers have become popular. Methyl *tert*-butyl ether (MTBE), for example, is popular and methyl tetrahydrofuran has recently been touted as an agreeable solvent [4]. Inexpensive natural products such as ethanol have the added advantage of being readily biodegradable and ethyl lactate, produced by combining two innocuous natural products, is currently being promoted as a solvent for chemical reactions.

It is worth noting that another contributing factor is the use of different solvents for the different steps in a multistep synthesis. Switching from one solvent to another inevitably leads to substantial wastage and chemists have a marked tendency to choose a different solvent for each step in a synthesis, But times are changing; as mentioned in Chapter 1, the new Pfizer process for sertraline uses ethanol for three consecutive steps, and ethyl acetate in the following step, obviating the need for the hexane, dichloromethane, tetrahydrofuran, and toluene used in the original process. The conclusion is clear: the problem with solvents is not so much their use but the seemingly inherent inefficiencies associated with their containment, recovery and reuse. Alternative solvents should therefore provide for their efficient removal from the product and reuse. The importance of alternative reaction media is underscored by the recent publication of an issue of Green Chemistry devoted to this topic [5].

The subject of alternative reaction media (neoteric solvents) also touches on another issue which is very relevant in the context of this book: recovery and reuse of the catalyst. This is desirable from both an environmental and an economic viewpoint (many of the catalysts used in fine chemicals manufacture contain highly expensive noble metals and/or (chiral) ligands.

7.1.3
Alternative Reaction Media and Multiphasic Systems

If a catalyst is an insoluble solid, that is, a heterogeneous catalyst, it can easily be separated by centrifugation or filtration. In contrast, if it is a homogeneous catalyst, dissolved in the reaction medium, this presents more of a problem. This offsets the major advantages of homogeneous catalysts, such as high activities and selectivities compared to their heterogeneous counterparts (see Table 7.1). However, as Blaser and Studer have pointed out [6], another solution is to develop a catalyst that, at least for economic reasons, does not need to be recycled.

Nonetheless, a serious shortcoming of homogeneous catalysis is the cumbersome separation of the (expensive) catalyst from reaction products and the quantitative recovery of the catalyst in an active form. Separation by distillation of reaction products from catalyst generally leads to heavy ends which remain in the catalyst phase and eventually deactivate it. In the manufacture of pharmaceuticals quantitative separation of the catalyst is important in order to avoid contamination of the product. Consequently there have been many attempts to heterogenize homogeneous catalysts by attachment to organic or inorganic supports or by separation of the products from the catalyst using a semipermeable membrane. However, these approaches have, generally speaking, not resulted in commercially viable processes, for a number of reasons, such as leaching of the metal, poor catalyst productivities, irreproducible activities and selectivities and degradation of the support.

This need for efficient separation of product and catalyst, while maintaining the advantages of a homogeneous catalyst, has led to the concept of liquid–liquid biphasic catalysis, whereby the catalyst is dissolved in one phase and the reactants and product(s) in the second liquid phase. The catalyst is recovered and recycled by simple phase separation. Preferably, the catalyst solution remains in the reactor and is reused with a fresh batch of reactants without further treatment or, ideally, it is adapted to continuous operation. Obviously, both solvents are subject to the same restrictions as discussed above for monophasic systems. The biphasic concept comes in many forms and they have been summarized by Keim in a recent review [7]:

Table 7.1 Homogeneous vs Heterogeneous Catalysis

	Homogeneous	Heterogeneous
Advantages	– Mild reaction conditions – High activity & selectivity – Efficient heat transfer	– Facile separation of catalyst and products – Continuous processing
Disadvantages	– Cumbersome separation & recycling of catalyst – Not readily adapted to a continuous process	– Heat transfer problems – Low activity and/or selectivity

- Two immiscible organic solvents
- Water (aqueous biphasic)
- Fluorous solvents (fluorous biphasic)
- Ionic liquids
- Supercritical CO_2

7.2
Two Immiscible Organic Solvents

Probably the first example of a process employing the biphasic concept is the Shell process for ethylene oligomerization in which the nickel catalyst and the ethylene reactant are dissolved in 1,4-butanediol, while the product, a mixture of linear alpha olefins, is insoluble and separates as a second (upper) liquid phase (see Fig. 7.1). This is the first step in the Shell Higher Olefins Process (SHOP), the largest single feed application of homogeneous catalysis [7].

Because of the resemblance to the 1,4-butane diol in the above example, it is worth mentioning that poly(ethylene glycol) (PEG) and poly(propyleneglycol) (PPG) have attracted interest as novel solvents for catalytic processes (see Fig. 7.2 for examples). They are both relatively inexpensive and readily available materials. They are essentially non-toxic (PPG is often used as a solvent for pharmaceutical and cosmetic preparations and both are approved for use in beverages) and have good biodegradability. Moreover, they are immiscible with water, non-volatile, thermally robust and can, in principle, be readily recycled after removal of the product.

For example, PEG-200 and PEG-400 (the number refers to the average molecular weight) were used as solvents for the aerobic oxidation of benzylic alcohols catalyzed by the polyoxometalate, $H_5PV_2Mo_{10}O_{40}$ [8]. Combination of the same polyoxometalate with Pd(II) was used to catalyze the Wacker oxidation of propyl-

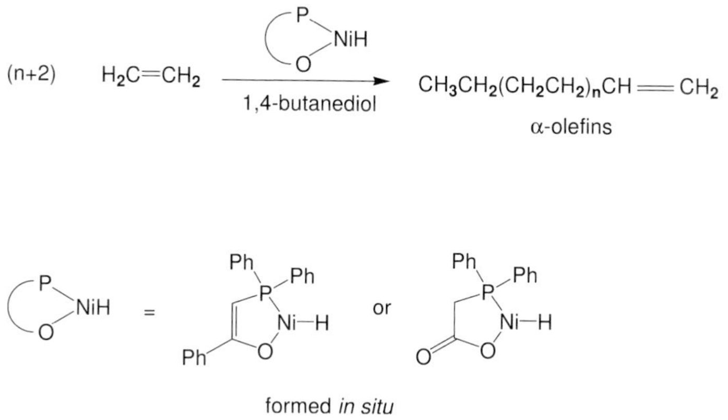

Fig. 7.1 The oligomerization step of the SHOP process.

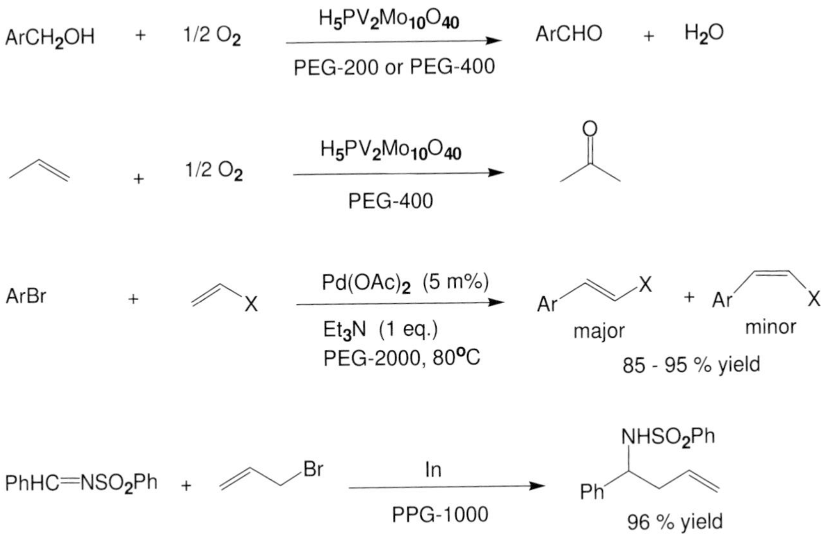

$$ ArCH_2OH \quad + \quad 1/2\ O_2 \quad \xrightarrow[\text{PEG-200 or PEG-400}]{H_5PV_2Mo_{10}O_{40}} \quad ArCHO \quad + \quad H_2O $$

$$ \quad + \quad 1/2\ O_2 \quad \xrightarrow[\text{PEG-400}]{H_5PV_2Mo_{10}O_{40}} $$

$$ ArBr \quad + \quad \text{(alkene-X)} \quad \xrightarrow[\substack{Et_3N\ (1\ eq.) \\ \text{PEG-2000, 80°C}}]{Pd(OAc)_2\ (5\ m\%)} $$

Ar⌁X (major) + Ar⌁X (minor)

85 - 95 % yield

$$ PhHC{=}NSO_2Ph \quad + \quad \text{(allyl-Br)} \quad \xrightarrow[\text{PPG-1000}]{In} $$

NHSO$_2$Ph

Ph⌁

96 % yield

Fig. 7.2 Catalytic reactions in PEG or PPG.

ene to acetone [8, 9]. Similarly, PEG-400 was used as a reusable solvent for the Heck reaction [10], and PPG-1000 for the indium mediated synthesis of homo-allylic amines from imines [11].

In the following sections we shall address the other forms of biphasic catalysis.

7.3
Aqueous Biphasic Catalysis

Water has several attractive features as a solvent and, as we have said elsewhere, the best solvent is no solvent, but if one has to use a solvent then let it be water. Water is the most abundant molecule on the planet and is, hence, readily available and inexpensive. It is nonflammable and incombustible and odorless and colorless (making contamination easy to spot). It has a high thermal conductivity, heat capacity and heat of evaporation, which means that exothermic reactions can be controlled effectively. It readily separates from organic solvents owing to its polarity, density and because of the hydrophobic effect [12], which makes it eminently suitable for biphasic catalysis. Indeed, water forms biphasic systems with many organic solvents, with fluorous solvents, some ionic liquids and with scCO$_2$ [13].

There are some disadvantages, however. Many organic compounds do not dissolve in water. Although the product should have a very low solubility in water the reactants must have some solubility since the catalyst is in the water phase. Reactants and or catalyst may not be stable or may be deactivated in water. If the water is eventually contaminated it may be difficult to purify. Hence, the amount of water used should be kept to a minimum and recycled, as a catalyst solution, as many times as possible before eventual clean-up is required owing to accumulation of impurities.

The numerous benefits of aqueous biphasic catalysis clearly outweigh any disadvantages and, consequently, it is the most studied and most widely applied biphasic technology. Almost all reactions involving organometallic catalysis [14–25], including asymmetric syntheses [26] and polymerizations [27] have been performed in this way. A comprehensive coverage of all these examples goes beyond the scope of this book and the reader is referred to the numerous books and reviews on the subject [14–27] for more details. In this section we shall cover some salient examples which particularly illustrate the industrial utility and the greenness aspects of the methodology.

A prerequisite for aqueous biphasic catalysis is that the catalyst should be soluble in water and this is generally achieved by incorporating hydrophilic moieties in the ligand(s). Literally hundreds of water soluble ligands have been designed for use in aqueous biphasic catalysis. A few examples are shown in Fig. 7.3 and the reader is referred to a review [17] on the subject for more examples. By far the most commonly used water soluble ligands are the sulfonated phosphines. The first water soluble, sulfonated phosphine, $Ph_2P(C_6H_4$-3-$SO_3Na)$ (tppms), was prepared in 1958 by Chatt and coworkers, by sulfonation of triphenyl phosphine [28]. The solubility of tppms in water is 80 g L^{-1} at 20 °C.

Pioneering studies of the use of water soluble noble metal complexes of sulfonated phosphines as catalysts in aqueous biphasic systems were performed in the early 1970s, by Joo and Beck in Hungary and Kuntz at Rhone-Poulenc in France. Joo and Beck studied catalytic hydrogenations and transfer hydrogenations using Rh or Ru complexes of tppms [24]. Kuntz, on the other hand, pre-

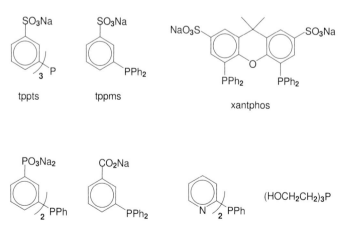

Fig. 7.3 Water soluble phosphines.

pared trisulfonated triphenylphosphine (tppts), by using longer sulfonation times, and it proved to be highly soluble in water (1100 g L^{-1} at 20 °C).

7.3.1
Olefin Hydroformylation

Kuntz subsequently showed that the RhCl (tppts)$_3$ catalyzed the hydroformylation of propylene in an aqueous biphasic system [29]. These results were further developed, in collaboration with Ruhrchemie, to become what is known as the Ruhrchemie/Rhone-Poulenc two-phase process for the hydroformylation of propylene to n-butanal [18, 19, 22, 30]. Ruhrchemie developed a method for the large scale production of tppts by sulfonation of triphenylphosphine with 30% oleum at 20 °C for 24 h. The product is obtained in 95% purity by dilution with water, extraction with a water insoluble amine, such as tri(isooctylamine), and pH-controlled re-extraction of the sodium salt of tppts into water with a 5% aqueous solution of NaOH. The first commercial plant came on stream in 1984, with a capacity of 100 000 tons per annum of butanal. Today the capacity is ca. 400 000 tpa and a cumulative production of millions of tons. Typical reaction conditions are $T = 120$ °C, $P = 50$ bar, CO/H$_2$ = 1.01, tppts/Rh = 50–100, [Rh] = 10–1000 ppm. The RhH(CO)(tppts)$_3$ catalyst is prepared in situ from e.g. rhodium 2-ethylhexanoate and tppts in water.

The RCH/RP process (see Fig. 7.4) affords butanals in 99% selectivity with a n/i ratio of 96/4. Rhodium carry-over into the organic phase is at the ppb level. The process has substantial economic and environmental benefits compared with conventional processes for the hydroformylation of propylene using Rh or Co complexes in an organic medium [31]:

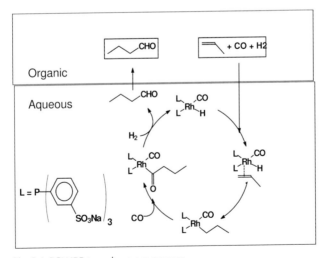

Fig. 7.4 RCH/RP two-phase oxo process.

- High catalyst activity and selectivity (less alkane production)
- High linearity (n/i) is obtained at lower ligand/Rh ratios
- Easy and essentially complete catalyst recovery
- Much simpler process engineering and the process is highly energy efficient (the plant is a net producer of heat)
- Virtual elimination of plant emissions and the avoidance of an organic solvent
- Conditions are less severe and heavy end formation is much less than in conventional processes where the product is separated from the catalyst by distillation. Moreover, the small amount of heavy ends (0.4%) formed is dissolved in the organic phase where it does not contaminate/deactivate the catalyst.

Based on the success of the RCH/RP process much effort was devoted to the development of new ligands that are even more efficient (see Fig. 7.5 for examples) [17]. However, to our knowledge they have not been commercialized, probably because they do not have a favorable cost/benefit ratio.

The aqueous biphasic hydroformylation concept is ineffective with higher olefins owing to mass transfer limitations posed by their low solubility in water. Several strategies have been employed to circumvent this problem [22], e.g. by conducting the reaction in a monophasic system using a tetraalkylammonium salt of tppts as the ligand, followed by separation of the catalyst by extraction into water. Alternatively, one can employ a different biphasic system such as a fluorous biphasic system or an ionic liquid/scCO$_2$ mixture (see later).

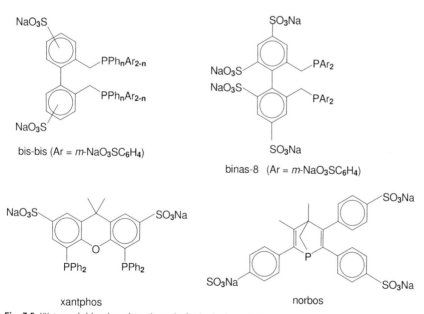

bis-bis (Ar = *m*-NaO$_3$SC$_6$H$_4$)

binas-8 (Ar = *m*-NaO$_3$SC$_6$H$_4$)

xantphos

norbos

Fig. 7.5 Water soluble phosphine ligands for hydroformylation.

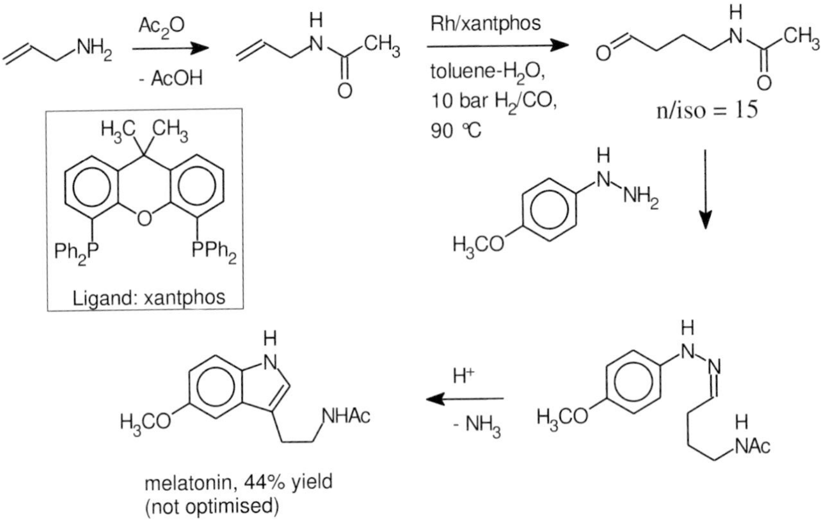

Fig. 7.6 Synthesis of melatonin using inverse aqueous phase catalysis.

What about when the substrate and product are water soluble? The problem of catalyst recovery in this case can be solved by employing *inverse aqueous biphasic catalysis*. An example is the hydroformylation of N-allylacetamide in an aqueous biphasic system in which the catalyst is dissolved in the organic phase and the substrate and product remain in the water phase. This formed the basis for an elegant synthesis of the natural product, melatonin, in which the aqueous solution of the hydroformylation product was used in the next step without work-up (Fig. 7.6) [32].

The success of aqueous biphasic hydroformylation stimulated a flurry of activity in the application of the concept to other reactions involving organometallic catalysis [14–27].

7.3.2
Hydrogenation

Following the pioneering work of Joo [24], the aqueous biphasic concept has been widely applied to catalytic hydrogenations and hydrogen transfer processes, using mainly complexes of Rh and Ru [15, 24, 25]. For example, Mercier et al., at Rhone-Poulenc, showed that hydrogenation of the unsaturated aldehyde, 3-methyl-2-buten-1-al, in the presence of Rh/tppts afforded the corresponding saturated aldehyde, while with Ru/tppts as the catalyst the unsaturated alcohol was obtained in high selectivity (Fig. 7.7) [33, 34]. Joo subsequently showed, using $RuCl_2(tppms)_2$ as the catalyst, that the chemoselectivity is pH dependent [35]. The active catalytic species for olefinic double bond hydrogenation is $RuHCl(tppms)_2$, formed in acidic solutions by dissociation of $RuHCl(tppms)_3$. In contrast, at basic pH the major species is $RuH_2 (tppms)_4$, which catalyzes

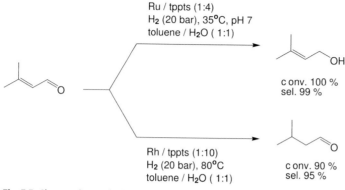

Fig. 7.7 Chemoselective hydrogenation of an unsaturated aldehyde.

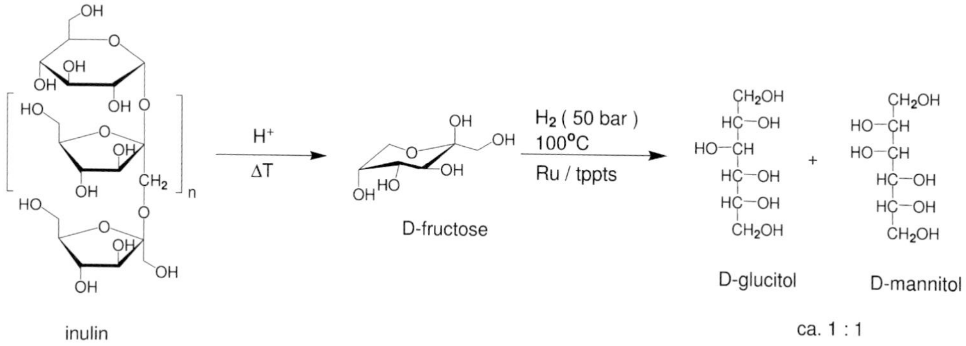

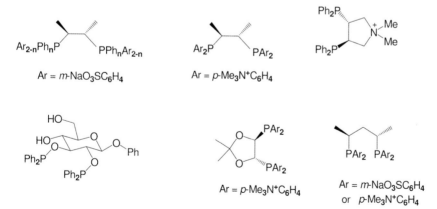

Fig. 7.8 Hydrogenation of carbohydrates in water.

Ar$_{2-n}$Ph$_n$P — PPh$_n$Ar$_{2-n}$

Ar = m-NaO$_3$SC$_6$H$_4$

Ar$_2$P — PAr$_2$

Ar = p-Me$_3$N$^+$C$_6$H$_4$

Ph$_2$P

Ph$_2$P

N$^+$ Me
Me

Ph$_2$P

O

HO
HO
Ph$_2$P
O
O
Ph
O
Ph$_2$P

O
O
O
O
PAr$_2$
PAr$_2$

Ar = p-Me$_3$N$^+$C$_6$H$_4$

PAr$_2$ PAr$_2$

Ar = m-NaO$_3$SC$_6$H$_4$
or p-Me$_3$N$^+$C$_6$H$_4$

Fig. 7.9 Chiral water soluble ligands.

the selective reduction of the carbonyl group. The coordinative saturation of the latter prevents the coordination of the olefinic double bond but allows the reduction of a carbonyl group by intermolecular hydride transfer [35].

The technique can also be used for the catalytic transfer hydrogenation of carbonyl groups using formate [34, 35] or isopropanol [36] as the hydrogen donor. It is also worth noting, in this context, that Ru/tppts can be used for the selective hydrogenation [37] or transfer hydrogenation [38] of carbohydrates in a monophasic aqueous system (Fig. 7.8).

When chiral water soluble ligands are used the technique can be applied to asymmetric hydrogenations [26]. Some examples are shown in Fig. 7.9.

7.3.3
Carbonylations

The majority of catalytic carbonylations employ palladium catalysts and the water soluble complex, Pd(tppts)$_3$, is easily prepared by reduction of PdCl$_2$/tppts with CO in water at room temperature [39]. Hence, this complex is readily generated *in situ*, under carbonylation conditions. It was shown to catalyze the carbonylation of alcohols [40, 41] and olefins [42–44], in the presence of a Bronsted acid cocatalyst (Fig. 7.10).

The reaction is proposed to involve the formation of an intermediate carbenium ion (hence the need for an acid cocatalyst) which reacts with the Pd(0) complex to afford an alkylpalladium(II) species (see Fig. 7.11) [44].

When a sulfonated diphosphine is used as the ligand, the complex formed with palladium(0) catalyzes the alternating copolymerization of ethylene and

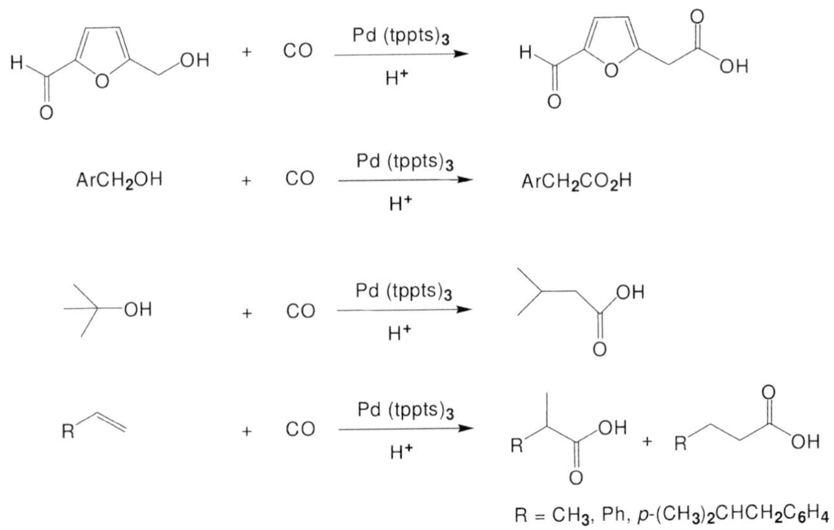

$$R = CH_3,\ Ph,\ p\text{-}(CH_3)_2CHCH_2C_6H_4$$

Fig. 7.10 Carbonylations catalyzed by Pd(tppts)$_3$ in water.

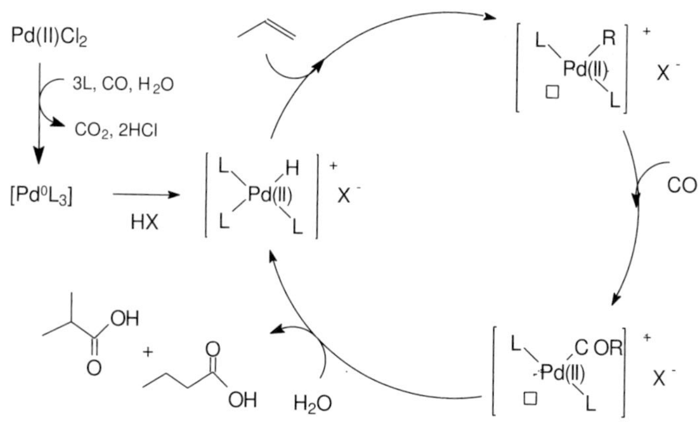

Fig. 7.11 Mechanism of Pd(tppts)$_3$ catalyzed carbonylations.

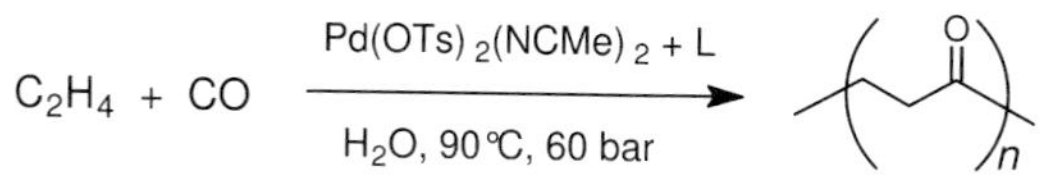

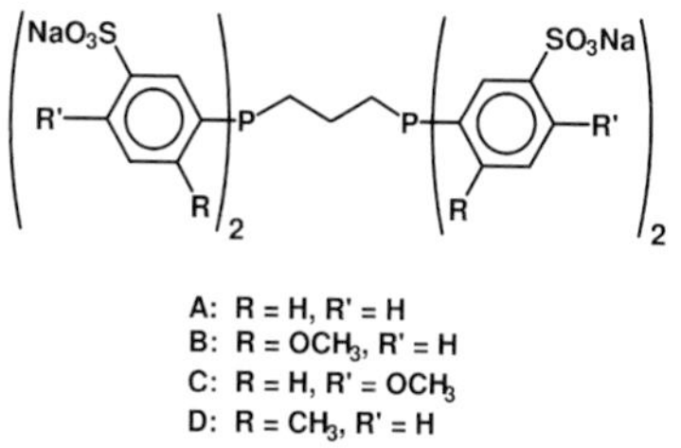

Ligand	Additive	Activity (kg/g Pd/h)
A	HOTs	4.0
B	HOTs	18.6
B	none	24.4
B	NaOH	0
C	HOTs	1.7
D	HOTs	1.7

A: R = H, R' = H
B: R = OCH$_3$, R' = H
C: R = H, R' = OCH$_3$
D: R = CH$_3$, R' = H

Fig. 7.12 Alternating copolymerization of ethylene and CO in water.

CO to give the engineering thermoplastic polyketone, Carilon [45, 46]. Indeed, when a well-defined complex was used (Fig. 7.12), exceptionally high activities were observed [46], with turnover frequencies (TOFs) higher than the conventional catalyst in methanol as solvent.

7.3.4
Other C–C Bond Forming Reactions

Many other C–C bond forming reactions involving organometallic catalysis have been successfully performed in an aqueous biphasic system. Examples are shown in Fig. 7.13 and include Heck [47, 48] and Suzuki couplings [48] and the Rhone-Poulenc process for the synthesis of geranylacetone, a key intermediate in the manufacture of vitamin E, in which the key step is Rh/tppts catalyzed ad-

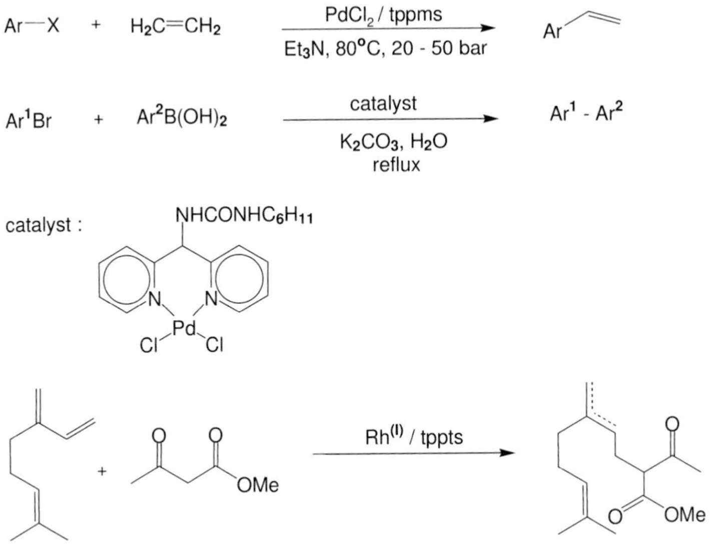
Ar—X + H2C=CH2
PdCl2 / tppms
Et3N, 80°C, 20 - 50 bar
Ar
Ar1Br + Ar2B(OH)2
catalyst
K2CO3, H2O
reflux
Ar1 - Ar2
catalyst :
NHCONHC6H11
N
N
Pd
Cl
Cl
+
O
O
OMe
Rh(I) / tppts
O
O
OMe

Fig. 7.13 Catalytic C–C bond formation in aqueous biphasic media.

dition of methyl acetoacetate to myrcene [49]. The intermediate beta-keto esters can also be used for the synthesis of pseudo-ionone, a key intermediate in the manufacture of vitamin A. Water soluble variants of the Grubbs Ru catalyst have been used in aqueous biphasic ring opening metathesis polymerization (ROMP) [50].

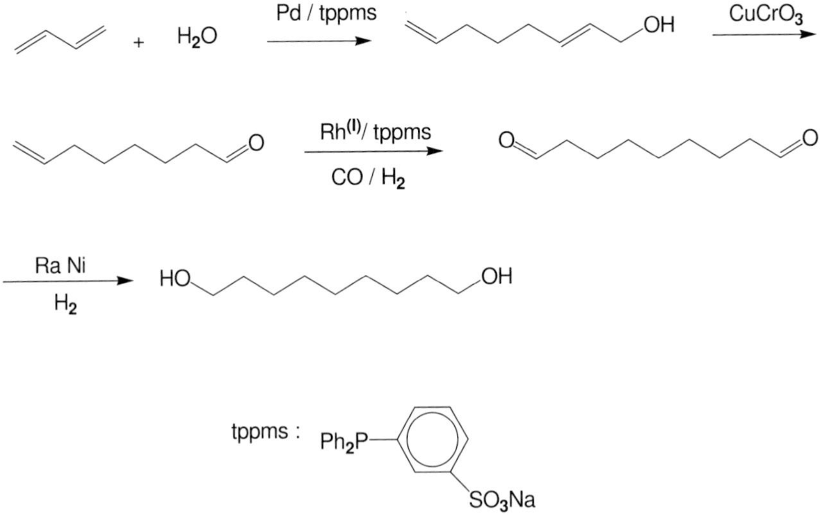
+ H2O
Pd / tppms
OH
CuCrO3
O
Rh(I)/ tppms
CO / H2
O
O
Ra Ni
H2
HO
OH
tppms :
Ph2P
SO3Na

Fig. 7.14 Kuraray process for nonane-1,9-diol.

In the Kuraray process for the production of nonane-1,9-diol two steps involve the use of aqueous biphasic catalysis: Pd/tppms catalyzed telomerization of butadiene with water as a reactant and Rh/tppms catalyzed hydroformylation (Fig. 7.14) [51].

7.3.5 Oxidations

The palladium(II) complex of sulfonated bathophenanthroline was used in a highly effective aqueous biphasic aerobic oxidation of primary and secondary alcohols to the corresponding aldehydes or carboxylic acids and ketones respectively (Fig. 7.15) [52, 53]. No organic solvent was necessary, unless the substrate was a solid, and turnover frequencies of the order of $100\ h^{-1}$ were observed. The catalyst could be recovered and recycled by simple phase separation (the aqueous phase is the bottom layer and can be left in the reactor for the next batch). The method constitutes an excellent example of a green catalytic oxidation with oxygen (air) as the oxidant, no organic solvent and a stable recyclable catalyst.

7.4 Fluorous Biphasic Catalysis

Fluorous biphasic catalysis was pioneered by Horvath and Rabai [54, 55] who coined the term 'fluorous', by analogy with 'aqueous', to describe highly fluorinated alkanes, ethers and tertiary amines. Such fluorous compounds differ markedly from the corresponding hydrocarbon molecules and are, consequently, immiscible with many common organic solvents at ambient temperature although they can become miscible at elevated temperatures. Hence, this provides a basis for performing biphasic catalysis or, alternatively, monophasic catalysis at elevated temperatures with biphasic product/catalyst separation at lower temperatures. A number of fluorous solvents are commercially available (see Fig. 7.16 for example), albeit rather expensive compared with common organic

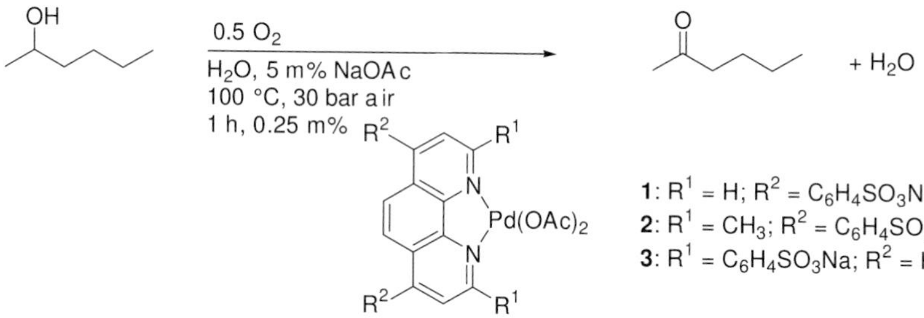

1: R^1 = H; R^2 = $C_6H_4SO_3Na$
2: R^1 = CH_3; R^2 = $C_6H_4SO_3Na$
3: R^1 = $C_6H_4SO_3Na$; R^2 = H

Fig. 7.15 Aqueous biphasic Pd catalyzed aerobic oxidation of alcohols.

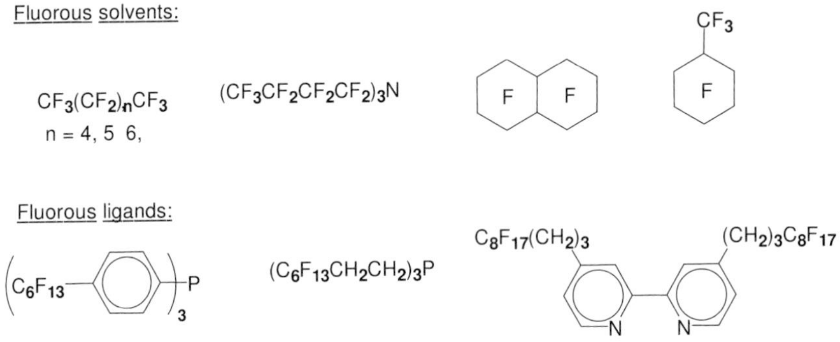

Fluorous solvents:

$CF_3(CF_2)_nCF_3$

n = 4, 5 6,

$(CF_3CF_2CF_2CF_2)_3N$

Fluorous ligands:

$\left(C_6F_{13} \!-\!\! \left\langle \bigcirc \right\rangle \!-\! \right)_{\!3} \!\! P$

$(C_6F_{13}CH_2CH_2)_3P$

Fig. 7.16 Some examples of fluorous solvents and ligands.

solvents (or water). Barthel-Rosa and Gladysz have published an extensive 'user's guide' to the application of fluorous catalysts and reagents [56].

In order to perform fluorous biphasic catalysis the (organometallic) catalyst needs to be solubilized in the fluorous phase by deploying "fluorophilic" ligands, analogous to the hydrophilic ligands used in aqueous biphasic catalysis. This is accomplished by incorporating so-called "fluorous ponytails".

The attachment of the strongly electron-withdrawing perfluoroalkyl groups would seriously reduce the electron density on, e.g. phosphorus, to an extent that it would no longer be an effective phosphine ligand. This problem is circumvented by placing an "organic" spacer between the perfluoroalkyl moiety and the coordinating atom, generally phosphorus. Some examples of such ligands are shown in Fig. 7.16. It should be noted however that these arguments do not apply to, for example, oxidation reactions, where a strongly electron-withdrawing ligand may actually increase the activity of the catalyst (see later).

7.4.1
Olefin Hydroformylation

As noted earlier, hydroformylation of higher olefins in an aqueous biphasic system is problematic owing to the lack of solubility of the substrate in the aqueous phase. On the other hand, hydroformylation in an organic medium presents the problem of separating the long-chain aldehydes from the catalyst. In contrast, this is not a problem with a fluorous biphasic system where at the elevated reaction temperature the mixture becomes a single phase. Cooling the reaction mixture to room temperature results in a separation into a fluorous phase, containing the catalyst, and an organic phase, containing the aldehyde products. This concept was applied by Horvath and Rabai, in their seminal paper [54], to the hydroformylation of 1-decene in a 1:1 mixture of $C_6F_{11}CF_3$ and toluene. The catalyst was prepared *in situ* from $Rh(CO)_2(acac)$ and $P[CH_2CH_2(CF_2)_5CF_3]_3$ (P/Rh=40) and the reaction performed at $100\,^\circ C$ and 10 bar CO/H_2. Upon completion of the reaction the reactor was cooled to room

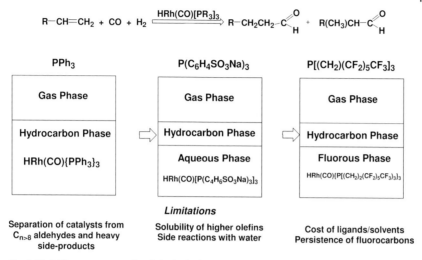

Fig. 7.17 Different concepts for olefin hydroformylation.

temperature when phase separation occurred. When the upper, organic phase was returned to the reactor, with fresh reactants, negligible reaction was observed, demonstrating that catalytically active rhodium species are not leached into the organic phase. It was subsequently shown [57, 58] that recycling of the catalyst phase, in nine consecutive runs, afforded a total turnover number (TON) of more than 35 000. The rhodium losses amounted to 4.2%, which constitutes ca. 1 ppm per mole of aldehyde. Unfortunately there was some leaching of the free ligand into the organic phase, resulting in a slight decrease in (n/i) selectivity (from ca. 92/8 to 89/11), which is dependent on the ligand/Rh ratio. However, the authors noted [57] that this could probably be improved by fine tuning of the solvent and/or ligand. They also emphasized that the system combined the advantages of a monophasic reaction with a biphasic separation. The three different concepts for olefin hydroformylation – organic solvent, aqueous biphasic and fluorous biphasic – are compared in Fig. 7.17.

7.4.2
Other Reactions

The successful demonstration of the fluorous biphasic concept for performing organometallic catalysis sparked extensive interest in the methodology and it has subsequently been applied to a wide variety of catalytic reactions, including hydrogenation [59], Heck and Suzuki couplings [60, 61] and polymerizations [62]. The publication of a special Symposium in print devoted to the subject [63] attests to the broad interest in this area.

Fluorous solvents would seem to be particularly suitable for performing aerobic oxidations, based on the high solubility of oxygen in fluorocarbons, a prop-

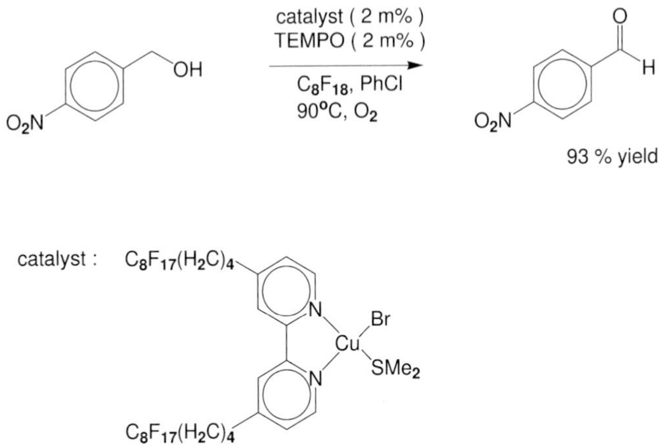

Fig. 7.18 Catalytic oxidation of alcohols in a fluorous medium.

erty which is exploited in their application as blood substitutes [64]. A few examples of catalytic oxidations in fluorous media have been reported [65, 66]. For example, the aerobic oxidation of alcohols was performed in a fluorous medium, using a copper complex of a bipyridine ligand containing perfluorinated ponytails (Fig. 7.18) [66].

Catalytic oxidations with hydrogen peroxide have also been performed in fluorous media (Fig. 7.19) [67].

Notwithstanding the seemingly enormous potential of the fluorous biphasic catalysis concept, as yet a commercial application has not been forthcoming. Presumably the cost of the solvents and ligands is a significant hurdle, and although the catalyst and products are well-partitioned over the two phases there is a finite solubility of the catalyst in the organic phase which has to be coped with. Perhaps an even more serious problem is the extremely long life-

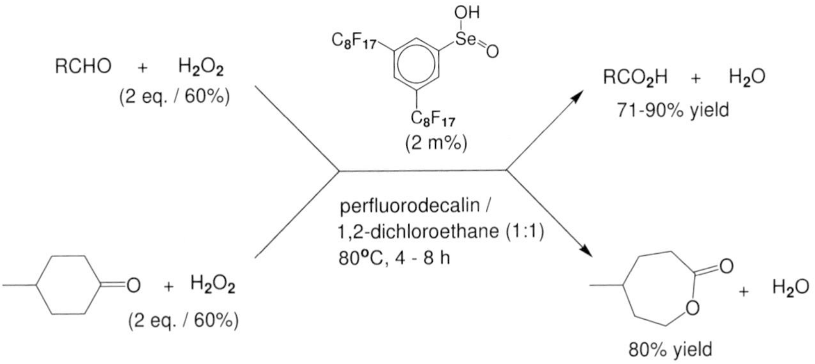

Fig. 7.19 Catalytic oxidations with hydrogen peroxide in fluorous media.

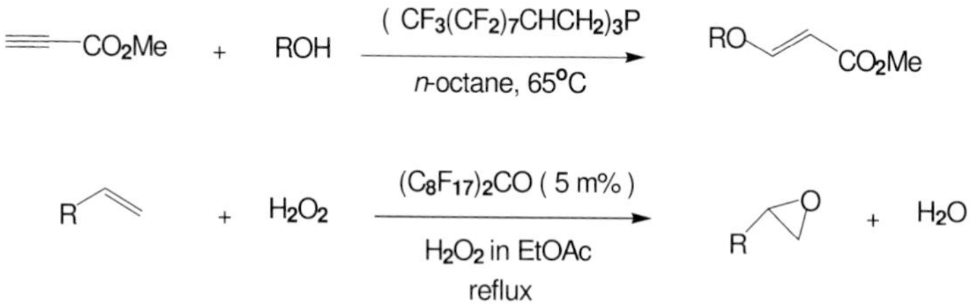

Fig. 7.20 Fluorous catalysis without fluorous solvents.

time of fluorocarbons in the environment which, even though they are chemically inert, essentially nontoxic and are not, in contrast to their cousins the CFCs, ozone-depleting agents, is still a matter for genuine concern.

In this context it is interesting to note the recent reports of *fluorous catalysis without fluorous solvents* [68]. The thermomorphic fluorous phosphines, $P[(CH_2)_m(CF_2)_7CF_3]_3$ ($m=2$ or 3) exhibit ca. 600-fold increase in n-octane solubility between –20 and 80 °C. They catalyze the addition of alcohols to methyl propiolate in a monophasic system at 65 °C and can be recovered by precipitation on cooling (Fig. 7.20) [68]. Similarly, perfluoroheptadecan-9-one catalyzed the epoxidation of olefins with hydrogen peroxide in e.g. ethyl acetate as solvent [69]. The catalyst could be recovered by cooling the reaction mixture, which resulted in its precipitation.

Presumably this technique can be applied to other examples of (organometallic) catalysis. We also note that catalysis can also be performed in supercritical carbon dioxide (scCO$_2$) as solvent (see next section).

7.5
Supercritical Carbon Dioxide

7.5.1
Supercritical Fluids

Supercritical fluids (SCFs) constitute a different category of neoteric solvents to the rest discussed in this chapter since they are not in the liquid state. The critical point (Fig. 7.21) represents the highest temperature and pressure at which a substance can exist as a vapor in equilibrium with a liquid. In a closed system, as the boiling point curve is ascended, with increasing pressure and temperature, the liquid becomes less dense, owing to thermal expansion, and the gas becomes denser as the pressure increases. The densities of the two phases converge and become identical at the critical point, where they merge and become a SCF. Hence the properties, e.g. density, viscosity and diffusivity, of a SCF are intermediate between a liquid and a vapor. SCFs also mix well with gases, making them potentially interesting media for catalytic reactions with hydrogen, carbon monoxide and oxygen.

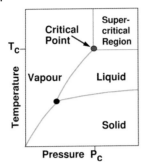

Fig. 7.21 A phase diagram illustrating the supercritical region.

7.5.2
Supercritical Carbon Dioxide

Several features of scCO$_2$ make it an interesting solvent in the context of green chemistry and catalysis. For carbon dioxide the critical pressure and temperature are moderate: 74 bar and 31 °C, respectively. Hence the amount of energy required to generate supercritical carbon dioxide is relatively small.

In addition, carbon dioxide is nontoxic, chemically inert towards many substances, nonflammable, and simple depressurization results in its removal.

It is miscible with, e.g. hydrogen, making it an interesting solvent for hydrogenation and hydroformylation (see below). Although it is a greenhouse gas its use involves no net addition to the atmosphere; it is borrowed as it were. Its main uses are as a replacement for VOCs in extraction processes. For example it is widely used for the decaffeination of coffee where it replaced the use of a chlorinated hydrocarbon. More recently, it has been commercialized as a replacement for trichloroethane in dry cleaning applications [70]. In the last decade attention has also been focused on the use of scCO$_2$ as a solvent for green chemistry and catalysis [71–74]. The pre-existence of an established SCF extraction industry meant that the necessary equipment was already available.

7.5.3
Hydrogenation

The most extensively studied reaction is probably hydrogenation. The miscibility of scCO$_2$ with hydrogen, as noted above, results in high diffusion rates and is a significant advantage. It provides the basis for achieving much higher reaction rates than in conventional solvents. The use of scCO$_2$ as a solvent for catalytic hydrogenations over supported noble metal catalysts was pioneered by Poliakoff [75–77]. For example, the Pd-catalyzed hydrogenation of a variety of functional groups, such as C=C, C=O, C=N and NO$_2$ was shown [75, 76] to proceed with high selectivities and reaction rates in scCO$_2$. The high reaction rates allowed the use of exceptionally small flow reactors. Chemoselectivities with multifunctional compounds could be adjusted by minor variations in reaction parameters.

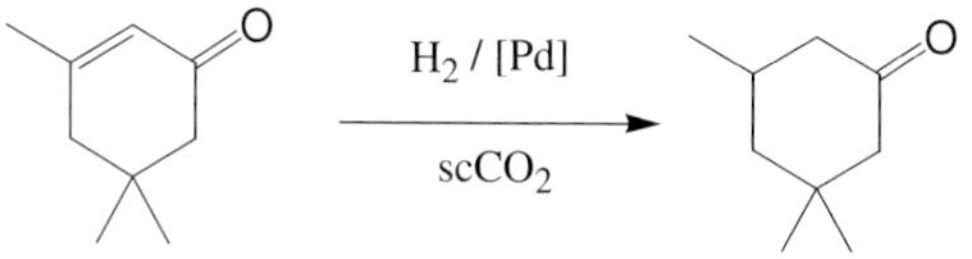

Fig. 7.22 Catalytic hydrogenation of isophorone in scCO$_2$.

The technology has been commercialized, in collaboration with the Thomas Swan company, for the manufacture of trimethyl cyclohexanone by Pd-catalyzed hydrogenation of isophorone (Fig. 7.22) [77].

The multi-purpose plant, which went on stream in 2002, has a production capacity of 1000 tpa (100 kg h^{-1}). The very high purity of the product and high productivities, together with the elimination of organic solvent, are particular benefits of the process. It is a perfect example of the successful transfer of green chemistry from academia to industry. Presumably it will stimulate further industrial application of the technology. Ikariya and coworkers [78], another pioneering group in the area of catalytic hydrogenations in scCO$_2$, recently reported that chemoselective hydrogenation of halogenated nitrobenzene (Fig. 7.23) proceeds very effectively over Pt/C catalysts in scCO$_2$. The rate is significantly enhanced compared to that with the neat substrate and competing dehalogenation is markedly suppressed, affording a higher chemoselectivity. The increased selectivity in scCO$_2$ was tentatively attributed to the generation of small amounts of CO which preferentially cover the more active sites on the platinum surface which are responsible for dehalogenation. Indeed, the addition of small amounts of CO to the neat reaction system caused a marked suppression of competing dehalogenation.

Catalytic asymmetric hydrogenations have also been performed in supercritical carbon dioxide [79–81]. For example, a-enamides were hydrogenated in high enantioselectivities comparable to those observed in conventional solvents, using a cationic rhodium complex of the EtDuPHOS ligand (Fig. 7.24) [79]. More recently, catalytic asymmetric hydrogenations have been performed in scCO$_2$ with

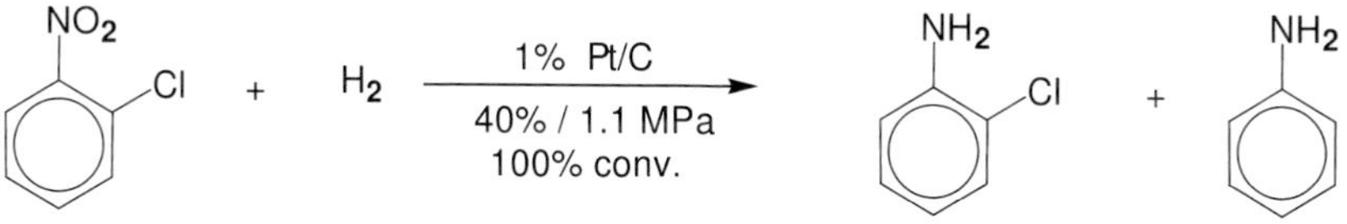

	Time (h)	Selectivity (%)	
		1	2
neat	5	95.6	2.1
sc CO$_2$	2.5	99.7	0.3

Fig. 7.23 Chemoselective hydrogenation of o-chloronitrobenzene in scCO$_2$.

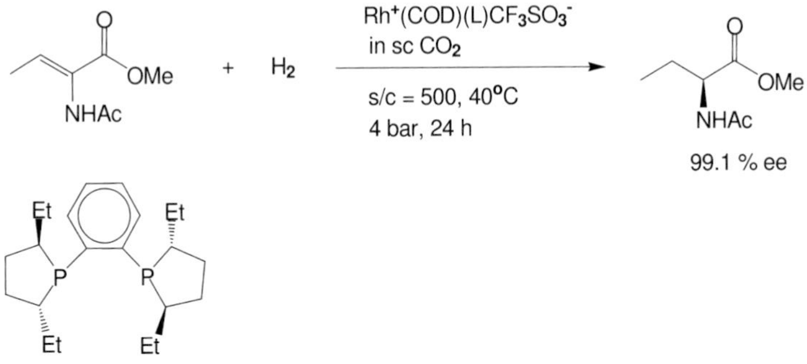

L = Et-DuPHOS

Fig. 7.24 Catalytic asymmetric hydrogenation in scCO$_2$.

Rh complexes containing monodentate perfluoroalkylated chiral ligands [80]. By using a chiral Rh complex immobilized onto alumina, via a phosphotungstic acid linker, asymmetric hydrogenations could be performed in continuous operation using scCO$_2$ as the solvent [81]. Jessop and coworkers [82] compared a variety of neoteric solvents – scCO$_2$, ionic liquids (ILs), ILs with cosolvents and CO$_2$-expanded ionic liquids (EILs) – in the asymmetric hydrogenation of prochiral unsaturated acids catalyzed by chiral ruthenium complexes. They concluded that the optimum solvent was dependent on the specific substrate used and no one solvent clearly outperforms all others for all substrates. Solvents thought to dissolve significant amounts of hydrogen gave good enantioselectivities for substrates known to be dependent on high H$_2$ concentrations. Solvents dissolving low amounts of hydrogen, e.g. ionic liquids, were ideal for substrates dependent on minimum H$_2$ concentrations for high enantioselectivities. Similarly, olefin hydroformylation has been conducted in scCO$_2$ using an immobilized Rh catalyst [83].

7.5.4
Oxidation

Supercritical carbon dioxide is, in principle, an ideal inert solvent for performing catalytic aerobic oxidations as it is nonflammable and completely miscible with oxygen. It is surprising, therefore, that there are so few studies in this area. A recent report describes the aerobic oxidation of alcohols catalyzed by PEG-stabilized palladium nanoparticles in a scCO$_2$/PEG biphasic system [84]. Recently, much interest has also been focused on catalytic oxidations with hydrogen peroxide, generated *in situ* by Pd-catalyzed reaction of hydrogen with oxygen, in scCO$_2$/water mixtures [85]. The system was used effectively for the direct epoxidation of propylene to propylene oxide over a Pd/TS-1 catalyst [86]. These reactions probably involve the intermediate formation of peroxycarbonic acid by reaction of H$_2$O$_2$ with CO$_2$ (Fig. 7.25).

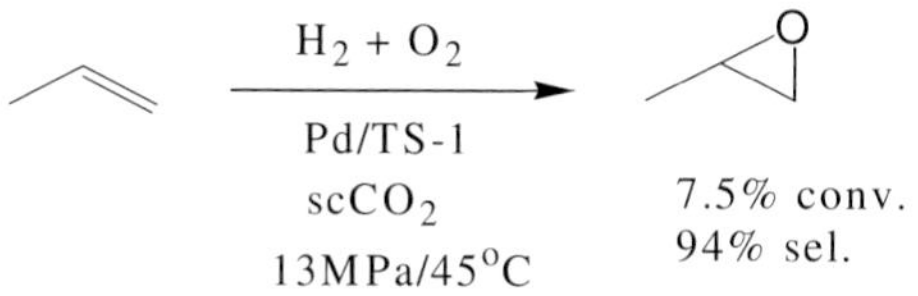

Fig. 7.25 Oxidation of propylene with H_2/O_2 in scCO$_2$.

7.5% conv.
94% sel.

7.5.5
Biocatalysis

scCO$_2$ is also an interesting solvent for performing bioconversions. The first reports of biocatalysis in scCO$_2$ date back to 1985 [87–89] and in the intervening two decades the subject has been extensively studied [90]. Enzymes are generally more stable in scCO$_2$ than in water and the *Candida antarctica* lipase (Novozym 435)-catalyzed resolution of 1-phenylethanol was successfully performed at temperatures exceeding 100 °C in this solvent [91]. Matsuda et al. found that the enantioselectivity of alcohol acylations catalyzed by Novozyme 435 in scCO$_2$ could be controlled by adjusting the pressure and temperature [92]. The same group recently reported a continuous flow system in scCO$_2$ for the enzymatic resolution of chiral secondary alcohols via Novozyme 435 catalyzed acylation with vinyl acetate (Fig. 7.26) [93]. For example, the kinetic resolution of 1-phenyl ethanol at 9 MPa CO$_2$ and 40 °C afforded the (R)-acetate in 99.8% *ee* and the (S)-alcohol in 90.6% *ee* at 48% conversion (E = 1800).

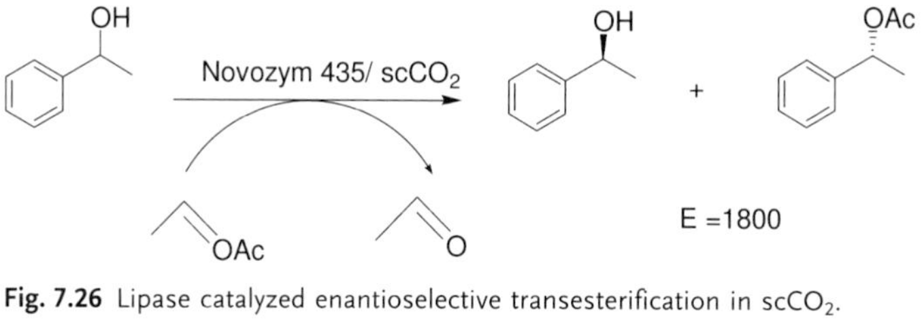

Fig. 7.26 Lipase catalyzed enantioselective transesterification in scCO$_2$.

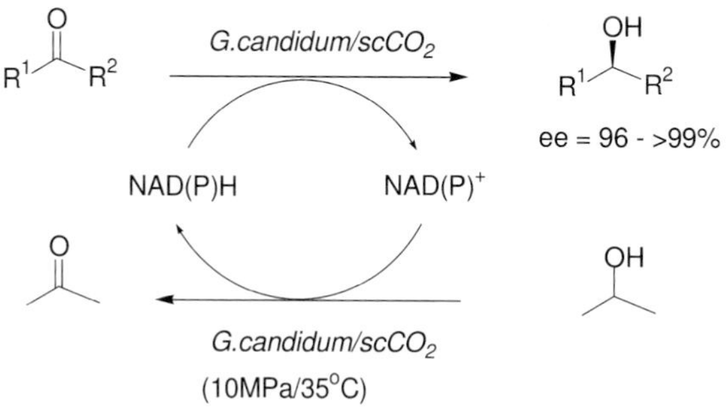

Fig. 7.27 Biocatalytic enantioselective reduction of ketones in scCO$_2$.

Similarly, the enantioselective reduction of prochiral ketones catalyzed by whole cells of *Geotrichum candidum* proceeded smoothly in scCO$_2$ in a semi-continuous flow system (Fig. 7.27) [94].

Enzyme catalyzed oxidations with O$_2$ have also been successfully performed in scCO$_2$ e.g. using cholesterol oxidase [95] and polyphenol oxidase [88]. The use of scCO$_2$ as a solvent for biotransformations clearly has considerable potential and we expect that it will find more applications in the future.

7.6
Ionic Liquids

Ionic liquids are quite simply liquids that are composed entirely of ions [96, 97]. They are generally salts of organic cations, e.g. tetraalkylammonium, alkylpyridinium, 1,3-dialkylimidazolium, tetraalkylphosphonium (Fig. 7.28). Room temperature ionic liquids exhibit certain properties which make them attractive media for performing green catalytic reactions. They have essentially no vapor pressure and are thermally robust with liquid ranges of e.g. 300 °C, compared to 100 °C for water. Polarity and hydrophilicity/hydrophobicity can be tuned by a suitable combination of cation and anion, which has earned them the accolade, 'designer solvents'.

Ionic liquids have been extensively studied in the last few years as media for organic synthesis and catalysis in particular [98]. For example, the hydroformylation of higher olefins, such as 1-octene, was performed in ionic liquids [99]. Good activities were observed with rhodium in combination with the water-soluble ligand, tppts, described above but the selectivity was low (n/iso ratio=2.6). In order to achieve both high activities and selectivities special ligands had to be designed (Fig. 7.29). No detectable (less than 0.07%) Rh leaching was observed and the IL phase containing the catalyst could be recycled after separating the product which formed a separate phase. However, the need for rather exotic ligands will presumably translate to higher costs for this process.

As would be expected, high rate accelerations can result when reactions proceeding through ionic intermediates, e.g. carbocations, are performed in ionic liquids. For example, Seddon and coworkers [100] studied the Friedel-Crafts acylation of toluene, chlorobenzene (Fig. 7.30) and anisole with acetyl chloride in [emim][Al$_2$Cl$_7$], whereby the ionic liquid is acting both as solvent and catalyst. They ob-

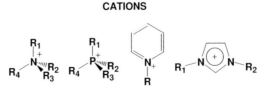

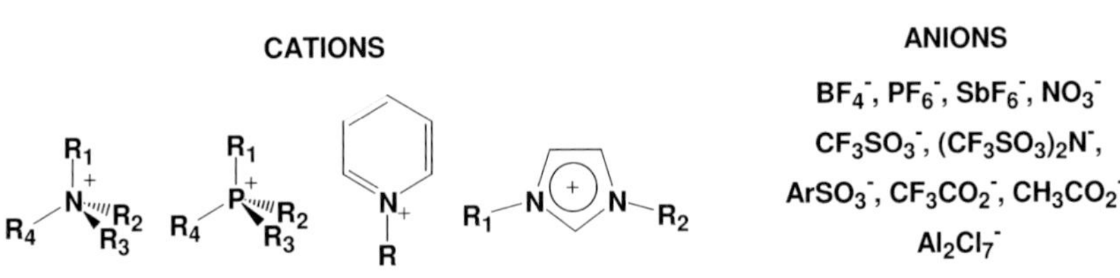

CATIONS

ANIONS

BF$_4^-$, PF$_6^-$, SbF$_6^-$, NO$_3^-$

CF$_3$SO$_3^-$, (CF$_3$SO$_3$)$_2$N$^-$,

ArSO$_3^-$, CF$_3$CO$_2^-$, CH$_3$CO$_2^-$,

Al$_2$Cl$_7^-$

Fig. 7.28 Structures of ionic liquids.

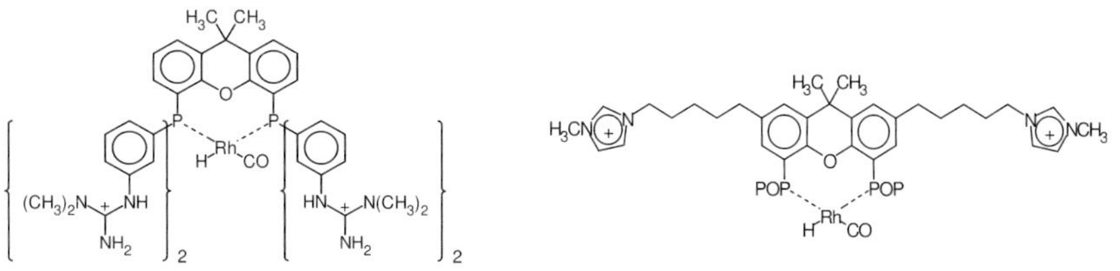

Catalyst	TOF(h⁻¹)	n/iso ratio

Catalyst	TOF(h^{-1})	n/iso ratio
(tppts)$_2$(CO)RhH	80	2.6
A	50	20
B	320	49

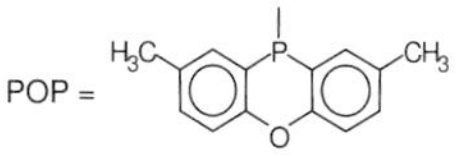

Fig. 7.29 Hydroformylation in ionic liquids.

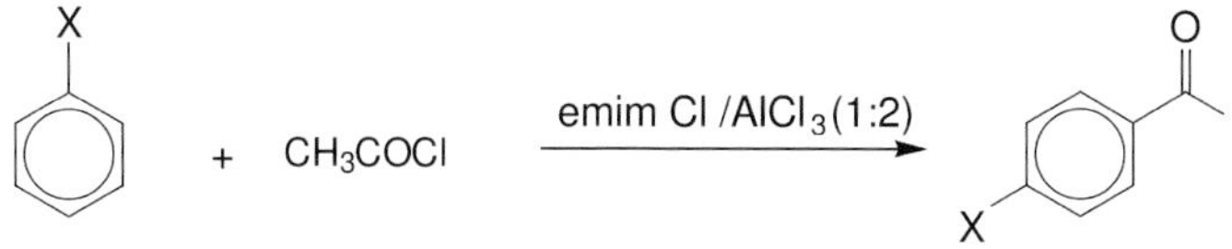

R	Temp (°C)	Time (h)	Yield (%)
MeO	-10	0.25	99
Me	20	1	98
Cl	20	24	97

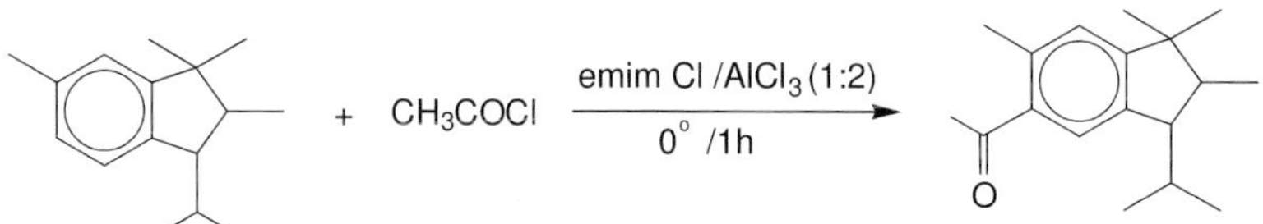

traseolide
99% yield

Fig. 7.30 Friedel-Crafts acylation in an ionic liquid.

tained dramatic rate enhancements and improved regioselectivities to the para-isomer compared to the corresponding reactions in molecular organic solvents. Similarly, the fragrance material, traseolide, was obtained in 99% yield as a single isomer (Fig. 7.30). It should be noted, however, that these reactions suffer from the same problems as conventional Friedel-Crafts acylations with regard to work-up and product recovery. The ketone product forms a strong complex with the chloroaluminate ionic liquid, analogous to that formed with aluminum chloride in molecular solvents. This complex is broken by water, leading to aqueous waste streams containing copious quantities of aluminum salts. More recently, the same group has described the use of chloroindate(III) ionic liquids as catalysts and solvents for Friedel-Crafts acylations [101]. Although indium is a much less active catalyst than aluminum, which limits the substrate scope, product recovery and catalyst recycling are simplified. Metal bistriflamide salts of e.g. Co(II), Mn(II) and Ni(II) were also shown to be effective catalysts (1 mol%) for Friedel-Crafts acylations, either neat or in ionic liquids as solvents [102].

Generally speaking, recycling of catalysts in ionic liquid systems is facilitated when they are ionic compounds (salts) as these are generally very soluble in the ionic liquid phase but insoluble in organic solvents, thus enabling extraction of the product and recycling of the catalyst in the ionic liquid phase. For example, olefin metathesis has been performed in an ionic liquid using a cationic ruthenium allenylidene complex as the catalyst (Fig. 7.31) [103]. It was demonstrated that the catalyst could be recycled in the IL phase but a markedly lower conversion was observed in the third run owing to decomposition of the catalyst, which is also observed in conventional organic solvents. The results were markedly dependent on the anion of the ionic liquid, the best results being obtained in [bmim] [CF$_3$CO$_2$].

In the last few years increasing attention has been devoted to conducting biocatalytic transformations in ionic liquids [104–107]. The first report of enzyme-(lipase-) catalyzed reactions in water-free ionic liquids dates from 2000 and involved transesterification, ammoniolysis and perhydrolysis reactions catalyzed by *Candida antarctica* lipase B (Fig. 7.32) [108].

The use of ionic liquids as reaction media for biotransformations has several potential benefits compared to conventional organic solvents, e.g., higher operational stabilities [109] and enantioselectivities [110] and activities are generally at least as high as those observed in organic solvents. They are particularly attractive for performing bioconversions with substrates which are very sparingly soluble in conventional organic solvents, e.g., carbohydrates [111] and nucleosides.

Notwithstanding the numerous advantages of ionic liquids as reaction media for catalytic processes widespread industrial application has not yet been forthcoming. The reasons for this are probably related to their relatively high prices and the paucity of data with regard to their toxicity and biodegradability. The replacement of conventional VOCs with ionic liquids is an obvious improvement with regard to atmospheric emissions but small amounts of ionic liquids will

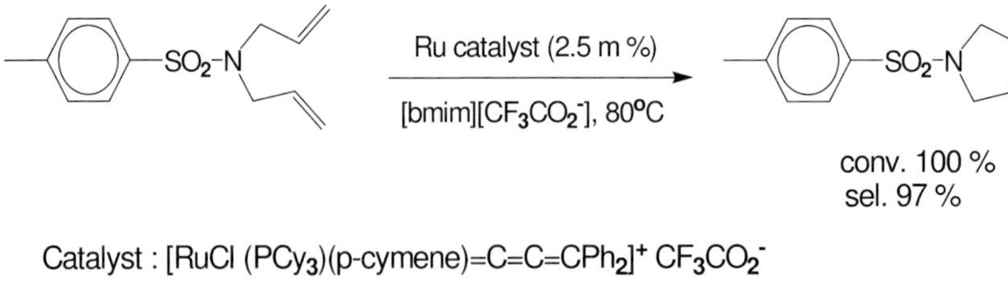

Ru catalyst (2.5 m %)

[bmim][CF$_3$CO$_2$], 80°C

conv. 100 %
sel. 97 %

Catalyst : [RuCl (PCy$_3$)(p-cymene)=C=C=CPh$_2$]$^+$ CF$_3$CO$_2^-$

Fig. 7.31 Olefin metathesis in an ionic liquid.

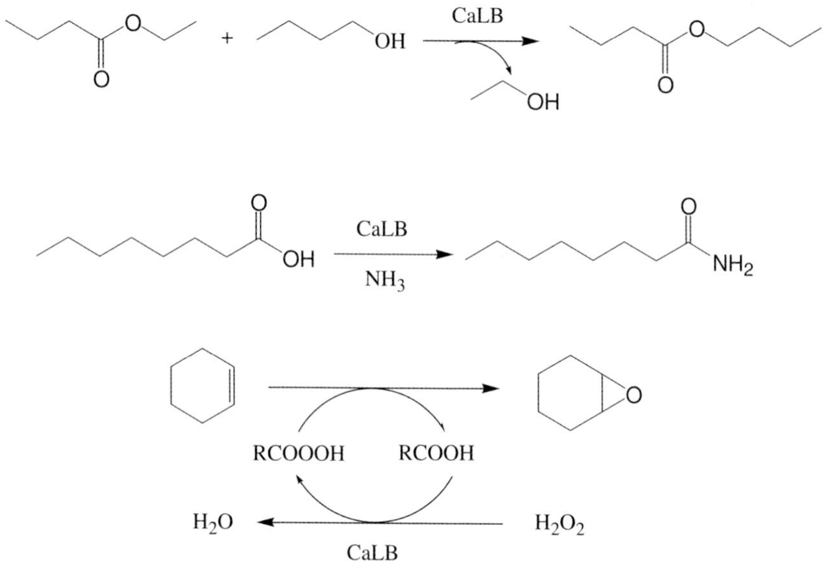

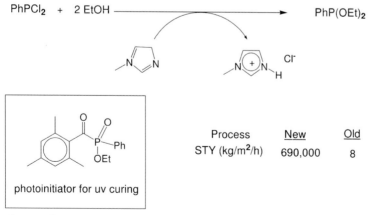

Fig. 7.32 Biotransformations in ionic liquids.

inevitably end up in the environment, e.g., in ground water. Consequently, it is important to establish their effect on the environment, e.g. with regard to their toxicity [112] and biodegradability [113]. Indeed, the current trend in ionic liquid research is towards the development of inexpensive, non-toxic, biodegradable ionic liquids, e.g. based on renewable raw materials [114].

One interesting application of an ionic liquid does not involve its use as a solvent. The company BASF used N-methylimidazole to scavenge the HCl formed in the reaction of dichlorophenyl phosphine with ethanol (Fig. 7.33). This results in the formation of N-methylimidazoliumchloride (Hmim-Cl), which has a

PhPCl$_2$ + 2 EtOH $\longrightarrow$ PhP(OEt)$_2$

Process	New	Old
STY (kg/m^2/h)	690,000	8

photoinitiator for uv curing

Fig. 7.33 The BASIL process.

melting point of 75 °C and is, hence, an ionic liquid under the reaction conditions. During the process, which has the acronym BASIL (biphasic scavenging utilizing ionic liquids), the ionic liquid separates as a second phase and is readily separated from the product and recycled [115]. The product is a precursor of UV photoinitiators which find a variety of applications. In the conventional process a solid amine hydrochloride is formed from e.g. triethylamine as the acid scavenger. This necessitates the use of large quantities of organic solvents and more complicated work-up. The methylimidazole also acts as a nucleophilic catalyst, resulting in a forty-fold rate enhancement. The new process is also performed in a continuous operation mode and the space time yield was increased from 8 kg m^{-2} h^{-1} to 690 000 kg m^{-2} h^{-1}.

7.7
Biphasic Systems with Supercritical Carbon Dioxide

One problem associated with the use of ILs is recovery of the product and recycling of the catalyst. If this is achieved by extraction with a volatile organic solvent then it is questionable what the overall gain is. An attractive alternative is to use scCO$_2$ as the second phase, whereby the catalyst remains in the IL phase and the product is extracted into the scCO$_2$ phase. This concept has been successfully applied to both homogeneous metal catalysis [116] and biocatalytic conversions [117]. We have recently applied the concept of using a 'miscibility switch' for performing catalytic reactions in IL/scCO$_2$ mixtures [118]. This takes advantage of the fact that, depending on the temperature and pressure, scCO$_2$ and IL mixtures can be mono- or biphasic. Hence, the reaction can be performed in a homogeneous phase and, following adjustment of the temperature and/or pressure, the product separated in the scCO$_2$ layer of the biphasic system.

Other combinations with scCO$_2$ have also been considered which dispense with the need for an ionic liquid altogether. For example, a biphasic water/scCO$_2$ system, whereby the catalyst, e.g. a metal complex of tppts, resides in the water phase and the product is removed in the scCO$_2$ phase [119, 120]. The system has its limitations: the catalyst needs to be water soluble and all reaction components must be stable towards the acidic conditions (pH 3) of carbonic acid. It is not effective for reactions in which the catalyst, substrate or reagent is insoluble in water or sensitive to the low pH of the aqueous phase [121]. More recently, an attractive system comprising a biphasic mixture of poly(ethylene glycol) (PEG) to dissolve the catalyst and scCO$_2$ as the extractive phase was used for the RhCl(Ph$_3$P)$_3$-catalyzed hydrogenation of styrene [122]. PEGs have the advantage over ILs that they are much less expensive and are nontoxic (analogous to CO$_2$, they are approved for use in foods and beverages). They are, moreover, miscible with common organic ligands and, in the above example, the catalyst was stable and recyclable in the PEG phase. Similarly, biocatalytic transformations have also been performed in a PEG/scCO$_2$ biphasic system (Fig. 7.34) [123].

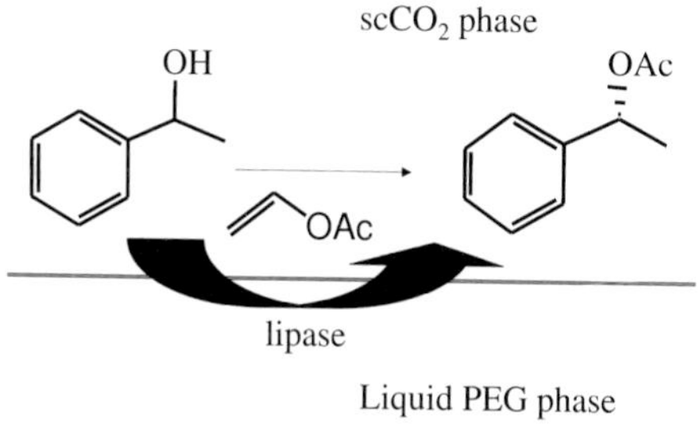

scCO$_2$ phase

OH

OAc

OAc

lipase

Liquid PEG phase

Fig. 7.34 Lipase-catalyzed transesterification in PEG/scCO$_2$.

7.8
Thermoregulated Biphasic Catalysis

Another approach to facilitating catalyst separation while maintaining the benefits of homogeneous catalysis involves the use of thermoregulated biphasic catalysis [124, 125], whereby the catalyst is dissolved in a particular solvent at one temperature and insoluble at another. For example, a diphosphine ligand attached to an ethylene oxide/propylene oxide block copolymer (Fig. 7.35) afforded rhodium complexes that are soluble in water at room temperature but precipitate on warming to 40 °C. The driving force for this inverted temperature dependence on solubility is dehydration of the ligand on heating. Hence, a rhodium-catalyzed reaction such as hydrogenation or hydroformylation can be performed at room temperature in a single phase and the catalyst separated by precipitation at a higher temperature. An added advantage is that runaway conditions are never achieved since the catalyst precipitates and the reaction stops on raising the temperature. This principle has also been applied to biotransformations by attaching enzymes to EO/PO block copolymers [126].

7.9
Conclusions and Prospects

The employment of catalytic methodologies – homogeneous, heterogeneous and enzymatic – in water or supercritical carbon dioxide as the reaction medium holds much promise for the development of a sustainable chemical manufacturing industry. Water is cheap, abundantly available, non-toxic and non-inflammable and the use of aqueous biphasic catalysis provides an ideal basis for recovery

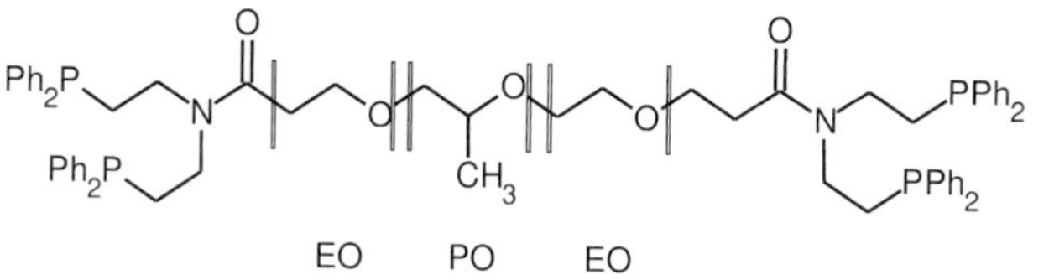

Ph$_2$P, Ph$_2$P, N, O, O, CH$_3$, O, O, N, PPh$_2$, PPh$_2$

EO   PO   EO

Fig. 7.35 Ethylene oxide/propylene oxide block copolymer.

and recycling of the (water-soluble) catalyst. Water is also the ideal solvent for many processes catalysed by Nature's catalysts, enzymes. Hence, the use of water as a reaction medium meshes well with the current trend towards a sustainable chemical industry based on the utilization of renewable raw materials rather than fossil fuels as the basic feedstock.

Supercritical carbon dioxide has many potential benefits in the context of sustainability. In common with water, it is cheap, abundantly available, non-toxic and non-inflammable. It is also an eminently suitable solvent for homogeneous, heterogeneous and biocatalytic processes and is readily separated from the catalyst and products by simple release of pressure. Reaction rates are very high in scCO$_2$, owing to its properties being intermediate between those of a gas and a liquid. Biphasic systems involving scCO$_2$ with, for example, an ionic liquid or polyethylene glycol also hold promise as reaction media for a variety of catalytic processes integrated with product separation and catalyst recycling.

The ultimate in sustainable catalytic processes is the integration of chemocatalytic and/or biocatalytic steps into catalytic cascade processes that emulate the metabolic pathways of the cell factory. It is an esthetically pleasing thought that, in the future, fuels, chemicals and polymers could be obtained from carbon dioxide and water as the basic raw materials via biomass, using sunlight as the external source of energy and water and supercritical carbon dioxide as solvents. The important difference between this bio-based scenario and the current oil-based one is the time required for renewal of the feedstocks.

References

1 K. Tanaka, F. Toda, *Chem. Rev.*, **2000**, *100*, 1025.
2 C.L. Hussey, in *Advances in Molten Salts*, Vol. 5, G. Mamantov and C. Mamantov (Eds), Elsevier, New York, **1983**, p. 185.
3 R.V. Ulijn, P.J. Halling, *Green Chem.*, **2004**, *6*, 488.
4 D.H. Brown Ripin, M. Vetelino, *Synlett*, **2003**, 2353.
5 *Green Chem.*, **2003**, *5*, pp, 105–284.
6 H.U. Blaser, M. Studer, *Green Chem.*, **2003**, *5*, 112.
7 W. Keim, *Green Chem.*, **2003**, *5*, 105; see also B. Driessen Holscher, P. Wasserscheid, W. Keim, *CatTech.*, **1998**, June, 47.
8 A. Haimov, R. Neumann, *Chem. Commun.*, **2002**, 876.
9 See also H. Alper, K. Januszkiewicz, D.J. Smith, *Tetrahedron Lett.*, **1985**, *26*, 2263.

10 S. Chandrasekhar, Ch. Narsihmulu, S.S. Sultana, N.K. Reddy, *Org. Lett.*, **2002**, *4*, 4399.
11 P.C. Andrews, A.C. Peatt, C. Raston, *Green Chem.*, **2004**, *6*, 119.
12 B. Widom, P. Bhimalapuram, K. Koga, *Phys. Chem. Chem. Phys.*, **2003**, *5*, 3085; R. Schmid, *Monatsh. Chem.*, **2001**, *132*, 1295; J. Kyte, *Biophys. Chem.*, **2003**, *100*, 192; R.L. Pratt, A. Pohorille, *Chem. Rev.*, **2002**, *102*, 2671.
13 D. Hancu, J. Green, E.J. Beckman, *Acc. Chem. Res.*, **2002**, *1*, 43.
14 B. Cornils, W.A. Herrmann (Eds.), *Aqueous Phase Organometallic Catalysis, Concepts and Applications*, 2nd edn., Wiley-VCH, Weinheim, 2004.
15 F. Joo, *Aqueous Organometallic Chemistry and Catalysis*, Kluwer, Dordrecht, 2001.
16 I.T. Horvath, F. Joo (Eds), NATO ASI 3/5, Kluwer, Dordrecht, 1995.

17 G. Papadogianakis, R. A. Sheldon, in *Catalysis. A Specialist Periodical Report,* Vol. 13. J. J. Spivey, Senior Reporter, Royal Society of Chemistry, London, 1997, pp. 114–193.

18 G. Papadogianakis, R. A. Sheldon, *New J. Chem.,* **1996**, *20,* 175.

19 B. Cornils, E. Wiebus, *Chemtech,* **1995**, *25,* 33; B. Cornils, *Org. Proc. Res. Dev.,* **1998**, *2,* 127.

20 P. A. Grieco (Ed.), *Organic Synthesis in Water,* Blackie, London, 1998.

21 N. Pinault, B. W. Duncan, *Coord. Chem. Rev.,* **2003**, *24,* 1.

22 P. W. N. M. van Leeuwen, P. C. J. Kamer, J. N. H. Reek, *Cattech,* **1999**, *3,* 164.

23 W. A. Herrmann, C. W. Kohlpaintner, *Angew. Chem. Int. Ed. Engl.,* **1993**, *32,* 1524.

24 F. Joo, *Acc. Chem. Res.,* **2002**, *35,* 738 and references cited therein.

25 T. Dwars, G. Oehme, *Adv. Synth. Catal.,* **2002**, *344,* 239.

26 D. Sinou, *Adv. Synth. Catal.,* **2002**, *344,* 221.

27 S. Mecking, A. Held, F. M. Bauers, *Angew. Chem. Int. Ed.,* **2002**, *41,* 545.

28 S. Ahrland, J. Chatt, R. N. Davies, A. A. Williams, *J. Chem. Soc.,* **1958**, 276.

29 E. G. Kuntz, *Chemtech.,* **1987**, 570 and references cited therein.

30 C. W. Kohlpaintner, R. W. Fischer, B. Cornils, *Appl. Catal. A: General,* **2001**, *221,* 219.

31 C. D. Frohning, C. W. Kohlpaintner, in *Applied Homogeneous Catalysis with Organometallic Compounds,* B. Cornils, W. Herrmann (Eds), VCH, Weinheim, 1996, pp. 29–104.

32 G. Verspui, G. Elbertse, F. A. Sheldon, M. A. P. J. Hacking, R. A. Sheldon, *Chem. Commun.,* **2000**, 1363.

33 J. M. Grosselin, C. Mercier, *J. Mol. Catal.,* **1990**, *63,* L25.

34 C. Mercier, P. Chabardes, *Pure Appl. Chem.,* **1994**, *66,* 1509.

35 F. Joo, J. Kovacs, A. Cs. Benyei, A. Katho, *Angew. Chem. Int. Ed. Engl.,* **1998**, *110,* 969; F. Joo, A. Cs. Benyei, A. Katho, *Catal. Today,* **1998**, *42,* 441.

36 A. N. Ajjou, J.-L. Pinet, *J. Mol. Catal. A: Chemical,* **2004**, *214,* 203.

37 A. W. Heinen, G. Papadogianakis, R. A. Sheldon, J. A. Peters, H. van Bekkum, *J. Mol. Catal. A: Chemical,* **1999**, *142,* 17.

38 V. Kolaric, V. Sunjic, *J. Mol. Catal. A: Chemical,* **1996**, *110,* 189 and *111,* 239.

39 G. Papadogianakis, L. Maat, R. A. Sheldon, *Inorg. Synth.,* **1998**, *32,* 25.

40 G. Papadogianakis, L. Maat, R. A. Sheldon, *J. Chem. Soc., Chem. Commun.,* **1994**, 2659.

41 G. Papadogianakis, L. Maat, R. A. Sheldon, *J. Chem. Tech. Biotechnol.,* **1997**, *70,* 83.

42 G. Verspui, G. Papadogianakis, R. A. Sheldon, *Catal. Today,* **1998**, *42,* 449.

43 G. Verspui, J. Feiken, G. Papadogianakis, R. A. Sheldon, *J. Mol. Catal. A: Chemical,* **1999**, *189,* 163.

44 G. Verspui, I. I. Moiseev, R. A. Sheldon, *J. Organometal. Chem.,* **1999**, *586,* 196.

45 G. Verspui, G. Papadogianakis, R. A. Sheldon, *J. Chem. Soc., Chem. Commun.,* **1998**, 401; G. Verspui, F. Schanssema, R. A. Sheldon, *Appl. Catal. A: General,* **2000**, *198,* 5.

46 G. Verspui, F. Schanssema, R. A. Sheldon, *Angew. Chem. Int. Ed.,* **2000**, *57,* 157.

47 J. P. Genet, E. Blart, M. Savignac, *Synlett,* **1992**, 715.

48 A. L. Casalnuovo, J. C. Calabrese, *J. Am. Chem. Soc.,* **1990**, *112,* 4324.

49 C. Mercier, P. Chabardes, *Pure Appl. Chem.,* **1994**, *66,* 1509.

50 D. M. Lynn, B. Mohr, R. H. Grubbs, *J. Am. Chem. Soc.,* **1998**, *120,* 1627.

51 N. Yoshimura, in [14], pp. 540–549.

52 G. J. ten Brink, I. W. C. E. Arends, R. A. Sheldon, *Science,* **2000**, *287,* 5458.

53 G. J. ten Brink, I. W. C. E. Arends, R. A. Sheldon, *Adv. Synth. Catal.,* **2002**, *344,* 355.

54 I. T. Horvath, J. Rabai, *Science,* **1994**, *266,* 72.

55 I. T. Horvath, *Acc. Chem. Res.,* **1998**, *31,* 641.

56 L. P. Barthel-Rosa, J. A. Gladysz, *Coord. Chem. Rev.,* **1999**, *190–192,* 587.

57 I. T. Horvath, G. Kiss, R. A. Cook, J. E. Bond, P. A. Stevens, J. Rabai, E. J. Mozeleski, *J. Am. Chem. Soc.,* **1998**, *120,* 3133.

58 See also D. F. Foster, D. Gudmunsen, D. J. Adams, A. M. Stuart, E. G. Hope, D. J. Cole-Hamilton, G. P. Schwarz, P. Pogorzelec, *Tetrahedron*, **2002**, *58*, 3901.

59 D. Rutherford, J. J. J. Juliette, C. Rocaboy, I. T. Horvath, J. A. Gladysz, *Catal. Today*, **1998**, *42*, 381.

60 J. Moineau, G. Pozzi, S. Quici, D. Sinou, *Tetrahedron Lett.*, **1999**, *40*, 7683; M. Moreno-Mañas, R. Pleixats, S. Villarroya, *Chem. Comm.*, **2002**, 60; C. Rocaboy, J. A. Gladysz, *Org. Lett.*, **2002**, *4*, 1993.

61 J. A. Gladysz C. Rocaboy, *Tetrahedron*, **2002**, *58*, 4007; D. Chen, F. Qing, Y. Huang, *Org. Lett.*, **2002**, *4*, 1003; C. C. Tzschucke, C. Markert, H. Glatz, W. Bannwarth, *Angew. Chem. Int. Ed.*, **2002**, *41*, 4500.

62 D. M. Haddleton, S. G. Jackson, S. A. F. Bon, *J. Am. Chem. Soc.*, **2000**, *122*, 1542.

63 J. A. Gladysz, (Ed.), *Tetrahedron*, **2002**, *58* (20), 3823–4131.

64 J. G. Riess, M. Le Blanc, *Angew. Chem. Int. Ed. Engl.*, **1978**, *17*, 621.

65 M. Cavazzini, S. Quici, G. Pozzi, *Tetrahedron*, **2002**, *58*, 3943 and references cited therein.

66 B. Betzemeier, M. Cavazzine, S. Quici, P. Knochel, *Tetrahedron Lett.*, **2000**, *41*, 4343.

67 G. J. ten Brink, J. M. Vis, I. W. C. E. Arends, R. A. Sheldon, *Tetrahedron*, **2002**, *58*, 3977 and references cited therein.

68 M. Wende, R. Meier, J. A. Gladysz, *J. Am. Chem. Soc.*, **2001**, *123*, 11490; M. Wende, J. A. Gladysz, *J. Am. Chem. Soc.*, **2003**, *125*, 5861.

70 J. M. DeSimone, *Science*, **2002**, *297*, 799.

71 P. G. Jessop, W. Leitner (Eds), *Chemical Synthesis Using Supercritical Fluids*, Wiley-VCH, Weinheim, 1999.

72 W. Leitner, *Top. Curr. Chem.*, **1999**, *206*, 107.

73 W. Leitner, *Acc. Chem. Res.*, **2002**, *35*, 746.

74 E. J. Beckman, *J. Supercrit. Fluid.*, **2004**, *28*, 121.

75 M. G. Hitzler, M. Poliakoff, *Chem. Commun.*, **1997**, 1667.

76 M. G. Hitzler, F. R. Smail, S. K. Ross, M. Poliakoff, *Org. Proc. Res. Dev.*, **1998**, *2*, 137.

77 P. Licence, J. Ke, M. Sokolova, S. K. Ross, M. Poliakoff, *Green Chem.*, **2003**, *5*, 99.

78 S. Ichikawa, M. Tada, Y. Inasawa, T. Ikariya, *Chem. Comm.*, **2005**, 924.

79 M. J. Burk, S. Feng, M. F. Gross, W. J. Tumas, *J. Am. Chem. Soc.*, **1995**, *117*, 8277.

80 D. J. Adams, W. Chen, E. G. Hope, S. Lange, A. M. Stuart, A. West, J. Xiao, *Green Chem.*, **2003**, *5*, 118.

81 P. Stephenson, P. Licence, S. K. Ross, M. Poliakoff, *Green Chem.*, **2004**, *6*, 521.

82 P. G. Jessop, R. R. Stanley, R. A. Brown, C. A. Eckert, C. L. Liotta, T. T. Ngo, P. Pollet, *Green Chem.*, **2003**, *5*, 123; see also J. Xiao, S. C. A. Nefkens, P. G. Jessop, T. Ikariya, R. Noyori, *Tetrahedron Lett.*, **1996**, *37*, 2813.

83 N. J. Meehan, A. J. Sandee, J. N. H. Reek, P. C. J. Kamer, P. W. N. M. van Leeuwen, M. Poliakoff, *Chem. Comm.*, **2000**, 1497.

84 Z. Hou, N. Theyssen, A. Brinkmann, W. Leitner, *Angew. Chem. Int. Ed.*, **2005**, *44*, 1346.

85 E. J. Beckmann, *Green Chem.*, **2003**, *5*, 332.

86 T. Danciu, E. J. Beckmann, D. Hancu, R. Cochran, R. Grey, D. Hajnik, J. Jewson, *Angew. Chem. Int. Ed.*, **2003**, *42*, 1140.

87 T. W. Randolph, H. W. Blanch, J. M. Prausnitz, C. R. Wilke, *Biotechnol. Lett.*, **1985**, *7*, 325.

88 D. A. Hammond, M. Karel, A. M. Klibanov, V. J. Krukonis, *Appl. Biochem. Biotechnol.*, **1985**, *11*, 393.

89 K. Nakamura, Y. M. Chi, Y. Yamada, T. Yano, *Chem. Eng. Commun.*, **1986**, *45*, 207.

90 For reviews see: A. J. Messiano, E. J. Beckmann, A. J. Russell, *Chem. Rev.*, **1999**, *99*, 623; H. Hartmann, E. Schwabe, T. Scheper, D. Combes, in *Stereoselective Biocatalysis*, R. N. Patel (Ed.), Marcel Dekker, New York, 2000, pp. 799–838.

91 A. Overmeyer, S. Schrader-Lippelt, V. Kascher, G. Brunner, *Biotechnol. Lett.*, **1999**, *21*, 65.

92 T. Matsuda, R. Kanamaru, K. Watanabe, T. Harada, K. Nakamura, *Tetrahedron Lett.*, **2001**, *42*, 8319; T. Matsuda, R. Kanamaru, K. Watanabe, T. Kamitanaka, T. Harada, K. Nakamura, *Tetrahedron Asymm.*, **2003**, *14*, 2087.

93 T. Matsuda, K. Watanabe, T. Harada, K. Nakamura, Y. Arita, Y. Misumi, S. Ichikawa, T. Ikariya, *Chem. Commun.*, **2004**, 2286.

94 T. Matsuda, K. Watanabe, T. Kamitanaka, T. Harada, K. Nakamura, *Chem. Comm.*, **2003**, 1198.

95 W. Randolph, D. S. Clark, H. W. Blanch, J. M. Prausnitz, *Science*, **1988**, *239*, 387.

96 R. D. Rogers, K. R. Seddon (Eds), *Ionic Liquids as Green Solvents; Progress and Prospects*, ACS Symp. Ser. 856, American Chemical Society, Washington DC, 2003; P. Wasserscheid, T. Welton, *Ionic Liquids in Synthesis*, Wiley-VCH, Weinheim, 2003.

97 S. A. Forsyth, J. M. Pringle, D. R. MacFarlane, *Aust. J. Chem.*, **2004**, *57*, 113; J. S. Wilkes, *J. Mol. Catal. A: Chemical*, **2004**, *214*, 11.

98 For recent reviews see: R. A. Sheldon, *Chem. Commun.*, **2001**, 2399; P. Wasserscheid, W. Keim, *Angew. Chem. Int. Ed.*, **2000**, *39*, 3772; J. Dupont, R. F. de Souza, P. A. Z. Suarez, *Chem. Rev.*, **2002**, *102*, 3667; C. E. Song, *Chem. Comm.*, **2004**, 1033–1043; D. Zhao, M. Wu, Y. Kou, E. Min, *Catal. Today*, **2002**, *74*, 157; C. M. Gordon, *Appl. Catal. A: General* **2001**, *222*, 101.

99 P. Wasserscheid, H. Waffenschmidt, P. Machnitzki, K. W. Kottsieper, O. Stelzer, *Chem. Commun.*, **2001**, 451; P. W. N. M. van Leeuwen, P. C. J. Kamer, J. N. H. Reek, P. Dierkes, *Chem. Rev.*, **2000**, *100*, 2741; R. P. J. Bronger, S. M. Silva, P. C. J. Kamer, P. W. N. M. van Leeuwen, *Chem. Commun.*, **2002**, 3044.

100 C. J. Adams, M. J. Earle, G. Roberts, K. R. Seddon, *Chem. Comm.*, **1998**, 2097.

101 M. J. Earle, U. Hakala, C. Hardacre, J. Karkkainen, B. J. McCauley, D. W. Rooney, K. R. Seddon, J. M. Thompson, K. Wahala, *Chem. Comm.*, **2005**, 901.

102 M. J. Earle, U. Hakala, B. J. McCauley, M. Nieuwenhuizen, A. Ramani, K. R. Seddon, *Chem. Comm.*, **2004**, 1368.

103 D. Semeril, H. Olivier-Bourbigou, C. Bruneau, P. H. Dixneuf, *Chem. Comm.*, **2002**, 146.

104 F. van Rantwijk, R. Madeira Lau, R. A. Sheldon, *Trends Biotechnol.*, **2003**, *21*, 131.

105 R. A. Sheldon, R. Madeira Lau, M. J. Sorgedrager, F. van Rantwijk, K. R. Seddon, *Green Chem.*, **2002**, *4*, 147–151.

106 U. Kragl, M. Eckstein, N. Kaftzik, *Curr. Opin. Biotechnol.*, **2002**, *13*, 565.

107 Z. Yang, W. Pan, *Enzyme Microb. Technol.*, **2005**, *37*, 19.

108 R. Madeira Lau, F. van Rantwijk, K. R. Seddon, R. A. Sheldon, *Org. Lett.*, **2000**, *2*, 4189.

109 T. de Diego, P. Lozano, S. Gmouth, M. Vaultier, J. L. Iborra, *Biotechnol. Bioeng.*, **2004**, *88*, 916; P. Lozano, T. de Diego, J. P. Guegan, M. Vaultier, J. L. Iborra, *Biotechnol. Bioeng.*, **2001**, *75*, 563; P. Lozano, T. de Diego, D. Carrie, M. Vaultier, J. L. Iborra, *Biotechnol. Lett.*, **2001**, *23*, 1529.

110 K. W. Kim, B. Song, M. Y. Choi, M. J. Kim, *Org. Lett.*, **2001**, *3*, 1507; S. H. Schöfer, N. Kaftzik, P. Wasserscheid, U. Kragl, *Chem. Comm.*, **2001**, 425.

111 Q. Liu, M. H. A. Janssen, F. van Rantwijk, R. A. Sheldon, *Green Chem.*, **2005**, *7*, 39.

112 B. Jastorff, K. Mölter, P. Behrend, U. Bottin-Weber, J. Filser, A. Heimers, B. Ondruschka, J. Ranke, M. Schaefer, H. Schroder, A. Stark, P. Stepnowski, F. Stock, R. Störmann, S. Stolte, U. Welz-Biermann, S. Ziegert, J. Thöming, *Green Chem.*, **2005**, *7*, 243 and references cited therein.

113 N. Gathergood, P. J. Scammells, *Aust. J. Chem.*, **2002**, *55*, 557; N. Gathergood, M. Teresa Gascia, P. J. Scammells, *Green Chem.*, **2004**, *6*, 166.

114 S. T. Handy, *Chem. Eur. J.*, **2003**, *9*, 2938; S. T. Handy, M. Okello, G. Dickenson, *Org. Lett.*, **2003**, *5*, 2513; see also J. H. Davis, P. A. Fox, *Chem. Comm.*, **2003**, 1209.

115 M. Freemantle, *C&EN*, March 31, 2003, p. 9; O. Huttenloch, M. Maase,

G. Grossmann, L. Szarvas, PCT Int. Appl. WO 2005019183 (2005) to BASF.

116 R.A. Brown, P. Pollet, E. McKoon, C.A. Eckhert, C.L. Liotta, P.G. Jessop, *J. Am. Chem. Soc.*, **2001**, *123*, 1254–1255; D.J. Cole-Hamilton, *Science*, **2003**, *299*, 1702.

117 P. Lozano, T. de Diego, D. Carrié, M. Vaultier, J.L. Iborra, *Chem. Comm.*, **2002**, 692 ; M.T. Reetz, W. Wiesenhofer, G. Francio, W. Leitner, *Chem. Comm.*, **2002**, 992; P. Lozano, T. de Diego, *Biotech. Progr.*, **2004**, *20*, 661.

118 M.C. Kroon, J. van Spronsen, C.J. Peters, R.A. Sheldon, G.-J. Witkamp, *Green Chem.* **2006**, *8*, 246.

119 B.M. Bhanage, Y. Ikushima, M. Shirai, M. Arai, *Chem. Comm.*, **1999**, 1277; *Tetrahedron Lett.* **1999**, *40*, 6427.

120 G.B. Jacobson, C.T. Lee, K.P. Johnston, *J. Am. Chem. Soc.*, **1999**, *121*, 11902.

121 R.J. Bonilla, B.R. James, P.G. Jessop, *Chem. Comm.*, **2000**, 941.

122 D.J. Heldebrant, P.G. Jessop, *J. Am. Chem. Soc.*, **2003**, *125*, 5600.

123 M. Reetz, W. Wiesenhofer, *Chem. Comm.*, **2004**, 2750.

124 D.E. Bergbreiter, *Chem. Rev.*, **2002**, *102*, 3345–3383.

125 Y. Wang, J. Jiang, Z. Jin, *Catal. Surveys Asia*, **2004**, *8* (2), 119.

126 A.F. Ivanov, E. Edink, A. Kumar, I.Y. Galaev, A.F. Arendsen, A. Bruggink, B. Mattiasson, *Biotechnol. Progr.*, **2003**, *19*, 1167.

8
Chemicals from Renewable Raw Materials

8.1
Introduction

For the last 70 years or so the chemical industry has been based on crude oil (petroleum) and natural gas as basic raw materials, hence the name petrochemicals. This may not be so for much longer, however. The chemical industry is currently on the brink of a new revolution, based on the switch from fossil resources to renewable agriculture-based raw materials. From a distance the production facility of Cargill in Blair, Nebraska looks very much like a small oil refinery or medium-sized petrochemicals plant. However, closer inspection reveals that it is a corn-processing plant; a biorefinery producing, *inter alia*, high-fructose corn syrup, ethanol and lactic acid. As James R. Stoppert, a senior executive of Cargill pointed out, the chemical industry is based on carbon and it does not matter if the carbon was fixed 2 million years ago or 6 months ago [1].

The necessity to switch from nonrenewable fossil resources to renewable raw materials, such as carbohydrates and triglycerides derived from biomass, was an important conclusion of the Report of the Club of Rome in 1972 [2]. It should be noted, however, that ca. 80% of the global production of oil is converted to thermal or electrical energy. If the world is facing an oil crisis it is, therefore, an energy crisis rather than a raw materials crisis for the chemical industry. Indeed, there are sufficient reserves of fossil feedstocks to satisfy the needs of the chemical industry for a long time to come.

Nonetheless, a (partial) switch to 'renewables' is desirable for other reasons, such as biocompatibility, biodegradability and lower toxicity, i.e. renewable raw materials leave a smaller 'environmental footprint' [3]. That the chemical industry has been slow to make the transition, in the three decades following the Report of the Club of Rome, is a consequence of the fact that oil and natural gas are excellent basic feedstocks and highly atom efficient, low waste, catalytic procedures are available for their conversion into commodity chemicals. The same cannot be said for the fine chemicals industry where processes are, generally speaking, much less efficient in many respects and there is considerable room for improvement.

Products based on renewable raw materials are derived from CO_2 and H_2O via photosynthesis and, following their use, are ultimately returned to the bio-

Green Chemistry and Catalysis. I. Arends, R. Sheldon, U. Hanefeld
Copyright © 2007 WILEY-VCH Verlag GmbH & Co. KGaA, Weinheim
ISBN: 978-3-527-30715-9

sphere as CO_2 and H_2O via biodegradation. In principle, they are CO_2 neutral and, hence, have a beneficial effect on greenhouse emissions. Furthermore, in an era of steeply rising oil and natural gas prices, they are becoming more dependable and relatively less expensive.

Renewable raw materials can contribute to the sustainability of chemical products in two ways: (i) by developing greener, biomass-derived products which replace existing oil-based products, e.g. a biodegradable plastic, and (ii) greener processes for the manufacture of existing chemicals from biomass instead of from fossil feedstocks. These conversion processes should, of course, be catalytic in order to maximize atom efficiencies and minimize waste (E factors) but they could be chemo- or biocatalytic, e.g. fermentation [3–5]. Even the chemocatalysts themselves can be derived from biomass, e.g. expanded corn starches modified with surface SO_3H or amine moieties can be used as recyclable solid acid or base catalysts, respectively [6].

What are renewable raw materials and how do their prices compare with oil, coal and natural gas? It has been estimated [7] that the global biomass production amounts to ca. 10^{11} tons per annum, only 3% of which is cultivated, harvested and used (food and non-food). It consists of 75% carbohydrates and 20% lignin, with the remaining 5% comprising oils, fats, proteins and terpenes. In a traditional agricultural economy farmers produce grains for food and feed. In a biobased economy farmers still produce these but the non-food parts of plants, e.g. wheat straw, corn stover and sugar cane bagasse, are used as raw materials for fuels and commodity chemicals.

The prices of some pertinent examples of agriculture-based raw materials (biomass) are compared with oil and coal in Table 8.1. It is obvious that the cheapest source of carbon is agricultural waste, i.e. waste plant biomass such as corn stover, wheat straw and sugar cane bagasse, which consists primarily of lignocellulose.

It has been estimated [3] that enough waste plant biomass is generated in the United States to produce all of the organic chemicals currently manufactured by the US chemical industry and supply a significant fraction of its liquid transportation fuel needs. For example, one ton of wheat straw affords ca. 600 kg of carbohydrates and ca. 200 kg of lignin. The former can be converted, by fermen-

Table 8.1 Prices of various raw materials.

Raw material	Average world market price (€/kg)
Crude oil	0.175 (0.400)
Coal	0.035
Corn	0.080
Wheat straw	0.020
Sugar	0.180
Ethanol	0.400
Ethylene	0.400

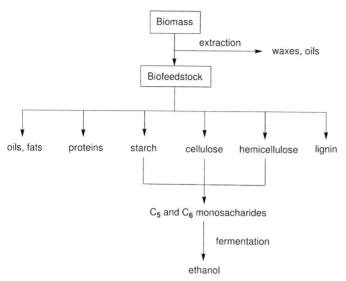

Fig. 8.1 The biorefinery.

tation, to 85 gallons (ca. 180 kg) of ethanol and the latter burned to generate electricity.

The transition to a biobased economy is currently in an intermediate phase where certain commodity chemicals, e.g. lactic acid, are being produced from corn starch. Ultimately, however, economically viable production of bulk chemicals and liquid fuels will only be possible from inexpensive lignocellulose, generated intentionally by cultivation of forage crops, e.g. hay, or derived from waste crop and forestry residues.

One can envisage the future production of liquid fuels and commodity chemicals in a 'biorefinery'. Biomass is first subjected to extraction to remove waxes and essential oils. Various options are possible for conversion of the remaining 'biofeedstock', which consists primarily of lignocellulose. It can be converted to synthesis gas (CO + H_2) by gasification, for example, and subsequently to methanol. Alternatively, it can be subjected to hydrothermal upgrading (HTU), affording liquid biofuels from which known transport fuels and bulk chemicals can be produced. An appealing option is bioconversion to ethanol by fermentation. The ethanol can be used directly as a liquid fuel and/or converted to ethylene as a base chemical. Such a 'biorefinery' is depicted in Fig. 8.1.

Recent advances in molecular genetics and metabolic engineering have laid the foundations for the transition from an oil-based to a biobased economy. The one remaining bottleneck, where technological breakthroughs are still needed, is the efficient conversion of lignocellulose to lignin and (hemi)cellulose [3].

8.2
Carbohydrates

According to an authorative estimate, ca. 10^{11} tons per annum of carbohydrates, mainly in the form of lignocellulose, grow and decay every year. The annual wood harvest is small by comparison and has been estimated at ca. 2×10^9 tons per annum. Approximately 5% of the wood harvest is converted into pulp and a minor fraction (5×10^6 tons per annum) of the latter into cellulose. Approximately 60% of this is processed into fibers and films or converted into cellulose esters and ethers [8]. Hence, there would be sufficient lignocellulose available to satisfy all of the raw materials needs many times over. It is, moreover, the cheapest source of biomass-derived energy [9] and a considerable research effort, dating back to the first energy crisis, has been devoted to the possible production of ethanol from lignocellulose [10, 11].

The native structure of lignocellulose, which is very complex, renders it resistant to enzymatic hydrolysis. The major component is cellulose, a $\beta 1 \rightarrow 4$ polymer of glucose (see Fig. 8.2 a), which accounts for 35–50% of the mass. Lignocellulose also contains 20–35% hemicellulose, a complex polymer of pentoses and hexoses and 10–25% lignin [10, 12]. Techniques to convert lignocellulose into a fermentable sugar mixture have been under development for over 20 years and are now entering use.

Digestion of lignocellulose, to render it susceptible to hydrolysis, can be performed in a number of ways [10, 13, 14] but few of these satisfy the green commandments. There seems to be a trend towards aqueous pretreatment, such as the steam explosion that is used by Iogen (Canada) [15, 16], followed by enzy-

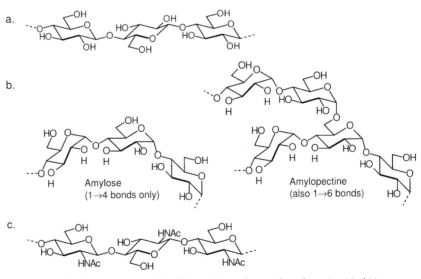

Fig. 8.2 Carbohydrate structures: (a) cellulose, (b) amylose and amylopectin, (c) chitin.

matic hydrolysis in the presence of cellulases. Due to the presence of hemicellulose, the resulting sugar mixture contains pentoses, which resist fermentation by *Saccharomyces* species (see later).

Starch is the other carbohydrate-based feedstock. Approximately 10 Mt is produced annually from corn (maize), wheat and potato, out of a total agricultural production of 1.6 Gt a^{-1} carbohydrate equivalents. A minor fraction of starch is amylose, a linear $a1 \rightarrow 4$ polymer of glucose (Fig. 8.2b). The native structure of amylose is helical; loose random coils are formed upon dissolution in water. The branched glucose polymer amylopectin is the major (approximately 75%) component of starch.

The industrial processing of starch into glucose, which provides the cheapest access to the latter, is very well developed industrially [17]. Acidic as well as enzymatic hydrolysis is employed. The development of enzymes that maintain their activity under the preferred process conditions is a major activity of the large suppliers of industrial enzymes.

A minor source of carbohydrates that should be mentioned, nevertheless, is chitin, a polymer of *N*-acetyl-D-glucosamine (see Fig. 8.2c) that resembles cellulose in its structure and general behavior. Chitin is extracted at a 150 kt a^{-1} scale from shellfish waste [18], which is a tiny fraction of the amount that grows and decays every year.

Non-food applications of poly- and oligosaccharides and glucose and their derivatives are many, but few of these involve the use of such carbohydrates as a chemical. The reason is obvious: the complex structure of carbohydrates and the large number of reactive groups make carbohydrate-based organic chemistry highly complicated. Nature has developed its own (also highly complex) ways to deal efficiently with structural complexity. Hence, it is not surprising that nearly all examples of manufacture of chemicals from carbohydrates build on Nature's achievements, using fermentation.

8.2.1
Chemicals from Glucose via Fermentation

Fermentation is the reproduction of microorganisms in the presence of a source of carbon and energy (such as sugar) and various nutrients. The products of fermentation fall into five major categories:
1. More microbial cells
2. Macromolecular products excreted by the cells, such as enzymes and polysaccharides
3. Primary metabolites, such as acetic acid, acetone, amino acids, citric acid, vitamins, that are essential for cellular growth or result from the conversion of glucose into energy for the living cell
4. Secondary metabolites, such as antibiotics, dyes, pigments, aroma compounds, poisons, that are not essential for cellular growth but serve the survival of the species

5. Microbial transformations of foreign substrates, often referred to as precursor fermentation. It may be preferred to perform such transformations with whole cells rather than isolated enzymes when the latter approach would involve the recycling of expensive cofactors. Hence, precursor fermentations are often preferred for conducting redox transformations (see Chapter 6).

Some primary metabolites have been produced via anaerobic fermentation in the past, before the onset of the petrochemical age. Thus, acetone was produced via acetone–butanol fermentation in *Clostridium acetobutylicum* from 1914 onwards. Some chemicals are produced via fermentation without any competition from chemical procedures, owing to the efficiency of the former (citric acid, L-glutamic acid, L-proline) or because the complexity of the product renders chemical synthesis unattractive (β-lactam antibiotics, vitamin B_{12}). With the present-day fermentation technology it seems a fair estimate that a fermentative process, to compete with a chemical one, should replace at least four to five chemical steps of moderate complexity. A much wider adoption of the biotechnological production of chemicals is hampered by two factors that put it at a competitive disadvantage compared with chemical processes: a low space-time yield and high operating costs.

Space-time yield (STY) sets the capital costs of the production facility and an STY of 100 g L^{-1} d^{-1} has often been mentioned as the minimum for the profitable production of a building block of intermediate complexity. Fermentation processes are often much less productive, due to regulation (inhibition) and toxicity problems.

Regulation is a blanket concept that covers all negative (i.e. stabilizing) feedback mechanisms of a product on its biosynthesis, such as inhibition of an enzyme by one of the downstream products or transcriptional control, which acts at the level of the gene.

Strain improvement, to increase the stoichiometric yield, the product concentration and the STY, was, until the early 1990s, generally achieved via classical techniques, i.e. by putting the species under selective pressure. Because the progress depends on the natural mutation frequency, the latter is routinely augmented by the application of mutagenic chemicals, UV radiation etc.

Classical strain improvement can be used, for example, to address the regulation of an amino acid by selecting mutants that are able to grow in the presence of a mimic that has similar regulation characteristics. The classical approach involves random mutagenesis and addresses the whole genome. Hence, the phenotype is primarily targeted and the actual mechanism is not easily subjected to rational control.

A recently introduced technique that also targets the phenotype is whole genome shuffling. This latter approach involves the amplification of the genetic diversity within a population through genetic recombination [19, 20]. Next, similar to the traditional procedure, the newly created library is screened for mutants with the desired properties. Its main advantage is that improvement is much faster than is possible with classical mutation and selection as described above.

Metabolic pathway engineering, in contrast, is a rational approach that targets selected enzymes or genes [21, 22]. It has eclipsed the classical approach in the course of the 1990s due to its much better rate of improvement. Metabolic pathway engineering involves the reprogramming of the metabolic network, based on recombinant DNA technology and knowledge of whole genomes. The objective is to optimize the flow of carbon into the biosynthesis of the desired product. Commonly followed basic approaches are:

1. The removal of kinetic bottlenecks by removing regulation and/or by amplifying the genes that code for the rate-limiting enzymes. Alternatively, the WT enzymes may be replaced by mutated or heterologous ones with a different control architecture.
2. Directing metabolic flow from a common biosynthetical intermediate to the desired product by overexpressing the branchpoint enzyme.
3. Reprogramming the central metabolism, if necessary, to supply the required redox equivalents and metabolic energy, usually in the form of ATP.
4. Fine-tuning of the enzymes in the biosynthetic pathway to prevent the accumulation of intermediates.

In the cases described below we will generally focus on metabolic pathway engineering rather than on classical strain improvement, for obvious reasons. One should be aware, however, that the classical approach has strengths that can make it a powerful partner of the rational approach. Thus, regulation problems have been addressed by the development of a feedback-resistant enzyme, using selective pressure and random mutagenesis as described above, in a research species, followed by introduction of the altered gene in the production species via recombinant techniques.

High operating costs put microbial synthesis of chemicals at a competitive disadvantage that is mainly due to the downstream processing (DSP), which is inherently more complex than the work-up of a chemical transformation. In the DSP of a microbial transformation, biomass is separated and the – often dilute – product is isolated from the aqueous supernatant and subsequently purified to the desired grade. The necessary precipitation, extraction and/or distillation steps may consume organic solvents, generate salts and consume energy on a scale that defeats the green purpose of a fermentative process.

In the examples presented below fermentation is compared with traditional chemical manufacturing processes, emphasizing the strengths and weaknesses of fermentative production techniques from a Green Chemistry viewpoint.

8.2.2
Ethanol

The anaerobic fermentation of ethanol from sugar (Fig. 8.3) goes back to the Stone Age. In 1997, the fermentation of ethanol, mainly from sugar cane, molasses (Brazil) and corn (USA), amounted to 24 Mt worldwide, dwarfing the chemical production of $2.6 \, \mathrm{Mt \, a^{-1}}$ [23]. Iogen (Canada) produces ethanol from

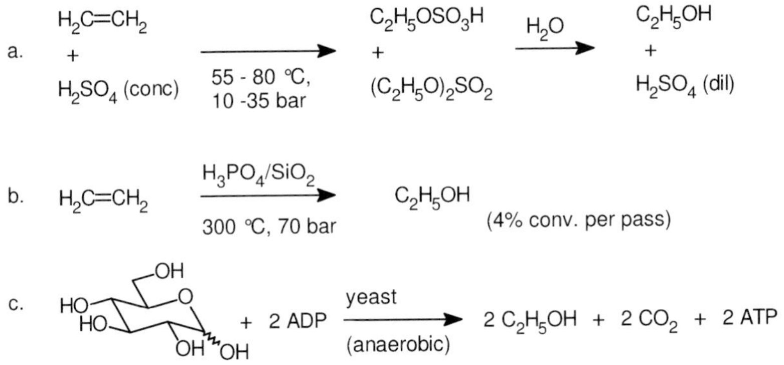

a.
$$H_2C{=}CH_2$$
$+$
$$H_2SO_4 \text{ (conc)}$$
→ (55 - 80 °C, 10 -35 bar)
$$C_2H_5OSO_3H$$
$+$
$$(C_2H_5O)_2SO_2$$
→ (H_2O)
$$C_2H_5OH$$
$+$
$$H_2SO_4 \text{ (dil)}$$

b.
$$H_2C{=}CH_2$$
→ (H_3PO_4/SiO_2, 300 °C, 70 bar)
$$C_2H_5OH$$
(4% conv. per pass)

c.
[glucose structure with OH, HO, HO, OH, OH]
$+$ 2 ADP
→ (yeast (anaerobic))
$2\,C_2H_5OH + 2\,CO_2 + 2\,ATP$

Fig. 8.3 Production of ethanol: (a), (b) chemically, from ethene; (c) via anaerobic fermentation.

straw at pilot-scale (approx. $800\,t\,a^{-1}$); a $130\,kt\,a^{-1}$ plant is expected to be completed in 2007 [15, 16].

The chemical production of ethanol involves acid-catalyzed hydration of ethene using either sulfuric acid (Fig. 8.3 a) [23] or a solid catalyst, such as H_3PO_4/SiO_2, in a recycle process (Fig. 8.3 b). A major disadvantage of these processes is the low conversion per pass of 4% [23], which is a consequence of the short contact time that is maintained to limit the formation of diethyl ether and ethene oligomers.

Under anaerobic conditions, many aerobic microbes switch their metabolism and excrete partially oxidized intermediates, such as ethanol or lactic acid, to maintain redox balance. The pathway involved, the glycolysis pathway, plays a leading role in all fermentations and is outlined in Fig. 8.4.

Glucose (Glc) is taken up and phosphorylated into glucose-6-phosphate (Glc6P), with consumption of ATP. Isomerization and phosphorylation afford fructose-1,6-bisphosphate (Fru1,6P$_2$), which is cleaved into two triose molecules: D-glyceraldehyde-3-phosphate (GA3P) and dihydroxyacetone monophosphate (DHAP). These are equilibrated by triose phosphate isomerase as only GA3P is metabolized further, except approximately 5 mol% of DHAP that leaks out of the pathway via reduction to glycerol, which is excreted as a side-product.

GA3P is subsequently converted, in a number of steps, into pyruvate (PYR), which is the branch-point between fermentation and respiration. *Saccharomyces* species are particularly well adapted to the anaerobic production of ethanol, via decarboxylation and reduction of PYR, to the near-exclusion of other metabolites. On account of this latter characteristic, as well as its high ethanol tolerance, *Saccharomyces* is the preferred organism to produce ethanol from hexoses.

The pentoses, such as xylose (Xyl), that result from the hydrolysis of lignocellulose (see above) resist fermentation by *Saccharomyces*, because it lacks an efficient mechanism to convert Xyl into xylulose (Xlu). The isomerization redox interconversion pathway of Xyl and Xlu, via xylitol, that is native to *Saccharomyces*, is inefficient due to a cofactor incompatibility (see Fig. 8.5) and results in a redox imbalance and the accumulation of xylitol [24]. Many bacteria, in contrast,

a   b

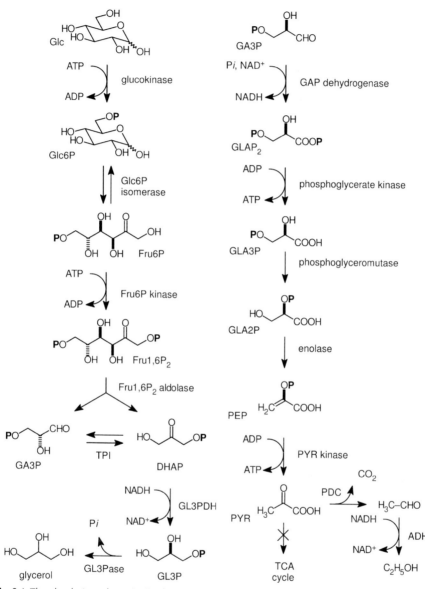

Fig. 8.4 The glycolysis pathway in *Saccharomyces* and the biosynthesis of ethanol and glycerol. Compounds: DHAP, dihydroxyl-acetone monophosphate; Fru6P, D-fructose-6-phosphate; Fru1,6P$_2$, D-fructose-1,6-bisphosphate; GA3P, D-glyceraldehyde-3-phosphate; GL3P, *sn*-glycerol-3-phosphate; GLAP$_2$, phosphoglycerate-3-phosphate; GLA3P, glycerate-3-phosphate; GLA2P, glycerate-2-phosphate; Glc, D-glucose; Glc6P, D-glucose-6-phosphate; PEP, phosphoenol-pyruvate; PYR, pyruvate. Enzymatic activities: GL3Pase, glycerol-3-phosphatase; ADH, alcohol dehydrogenase; PDC, pyruvate decarboxylase; TPI: triose phosphate isomerase.

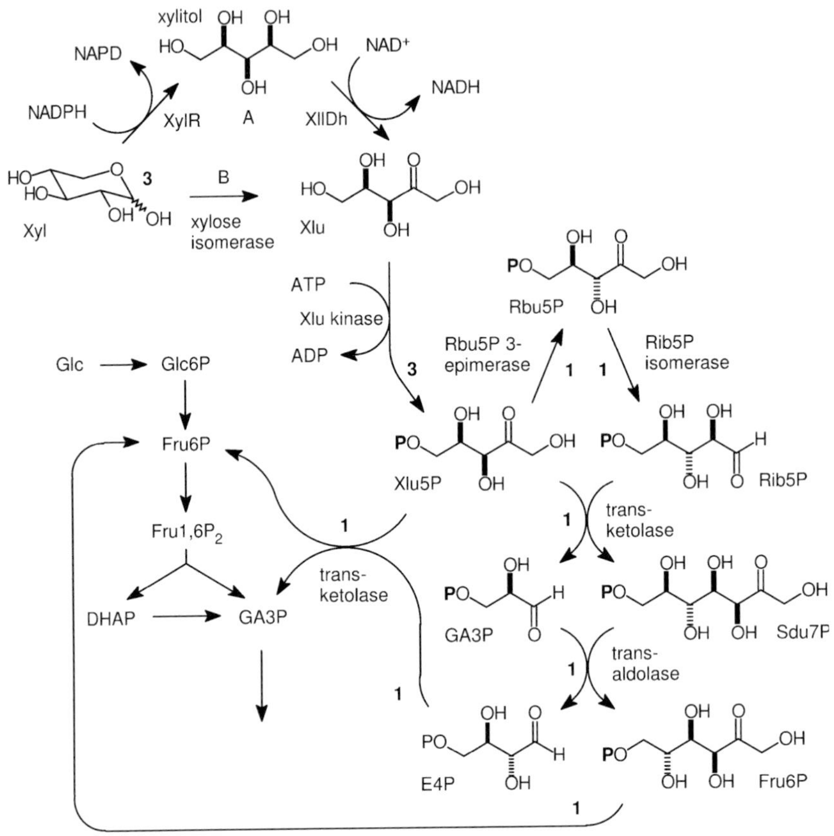

Fig. 8.5 Biochemical pathways for the conversion of xylose into xylulose (A: in yeasts; B: in bacteria) and the pentose phosphate pathway showing the idealized carbon flow in a xylose-fermenting organism. Compounds: E4P, D-erythrose-4-phosphate; Rib5P, D-ribose-5-phosphate; Rbu5P, D-ribulose-5-phosphate; Sdu7P, D-sedoheptulose-7-phosphate; Xlu, D-xylulose; Xlu5P: D-xylulose-5-phosphate; Xyl, D-xylose. Enzymatic activities: XylR, xylose reductase; XllDH, xylitol dehydrogenase.

harbor a Xyl isomerase which converts Xyl into Xlu, but these organisms generally are inefficient ethanol producers.

Xlu is subsequently phosphorylated into xylulose-5-phosphate (Xlu5P), which is the entrance point into the pentose phosphate pathway. This latter metabolic pathway, which is interconnected with the upstream half of the glycolysis pathway (see Fig. 8.5), can metabolize Xlu5P into GA3P.

8.2.2.1 Microbial Production of Ethanol

Raw materials for the fermentation of ethanol are sugar molasses (Brazil), corn steep liquor and corn starch hydrolysate (USA). Industrial ethanol fermentation is highly developed and the stoichiometric yield can be as high as 1.9 mol mol^{-1} [25, 26]. The

STY is high for a fermentative procedure and ranges from 140 g L^{-1} d^{-1} for a continuous tank reactor to 1.2 kg L^{-1} d^{-1} in a continuous tower reactor with cell recycle. Depending on the ethanol tolerance of the production species, ethanol is produced to a concentration of 12–20%. The ethanol is traditionally recovered from the fermentation broth via an energy-intensive distillation step, but it is sought to replace the latter by pervaporation or reversed osmosis [25].

It has been predicted that lignocellulose conversion will, eventually, provide the cheapest route to ethanol [27], but its adoption has been delayed by the expensive pretreatment and the inefficient fermentation of the pentose-rich cellulose-derived sugars. Improvement has been sought via metabolic engineering in two ways. The first was to introduce the ethanol-forming enzymes PDC and ADH (see Fig. 8.4) in non-ethanologenic pentose-fermenting microbes (such as *Klebsiella oxytoca*). Alternatively, it has been attempted to engineer the capability to ferment pentoses in ethanol-producing yeasts, mainly *S. cerevisiae* and *Zymomonas mobilis* [24]. Both strategies were successful in principle, but the ethanol concentrations (<5%) remained undesirably low [24, 28].

This intractable problem may now be close to being solved. A *Saccharomyces* species that expressed the xylose isomerase gene from an anaerobic fungus was found to grow slowly on pentoses [29]. Improvement resulted from a combination of rational engineering – overexpression of the pentose phosphate-converting enzymes (see Fig. 8.5) – and classical strain improvement [30]. The authors conclude: "The kinetics of xylose fermentation are no longer a bottleneck in the industrial production of ethanol with yeast."

8.2.2.2 Green Aspects

A major part of the bioethanol is used as transport fuel, with the triple objectives to decrease the use of fossil fuels, reduce traffic-generated CO_2 emissions and improve the quality of the exhaust gas by adding an oxygenate to the fuel. The option to produce ethene from ethanol is still being considered. An energy analysis of the production of ethanol from corn [31] concluded that the energy gain (from sun to pump) ranges from 4% to 46% (taking the energy *input* as 100%). An energy gain of 75% would be possible by applying the best available technology in every step, which still means that 57% of the energy value of the output is required to drive the process.

In contrast, the energy gain of ethanol fermentation from a cellulose-based crop was estimated at only 10% [31]. A life cycle assessment of bioethanol from wood came to a similar conclusion [32]. This unsatisfactory outcome mainly results from the energy-intensive pretreatment with steam explosion, such as is used by Iogen [16]. The replacement of the latter by CO_2 explosion [33] may redress the energetic balance.

The unfavorable energy balance notwithstanding, the fermentation of ethanol, as well as other chemicals, from lignocellulose hydrolysate is considered highly desirable by some to avoid competition with food production. We note, however, that the large scale processing of waste lignocellulose runs counter to sensible agricultural practice to return plant material to the soil for the preservation of long-term soil structure [16].

8.2.3
Lactic Acid

The industrial production of lactic acid [34, 35], which dates back to 1881, is undergoing a remarkable transition. Lactic acid used to be a fairly mature fine chemical that was produced, in the mid 1990s, at a volume of $50-70 \, kt \, a^{-1}$ worldwide. A major share ($25 \, kt \, a^{-1}$, including simple esters etc.) is used in the food industry.

The potential of lactic acid as a renewable, broadly applicable intermediate and a building block for green polymers, solvents and plasticizers was recognized over 10 years ago [36]. The excellent balance of properties of lactic acid polymers (polylactic acid, PLA) [37, 38] and copolymers would make these materials ideal renewable and biodegradable replacements for, e.g., polycarbonate and polystyrene. The production of PLA was too expensive for such large-scale applications, however, and its use remained limited to niche markets, such as surgical sutures. As lactic acid was generally regarded as a relatively mature fine chemical that lacked any incentive, until the mid-1980s, to obtain cost reductions through innovation, the price was kept high by a lack of high-volume applications and vice versa.

This is now changing, as a new and very big player has entered the field. Cargill (USA), which in the late 1990s started to push fermentatively produced lactic acid as an emerging commodity [39], has opened a production facility with a capacity of $140 \, kt \, a^{-1}$ of polylactic acid in 2002 [40]; a market potential of $500 \, kt \, a^{-1}$ of lactic acid products in 2010 is anticipated.

Lactic acid is produced chemically from acetaldehyde, by hydrocyanation, followed by acid hydrolysis of the cyanohydrin (Fig. 8.6 a). The crude lactic acid is purified via esterification with methanol, distillation of the ester and hydrolysis with recycling of the methanol [34]. Major drawbacks are the production of an equivalent of ammonium sulfate and the cumbersome purification procedure that is required to obtain food-grade lactic acid.

The fermentative production of lactic acid from carbohydrates has repeatedly been reviewed recently [36, 41, 42]. Two classes of lactic acid producers are discerned: the homofermentative lactic acid bacteria, which produce lactic acid as the sole product, and the heterofermentative ones, which also produce ethanol, acetic acid etc. [43]. Recently, the focus has been on (S)-L-lactic acid producing, homofermentative *Lactobacillus delbrueckii* subspecies [42].

The anaerobic fermentation of lactic acid is traditionally performed at up to 50 °C over 2–8 d at pH 5.5–6.5 (lactic acid bacteria are highly sensitive to acid). The pH is maintained by titration with a base, usually calcium carbonate. The product concentration is kept below approx. $100 \, g \, L^{-1}$ to prevent precipitation of calcium lactate, as the separation of a precipitate from the biomass would be too elaborate. The stoichiometric yields are high, of the order of $1.7-1.9 \, mol \, mol^{-1}$ (85–95% of the theoretical yield) but the space–time yield, which is ap-

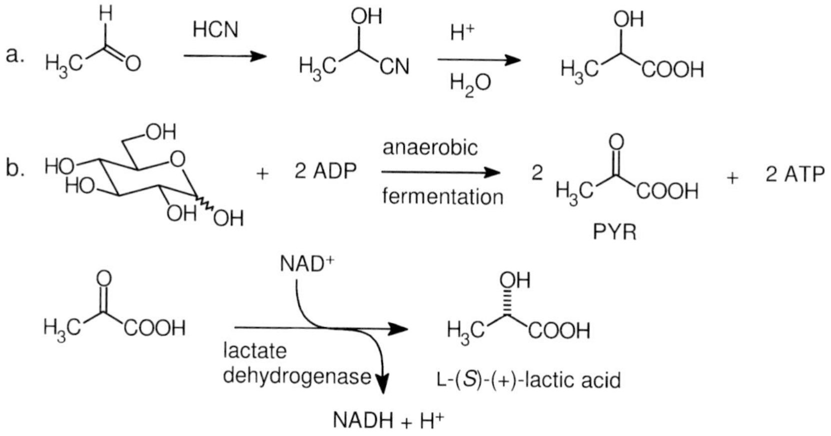

Fig. 8.6 Lactic acid and its production routes: (a) chemically, from acetaldehyde; (b) via fermentation.

prox. 100 g L^{-1} d^{-1}, is rather modest for a commodity and reflects the traditional nature of industrial lactic acid fermentation.

Acidification of the culture supernatant, filtration, treatment with active charcoal and concentration usually result in food-grade lactic acid, if the feedstock was pure glucose. Otherwise, more extensive purification may be required.

The fermentative process described above is not very atom-efficient – approx. 1 kg of calcium sulfate is formed per kg of lactic acid – and is expensive to operate due to the laborious DSP. It was good enough to compete, at times precariously, with the chemical manufacture of lactic acid [35] but extensive innovations were required to elevate lactic acid to commodity status.

The growing interest in renewable raw materials spurred, from the mid-1980s onwards, interest in the development of highly integrated, low-waste procedures for the fermentation, primary purification and processing of lactic acid. Continuous removal of lactic acid, via solvent extraction or otherwise, would obviate the need for adding base but is not feasible at pH 5.5–6.5, however, because only 2% of the lactic acid (pK_a 3.78) is uncharged at pH 5.5.

Acid-tolerant lactic acid bacteria would solve this latter problem and have indeed been obtained from screening [44] and via whole genome shuffling [45]. Alternatively, a lactate dehydrogenase gene could be inserted in a yeast, such as S. cerevisiae. These grow well at low pH and efficiently channel carbon into the glycolysis pathway under anaerobic conditions [46, 47]. Such constructs were indeed found to produce up to 55 g L^{-1} of lactic acid at pH 3.6 [46].

When the initial shortcomings of these engineered yeasts, such as low stoichiometric yield and productivity, have been ironed out, it may be expected that lactic acid fermentation in an acidic medium, combined with solvent extraction of the product, will evolve into a procedure of unprecedented efficiency.

Alternatively, the fermentation can be performed at pH 5.5–6.5 while employing a salt-free procedure to convert lactate into lactic acid [48]. Cargill has pat-

high molecular weight polymer

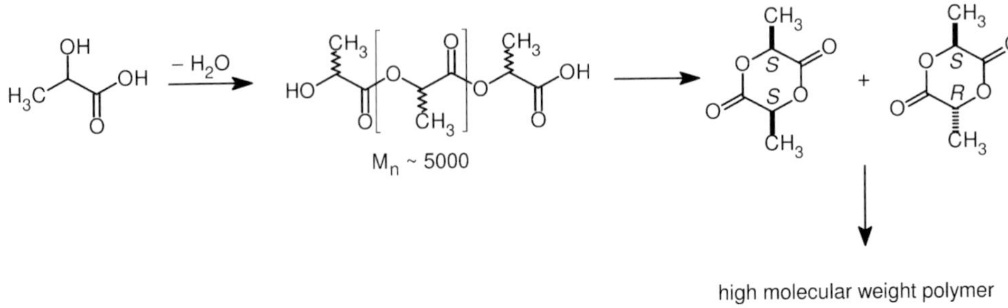

Fig. 8.7 Schematic representation of the solventless conversion of lactic acid into PLA.

ented a procedure, comprising acidification of the culture supernatant with CO_2 and extraction of the lactic acid into an organic phase [49], that may have been adopted.

Highly integrated procedures have been developed to convert lactic acid into dilactides and subsequently into PLA [38b, 38c, 50]. In the Cargill process, the aqueous lactic acid, which contains a preponderance of the (S)-enantiomer, is condensed into a mixture of oligomers, which is subsequently converted, in the presence of a tin catalyst [50], into the (S,S)- and (R,S)-dilactides (see Fig. 8.7) [51]. These are separated by vacuum distillation to tune (S,S)/(R,S) to the desired properties of the polymer [38c]. Subsequently, the lactide is subjected to ring-opening polymerization in the presence of a tin catalyst, such as tin octanoate, in a solventless procedure [38b, 50]. Major strengths of the procedure are the removal of water early in the procedure, the efficient separation of the dilactide diastereoisomers and the absence of solvents. The PLA is to be marketed under the trade names of Nature Works[TM] PLA by Cargill Dow Polymers and LACEA by Mitsui (Japan) [38c]. If the current price of lactic acid ($2 kg^{-1}) is translated into $3 kg^{-1} for PLA, the latter product should be able to compete with polycarbonate at $4.4 kg^{-1} [52]. If plans to produce lactic acid from cellulose hydrolysate [16] come to fruition, PLA could even become competitive with polystyrene ($2 kg^{-1}).

8.2.4
1,3-Propanediol

1,3-Propanediol (1,3PD) is also undergoing a transition from a small-volume specialty chemical into a commodity. The driving force is its application in poly(trimethylene terephthalate) (PTT), which is expected to partially replace poly(ethylene terephthalate) and polyamide because of its better performance, such as stretch recovery. The projected market volume of PTT under the tradenames CORTERRA (Shell) and Sorona[TM] 3GT (Dupont) is 1 Mt a^{-1} within a few years. In consequence, the production volume of 1,3PD is expected to expand from 55 kt a^{-1} in 1999 to 360 kt a^{-1} in the near future. 1,3PD used to be synthesized from acrolein by Degussa and from ethylene oxide by Shell (see Fig. 8.8) but a fermentative process is now joining the competition.

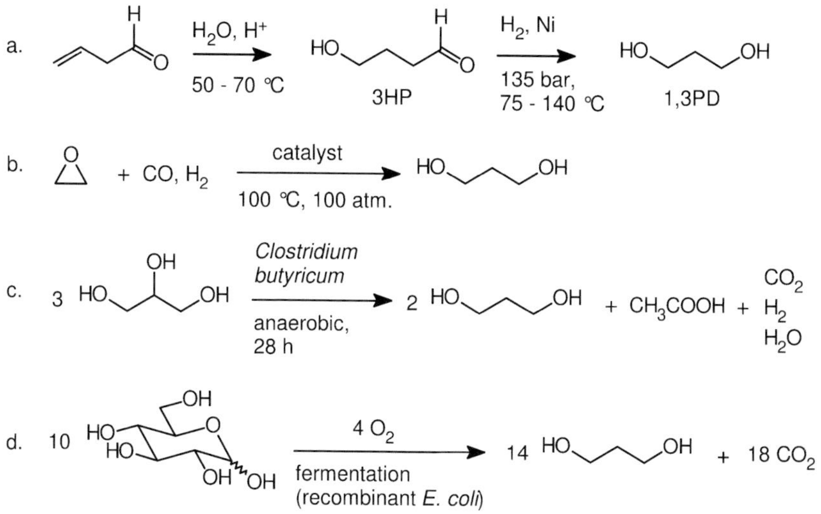

Fig. 8.8 Processes for 1,3PD: (a) from acrolein (Degussa); (b) from ethylene oxide (Shell); (c) from glycerol, via anaerobic fermentation (Henkel); (d) from glucose, using an E. coli cell factory (Dupont-Genencor). Compounds: 3HP, 3-hydroxy-propanal; 1,3PD, 1,3-dihydroxypropane.

The Degussa process (now owned by Dupont) starts from acrolein, which is hydrated in the presence of an acidic ion exchanger into 3-hydroxypropanal (3HP, Fig. 8.8a). The latter is subsequently extracted into isobutyl alcohol and hydrogenated over a Ni catalyst [53]. The overall yield does not exceed 85%, due to competing water addition at the 2-position and ether formation in the initial step. It has been announced that Degussa will supply up 10 kt a^{-1} to Dupont until the fermentative process of the latter company (see below) comes on stream [54].

Shell produces 1,3PD from ethylene oxide via hydroformylation with synthesis gas (Fig. 8.8b). The transformation required two separate steps in the past [55], but has been improved [56], which made the large-volume use of 1,3PD in poly(trimethylene terephthalate) economically viable, and the two steps have been telescoped into one [57, 58]. Shell has a capacity to 70 kt a^{-1} [59].

The fermentation of 1,3PD from glycerol (see Fig. 8.8c) was discovered in the late 19th century [60]. It has since been found that a considerable number of bacteria can use glycerol as a source of carbon and energy under anaerobic conditions and the reaction pathways have been elucidated [61]. Out of every three molecules of glycerol, one is oxidized, phosphorylated into DHAP and subsequently metabolized via the glycolysis pathway and the TCA cycle; the other two are converted into 1,3PD, to maintain redox balance (see Fig. 8.9), via dehydration and NADH-driven reduction.

There has been a long-standing interest in the possible microbial production of 1,3PD from glycerol in *Citrobacter*, *Klebsiella* and *Clostridia* species [61–63], to

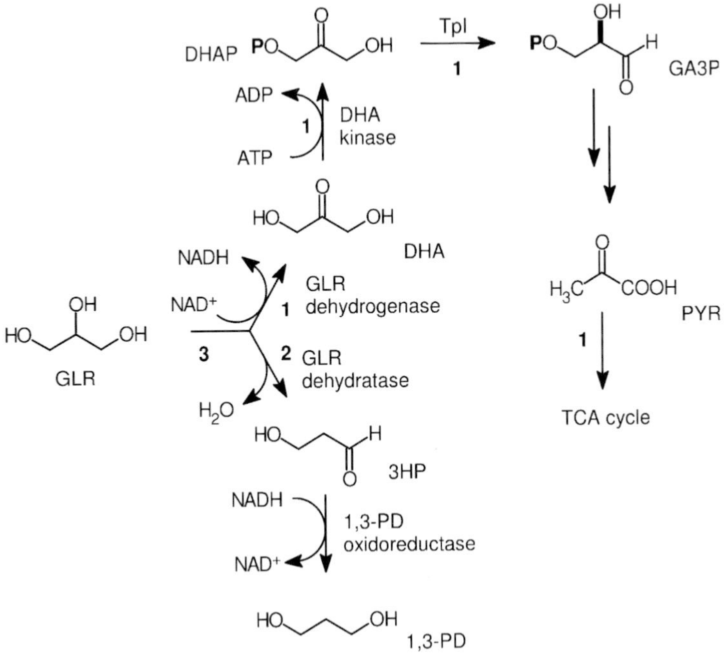

DHAP

Tpl

GA3P

ADP

DHA
kinase

ATP

1

DHA

NADH

GLR
dehydrogenase

NAD+

1

GLR

H₃C　COOH

PYR

3

2 GLR
dehydratase

1

GLR

H₂O

TCA cycle

3HP

NADH

1,3-PD
oxidoreductase

NAD+

1,3-PD

Fig. 8.9 Anaerobic fermentation of glycerol [61].
Compounds: GLR, glycerol; DHA, dihydroxyacetone.

relieve a projected oversupply of glycerol. The stoichiometric yields are close to the theoretical maximum of 0.67 mol/mol but the product titer has remained limited to 70–78 g L^{-1} (STY 75 g L^{-1} d^{-1}), in spite of metabolic pathway engineering studies [61].

Genencor has taken a radically different approach [22] by engineering the central metabolism of *E. coli* for the production of 1,3PD from glucose [64]. To this end, the enzymes from *S. cerevisiae* that convert DHAP into glycerol (see Fig. 8.4), as well as the glycerol dehydratase complex from *K. pneumonia*, have been cloned into the production organism. Hence, the latter combines functionalities that in Nature require two very different organisms. The final reduction step, of 3HP into 1,3PD, is taken care of by a non-specific NADPH-dependent alcohol dehydrogenase that is native to *E. coli*.

Efficiency required that all glycerol should flow into the 1,3PD pathway. Hence, glycerol was prevented from re-entering the central metabolism [64]. A second, more radical modification was made to the glucose uptake and phosphorylation mechanism, which in *E. coli* natively occurs via the PEP-consuming phosphate transfer system (PTS) with formation of PYR, which enters the oxidative branch. The consumption of 1 mol of PEP per mol of glucose puts an artificial ceiling on the stoichiometric yield of 1.0 mol mol^{-1}. Replacing the PTS by an ATP-dependent system [64] effectively removed the restriction of the product yield.

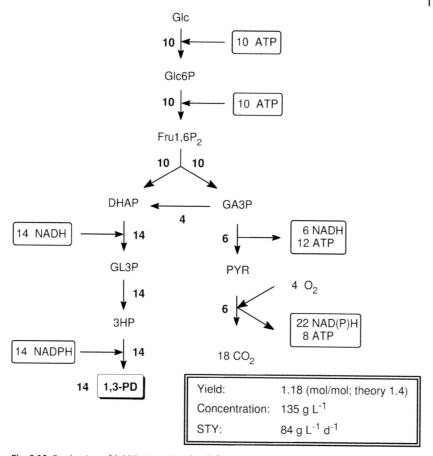

Fig. 8.10 Production of 1,3PD in an *E. coli* cell factory [64].

As shown in Fig. 8.10, energy and redox equivalents are required to drive the transformation of glucose into 1,3PD. These are provided by converting some of the glucose all the way into CO_2, which restricts the stoichiometric yield on glucose to 1.4 mol mol^{-1} maximum. A stoichiometric yield of 1.18 mol mol^{-1} of 1,3PD (50% by weight) has been obtained in practice [64], which corresponds with 85% of the theoretical yield, at a product concentration of 135 g L^{-1}.

It should be noted that 1,3PD represents only 36% of the molecular mass of the polymer; hence, if the fermentative route to 1,3PD is the winner its contribution to sustainability still remains limited. Unless, of course, the terephthalic acid building block also is replaced by a renewable one, such as furan-1,5-dicarboxylic acid (see Section 8.3.5).

8.2.5
3-Hydroxypropanoic Acid

3-Hydroxypropanoic acid (3HPA) is under development as a future platform chemical and monomer derived from biomass. It is, at the present time, not produced on an industrial scale, either chemically or biotechnologically. 3HPA could be a key compound for the production of biomass-derived C_3 intermediates, such as acrylic acid, acrylic amide and malonic acid (see Fig. 8.11). Hydrogenation of 3HPA would provide a competing procedure for the production of 1,3PD (see Section 8.2.4) that could be more economical than the DuPont and Shell processes [65].

There is no microbe known to produce 3HPA. Various metabolic routes from glucose to 3HPA can be designed [66] but only two of these, via lactic acid and 3-aminopropionic acid, respectively (Fig. 8.12), have been developed in any depth, mainly by Cargill. The lactic acid route has two major disadvantages: it passes through a local energy maximum (acrylCoA) and there is hardly any driving force because the heats of formation and ionization constants of lactic acid and 3HP are very close [67].

The alternate route, via pyruvic acid and 3-aminopropanoate, is exothermic for nearly the whole way and is anaerobic with production of ATP. The transformation of L-alanine into 3-aminopropanoate is a major hurdle, as this activity is not known in Nature and was not found upon screening. A corresponding lysine mutase is known, however, and was engineered into an alanine mutase, in collaboration with Codexis [67, 68]. The selection method was based on the disruption of the *panD* gene (see Section 8.2.9), which puts the organism under selective pressure to evolve an alternative route to 3-aminopropionate to satisfy its requirement for (R)-pantoic acid. Furthermore, all competing NAD^+-forming

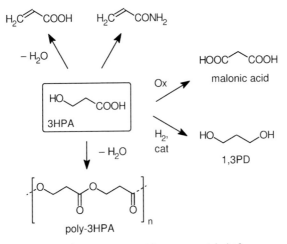

Fig. 8.11 3-Hydroxypropanoic acid as a potential platform chemical. Compound: 3HPA, 3-hydroxypropanoic acid.

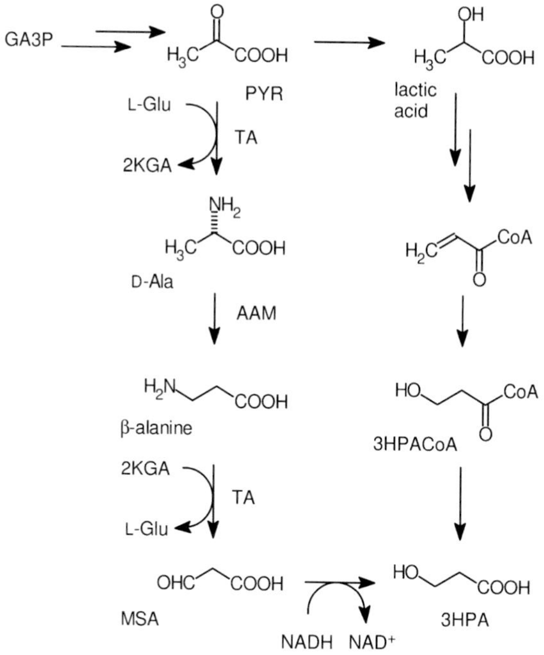

Fig. 8.12 Microbial production routes to 3-hydroxypropanoic acid. Compounds: 3HPCoA, 3-hydroxypropanoyl coenzyme A ester; 2KGA, 2-ketogluconic acid; MSA, malonate semialdehyde. Enzymatic activities: AAM, alanine-2,3-aminomutase; TA, transaminase.

pathways, such as production of ethanol, acetate, lactate etc., were switched off to force the organism into producing 3HPA to maintain redox balance [67]. Cargill has indicated that commercialization of the microbial production of 3HPA is still some years away.

8.2.6
Synthesizing Aromatics in Nature's Way

Microbes and plants synthesize aromatic compounds to meet their needs of aromatic amino acids (L-Phe, L-Tyr and L-Trp) and vitamins. The biosynthesis of these aromatics [69] starts with the aldol reaction of D-erythrose-4-phosphate (E4P) and phosphoenolpyruvate (PEP), which are both derived from glucose via the central metabolism, into DAHP (see Fig. 8.13). DAHP is subsequently converted, via a number of enzymatic steps, into shikimate (SA) and eventually into chorismate (CHA, see later), which is the common intermediate in the biosynthesis of the aromatic amino acids [70] and vitamins.

DAHP synthase combines E4P and PEP, which are both derived from the central metabolism (see Figs. 8.4 and 8.5 for more details). The biosynthesis of

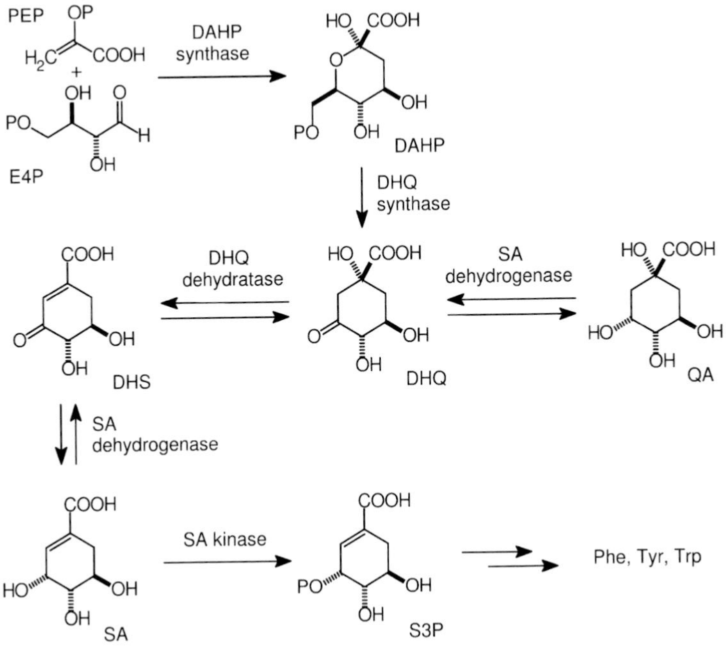

Fig. 8.13 Biosynthesis of aromatic compounds in *E. coli* [71]. Compounds: E4P, D-erythrose-4-phosphate; PEP, phosphoenol-pyruvate; DAHP, 3-deoxy-D-*arabino*-heptulo-sonate-7-phosphate; DHQ, 3-dehydroquinate; QA, quinate; DHS, 3-dehydroshikimate; SA, shikimate; S3P, shikimate-3-phosphate.

E4P and PEP has been studied in much detail; the extensive mutagenesis that has been applied to increase the bioavailability of these latter compounds has been reviewed [70–72]. It should be noted that the PTS, which mediates the up-take and phosphorylation of glucose, limits the yield of DAHP from glucose at 0.43 mol mol^{-1} in a situation that is similar to that described above in Section 8.2.4. If all PEP could be channeled into the aromatic pathway the stoichio-metric yield of DAHP could be as high as 0.86 mol mol^{-1}, at least in theory [73]. Various approaches to PEP conservation have been demonstrated [72, 74, 75].

The common pathway is regulated by several mechanisms, depending on the organism. In *E. coli*, which is the best investigated organism in this respect, me-tabolic engineering usually starts with the introduction of a feedback resistant DAHP synthase gene [70]. Overexpression of the subsequent enzymes in the common route, DHQ synthase and SA dehydrogenase, alleviates the kinetic bot-tlenecks caused by product inhibition.

SA has recently emerged as the preferred intermediate in an industrial pro-cess for the antiviral medicine Tamiflu® [76]. Because SA is only (sparingly) available from the fruit of the *Illicium* plants, there is much interest in its mi-crobial production, mainly in rationally designed *E. coli* mutants [77].

An *E. coli* in which further conversion of SA was blocked by disruption of SA kinase, combined with mutagenesis for increased SA production, excreted SA into the culture medium [78]. 27 g L^{-1} of SA was produced, accompanied by DHS (4.4 g L^{-1}) and QA (12.6 g L^{-1}) [78]. This latter compound, which is particularly troublesome in the subsequent purification of SA, arises from the conversion of DHQ into QA, due to incomplete selectivity of SA dehydrogenase. The formation of QA could eventually be suppressed by performing the culture under glucose-rich conditions [79]. When combined with PEP conservation, SA was produced at a level of 84 g L^{-1} in a fermentation on the 10 L scale, which corresponds with a yield of 0.33 mol mol^{-1} glucose [75], along with DHS (10 g L^{-1}) and QA (2 g L^{-1}). Hence, the total yield on hydroaromatic products was 0.38 mol mol^{-1}.

There also is some interest in QA as a chiral building block and a food acidulant. The compound is not naturally produced by *E. coli* but, as noted above, an *E. coli* engineered for the production of SA fortuitously produced modest amounts of QA [78, 79]. A very similar mutant, which lacked DHQ dehydratase (see Fig. 8.13), produced QA in copious amounts (49 g L^{-1}), along with a small amount (3.3 g L^{-1}) of DHQ [80].

The Frost group [80, 81] has been particularly active in devising microbial and chemical-microbial procedures for the benzene-free synthesis of phenolic compounds from glucose via the hydroaromatic compounds DHS, QA and SA. As has been noted previously, such products can be synthesized more cheaply from coal via the methanol route, if oil should become too expensive [82].

The same group has demonstrated the microbial synthesis of vanillin in an *E. coli* mutant [83] but the yield and titer (6.2 g L^{-1}) were too low to be of immediate practical value. A completely microbial route to gallic acid [84], which is currently isolated from natural sources [85], produced 20 g L^{-1} of the desired product in the presence of considerable amounts of side-products [84].

8.2.7
Aromatic α-Amino Acids

The market volume of *L-phenylalanine* (L-Phe) has rapidly increased in recent years to 14 kt in 2002, according to an authorative estimate [86], which is mainly due to the commercial success of the artificial sweetener aspartame. L-Phe has been produced by a number of chemoenzymatic routes in the past [87] (see Fig. 8.14) and many more have been developed [88] but never commercialized. Fermentation of L-Phe is now so efficient that it has rendered all of the chemoenzymatic procedures obsolete.

L-Phe can be prepared via the enantioselective hydrolysis of N-acetyl-D,L-Phe and microbial reductive amination of phenylpyruvate (see Fig. 8.14) [87]. The stoichiometric yields of these processes were high but the precursors required 3–4 synthetic steps from the basic starting materials in most cases. The phenylammonia lyase route, in contrast [89], provided L-Phe in only two steps from the basic chemicals benzaldehyde and acetic anhydride [90]. The enzymatic step

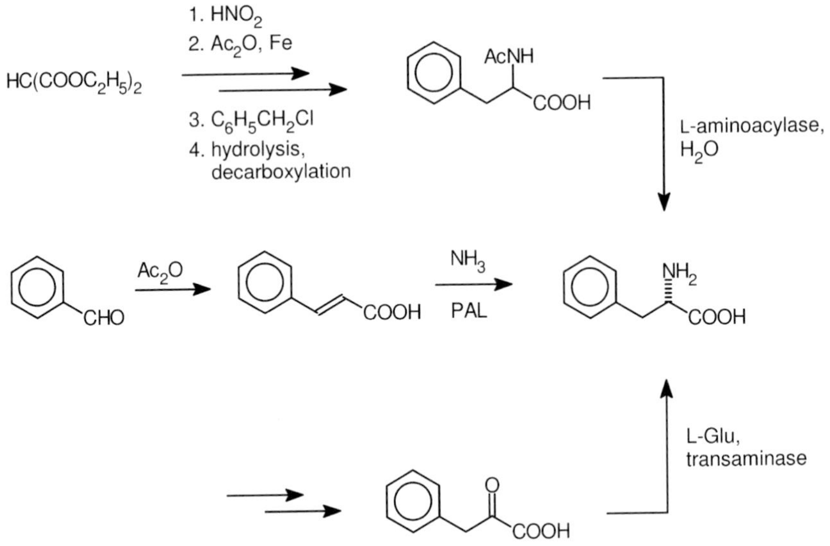

Fig. 8.14 Chemoenzymatic processes for L-phenylalanine.

afforded up to 50 g L^{-1} of L-Phe in 83% yield [89c]; an STY of 34 g L^{-1} d^{-1} has been claimed [89b]. The procedure was used to provide the L-Phe for the first production campaign of aspartame by Searle [87].

The biosynthetic pathway from SA into L-Phe [69, 70] is shown in Fig. 8.15. The synthesis of chorismate (CHA), the common intermediate in the biosynthesis of the aromatic amino acids, requires an extra equivalent of PEP, which limits the yield of L-Phe from glucose to 0.30 mol mol^{-1} if PEP is not conserved [91]. The further transformation of CHA into phenylpyruvic acid (PPY) suffers from inhibition by L-Phe and is also subject to transcriptional control [69, 92]. The final step is a reductive amination of PPY into L-Phe with consumption of L-Glu.

In the early years, L-Phe was microbially produced with *Corynebacterium gluta-micum* and *E. coli* strains which had been deregulated with respect to the end product via classical strain improvement. More recently, metabolic engineering has been employed to address nearly all aspects of the biosynthesis of L-Phe; the work has been reviewed [70, 93].

Little is known about the current production levels of L-Phe, but in the early days of metabolic engineering 50 g L^{-1} of L-Phe, with a yield on glucose of 0.27 mol mol^{-1}, was produced by an engineered *E. coli* [92]. This is near the theoretical limit [91] and one would surmise that in the starting species used by these authors some PEP-conserving mutations had been introduced via classical strain improvement.

Few details have been disclosed on the DSP of L-Phe fermentation, which may indicate that it is fairly traditional. A procedure for integrated product removal, which also resulted in increased productivity and yield on glucose, has recently been published [94].

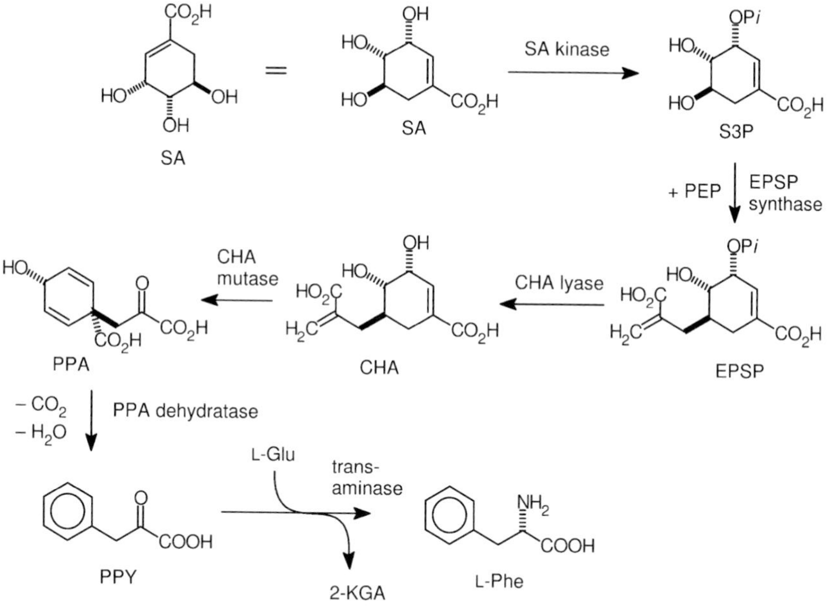

Fig. 8.15 Biosynthesis of L-Phe. Compounds: EPSP, 5-*enol*-pyruvoylshikimic acid-3-phosphate; CHA, chorismic acid; PPA, prephenic acid; PPY, phenylpyruvic acid; S3P, shikimic acid 3-phosphate.

The fermentation of L-Phe has obviously emerged victorious, on purely economic grounds and is now so efficient that even D,L-Phe, which is used by the Holland Sweetener Company in the enzymatic production of aspartame, is obtained by racemization of L-Phe rather than via a chemical procedure.

L-Tryptophan (L-Trp) was produced, mainly by Japanese companies, on a scale of 500–600 t a^{-1} in 1997 [70]. It is an essential amino acid that is used as a food and feed additive and in medical applications. L-Trp is, at US$ 50 kg^{-1} (feed quality), the most expensive aromatic amino acid and it is thought that the market for L-Trp could expand drastically if the production costs could be brought down. There is no chemical process for L-TrP and enzymatic procedures starting from indole, which were very efficient, could not compete with fermentation [95]. L-Trp has been produced by precursor fermentation of anthranilic acid (ANT, see Fig. 8.16), but the serious effects of minor by-products caused the process to be closed down. Since the mid-1990s all L-Trp is produced by *de novo* fermentation.

The biosynthesis of L-Trp from CHA is outlined in Fig. 8.16. The complex transformation of CHA into phosphoribosyl anthranilate (PRAA) is, in *E. coli* and *C. glutamicum*, catalyzed by a protein aggregate that is inhibited by the L-Trp end product. The final two steps are performed by a single protein; indole, which is toxic to the cell, is channeled directly into the active site where the fi-

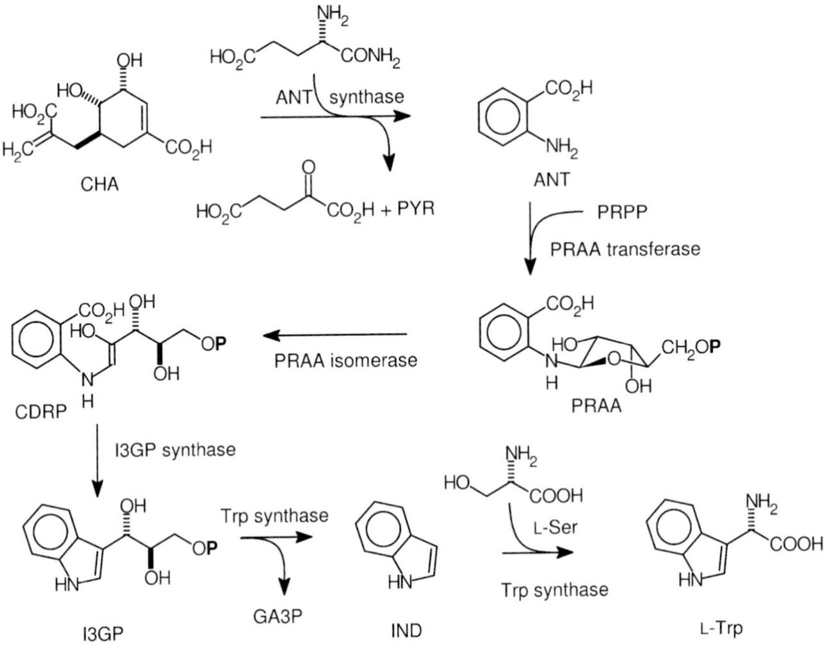

Fig. 8.16 Biosynthesis of L-Trp. Compounds: ANT, anthranilate; CDRP, 1-(o-carboxyphenylamino)-1-deoxyribulose-5-phosphate; I3GP, indole-3-glycerolphosphate; IND, indole; PRAA, phosphoribosyl anthranilate; PRPP, 5-phosphoribosyl-α-pyrophosphate.

nal conversion takes place [69]. The genes that code for the Trp pathway are tightly clustered on the *trp* operon, which is subject to transcriptional control by a single repressor protein.

In comparison with L-Phe, extra glucose is required to provide the ribose phosphate derivative and the maximum yield of L-Trp from glucose, without PEP-conserving modifications, has been estimated at 0.20 mol mol^{-1} [96].

C. glutamicum and *E. coli*, which share very similar biosynthetic pathways and control architectures, have been subjected to pathway engineering for the production of L-Trp. Modifications that have been reported include the, now familiar, feedback-resistant DAHP synthase and the enzymes in the Trp pathway were freed from regulation [72, 97] and overexpressed.

A shortage of L-Ser, which caused accumulation of indole and cell death, was remedied by overexpression of the first enzyme in the Ser pathway [98]. Additionally, engineering of the central metabolism to increase the availability of E4P (see Fig. 8.5) increased the production of L-Trp in *C. glutamicum* [99].

The best published L-Trp production levels with engineered *C. glutamicum* and *E. coli* species range from 45–58 g L^{-1} (STY 17–20 g L^{-1} d^{-1}) [72, 99]; a 0.2 mol mol^{-1} yield on Glc, which is the theoretical maximum, has been reported [100].

8.2.7
Indigo: the Natural Color

The blue indigo dye has been used from prehistoric times onwards [101]. Traditionally, it was prepared from plants, such as *Indigofera tinctoria* (in the East) or *Isatis tinctoria* (woad, in Europe). These plants contain derivatives of indoxyl (see Fig. 8.17) from which the latter was liberated by laborious fermentation. Spontaneous aerobic oxidation of indoxyl yields indigo; the reaction seems to proceed via leucoindigo and generates 1 mol of H_2O_2 per mol of indoxyl oxidized [102]. Reduction of the insoluble indigo, by various methods, yields the soluble and colorless leucobase, which is applied to the textile and is oxidized back to indigo upon exposure to air.

A chemical process for indigo, developed by BASF in the late 19th century, rapidly pushed the natural material from the market [103]. Nowadays, 17 kt a^{-1} of indigo is produced chemically; most of this output is used to dye the 10^9 blue jeans that are produced annually. Amgen and later Genencor have been developing a fermentative process for indigo since the early 1980s. Remarkably, the EC now supports a research project that aims at a revival of the agricultural production of indigo [104].

Nearly all indigo is produced from N-phenylglycine (see Fig. 8.18) via fusion with potassium and sodium hydroxide, followed by treatment with sodamide [103]. The melt containing the dialkalimetal salt of indoxyl is subsequently dissolved in water, and indigo is formed by aerobic oxidation. Filtration and wash-

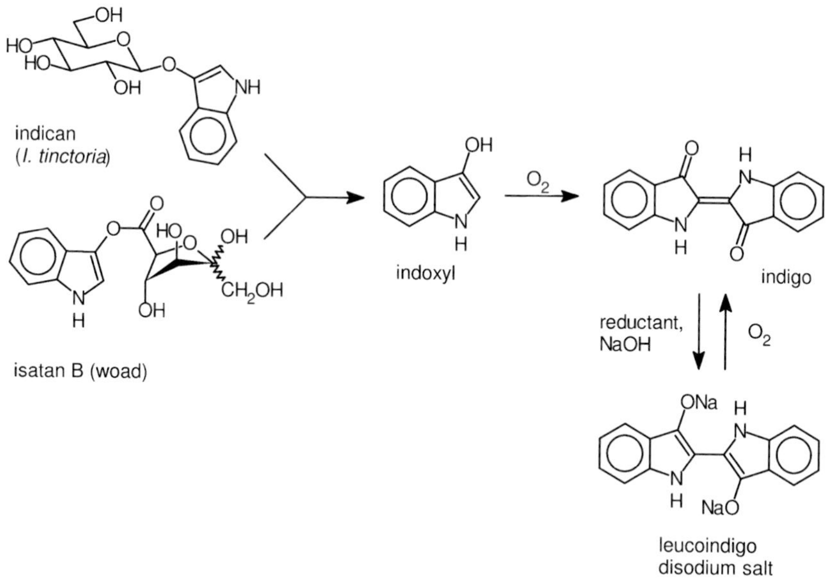

Fig. 8.17 Indigo, its basic chemistry and its natural sources.

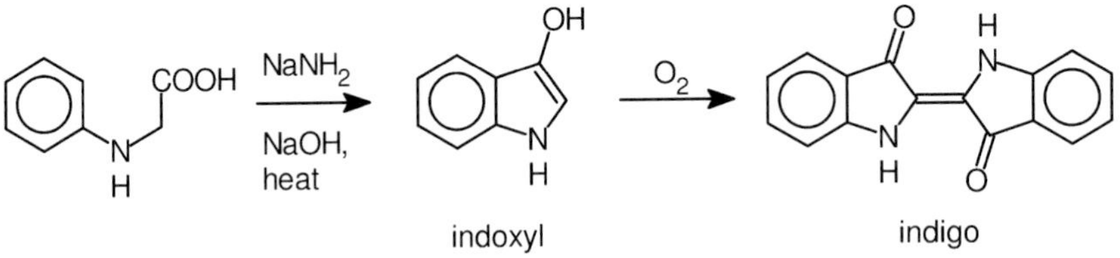

Fig. 8.18 Indigo production via the Heumann-Pfleger process (BASF).

ing affords nearly pure indigo; the alkaline filtrate is concentrated, filtered to re-move side-products and recycled [103].

The key to an industrially viable microbial synthesis of indigo, which already had been demonstrated in principle in 1927 [105], was the discovery that naphthalene dioxygenase (NDO) from *P. putida* was able to oxidize indole into the indole precursor *cis*-2,3-hydroxy-2,3-dihydroindole (see Fig. 8.19) [106]. Gen-encor has disclosed details of a prototype indigo-producing *E. coli* that is based on its L-Trp producing strain [107]. This latter species carried the genes for a feedback-resistant DAHP synthase (see Fig. 8.11) and ANT synthase to maxi-mize the carbon flow into the Trp pathway. The Trp synthase was engineered to allow indole to escape from the enzyme complex, which does not occur nor-mally [69]; the insertion of a *P. putida* gene for naphthalene dioxygenase re-sulted in the production of *cis*-2,3-dihydroxyindole (DHI), which reacts extracel-lularly into indigo (Fig. 8.19).

The indigo production was 50% lower than expected on the basis of the Trp productivity of the parent strain [107]. This latter problem was traced to inactiva-tion of DAHP synthase by exposure to indoxyl or its oxidation products; it was also found that PEP exerted a protecting effect on DAHP synthase. The produc-tion of indigo was improved to approx. $18\ \text{g L}^{-1}$ ($6.2\ \text{g L}^{-1}\ \text{d}^{-1}$) by increasing the gene dosage for DAHP synthase and mutations that increased the bioavailability of PEP [107].

A problem connected with the microbial production of indigo is the formation of a small amount of isatin, which could be partly suppressed by adjusting the O_2 concentration [108]. Isatin reacts with indoxyl to give indirubin, which gives the finished denim an undesirable red cast. The introduction of an isatin-de-grading enzyme (isatin hydrolase from *P. putida*) satisfactorily reduced the level of indirubin [107].

Subsequently, indigo was successfully produced via fermentation on the 300 000 L scale at a cost that was comparable with the price of chemical indigo [109]. Commercialization proved elusive, however, presumably because chemical indigo is marketed with a substantial profit margin. We note that it is common experience that the *total* costs of a new process, to compete, must be equal to (or lower than) the *production* cost of the existing process. In this particular case, it would seem that the STY of the fermentative process is too low to com-pete with a chemical procedure that encompasses only three steps from the ba-sic chemicals aniline and acetic acid.

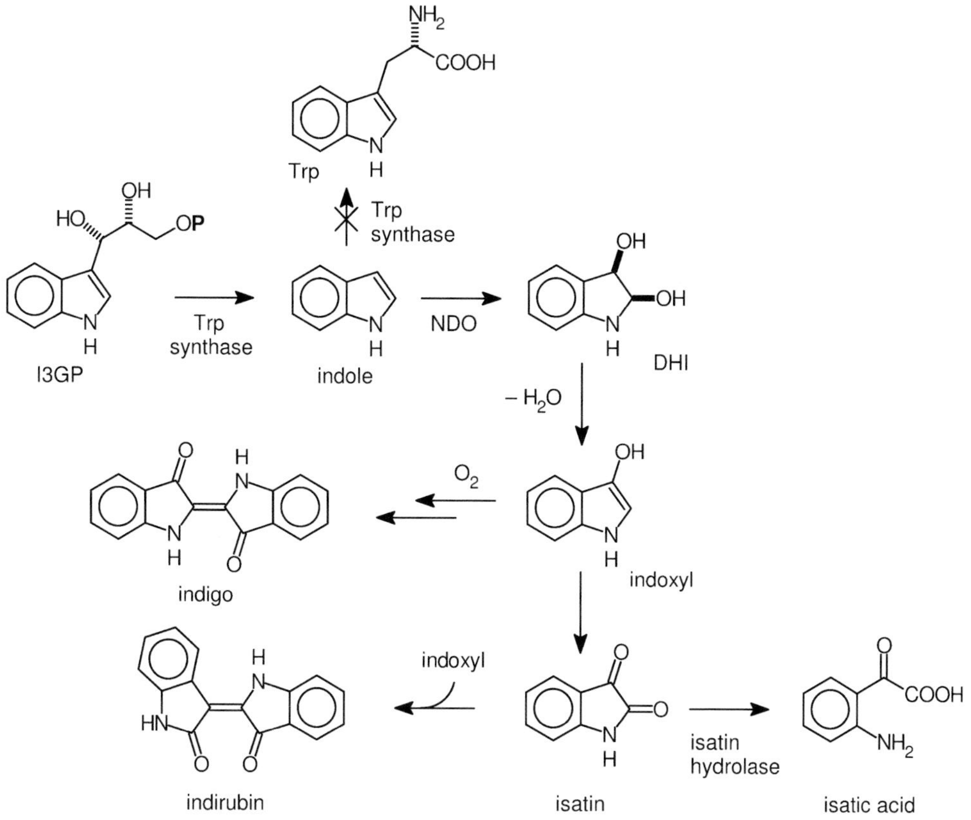

Fig. 8.19 Microbial production of indigo by an *E. coli* cell factory. Abbreviations: DHI, *cis*-2,3-dihydroxy-2,3-dihydroindole; NDO, naphthalene dioxygenase.

8.2.8
Pantothenic Acid

(*R*)-Pantothenic acid (vitamin B$_5$) is synthesized by microbes and plants, but not by mammals, who require it as a nutritional factor. Only the (*R*)-enantiomer is physiologically active. (*R*)-Pantothenic acid is produced as its calcium salt on a 6 kt a^{-1} scale, 80% of which is applied as an animal feed additive; major suppliers are Roche, Fuji and BASF. Pantothenic acid is produced via chemical methods [110] but a fermentative procedure has recently been commercialized.

The major industrial route to calcium pantothenate starts from isobutyraldehyde, which is condensed with formaldehyde. Hydrocyanation and hydrolysis affords the racemic pantolactone (Fig. 8.20). The resolution of pantolactone is carried out by diastereomeric crystallization with a chiral amine, such as (+)-2-aminopinane (BASF), 2-benzylamino-1-phenylethanol (Fuji) or (1*R*)-3-*endo*-aminonorborneol (Roche). The undesired enantiomer is racemized and recycled.

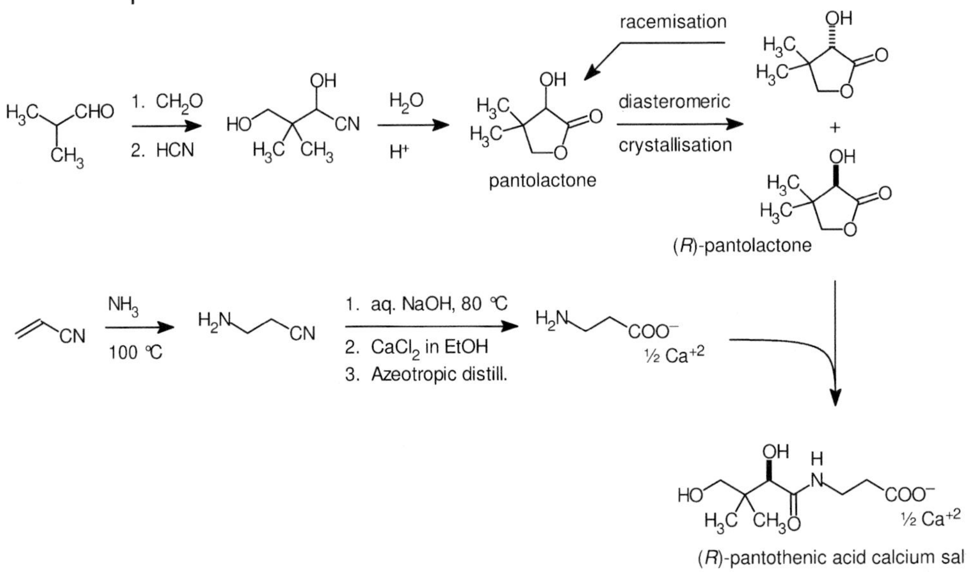

Fig. 8.20 Chemical production of (R)-pantothenic acid.

The reaction of (R)-pantolactone with calcium 3-aminopropionate (synthesized from acrylonitrile, see Fig. 8.20) affords calcium pantothenate [111].

The diastereomeric crystallization of pantolactone is laborious due to the need to recycle the resolving agent. Various schemes to replace this latter step with the chemical or microbial oxidation of pantolactone, followed by microbial reduction to the (R)-enantiomer, were unsuccessful because the productivity of the microbial step remained too low [110a].

The enantioselective hydrolysis of pantolactone into (R)-pantoic acid and (S)-pantolactone (Fig. 8.21), in the presence of (R)-pantolactone hydrolase from *Fusarium oxysporum* [110b], offers a better alternative. An alginate-entrapped

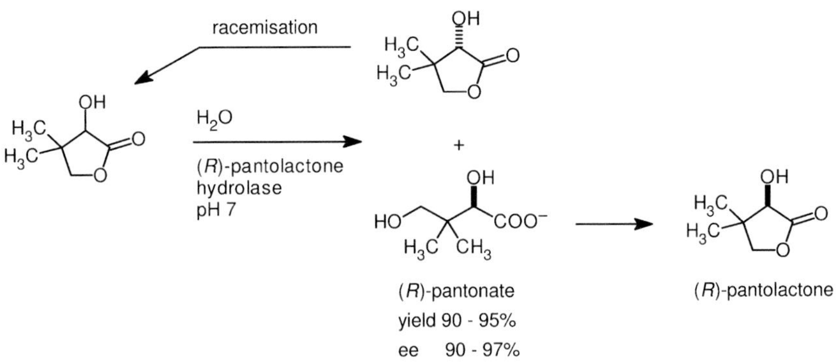

Fig. 8.21 Enzymatic resolution of pantolactone.

whole-cell preparation maintained its activity for 180 hydrolysis cycles at $350\ g\ L^{-1}$ pantolactone. The enzymatic resolution of pantolactone does not, however, remedy the major drawback of the traditional production of pantothenic acid, which is the laborious purification of (R)-pantolactone.

The biosynthesis of (R)-pantothenate in E. coli [112] (see Fig. 8.22) and Corynebacterium glutamicum [113] has been elucidated. 3-Methyl-2-oxobutyrate, an intermediate in the L-Val biosynthesis pathway, is successively hydroxymethylated and reduced to (R)-pantoate. The latter intermediate is coupled, in an ATP-requiring reaction, with 3-aminopropionate that is derived from L-aspartate via decarboxylation. The corresponding genes have been identified [114].

(R)-pantothenic acid is an obvious candidate to be produced via fermentation, because all microorganisms synthesize the vitamin to meet their own requirements. Takeda Chemical Industries has developed a microbial partial synthesis of (R)-pantothenate in an E. coli mutant with enhanced expression of the panB, panC and panD genes [115]. High levels of (R)-pantothenate, $60\ g\ L^{-1}$ [116], which corresponds with $30\ g\ L^{-1}\ d^{-1}$, were obtained when 3-aminopropionate was fed to the culture. Presumably, fermentation of (R)-pantothenate with suppletion of 3-aminopropionate is used by Degussa in the production of Biopan®.

Attempts at de novo fermentation of (R)-pantothenate in organisms such as E. coli [117], Bacillus species [118] and C. glutamicum [113, 119] met with low production levels of $1–2\ g\ L^{-1}$ [113], which is too low to be of immediate practical value.

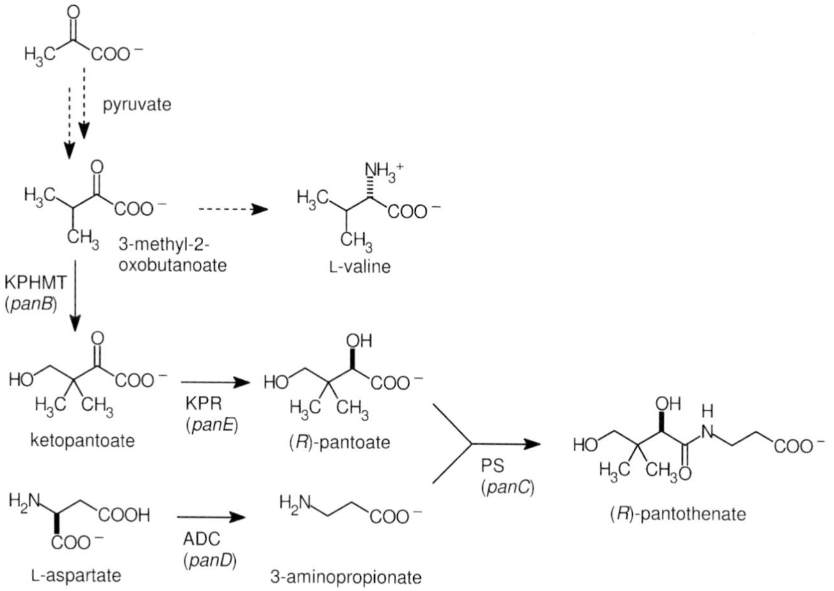

Fig. 8.22 The biosynthesis of (R)-pantothenate in E. coli [112]. Enzymatic activities: ADC, L-aspartate-1-decarboxylase; KPHM, a-ketopantoate hydroxymethyltransferase; KPR, a-ketopantoate reductase; PS, pantothenate synthase.

In conclusion, at least one producer of (R)-pantothenate is operating a fermentative process, with suppletion of 3-aminopropionate. Total microbial synthesis is much more challenging but, considering the pace of metabolic engineering, there is little doubt that one will be commercialized in the near future. Alternatively, the 3-aminopropionate could be derived from L-Ala, using the cellular machinery developed for the production of 3HPA (see Section 8.2.5) [67, 68].

8.2.9
The β-Lactam Building Block 7-Aminodesacetoxycephalosporanic Acid

Penicillin G is produced on a scale of > 20 kt a^{-1} via fermentation in *Penicillium chrysogenum*; there is no competition from chemistry as production by chemical means would be far too elaborate. A relentless competition has driven the continuous improvement of penicillin G production [120]. Performance figures of the penicillin G culture are well-kept secrets but can be estimated at 15–20 g L^{-1} d^{-1}.

Penicillin G as well as approx. 4 kt a^{-1} of penicillin V are enzymatically hydrolyzed into the β-lactam nucleus 6-aminopenicillanic acid (6-APA), which is the building block for the semisynthetic penicillin antibiotics ampicillin and amoxycillin [121] (see Fig. 8.23). Part of the penicillin G is converted into 7-aminodesacetoxycephalosporanic acid (7-ADCA, 3 kt a^{-1}), which is the intermediate for the semisynthetic cephalosporins cephalexin and cefadroxil [122]. 7-ADCA has traditionally been produced from penicillin G via a chemical ring expansion but an all-bio process has recently been started up by DSM.

The chemical process for 7-ADCA dates back to the early days of cephalosporin chemistry [123] and involves the sequential sulfoxidation, esterification and dehydratation/expansion of the penicillin nucleus. The yields were low, initially, but improved as the result of a considerable research effort [124]. An optimized

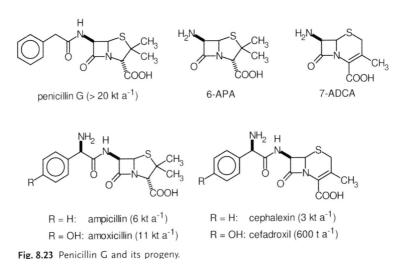

penicillin G (> 20 kt a^{-1})   6-APA   7-ADCA

R = H: ampicillin (6 kt a^{-1})   R = H: cephalexin (3 kt a^{-1})
R = OH: amoxicillin (11 kt a^{-1})   R = OH: cefadroxil (600 t a^{-1})

Fig. 8.23 Penicillin G and its progeny.

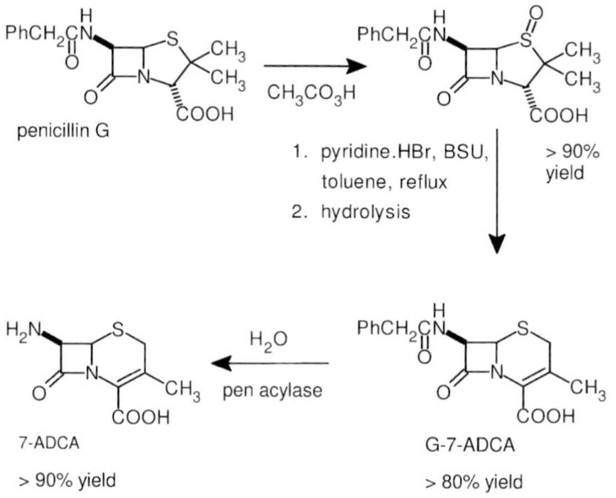

Fig. 8.24 Chemoenzymatic synthesis of 7-ADCA from penicillin G. Abbreviation: BSU, N,N'-bis(trimethylsilyl)urea.

procedure (Fig. 8.24) afforded G-7-ADCA in >80% yield [125, 126]; subsequent enzymatic hydrolysis affords 7-ADCA.

The chemical ring expansion process has been employed universally for over 20 years and has remained typical for the times in which it was conceived: although stoichiometrically efficient it depends on hazardous, highly active chemicals, which are difficult to recover and reuse, and it generates an excessive amount of waste.

The relationship between the biosynthesis of the penicillin and cephalosporin nuclei [127] is shown in Fig. 8.25. The common intermediate in the biosynthesis of penicillins and cephalosporins is isopenicillin N (IPN), which in *Penicillium* is converted into penicillin G by replacement of the L-2-aminoadipyl sidechain with externally supplied phenylacetic acid, mediated by IPN acyl transferase (IPN AcT). In the cephalosporin-producing *Acremonium chrysogenum*, IPN is subjected to an enzymatic ring expansion.

The replacement of the traditional, chemoenzymatic process for 7-ADCA by a biotransformation, preferably in the course of the fermentation, was greatly desired [122] but far from trivial to accomplish. It should be noted that the enzymatic ring expansion step (see Fig. 8.25) is not very selective, and the productivity is modest even in natural cephalosporin producing organisms.

Attempts to perform an enzymatic ring expansion on penicillin G were not successful, due to the strict selectivity of desacetoxycephalosporinase (expandase) [127a], even when genetically modified [128]. A solution was found by feeding, instead of phenylacetic acid, adipic acid to a culture of a transgenic *P. chrysogenum* that expressed the gene for expandase, resulting in the formation of adipyl-7-ADCA (Fig. 8.26) [121, 129]. The production level of adipyl-7-ADCA was improved via site-directed mutagenesis of the expandase [130] and by DNA shuffling to relieve suspected kinetic bottlenecks [131].

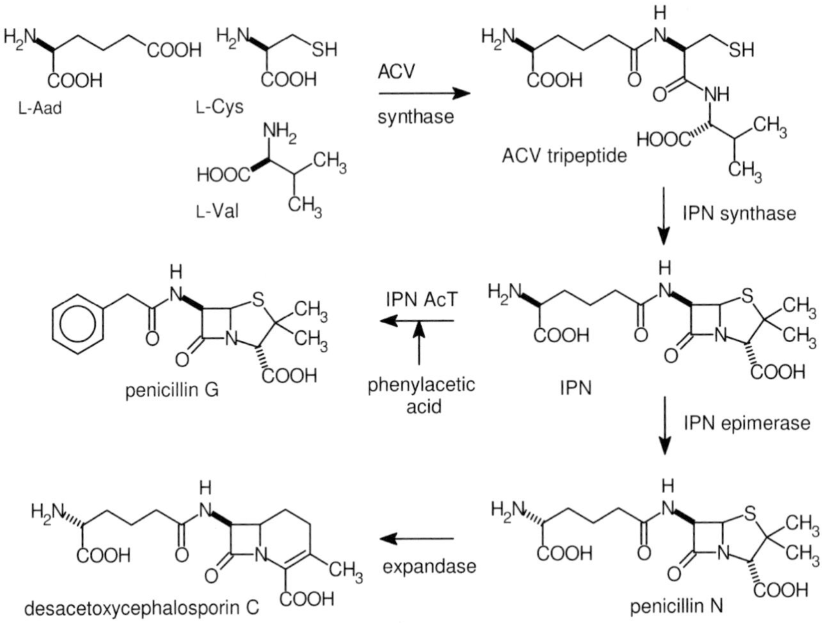

Fig. 8.25 Biosynthesis of the penicillin and cephalosporin nuclei. Compounds: Aad, L-2-aminoadipate; IPN, isopenicillin N. Enzymatic activity: IPNAcT, IPN acyltransferase.

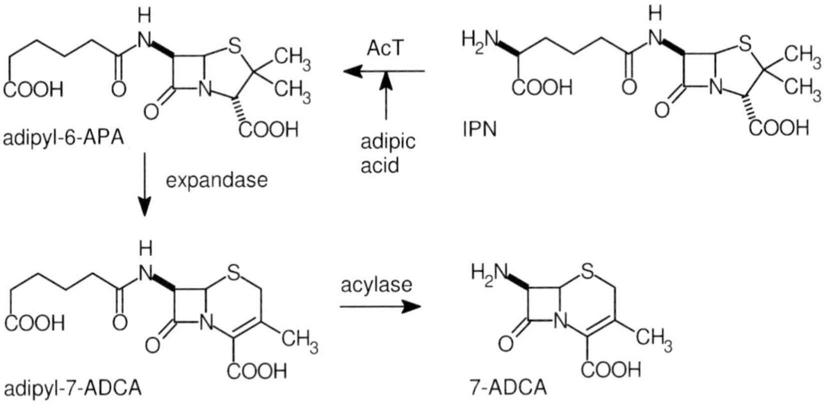

Fig. 8.26 An all-bio process for 7-ADCA: biosynthesis of adipyl-7-ADCA in a recombinant P. chrysogenum and enzymatic hydrolysis of the adipyl group.

The enzymatic ring expansion is neither complete nor selective, necessitating product isolation via chromatography in a simulated moving bed system. Because adipyl-7-ADCA rather than G-7-ADCA is produced, a new enzyme, adipyl-7-ADCA acylase, was developed to remove the side chain from adipyl-7-ADCA [132, 133], as the latter is not a substrate for penicillin G acylase.

DSM now produces green 7-ADCA in a new production facility in Delft and has closed down the installations in Delft and Matosinhos (Portugal) that employed the traditional technology. In conclusion, the key β-lactam nucleus 7-ADCA is now synthesized via biotransformations at great savings in chemicals, solvents and energy [134].

8.2.9
Riboflavin

Riboflavin (vitamin B_2) is an essential nutritional factor for humans (0.3–1.8 mg d^{-1}) and animals (1–4 mg (kg diet)$^{-1}$), who need it as a precursor for flavoproteins [135]. It is produced at a volume of approx. 3 kt a^{-1}, mainly as an animal feed additive. Approx. 300 t a^{-1} is used as a food additive and food colorant (E-101) and the remainder (500 t a^{-1}) is used in pharmaceutical applications. Major producers are Roche (Switzerland), BASF (Germany), Archer-Daniels-Midland (USA) and Takeda (Japan). Microbial and chemical production have coexisted for many years but the latter has recently been phased out [136].

For many years, riboflavin has been produced from D-ribose (Rib) and 3,4-xylidine, via the Karrer-Tishler process (see Fig. 8.27) [135], which pushed the existing fermentative procedures from the market in the late 1960s [137], although the laborious synthesis of Rib [135] was a serious drawback. Later, a fermentative process for Rib [138] was adopted. The Karrer-Tishler process is, from the green standpoint, not one of the worst as the overall yield is 60%, relatively little organic solvents are used and few auxiliary groups are discarded.

The biosynthesis of riboflavin, from the nucleotide guanosine triphosphate (GTP), requires at least six enzymatic activities and is subject to a complex regulation architecture [136, 139]. The genes that encode the enzymes have been identified and cloned [136].

The biosynthetic pathway of riboflavin in *B. subtilis* is outlined in Fig. 8.28. The starting compound GTP is, as well as the other purines, tightly regulated as these normally are not present in the cell in appreciable amounts [140]. The final step is a dismutation of DRL into riboflavin and ArP. The six enzymatic activities involved (not counting the phosphatase) are encoded by four genes, which are closely grouped on the *rib* operon [139].

The aerobic microbial production of riboflavin in various microbes dates back to the 1930s [141] but was discontinued in the late 1960s because the chemical process described above was more profitable. Fermentation soon (Merck, 1974) made a comeback, however, and subsequently has steadily eroded the position of the Karrer-Tishler process. Around 1990, production levels of 15–20 g L^{-1} had been achieved [136], up from 2 g L^{-1} in 1940 [141], causing the product to crys-

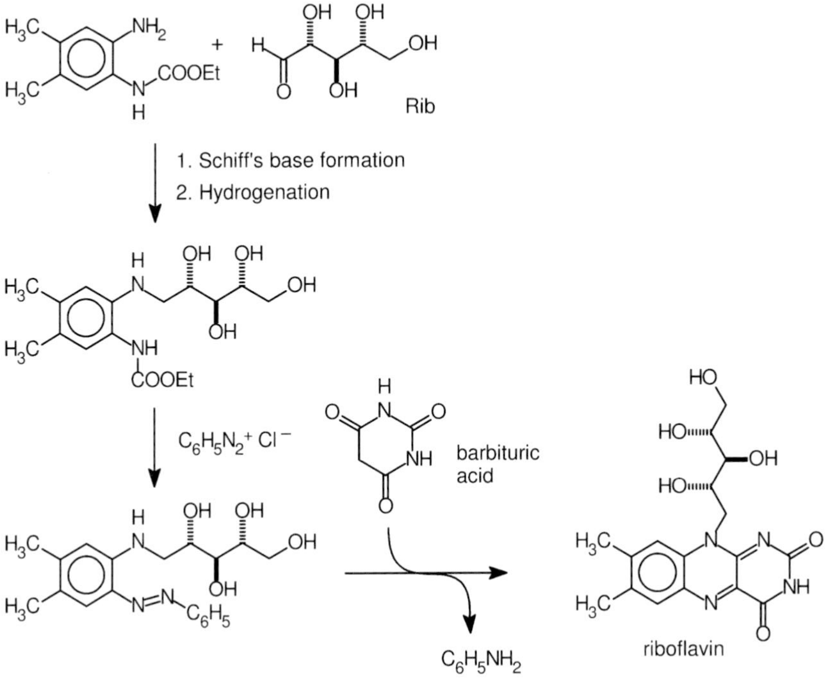

Fig. 8.27 The Karrer-Tishler industrial process for riboflavin. Abbreviation: Rib, D-ribose.

tallize from the fermentation medium. These "natural riboflavin overproducing" organisms mainly resulted from classical strain improvement as metabolic design studies have been initiated only recently [136]. BASF (Germany) started to ferment riboflavin in 1990 and closed down the chemical production in 1996 [136].

Roche (Switzerland), in collaboration with Omnigene (USA), has taken a different approach by engineering a *Bacillus subtilis*, although the latter organism is not a natural overproducer of riboflavin, via a combination of classical mutant selection and fermentation improvement with genetic engineering. The key to success was gene deregulation, replacement of native promoters by constitutive ones and increasing gene copy numbers, based on detailed knowledge of *rib* operon and its control architecture [139, 142].

Downstream processing of riboflavin from fermentation is straightforward, as the product precipitates from the fermentation broth. The crystals are collected by differential centrifuging and a pure product is obtained after repeated crystallization [143]. Roche has replaced its chemical production of riboflavin by the biotechnological process in 2000, with a 50% savings in production costs [54] as well as a 75% reduction in the use of non-renewable materials and very significantly reduced emissions into the environment [144].

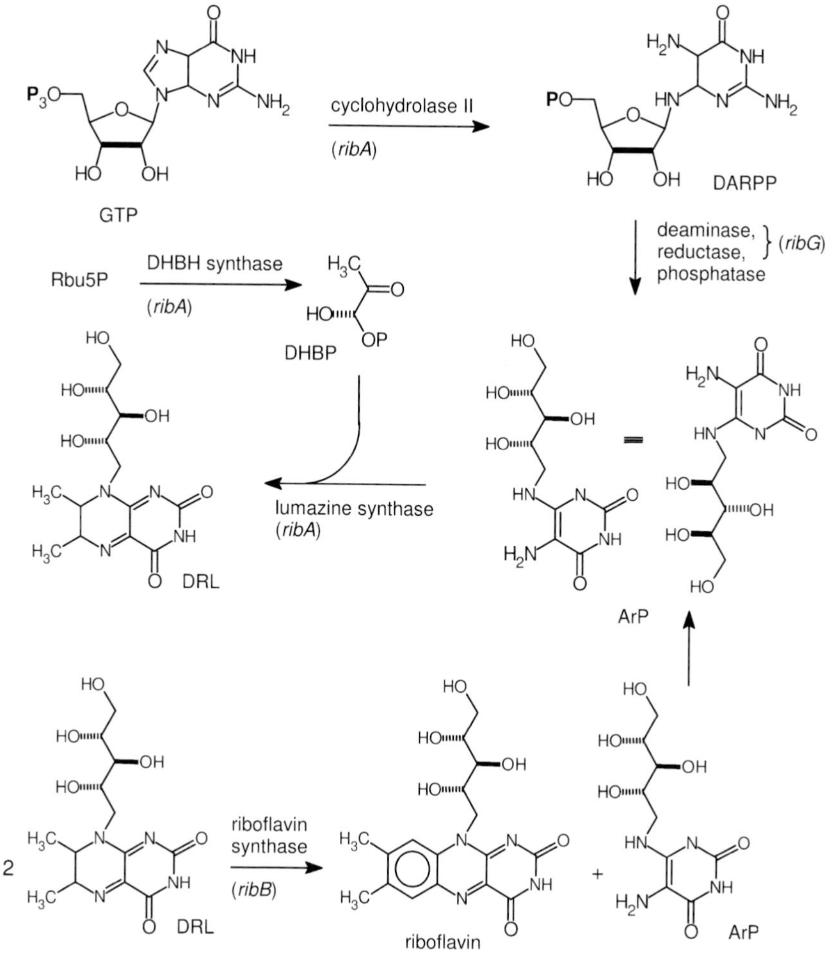

Fig. 8.28 Biosynthesis of riboflavin in bacteria. Compounds: ArP, 5-amino-6-(ribityl-amino)-2,4-(1H,3H)-pyrimidinedione; DARPP, 2,5-diamino-6-ribosylamino)-4(3H)-pyrimidinedione-5′-phosphate; DHBP, L-(S)- 3,4-dihydroxy-2-butanone-4-phosphate; DRL, 6,7-dimethyl-8-ribityllumazine; GTP, guanidine triphosphate; Rbu5P: D-ribulose-5-phosphate.

8.3
Chemical and Chemoenzymatic Transformations of Carbohydrates into Fine Chemicals and Chiral Building Blocks

We have noted in Section 8.2.1 that the structural complexity of carbohydrates is an obstacle to their application as a feedstock for chemistry. Nonetheless, the complexity of carbohydrates can be exploited in either of two ways. The first is to match the starting material to the structure of the product; L-ascorbic acid is the best-known example of this approach. The other is to break down the carbohydrate into enantiomerically pure C_3 and C_4 units that are broadly applicable building blocks.

8.3.1
Ascorbic Acid

L-Ascorbic acid (vitamin C, ASA) is produced on a scale of 80 kt a^{-1} worldwide; it is used in food supplements, pharmaceutical preparations, cosmetics, as an antioxidant in food processing and a farm animals feed supplement [145]. It is synthesized *in vivo* by plants and many animals, but not by primates, including Man, or microbes.

The industrial production of ASA has been dominated by the Reichstein-Grussner process (Fig. 8.29) [146] since the mid 1930s. Although the intermedi-

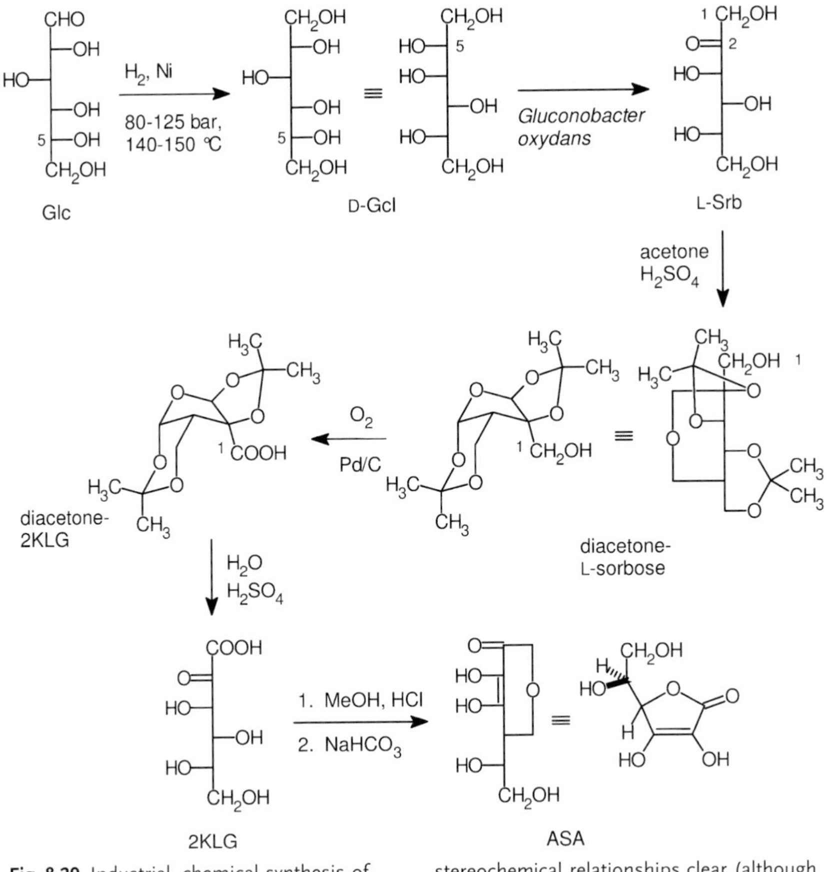

Fig. 8.29 Industrial, chemical synthesis of L-ascorbic acid. Experimental conditions have been taken from [145]. Structures (except those of the acetone derivatives) are in open-chain Fischer projection to make the stereochemical relationships clear (although such structures do not exist in aqueous solution). Compounds: ASA, L-ascorbic acid; Glc, D-glucose; D-Gcl, D-glucitol (sorbitol); 2KLG, 2-keto-L-gulonate; L-Srb, L-sorbose.

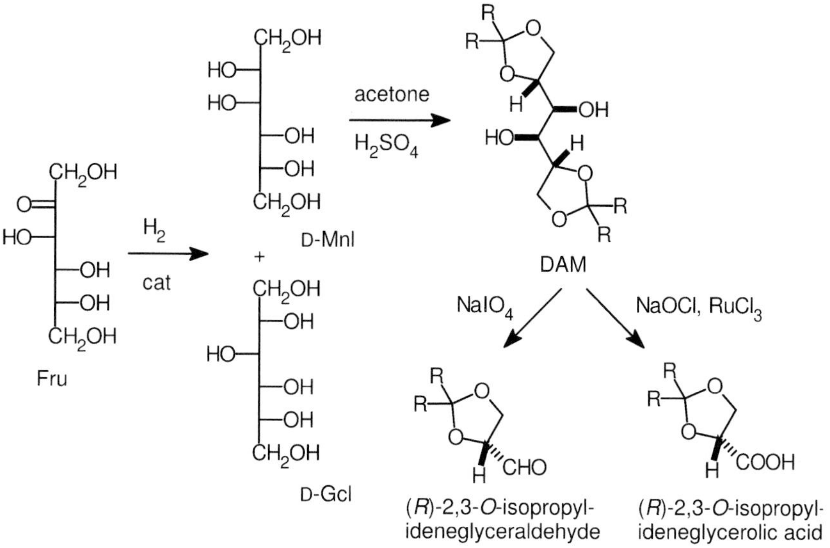

Fig. 8.32 Synthetic routes to enantiomerically pure (*R*)-C₃ building blocks. Compounds: Fru, D-fructose; D-Gcl, D-glucitol; D-Mnl, D-mannitol; DAM, 1,2:5,6-di-*O*-isopropylidene-D-mannitol.

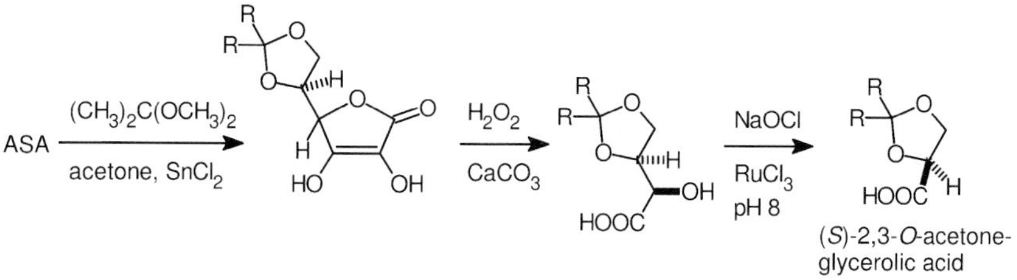

Fig. 8.33 Conversion of L-ascorbic acid into an (*S*)-glycerolic acid derivative. Compound ASA, L-ascorbic acid.

of *S*3HBL from maltose is only 33% maximum (28% in practice), because the non-reducing carbohydrate unit is sacrificed. Much higher yields, up to a theoretical 70% (w/w), can be obtained from polysaccharides, as each elimination step generates a new reducing end but, due to low concentration of the latter, the reaction will be slow. It has been suggested that maltodextrins with dp 10 would strike a good balance between yield and reaction rate [182].

The configuration at C-3 in *S*3HBL is derived from C-5 in the hexose system, which is D in all readily available oligo- and polysaccharides. Hence, the starting material for *R*3HBL is restricted to L-monosaccharides such as L-arabinose, which is available in abundance from sugar beet pulp, but requires a protection/activation strategy at C-4. It was accordingly shown that the readily availa

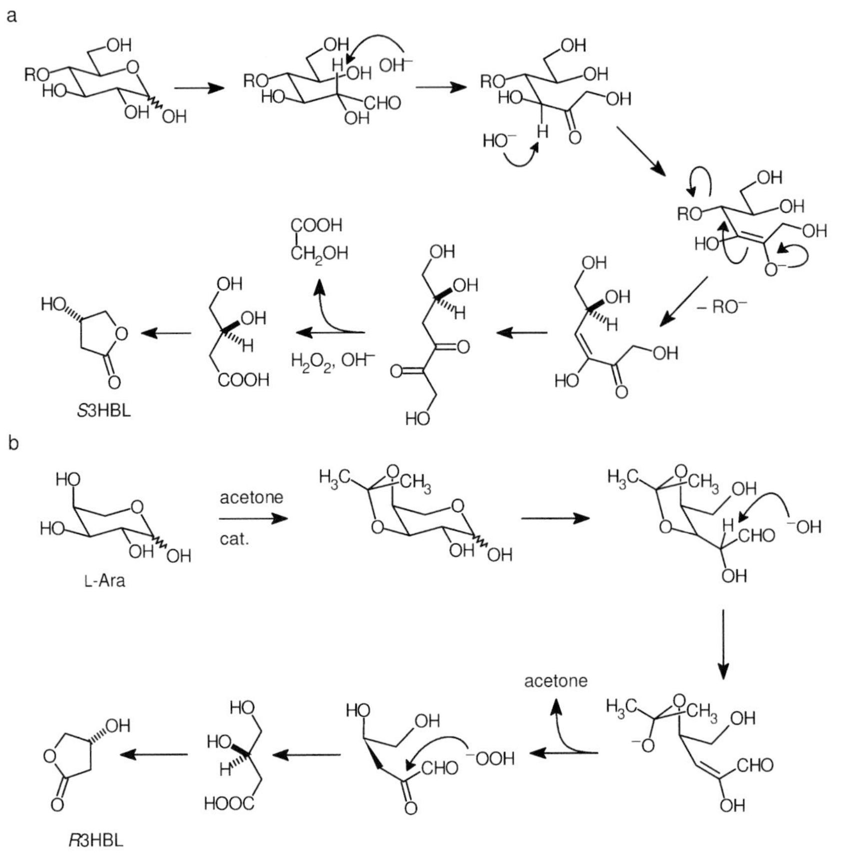

b

Fig. 8.34 Synthesis of (R)- and (S)-3-hydroxybutyrolactone from carbohydrate starting materials. Compounds: L-ARA, L-arabinose, R3HBL, (R)-3-hydroxybutyrolactone, S3HBL, (S)-3-hydroxybutyrolactone.

able derivative 3,4-O-isopropylidene-L-arabinose could be converted, via a similar sequence as described above (Fig. 8.34 b), into R3HBL in 96% yield [183].

8.3.3
5-Hydroxymethylfurfural and Levulinic Acid

It has often been proclaimed that 5-hydroxymethylfurfural (HMF, Fig. 8.35) could be an ideal cross-over compound between carbohydrates and petrochemistry [184], as it is a bifunctional heteroaromatic compound that is accessible from fructose in one step. It was expected that HMF could be developed into a valuable synthetic building block and that its derivatives, such as furan-2,5-dicarboxylic acid (FDA), would be able to compete with fossil-derived monomers for use in thermostable polyesters and polyamides.

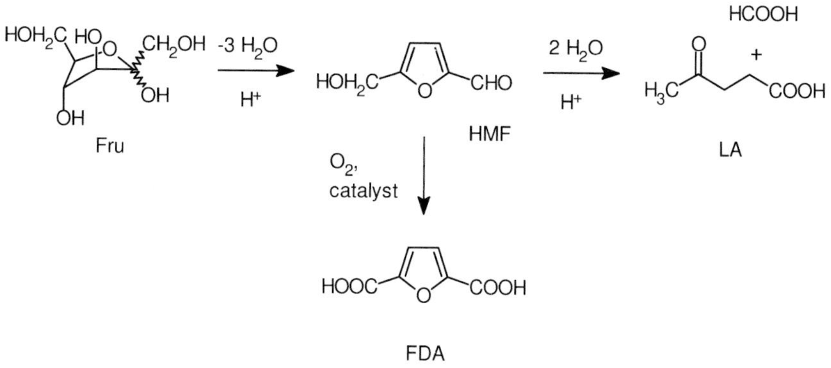

Fig. 8.35 Transformation of D-fructose into 5-hydroxymethyl-furfural and furan-2,5-dicarboxylic acid. Compounds: Fru, D-fructose; HMF, 5-hydroxymethylfurfural; FDA, furan-2,5-dicarboxylic acid, LA, levulinic acid.

Fructose, or a fructose-containing polysaccharide, can be transformed into HMF by aqueous sulfuric acid at 150 °C via a number of isomerization and dehydration steps [185]. Presumably, such a procedure was at the basis of the pilot-scale production of HMF by Südzucker (Germany) [186]. Major problems are the competing rehydration of HMF into levulinic acid (4-oxopentanoic acid, LA, see Fig. 8.35) and the formation of polymeric side-products, which necessitate chromatographic purification. Yields have stayed modest (approx. 60%), in consequence.

Various strategies, such as the use of solid acids, reaction in biphasic media or in anhydrous DMSO, have been attempted to improve the yield and selectivity [187]. Recently, HMF has been prepared in 1,2-dimethoxyethane, combined with water removal [188], in ionic liquid media [189] and in supercritical acetone [190] but the selectivity for HMF could not be improved beyond approx. 80%, which is undesirably low.

In conclusion, the selective conversion of biomass into HMF is still a formidable obstacle after 50 years of study. Due to the modest selectivity, the DSP is too elaborate for a commodity chemical and the target price of $ 2000 t^{-1} [190] is still elusive.

The interest in FDA arises from its possible application as a renewable-derived replacement for terephthalic acid in the manufacture of polyesters. A multitude of oxidation techniques has been applied to the conversion of HMF into FDA but, on account of the green aspect, platinum-catalyzed aerobic oxidation (see Fig. 8.35), which is fast and quantitative [191], is to be preferred over all other options. The deactivation of the platinum catalyst by oxygen, which is a major obstacle in large-scale applications, has been remedied by using a mixed catalyst, such as platinum–lead [192]. Integration of the latter reaction with fructose dehydration would seem attractive in view of the very limited stability of HMF, but has not yet resulted in an improved overall yield [193].

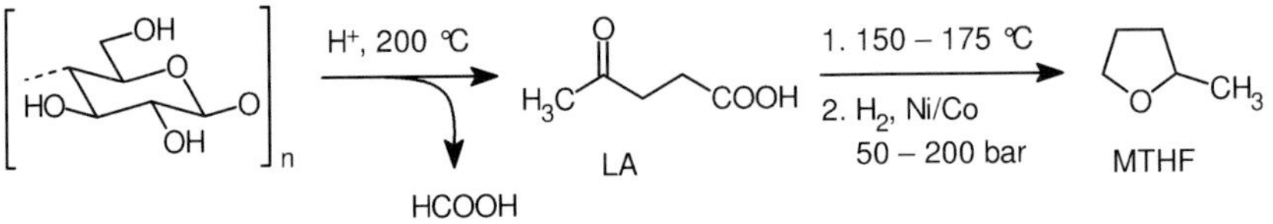

Fig. 8.36 Production of levulinic acid and transformation into 2-methyltetrahydrofuran. Compounds: LA, levulinic acid; MTHF, 2-methyltetrahydrofuran.

We note that, even if the price of HMF would drop to $ 2000 t^{-1}, FDA derived from the latter would be too expensive by a factor of eight (at least) to compete with terephthalic acid at $ 400 t^{-1}. Hence, there is little prospect that such a substitution will be realized in practice.

LA, which has already been mentioned as an undesirable side-product in the synthesis of HMF, is also a fine chemical, with a wide range of small-scale applications, that has been produced since 1870 [194].

The possible development of LA into a cross-over chemical between carbohydrates and petrochemicals has recently spurred the development of more efficient procedures for its production. Most start from cheap starting materials such as lignocellulose residues and waste paper in acidic medium at approx. 200 °C; the theoretical yield of such a procedure is 0.71 kg kg^{-1} (see Fig. 8.36). In a patent application for a two-stage procedure the claimed yields were 62–87% of the theory, depending on the raw material [195]. A much simpler, extrusion-based procedure has been described but even when fitted with a second stage the yield was not better than 66% of the theory [196]. Efficient DSP is not trivial and the chromatographic separation that has been described [197] is obviously not compatible with the aimed-for commodity status of LA. Alternatively, the LA can be esterified *in situ* [197].

One potential very large scale application of LA is its transformation, via ring closure and hydrogenation, into 2-methyltetrahydrofuran [197], a possible motor fuel additive. One of the companies involved, Biofine Inc. (now BioMetics) has received the 1999 Presidential Green Chemistry Award for its work on LA.

8.4
Fats and Oils

Naturally occurring oils and fats constitute another important source of renewable raw materials [198]. Whether they are referred to as fats or oils depends on whether they are solid or liquid at room temperature, respectively. They are composed primarily of triglycerides (triesters of glycerol) together with small amounts of free fatty acids, phospholipids, sterols, terpenes, waxes and vitamins. Oils and fats are either of vegetable or animal origin and are produced in the approximate proportions: 55% vegetable oils, 40% land-animal fats and 5% marine oils [199].

8.4.1   Biodiesel

Considerable attention is currently being focused on the use of renewable vege-table oils as feedstocks for the production of biodiesel. The latter has obvious benefits in the context of green chemistry and sustainability: (i) since it is plant-derived its use as a fuel is CO_2-neutral, (ii) it is readily biodegradable, (iii) its use results in reduced emissions of CO, SO_x, soot and particulate matter.

Vegetable oils can be used directly, as the triglycerides, to replace conventional diesel fuel but their much higher viscosities make their use rather impractical. Consequently, biodiesel generally consists of a mixture of fatty acid methyl es-ters (FAMEs), produced by transesterification of triglycerides with methanol (Fig. 8.37). It is worth noting, however, that fatty acid ethyl esters, produced from triglycerides and bioethanol, would constitute a truly green fuel.

The reaction is catalyzed by a variety of both acids and bases but simple bases such as NaOH and KOH are generally used for the industrial production of bio-diesel [200, 201]. The vegetable oil feedstock, usually soybean or rapeseed oil, needs to be free of water (<0.05%) and fatty acids (<0.5%) in order to avoid catalyst con-sumption. This presents a possible opportunity for the application of enzymatic transesterification. For example, lipases such as *Candida antarctica* B lipase have been shown to be effective catalysts for the methanolysis of triglycerides. When the immobilized form, Novozyme 435, was used it could be recycled 50 times with-out loss of activity [201, 202]. The presence of free fatty acids in the triglyceride did not affect the enzymes' performance. The methanolysis of triglycerides catalyzed by Novozyme 435 has also been successfully performed in scCO$_2$ as solvent [203].

Alkali-catalyzed transesterifications have several drawbacks in addition to the problem of free fatty acids and water in the feedstock. They are energy inten-sive, recovery of the glycerol is difficult, the basic catalyst has to be removed from the product and the alkaline waste water requires treatment. These disad-vantages could be circumvented by employing a lipase catalyst. But, in order to be economically viable, the enzyme costs have to be minimized through effec-tive immobilization and recycling.

A spin-off effect of the recent enormous increase in biodiesel production is that the coproduct, glycerol, has become a low-priced commodity chemical. Con-sequently, there is currently considerable interest in finding new applications of glycerol [204]. One possibility is to use glycerol as the feedstock for fermentative production of 1,3-propanediol (see earlier).

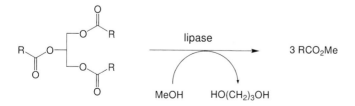

Fig. 8.37 Biodiesel production by transesterification of a triglyceride.

8.4.2
Fatty Acid Esters

Esters of fatty acids with monohydric alcohols find applications as emollients in cosmetics. They are prepared by acid- or base-catalyzed (trans)esterifications [200, 205]. As with biodiesel production, the use of enzymatic catalysis offers potential benefits but in the case of these specialty fatty acid esters there is a special advantage: the products can be labelled as 'natural'. Consequently, they command a higher price in personal care products where 'natural' is an important customer-perceived advantage. Examples include the synthesis of isopropylmyristate by CaLB-catalyzed esterification [206] and n-hexyl laurate by *Rhizomucor miehei* lipase (Lipozyme IM-77)-catalyzed esterification [207] (see Fig. 8.38).

Partial fatty acid esters of polyhydric alcohols such as glycerol, sorbitol and mono- and disaccharides constitute a group of nonionic surfactants and emulsifying agents with broad applications in pharmaceuticals, food and cosmetics. Traditionally, such materials have been prepared using mineral acids or alkali metal hydroxides as catalysts but more recently emphasis has been placed on recyclable solid acids and bases as greener alternatives [200, 205, 207]. For example, sucrose fatty acid esters are biodegradable, nontoxic, nonirritating materials that are applied as 'natural' surfactants and emulsifiers in food and personal care products. They are currently produced by base-catalyzed transesterification but lipase-catalyzed acylations are potentially greener alternatives [208, 209]. Base-catalyzed acylations require high temperatures and produce a complex mixture of mono-, di- and polyesters together with colored byproducts. In contrast, enzymatic acylation can be performed under mild conditions and affords only a mixture of two mono-esters [208, 209].

The major problem associated with the enzymatic acylation of sucrose is the incompatibility of the two reactants: sucrose and a fatty acid ester. Sucrose is hydrophilic and readily soluble in water or polar aprotic solvents such as pyridine and dimethylformamide. The former is not a feasible solvent for (trans)esterifications, for obvious thermodynamic reasons, and the latter are not suitable for the manufacture of food-grade products. The selective acylation of sucrose, as a suspension in refluxing *tert*-butanol, catalyzed by *C. antarctica* lipase B, afforded a 1:1 mixture of the 6 and 6′ sucrose monoesters (Fig. 8.39) [208]. Unfortunately, the rate was too low (35% conversion in 7 days) to be commercially useful.

$$R^1CO_2H \ + \ R^2OH \ \xrightarrow{\text{lipase}} \ R^1CO_2R^2 \ + \ H_2O$$

$$R^1 = n\text{-}C_{14}H_{29}; \ n\text{-}C_{16}H_{33}$$

$$R^2 = i\text{-}C_3H_7; \ CH_3(CH_2)_3CH(C_2H_5)CH_2\text{---}$$

Fig. 8.38 Lipase-catalyzed synthesis of fatty acid esters.

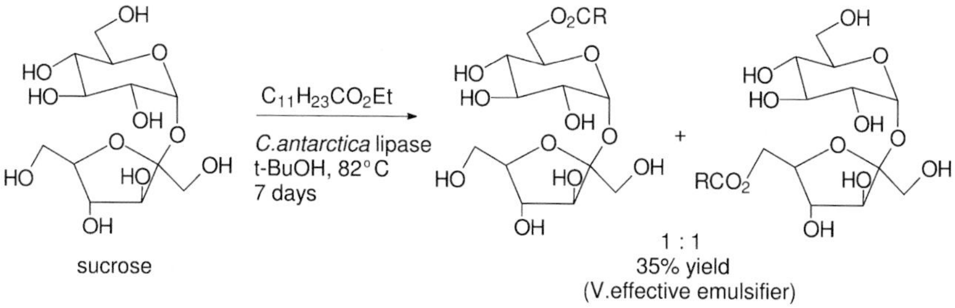

Fig. 8.39 CaLB-catalyzed acylation of sucrose.

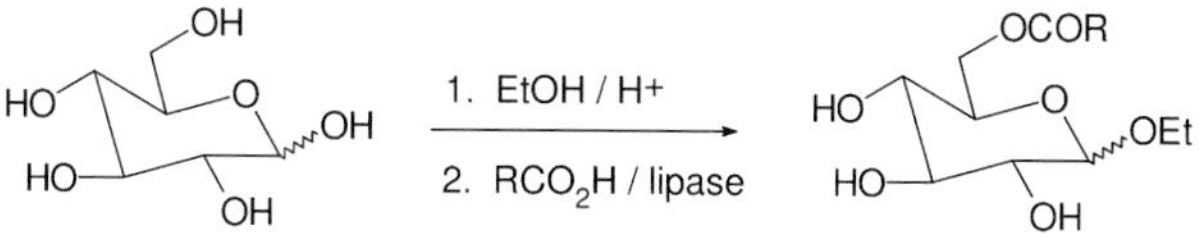

Fig. 8.40 Synthesis of alkylpolyglycosides and their fatty acid esters.

A possible solution to this problem of low rates is to use ionic liquids as solvents. Certain ionic liquids, e.g. those containing the dicyanamide anion, $(NC)_2N^-$, have been shown to dissolve sucrose in concentrations of several hundred grams per litre [210]. The lipase-catalyzed acylation of glucose with fatty acid esters has been shown to occur in a mixture of an ionic liquid, [bmim][BF$_4$], and *tert*-butanol [211].

Another interesting application involving both carbohydrate and oleochemical feedstocks is the production of alkyl polyglycosides (APGs), for use as biodegradable nonionic surfactants in laundry detergents, by acid-catalyzed reaction of a sugar, e.g. glucose, xylose, lactose, etc., with a fatty alcohol. A greener alternative to conventional mineral acids involves the use of solid acids, such as sulfonic acid resins, clays, zeolites and mesoporous aluminosilicates, as catalysts for glycosidation [212]. Further modification by lipase-catalyzed acylation with fatty acid esters affords surfactants with specialty applications, e.g. in cosmetics (Fig. 8.40).

8.5
Terpenes

A wide variety of terpenes are renewable, sustainable feedstocks for the fine chemical industry [213–215]. For example, they are an important source of ingredients and intermediates for flavors and fragrances and vitamins A and E. The major bulk terpenes are shown in Fig. 8.41.

World production of turpentine oils was ca. 330 000 tons in 1995 [213]. Its major constituents a- and β-pinene are obtained in a pure state by fractional distillation. The flavor and fragrance industry consumes around 30 000 t a^{-1} of pinenes [213].

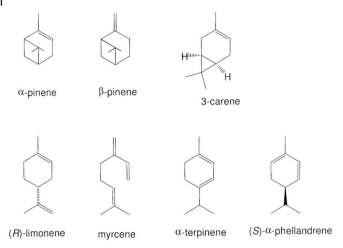

α-pinene   β-pinene

3-carene

(*R*)-limonene   myrcene   α-terpinene   (*S*)-α-phellandrene

Fig. 8.41 Major terpene feedstocks.

Another major feedstock, limonene, is derived from citrus oils, e.g. as a byproduct of orange juice production, to the extent of around $30\,000\ t\ a^{-1}$ [213].

Catalytic transformations of terpenes are well documented [213–215], comprising a wide variety of reactions: hydrogenation, dehydrogenation, oxidation, hydroformylation, carbonylation, hydration, isomerization and rearrangement, and cyclization.

For example, hydrogenation of α- and/or β-pinene, over nickel or palladium catalysts, affords *cis*-pinane. Autoxidation of *cis*-pinane (in the absence of a catalyst) gives the tertiary hydroperoxide (see Fig. 8.42) which is hydrogenated to the corresponding alcohol. Thermal rearrangement of the latter affords the important flavor and fragrance and vitamin A intermediate, linalool (Fig. 8.42) [213].

Thermal rearrangement of β-pinene affords myrcene (Fig. 8.43) which is the raw material for a variety of flavor and fragrance compounds, e.g. the Takasago process for the production of optically pure L-menthol (see Chapter 1). Dehydro-

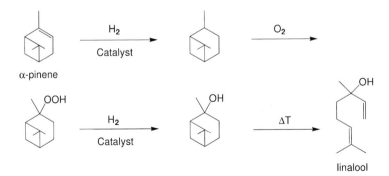

Fig. 8.42 Conversion of α-pinene to linalool

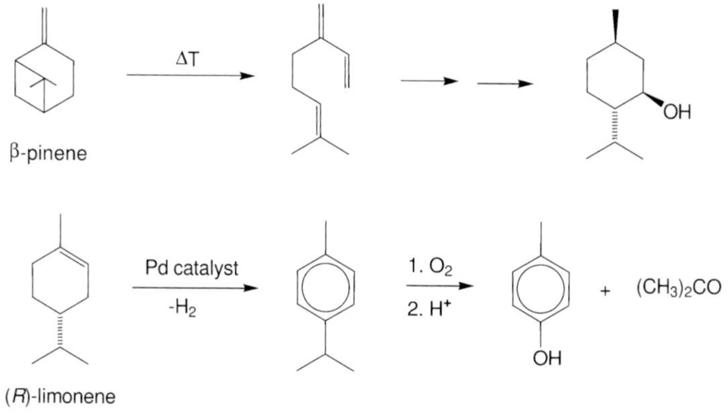

β-pinene

(R)-limonene

Fig. 8.43 Myrcene and p-cymene as key intermediates.

genation of limonene over a palladium catalyst affords p-cymene (Fig. 8.43) [213] which is the raw material for p-cresol production, via autoxidation and rearrangement, analogous to phenol production from cumene.

Terpenes can also be converted to the corresponding epoxides by reaction with hydrogen peroxide in the presence of a variety of catalysts based on e.g. tungstate, manganese and ruthenium (see Chapter 4 for a discussion of metal-catalyzed epoxidations) [213]. Lewis acid-catalyzed rearrangements of terpene epoxides lead to valuable flavor and fragrance intermediates. Such reactions were traditionally performed with e.g. ZnCl$_2$ but more recently solid Lewis acids such as Ti-beta have been shown to be excellent catalysts for these reactions (see also Chapter 2) [215]. For example, α-pinene can be converted to the corresponding epoxide which undergoes rearrangement in the presence of a Ti-beta catalyst to give the flavor and fragrance intermediate, campholenic aldehyde (Fig. 8.44) [216]. Carvone which has applications in the flavor and fragrance industry and

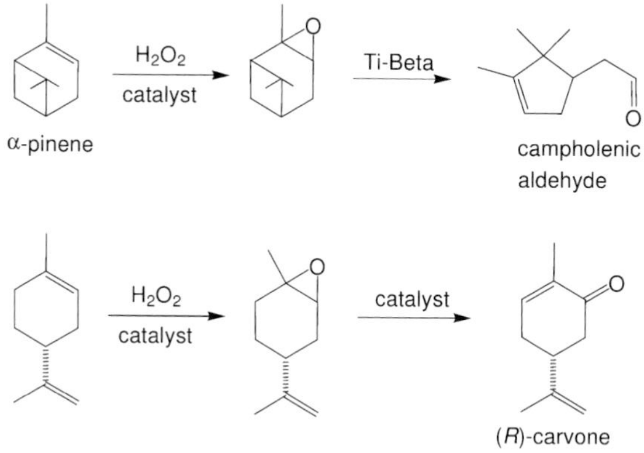

α-pinene

campholenic aldehyde

(R)-carvone

Fig. 8.44 Synthesis of campholenic aldehyde and carvone.

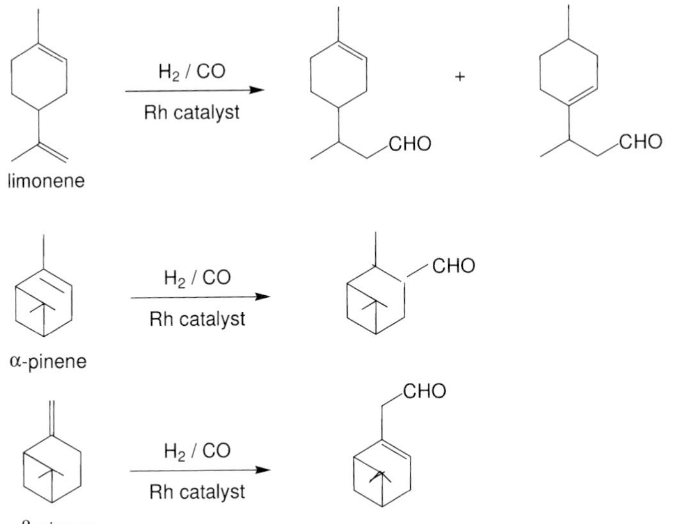

limonene

α-pinene

β-pinene

Fig. 8.45 Hydroformylation of terpenes.

as a potato sprouting inhibitor and antimicrobial agent [217], can similarly be prepared by rearrangement of limonene epoxide (Fig. 8.44) [215].

Hydroformylation of terpenes, catalyzed by rhodium complexes (see Chapter 5), is an important route to aldehydes of interest in the perfume industry [215]. Some examples are shown in Fig. 8.45.

Various condensation reactions of terpene-derived aldehydes and ketones, catalyzed by solid bases such as hydrotalcites [215–217], have been described in Chapter 3.

8.6
Renewable Raw Materials as Catalysts

Renewable resources can not only be used as raw materials for green catalytic processes in the chemical industry, but also can be employed as a source of the catalysts themselves. This is surely the ultimate in sustainability. For example, a new solid acid catalyst, for producing biodiesel, has been prepared from sulfonated burnt sugar [218]. The solid, recyclable catalyst is stable at temperatures up to 180 °C and is up to eight times more active than conventional catalysts such as Nafion (see Chapter 2). A carbohydrate, such as sucrose, starch or cellulose, is carbonized to produce polycyclic aromatic sheets which are subsequently treated with sulfuric acid to generate surface sulfonic acid groups. The resulting black amorphous powder can be processed into hard pellets or thin films for use in large-scale biodiesel production.

Another recent innovation is the use of expanded starch as a novel catalyst support [219, 220]. For example, a highly active supported palladium catalyst,

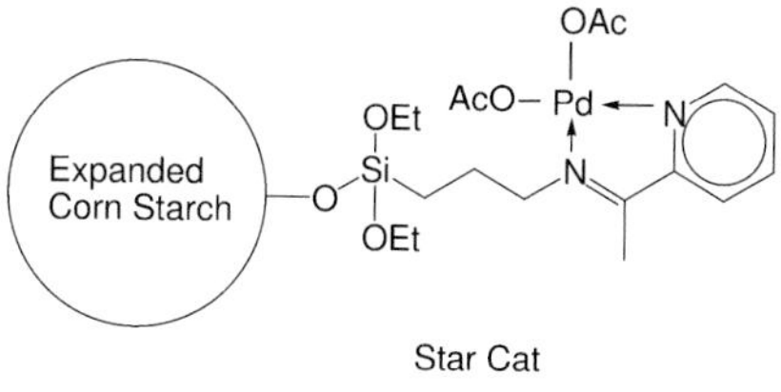

Fig. 8.46 Expanded starch-supported palladium catalyst.

Star Cat

for Heck, Suzuki and Sonagashira reactions, was produced by surface modification of expanded starch by sequential treatment with 3-aminopropyltriethoxysilane and 2-acetylpyridine and palladium acetate (Fig. 8.46). The resulting material was shown to be an effective recyclable catalyst for palladium-mediated C–C bond-forming reactions.

8.7
Green Polymers from Renewable Raw Materials

Renewable feedstocks can also be used as the raw materials for the synthesis of green, biodegradable polymers. A pertinent example is polylactate, derived from lactic acid which is produced by fermentation (see earlier). Another recent example is the production of polycarbonates by reaction of CO_2 with (R)-limonene oxide in the presence of a zinc catalyst (Fig. 8.47) [221].

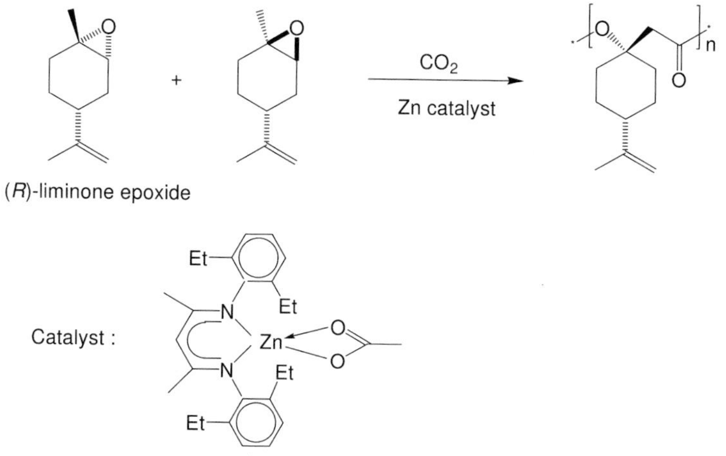

(R)-liminone epoxide

Catalyst :

Fig. 8.47 Polycarbonate production from (R)-limonene epoxide.

8.8
Concluding Remarks

The transition from a traditional fossil fuel based economy to a sustainable bio-based economy which utilizes renewable raw materials is surely one of the major scientific and technological challenges of the 21st century. It encompasses both the production of liquid fuels, e.g. bioethanol and biodiesel, bulk chemicals, polymers and specialty chemicals. Here again, the key to sustainability will be the development of economically viable, green processes involving the total spectrum of catalysis: heterogeneous, homogeneous and biocatalysis.

References

1 M. McCoy, C&EN, Dec. 5, 2003, pp. 17–18.
2 D. H. Meadows, D. I. Meadows, J. Randers, W. W. Behrens, Limits to Growth, New American Library, New York, 1972.
3 B. E. Dale, J. Chem. Technol. Biotechnol. 2003, 78, 1093.
4 C. Okkerse, H. van Bekkum, Green Chem. 1999, 1, 107.
5 J. R. Rostrup-Nielsen, Catal. Rev. 2004, 46, 247.
6 S. Doi, J. H. Clark, D. J. MacQuarrie, K. Milkowski, Chem. Commun. 2002, 2632.
7 M. Eggersdorfer, J. Meijer, P. Eckes, FEMS Microbiol. Rev. 1992, 103, 355.
8 For a recent review see: D. Klemm, B. Heublein, H.-P. Fink, A. Bohn, Angew. Chem. Int. Ed. 2005, 44, 3358–3393.
9 L. R. Lynd, C. E. Wyman, T. U. Gerngross, Biotechnol. Progr. 1999, 15, 777–793.
10 C. S. Gong, N. J. Cao, J. Du, G. T. Tsao, Adv. Biochem. Eng. Biotechnol. 1999, 65, 207–241.
11 T. W. Jeffries, Y.-S. Jin, Adv. Appl. Microbiol. 2000, 47, 221–268.
12 B. C. Saha, R. J. Bothast in: Fuels and Chemicals from Biomass, B. C. Saha, J. Woodward (Eds), ACS Symposium Series, Vol. 666, ACS, Washington, 1997, pp. 46–56.
13 M. Galbe, G. Zacchi, Appl. Microbiol. Biotechnol. 2002, 59, 618–628.
14 For a review of lignocellulose processing see: M. E. Himmel, W. S. Adney, J. O. Baker, R. Elander, J. D. MacMillan, R. A. Nieves, J. J. Sheehan, S. R. Thomas, T. B.

Vinzant, M. Zhang, in Fuels and Chemicals from Biomass, B. C. Saha, J. Woodward (Eds), ACS Symposium Series, Vol. 666, ACS, Washington, 1997, pp. 2–45.
15 OECD, The Application of Biotechnology to Industrial Sustainability, OECD Publications, Paris (Fr), 2001, pp. 133–136.
16 S. K. Ritter, Chem. Eng. News 2004 (May 31), 82, 31–34.
17 G. M. A. van Beynum, J. A. Roels, Starch Conversion Technology, Dekker, New York, 1985.
18 S. Hirano, in Ullmann's Encyclopaedia of Industrial Chemistry, B. Elvers, S. Hawkins, W. Russey, G. Schulz (Eds), VCH, Weinheim, 1993, Vol. A6, pp. 231–232.
19 Y.-X. Zhang, K. Perry, V. A. Vinci, K. Powell, W. P. C. Stemmer, S. B. del Cardayré, Nature 2002, 415, 644–646.
20 G. Stephanopoulos, Nature Biotechnol. 2002, 20, 666–668.
21 S. Y. Lee (Ed.), Metabolic Engineering, Marcel Dekker, New York, 1999.
22 G. Chotani, T. Dodge, A. Hsu, M. Kumar, R. LaDuca, D. Trimbur, W. Weyler, K. Sanford, Biochim. Biophys. Acta 2000, 1543, 434–455.
23 K. Weissermehl, H.-J. Arpe, Industrial Organic Chemistry, 4th edn, Wiley-VCH, Weinheim, 2003, pp. 193–198.
24 A. Arisidou, M. Penttilä, Curr. Opin. Biotechnol. 2000, 11, 187–198.
25 N. Kosaric, Z. Duvnjak, A. Farkas, H. Sahm, S. Bringer-Meyer, O. Goebel, D. Mayer, in Industrial Organic Chemicals: Starting Materials and Intermediates; an

Ullmann Encyclopedia, Vol. 4, Wiley-VCH, Weinheim, **1999**, pp. 2047–2147].

26 For an extensive review of the industrial fermentation of ethanol see: N. Kosaric, in *Biotechnology Vol. 6. Products of Primary Metabolism*, M. Roehr (Ed.), 2nd edn, Wiley-VCH, Weinheim, 1996, pp. 121–203.

27 R. Wooley, M. Ruth, D. Glassner, J. Sheehan, *Biotechnol. Progr.* **1999**, *15*, 794–803.

28 See, for example, [5], p. 233.

29 M. Kuyper, H. R. Harangi, A. K. Stave, A. A. Winkler, M. S. M. Jetten, W. T. A. M. de Laat, J. J. J. den Ridder, H. J. M. Op den Camp, J. P. van Dijken, J. T. Pronk, *FEMS Yeast Res.* **2003**, *4*, 69–78.

30 M. Kuyper, M. M. P. Hartog, M. J. Toirkens, M. J. H. Winkler, J. P. van Dijken, J. T. Pronk, *FEMS Yeast Res.* **2005**, *5*, 399–409; M. Kuyper, M. J. Toirkens, J. A. Diderich, A. A. Winkler, J. P. van Dijken, J. T. Pronk, *FEMS Yeast Res.* **2005**, *5*, 925–934.

31 D. Lorenz, D. Morris, *How Much Energy Does it Take to Make a Gallon of Ethanol?*, Institute for Local Self-Reliance, Minneapolis, 1995 [available from: http://www.ilsr.org].

32 OECD, *The Application of Biotechnology to Industrial Sustainability*, OECD Publications, Paris, 2001, pp. 137–141.

33 G. P. Philippidis, in *Enzymatic Conversion of Biomass for Fuels Production*, M. E. Himmel, J. O. Baker, R. P. Overend (Eds), ACS Symposium Series Vol. 566, ACS, Washington, 1994, pp. 188–217.

34 R. Datta in *Kirk-Othmer Encyclopaedia of Chemical Technology*, 4th edn, Vol. 13, J. I. Kroschwitz, M. Howe-Grant (Eds), Wiley-Interscience, New York, 1995, pp. 1042–1062.

35 H. Benninga, *A History of Lactic Acid Making*, Kluwer Academic Publishers, Dordrecht, 1990.

36 R. Datta, S.-P. Tsai, P. Bonsignore, S.-H. Moon, J. R. Frank, *FEMS Microbiol. Rev.* 1995, *16*, 221–231.

37 D. Garlotta, *J. Polym. Environ.* **2001**, *9*, 63–84.

38 (a) H. Tsuji, in *Biopolymers. Volume 4. Polyesters III, Applications and Commercial Products*, Y. Doi, A. Steinbüchel (Eds),

Wiley-VCH, Weinheim, 2002, pp. 129–177; (b) P. Gruber, M. O'Brien, in *Biopolymers. Vol. 4. Polyesters III, Applications and Commercial Products*, Y. Doi, A. Steinbüchel (Eds), Wiley-VCH, Weinheim, 2002, pp. 235–250; (c) N. Kawashima, S. Ogawa, S. Obuchi, M. Matsuo, T. Yagi in: *Biopolymers. Vol. 4. Polyesters III, Applications and Commercial Products*, Y. Doi, A. Steinbüchel (Eds), Wiley-VCH, Weinheim, 2002, pp. 251–274.

39 OECD, *The Application of Biotechnology to Industrial Sustainability*, OECD Publications, Paris, 2001, pp. 87–90.

40 Cargill-Dow, Press release April 2, 2002.

41 J. H. Litchfield, *Adv. Appl. Microbiol.* **1996**, *2*, 45-95.

42 J. Kăščák, J. Komínek, M. Roehr, in *Biotechnology Vol. 6. Products of Primary Metabolism*, M. Roehr (Ed.), Wiley-VCH, Weinheim, 1996, pp. 293–306.

43 For a functional classification of the lactic acid bacteria see 36], p. 95.

44 T. L. Carlson, E. M. Peters, Jr. (Cargill, Inc.) *WO* 99/19503, **1999**.

45 R. Patnaik, S. Loie, V. Gavrilovic, K. Perry, W. P. C. Stemmer, C. M. Ryan, S. del Cardayré, *Nature Biotechnol.* **2002**, *20*, 707–712.

46 S. Colombié, S. Dequin, J. M. Sablayrolles, *Enzyme Microb. Technol.* **2003**, *33*, 38–46.

47 B. Hause, V. Rajgarhia, P. Suominen (Cargill Dow Polymers LLC), *WO* 03/102152, **2003**.

48 Procedures for the (in-process extraction) of lactate *ion* (see [35], pp. 84–85) are not salt-free, because the subsequent recovery of lactic acid still requires an equivalent of acid.

49 A. M. Baniel, A. M. Eyal, J. Mizrahi, B. Hazan, R. R. Fisher, J. J. Kolstad, B. F. Steward (Cargill, Inc.), *WO* 95/24496, **1995**.

50 R. E. Drumright, P. R. Gruber, D. E. Henton, *Adv. Mater.* **2000**, *12*, 1841–1846.

51 P. R. Gruber, J. J. Kolstad, M. L. Iwen, R. D. Benson, R. L. Borchardt (Cargill, Inc.), *US* 5258488, **1993**.

52 Price quotations are from the Chemical Market Reporter of Aug 21, 2006.

53 C. Brossmer, D. Arntz (E.I. du Pont de Nemours and Co.), *US* 6140543, **2000**.

54 M. McCoy, *Chem. Eng. News* **1998**, June 22, *76*, 13–19.

55 C. J. Sullivan, in *Ullmann's Encyclopaedia of Industrial Chemistry*, B. Elvers, S. Hawkins, W. Russey, G. Schulz (Eds), VCH, Weinheim, 1993, Vol. A22, pp. 163–171.

56 Shell Chemical Company, Press Release May 9, 1995.

57 M. A. Murphy, B. L. Smith, A. Aguillo, K. D. Tau (Hoechst Celanese Corp.), *US* 4873378, **1989**.

58 (a) J. F. Knifton, T. G. James, K. D. Allen (Shell Oil Company), *US* 6576802 **2003**; (b) J. F. Knifton, T. G. James, K. D. Allen, P. R. Weider, J. B. Powell, L. H. Slaugh, T. Williams (Shell Oil Company), *US* 6586643 **2003**.

59 A. Tullo, *Chem. Eng. News* **2004**, *82*, 14.

60 A. Freund, *Monatsh. Chem.* **1881**, *2*, 636–641.

61 A.-P. Zeng, H. Biebl, *Advan. Biochem. Eng. Biotechnol.* **2002**, *74*, 239–259.

62 W.-D. Deckwer, *FEMS Microbiol. Rev.* **1995**, *16*, 143–149.

63 H. Biebl, A. P. Zeng, K. Menzel, W.-D. Deckwer, *Appl. Microbiol. Biotechnol.* **1998**, *50*, 24–29.

64 C. E. Nakamura, G. M. Whited, *Curr. Opin. Chem. Biol.* **2003**, *14*, 454–459.

65 M. McCoy, *Chem. Eng. News* **2003**, *81*, 17–18.

66 See: A. J. J. Straathof, S. Sie, T. T. Franco, L. A. M. van der Wielen, *Appl. Microbiol. Biotechnol.* **2005**, *67*, 727–734.

67 D. Cameron (Cargill, Inc.), lecture presented at Biotrans 2005, Delft, July 3–8, **2005**.

68 H. H. Liao, R. R. Gokarn, S. J. Gort, H. J. Jessen, O. Selifonova (Cargill, Inc.), *WO* 03/062173, **2003**.

69 A. J. Pittard in *Escherichia coli and Salmonella, Cellular and Molecular Biology*, 2nd edn, Vol. I, R. Curtiss III, J. L. Ingraham, F. C. Neidhart (Eds), ASM Press, Washington D.C., **1996**, pp. 458–484.

70 J. Bongaerts, M. Krämer, U. Müller, L. Raeven, M. Wubbolts, *Metabol. Eng.* **2001**, *3*, 289–300.

71 J. W. Frost, K. M. Draths, *Annu. Rev. Microbiol.* **1995**, *49*, 557–559.

72 A. Berry, *Trends Biotechnol.* **1996**, *14*, 250–256.

73 R. Patnaik, J. C. Liao, *Appl. Environ. Microbiol.* **1994**, *60*, 3903–3908.

74 N. J. Grinter, *Chemical Technol.* **1998**, July, 33–37.

75 S. S. Chandran, J. Yi, K. M. Draths, R. von Daeniken, W. Weber, J. W. Frost, *Biotechnol. Progr.* **2003**, *19*, 808–814.

76 M. Federspiel, R. Fischer, M. Hennig, H.-J. Mair, T. Oberhauser, G. Rimmler, T. Albiez, J. Bruhin, H. Estermann, C. Gandert, V. Göckel, S. Götzö, U. Hoffmann, G. Huber, G. Janatsch, S. Lauper, O. Röckel-Stäbler, R. Trussardi, A. G. Zwahlen, *Org. Proc. Res. Dev.* **1999**, *3*, 266–274.

77 M. Krämer, J. Bongaerts, R. Bovelberg, S. Kremer, U. Müller, S. Orf, M. Wubbolts, L. Raeven, *Metabol. Eng.* **2003**, *5*, 277–283.

78 K. M. Draths, D. R. Knop, J. W. Frost, *J. Am. Chem. Soc.* **1999**, *121*, 1603–1604.

79 D. R. Knop, K. M. Draths, S. S. Chandran, J. L. Barker, R. von Daeniken, W. Weber, J. W. Frost, *J. Am. Chem. Soc.* **2001**, *123*, 10173–10182.

80 N. Ran, D. R. Knop, K. M. Draths, J. W. Frost, *J. Am. Chem. Soc.* **2001**, *123*, 10927–10934.

81 J. M. Gibson, P. S. Thomas, J. D. Thomas, J. L. Barker, S. S. Chandran, M. K. Harrup, K. M. Draths, J. W. Frost, *Angew. Chem. Int. Ed. Engl.* **2001**, *40*, 1945–1948.

82 S. Borman, *Chem. Eng. News* **1992**, Dec. 14, *70*, 23–29.

83 K. Li, J. W. Frost, *J. Am. Chem. Soc.* **1998**, *120*, 10545–10546.

84 S. Kambourakis, K. M. Draths, J. W. Frost, *J. Am. Chem. Soc.* **2000**, *112*, 9042–9043.

85 G. Leston, in *Kirk-Othmer Encyclopedia of Chemical Technology*, 5th edn, J. I. Kroschwitz, M. Howe-Grant (Eds), Wiley-Interscience, New York, 1996, Vol. 19, p. 778.

86 A. Budzinski, *Chem. Rundschau* **2001**, *6*, 10.

87 W. Leuchtenberger, in *Biotechnology Vol. 6. Products of Primary Metabolism*, 2nd edn, M. Roehr (Ed.), VCH, Weinheim, 1996, pp. 465–502 [p. 473].

88 A. Kleemann, W. Leuchtenberger, B. Hoppe, H. Tanner, in *Industrial Organic Chemicals: Starting Materials and Intermediates; an Ullmann Encyclopedia*, Vol. 1, Wiley-VCH, Weinheim, **1999**, pp. 536–599 [pp. 558–559].

89 (a) W. E. Swann (Genex Corporation), *GB* 2127821, **1984**; (b) P. J. Vollmer, J. J. Schruben, J. P. Montgomery, H.-H. Yang (Genex Corporation), *US* 4584269, **1986**; (c) C. T. Evans, C. Choma, W. Peterson, M. Misawa, *Biotechnol. Bioeng.* **1987**, *30*, 1067–1072.

90 R. G. Eilerman in *Kirk-Othmer Encyclopedia of Chemical Technology*, 4th edn, Vol. 6, J. I. Kroschwitz, M. Jowe-Grant (Eds), Wiley-Interscience, New York, **1993**, pp. 344–353.

91 C. Förberg, T. Eliaeson, L. Häggström, *J. Biotechnol.* **1988**, *7*, 319–332.

92 K. Backmann, M. J. O'Connor, A. Maruya, E. Rudd, D. McKay, V. DiPasquantonio, D. Shoda, R. Hatch, K. Venkasubramanian, *Ann. N. Y. Acad. Sci.* **1990**, *589*, 16–24.

93 M. Ikeda, *Adv. Biochem. Eng./Biotechnol.* **2003**, *79*, 1–35.

94 N. Rüffer, U. Heidersdorf, I. Kretzers, G. A. Sprenger, L. Raeven, R. Takors, *Bioprocess Biosyst. Eng.* **2004**, *26*, 239–248.

95 W. Leuchtenberger, in: *Biotechnology Vol. 6. Products of Primary Metabolism*, 2nd edn, M. Roehr (Ed.), VCH, Weinheim, 1996, pp. 465–502 [p. 484].

96 J. W. Frost, J. Lievense, *New J. Chem.* **1994**, *18*, 341–348 and pertinent references cited therein.

97 R. Katsumata, M. Ikeda, *Biotechnology* **1993**, *11*, 921–925.

98 M. Ikeda, K. Nakanishi, K. Kino, R. Katsumata, *Biosci. Biotechnol. Biochem.* **1994**, *58*, 674–678.

99 M. Ikeda, R. Katsumata, *Appl. Environ. Microbiol.* **1999**, *65*, 2497–2509.

100 T. C. Dodge, J. M. Gerstner, *J. Chem. Technol. Biotechnol.* **2002**, *77*, 1238–1245.

101 M. Seefelder, *Indigo, Kultur, Wissenschaft und Technik*, 2nd edn, Ecomed, Landsberg, 1994.

102 S. Cotson, S. J. Holt, *Proc. R. Soc. (London), Ser. B* **1958**, *148*, 506–519.

103 E. Steingruber, in *Ullmann's Encyclopedia of Industrial Chemistry*, 5th edn, Vol. A14, B. Elvers, S. Hawkins, M. Ravenscroft, G. Schulz (Eds), VCH, Weinheim, 1993, pp. 149–156.

104 The SPINDIGO project [http://www.spindigo.net/].

105 P. H. H. Gray, *Proc. R. Soc. (London)* **1928**, *B102*, 263-280.

106 B. D. Ensley, B. J. Ratzkin, T. D. Osslund, M. J. Simon, L. P. Wackett, D. T. Gibson, *Science* **1983**, *222*, 167–169.

107 A. Berry, T. C. Dodge, M. Pepsin, W. Weyler, *J. Ind. Microbiol. Biotechnol.* **2002**, *28*, 127–133.

108 H. Bialy, *Nat. Biotechnol.* **1997**, *15*, 110.

109 M. G. Wubbolts, C. Bucke, S. Nielecki, in *Applied Biocatalysis*, 2nd edn, A. J. J. Straathof, P. Adlercreutz (Eds), Harwood Academic, Australia, 2000, pp. 153–211 [p. 178].

110 T. R. Rawalpally in *Kirk-Othmer Encyclopedia of Chemical Technology*, 4th edn, Vol. 15, J. I. Kroschwitz, M. Howe-Grant (Eds), Wiley-Interscience, New York, 1998, pp. 99–116.

111 J. F. Verbeek (Thompson-Hayward Chem. Co.), *US* 3935216, **1976**.

112 S. Jackowski in Escherichia coli *and* Salmonella, *Cellular and Molecular Biology*, 2nd edn, Vol. I, R. Curtiss III, J. L. Ingraham, F. C. Neidhart (Eds), ASM Press, Washington D.C., **1996**, pp. 465–477.

113 A. T. Hüser, C. Cassagnole, N. D. Lindley, M. Merkamm, A. Guyonvarch, V. Elišáková, M. Pátek, J. Kalinowski, I. Brune, A. Pühler, A. Tauch, *Appl. Environm. Microbiol.* **2005**, *71*, 3255–3268.

114 (a) M. K. B. Berlyn, K. B. Low, K. E. Rudd in Escherichia coli *and* Salmonella, *Cellular and Molecular Biology*, 2nd edn, Vol. I, R. Curtiss III, J. L. Ingraham, F. C. Neidhart (Eds), ASM Press, Washington D.C., **1996**, pp. 1715–1902 (p. 1787); (b) F. Elischewski, A. Pühler, J. Kalinowski, *J. Biotechnol.* **1999**, *75*, 135–146.

115 Y. Hikichi, T. Moriya, H. Miki, T. Yamaguchi (Takeda Chemical Industries, Ltd), *EP* 0590857, **1994**.

116 T. Moriya, Y. Hikichi, Y. Moriya, T. Yamaguchi (Takeda Chemical Industries, Ltd), *WO* 97/10340, **1997**.

117 T. Hermann, A. Marx, W. Pfefferle, M. Rieping (Degussa AG), *WO* 02/064806, **2002**.

118 (a) D. Kruse, G. Thierbach (Degussa AG), *WO* 03/004673, **2003**; (b) D. Kruse, G. Thierbach (Degussa AG), *WO* 03/006664, **2003**.

119 (a) N. Dusch, G. Thierbach (Degussa-Hüls AG; Institut für Innovation an der Universität Bielefeld GmbH), *EP* 1083225, **2001**; (b) N. Dusch, G. Thierbach (Degussa AG), *EP* 1167520, **2002**; (c) L. Eggeling, H. Sahm (Forschungszentrum Jülich GmbH), *WO* 02/055711, **2002**; (d) N. Dusch, A. Marx, W. Pfefferle, G. Thierbach (Degussa AG), *EP* 1247868, **2002**.

120 J. Thykaer, J. Nielsen, *Metab. Eng.* **2003**, 5, 56–69.

121 For a review see: M.A. Wegman, M.H.A. Janssen, F. van Rantwijk, R.A. Sheldon, *Adv. Synth. Catal.* **2001**, 343, 559–576.

122 E.J.A.X. van de Sandt, E. de Vroom, *Chimica Oggi* **2000**, 18, 72–75.

123 R.B. Morin, B.G. Jackson, R.A. Mueller, E.R. Lavagnino, W.B. Scanlon, S.L. Andrews, *J. Am. Chem. Soc.* **1963**, 85, 1896–1897.

124 For reviews see: (a) R.D.G. Cooper, L.D. Hatfield, D.O. Spry, *Acc. Chem. Res.* **1973**, 6, 32–40; (b) P.G. Sammes, *Chem. Rev.* **1976**, 76, 113–155.

125 (a) J.J. de Koning, H.J. Kooreman, H.S. Tan, J. Verweij, *J. Org. Chem.* **1975**, 40, 1346–1347; (b) J. Verweij, H.S. Tan, H.J. Kooreman (Gist-Brocades N.V.), *US* 4003894, **1977**.

126 A 97% yield has erroneously been cited in: (a) C.A. Bunnell, W.D. Luke, F.M. Perry, Jr. in *Beta-Lactam Antibiotics for Clinical use*, S.F. Queener, J.A. Webber, S.W. Queener (Eds), Marcel Dekker, New York, 1986, pp. 255–283 [p. 261]; (b) J. Roberts in: *Kirk-Othmer Encyclopedia of Chemical Technology*, 4th edn, Vol. 3, J.I. Kroschwitz, M. Howe-Grant (Eds), Wiley-Interscience, New York, 1992, pp. 28–82 [p. 57].

127 For reviews see: (a) J.E. Baldwin, E. Abraham, *Nat. Prod. Rep.* **1988**, 5, 129–145; (b) J.F. Martin, *Appl. Microb. Biotechnol.* **1998**, 50, 1–15.

128 C.A. Cantwell, R.J. Beckmann, J.E. Dotzlaf, D.L. Fisher, P.L. Skatrud,

W.-K. Yeh, S.W. Queener, *Curr. Genet.* **1990**, 17, 213–221.

129 (a) M.J. Conder, L. Crawford, P.C. McAda, J.A. Rambosek (Merck & Co, Inc.), *EP* 532341, **1992**; (b) L. Crawford, A.M. Stepan, P.C. McAda, J.A. Rambosek, M.J. Conder, V.A. Vinci, C.D. Reeves, *Biotechnology* **1995**, 13, 58–62.

130 M. Nieboer, R.A.L. Bovenberg (DSM N.V.), *WO* 99/60102, **1999**.

131 Press release by Codexis, Inc. in *Chem. Eng. News* **2003**, Nov. 10, 45, 4.

132 (a) C.F. Sio, A.M. Riemens, J.-M. van der Laan, R.M.D. Verhaert, W.J. Quax, *Eur. J. Biochem.* **2002**, 269, 4495–4504; (b) L.G. Otten, C.J. Sio, J. Vrielink, R.H. Cool, W.J. Quax, *J. Biol. Chem.* **2002**, 277, 42121–42127.

133 For a process design study of adipyl-7-ADCA hydrolysis see: C.G.P.H. Schroën, S. van de Wiel, P.J. Kroon, E. de Vroom, A.E.M. Janssen, J. Tramper, *Biotechnol. Bioeng.* **2000**, 70, 654–661.

134 OECD, *The Application of Biotechnology to Industrial Sustainability*, OECD, 2001, pp. 59–62.

135 R. Kurth, W. Paust, W. Hähnlein in: *Ullmann's Encyclopedia of Industrial Chemistry*, 5th edn, Vol. A27, B. Elvers, S. Hawkins (Eds), VCH, Weinheim, 1996, pp. 521–530.

136 K.-P. Stahmann, J.L. Revuelta, H. Seulberger, *Appl. Microbiol. Biotechnol.* **2000**, 53, 509–516.

137 B.D. Lago, L. Kaplan in: *Advances in Biotechnology Vol. III. Fermentation Products*, C. Vezina, K. Singh (Eds), Pergamon, Toronto, 1981, pp. 241–246.

138 P. de Wulf, E.J. Vandamme, *Adv. Appl. Microbiol.* **1997**, 44, 167–214.

139 J.B. Perkins, A. Stoma, T. Hermann, K. Theriault, E. Zachgo, T. Erdenberger, N. Hannett, N.P. Chatterjee, V. Williams II, G.A. Rufo, Jr., R. Hatch, J. Pero, *J. Ind. Microbiol. Biotechnol.* **1999**, 22, 8–18 and references cited therein.

140 H. Zalkin, P. Nygaard in Escherichia coli and Salmonella, *Cellular and Molecular Biology*, 2nd edn, Vol. I, R. Curtiss III, J.L. Ingraham, F.C. Neidhart (Eds), ASM Press, Washington D.C., **1996**, pp. 561–579

141 D. Perlman in: *Microbial Technology.*
Vol. I. Microbial Processes, H. J. Peppler,
D. Perlmann (Eds), Academic Press,
New York, 1979, pp. 521–627.

142 M. Hümbelin, V. Griessner, T. Keller,
W. Schurter, M. Haiker, H.-P. Hoh-
mann, H. Ritz, G. Richter, A. Bacher,
A. P. G. M. van Loon, *J. Ind. Microbiol.
Biotechnol.* **1999**, *22*, 1–7.

143 W. Bretzel, W. Schurter, B. Ludwig, E.
Kupfer, S. Doswald, M. Pfister,
A. P. G. M. van Loon, *J. Ind. Microbiol.
Biotechnol.* **1999**, *22*, 19–26.

144 OECD, *The Application of Biotechnology
to Industrial Sustainability*, OECD, 2001,
pp. 51–53.

145 B. Oster, U. Fechtel in: *Ullmann's Ency-
clopedia of Industrial Chemistry*, 5th edn,
Vol. A27, B. Elvers, S. Hawkins (Eds),
VCH, Weinheim, 1996, pp. 547–559.

146 T. Reichstein, A. Grussner, *Helv. Chim.
Acta* **1934**, *17*, 311–328.

147 T. C. Crawford, S. A. Crawford, *Advan.
Carbohydr. Chem. Biochem.* **1980**, *37*,
79.

148 V. Delić, D. Šunić, D. Vlašić in: *Biotechnology of Vitamins, Pigments and
Growth Factors*, E. J. Vandamme (Ed.),
Elsevier Applied Science, London,
1989, pp. 299–334.

149 R. D. Hancock, R. Viola, *Trends Biotechnol.* **2002**, *20*, 299–305.

150 The trivial name of ᴅ-glucitol is sorbitol, without a configuration prefix. The
often used name ᴅ-sorbitol is completely erroneous, as the compound is
derived from ʟ-, not ᴅ-sorbose. We note
that deriving the name of a sugar alcohol from a ketosugar is ambiguous
anyway; hydrogenation of ʟ-sorbose, for
example, would yield a mixture of ʟ-gu-
litol and ʟ-iditol.

151 J.-C. M.-P. G. de Troostembergh, I. A.
Debone, W. R. Obyn, C. G. M. Peuzet
(Cerestar Holding B.V.), *WO 03/
016508*, **2003**.

152 Synthetic organic chemists of a pre-
vious generation felt that Reichstein
had cheated by including a microbial
transformation.

153 Green zealots often object to the hydrogenation step, because of the elevated
temperature and pressure. The authors

wonder which procedure consumes
more energy: a hydrogenation at 140 °C
and 100 bar that is finished in a few h,
or a fermentation in a 10 times larger
vessel (because of the modest STY)
that runs for 5 days at 40 °C.

154 The survival of the Reichstein process
has been prolonged further by a price-
fixing cartel of the major producers of
ASA, see: Competition Commission,
*BASF AG and Takeda Chemical Indus-
tries Ltd: a Report on the Acquisition by
BASF AG of Certain Assets of Takeda
Chemical Industries Ltd*, London, 2001
[available from: http://www.competi-
tion-commission.org.uk].

155 W. Ning, Z. Tao, C. Wang, S. Wang, Z.
Yan, Q. Yin (Institute of Microbiology,
Academia Sinica), *EP 0278447*, **1988**.

156 See [149], p. 164 and [143], p. 311.

157 See [149], p. 164.

158 Recent Chinese work in this field un-
fortunately is not readily accessible,
see: (a) B. Zhou, Y. Li, Y. Liu, Z. Zang,
K. Zhu, D. Liao, Y. Gao, *Yingyong
Shengtai Xuebao* **2003**, *13*, 1452–1454
[*Chem. Abstr.* **2003**, *138*, 302693]; (b) W.
Lin, Q. Ye, C. Qiao, G. Yin, *Weisheng-
wuxue Tongbao* **2002**, *29*, 37–41 [*Chem.
Abstr.* **2003**, *138*, 336468].

159 H. Liaw, R. Kowzic, J. M. Eddington,
Y. Yang (Archer-Daniels-Midland Co.),
WO 01/83798, **2001**.

160 See document claimed in [149], p. 169.

161 T. Shibata, C. Ichikawa, M. Matsuura,
Y. Takata, Y. Noguchi, Y. Saito, M. Yamashita, *J. Biosci. Bioeng.* **2000**, *90*,
223–225.

162 R. A. Lazarus, J. L. Seymour, R. K. Staf-
ford, C. B. Marks, S. Anderson in: *Ge-
netics and Molecular Biology of Industrial
Microorganisms*, C. L. Hershberger, S. W.
Queener, G. Hegeman (Eds), American
Society for Microbiology, Washington,
D.C., 1989, pp. 187–193.

163 Y. Saito, Y. Ishii, H. Hayashi, K. Yoshikawa, Y. Noguchi, S. Yoshida, S. Soeda,
M. Yoshida, *Biotechnol. Bioeng.* **1998**,
58, 309–315.

164 T. Soyonama, H. Tani, K. Matsuda, B.
Kageyama, M. Tanimoto, K. Kobayashi,
S. Yagi, H. Kyotani, K. Mitsushima,

Appl. Environ. Microbiol. **1982**, *43*, 1064–1069.

165 (a) S. Anderson, C. B. Marks, R. Lazarus, J. Miller, K. Stafford, J. Seymour, D. Light, W. Rastetter, D. Estell, *Science* **1985**, *230*, 144–149; (b) T. Soyonama, B. Kageyama, S. Yagi, K. Mitsushima, *Agric. Biol. Chem.* **1987**, *51*, 3039–3047.

166 T. Dodge, F. Valle (Genencor International, Inc.), *WO 02/081140*, **2002**.

167 M. Kumar (Genencor International, Inc. and Microgenomics, Inc.), *WO 02/12528*, **2002**.

168 M. G. Boston, B. A. Swanson (Genencor International, Inc.), *US 2002/0177198*, **2002**.

169 *Chem. Eng. News* **1999**, Aug. 23, *77*, 15.

170 K. M. Moore, A. Sanborn (Archer-Daniels-Midland Co.), *WO 01/09074*, **2001**.

171 N. A. Collins, M. R. Shelton, G. W. Tindall, S. T. Perri, R. S. O'Meadhra, C. W. Sink, B. K. Arumugam, J. C. Hubbs (Eastman Chemical Co.), *WO 01/66508*, **2001**.

172 M. Rauls, H. Voss, T. Faust, T. Domschke, M. Merger (BASF AG), *WO 03/033448*, **2003**.

173 B. K. Arumugam, L. W. Blair, W. B. Brendan, N. A. Collins, D. A. Larkin, S. T. Perri, C. W. Sink (Eastman Chemical Co.), *US 6518454*, **2003**.

174 B. K. Arumugam, S. T. Perri, E. B. Mackenzie, L. W. Blair, J. R. Zoeller (Eastman Chemical Co.), *US 6476239*, **2002**.

175 See also: (a) B. K. Arumugam, N. Collins, T. Macias, S. Perri, J. Powell, C. Sink, M. Cushman (Eastman Chemical Co.) *WO 02/051826*, **2002**; (b) J. R. Zoeller, A. L. Crain (Eastman Chemical Co.) *WO 03/018569*, **2003**.

176 M. Kumar (Genencor International, Inc.) *US 2002/0090689*, **2002**.

177 R. I. Hollingsworth, G. Wang, *Chem. Rev.* **2000**, *100*, 4267–4282.

178 (a) R. Vogel in: *Ullmann's Encyclopedia of Industrial Chemistry*, 5th edn, Vol. A25, B. Elvers, S. Hawkins, W. Russey (Eds), VCH, Weinheim, 1996, pp. 418–423 [420]; (b) E. Schwartz in: *Ullmann's Encyclopedia of Industrial Chemistry*, 5th edn, Vol. A25, B. Elvers, S. Hawkins, W. Russey (Eds), VCH, Weinheim, 1996, pp. 423–426.

179 C. H. H. Emons, B. F. M. Kuster, J. A. J. M. Vekemans, R. A. Sheldon, *Tetrahedron: Asym.* **1991**, *2*, 359–362.

180 G. Huang, R. I. Hollingsworth, *Tetrahedron* **1998**, *54*, 1355–1360.

181 (a) R. I. Hollingsworth, G. Wang, *Chimica Oggi* **2002** (Sept.), 39–42; (b) R. I. Hollingsworth, *FarmaChem.* **2002** (March), 20–23.

182 R. I. Hollingsworth, *US 5292939*, **1994**.

183 R. I. Hollingsworth, *Org. Chem.* **1999**, *64*, 7633–7634.

184 (a) H. Schiweck, M. Munir, K. M. Rapp, B. Schneider, M. Vogel, in *Carbohydrates as Organic Raw Materials*, F. W. Lichtenthaler (Eds), VCH, Weinheim, 1991, pp. 57–94 [80–81]; (b) K. M. Rapp, J. Daub, in *Nachwachsende Rohstoffe: Perspektiven für die Chemie*, M. Eggersdorf, S. Warwel, G. Wulff (Eds), VCH, Weinheim, 1993, pp. 183–196.

185 (a) M. J. Anal, W. S. L. Mok, G. N. Richards, *Carbohydr. Res.* **1990**, *199*, 91–109; (b) A. T. W. J. de Goede, F. van Rantwijk, H. van Bekkum, *Starch* **1995**, *47*, 233–237.

186 K. M. Rapp. (Süddeutsche Zucker AG), *EP 0230250*, **1987**.

187 B. F. M. Kuster, *Starch* **1990**, *42*, 314–321.

188 E. J. M. Mombarg, *Catalytic Modifications of Carbohydrates*, Ch. 6., Thesis Delft University of Technology, 1997.

189 C. Lansalot-Matras, C. Moreau, *Catal. Commun.* **2003**, *4*, 517–520.

190 M. Bicker, J. Hirth, H. Vogel, *Green Chem.* **2003**, *5*, 280–284.

191 P. Vinke, H. E. van Dam, H. van Bekkum, *Stud. Surf. Sci. Catal.* **1990**, *55*, 147.

192 A. Gaset, L. Rigal, M. Delmas, N. Merat, P. Verdeguer (Furchim), *FR 2669634*, **1992**.

193 M. Kröger, U. Prüße, K.-D. Vorlop, *Top. Catal.* **2000**, *13*, 237–242.

194 F. D. Klinger, W. Ebertz, Oxocarboxylic Acids in: *Ullmann's Encyclopaedia of Industrial Chemistry*, B. Elvers, S. Hawkins, W. Russey, G. Schulz (Eds), VCH, Weinheim, 1993, Vol. A18, pp. 313–319.

195 S. W. Fitzpatrick (Biofine, Inc.), *US* 5608105, **1997**.

196 J. Y. Cha, M. A. Hana, *Ind. Crops Prod.* **2002**, *16*, 109–118.

197 W. A. Farone, J. E. Cuzens (Arkenol, Inc.), *US* 6054611, **2000**.

198 U. Biermann, W. Friedt, S. Lang, W. Lühs, G. Machmüller, J. O. Metzger, M. Rüschgen-Klaas, H. J. Schäfer, M. P. Schneider, *Angew. Chem. Int. Ed.* **2002**, *39*, 2206–2224.

199 A. Thomas, *Ullmann's Encyclopedia of Industrial Chemistry*, 5th edn, Vol. A10, VCH, Weinheim, 1987, pp. 173–243.

200 H. E. Hoydonckx, D. E. DeVos, S. A. Chavan, P. A. Jacobs, *Top. Catal.* **2004**, *27*, 83–96.

201 H. Fukuda, A. Kondo, H. Noda, *J. Biosci. Bioeng.* **2001**, *92*, 405–416.

202 T. Samukawa, M. Kaieda, T. Matsumoto, K. Ban, A. Kondo, Y. Shimada, H. Noda, H. Fukuda, *J. Biosci. Bioeng.* **2000**, *90*, 180.

203 M. A. Jackson, J. W. King, *J. Am. Oil Chem. Soc.* **1996**, *73*, 353.

204 See, for example, S. Carrettin, P. McMorn, P. Johnston, K. Griffin, C. J. Kiely, G. A. Attard, G. J. Hutchings, *Top. Catal.* **2004**, *27*, 131–136; J. Barrault, J. M. Clacens, Y. Pouilloux, *Top. Catal.* **2004**, *27*, 137–142.

205 P. Bondioli, *Top. Catal.* **2004**, *27*, 77–82.

206 O. Kirk, M. W. Christensen, *Org. Proc. Res. Dev.* **2002**, *6*, 446–451.

207 C. Marquez-Alvarez, E. Sastre, J. Perez-Pariente, *Top. Catal.* **2004**, *27*, 105–117.

208 M. Woudenberg-van Oosterom, F. van Rantwijk, R. A. Sheldon, *Biotechnol. Bioeng.* **1996**, *49*, 328–333.

209 D. Reyes-Duarte, N. Lopez-Cortes, M. Ferrer, F. J. Plou, A. Ballesteros, *Biocat. Biotrans.* **2005**, *23*, 19–27.

210 Q. Liu, M. H. A. Janssen, F. van Rantwijk, R. A. Sheldon, *Org. Lett.* **2005**, *7*, 327–329.

211 F. Ganske, U. T. Bornscheuer, *J. Mol. Catal. B: Enzymatic* **2005**, *36*, 40–42.

212 A. Corma, in *Fine Chemicals through Heterogeneous Catalysis*, R. A. Sheldon, H. van Bekkum (Eds), Wiley-VCH, Weinheim, 2002, pp. 257–274.

213 K. A. D. Swift, *Top. Catal.* **2004**, *27*, 143–155.

214 N. Ravasio, F. Zaccheria, M. Guidotti, R. Psaro, *Top. Catal.* **2004**, *27*, 157–168.

215 J. L. F. Monteiro, C. O. Veloso, *Top. Catal.* **2004**, *27*, 169–180.

216 P. J. Kunkeler, J. C. van der Waal, J. Bremmer, B. J. Zuurdeeg, R. S. Downing, H. van Bekkum, *Catal. Lett.* **1998**, *53*, 135.

217 C. C. C. R. de Carvalho, M. M. R. da Fonseca, *Food Chem.* **2006**, *95*, 413–422.

218 M. Toda, A. Takagaki, M. Okumura, J. N. Kondo, S. Hayashi, K. Domen, M. Hara, *Nature* **2005**, *438*, 178.

219 S. Doi, J. H. Clark, D. J. Macquarrie, K. Milkowski, *Chem. Commun.* **2002**, 2632.

220 M. J. Gronnow, R. Luque, D. J. Macquarrie, J. H. Clark, *Green Chem.* **2005**, *7*, 552–557.

221 C. M. Byrne, S. D. Allen, E. B. Lobkovsky, G. W. Coates, *J. Am. Chem. Soc.* **2004**, *126*, 11404–11405.

9
Process Integration and Cascade Catalysis

9.1
Introduction

As noted in Chapter 1, the key to success in developing green, sustainable processes is the effective integration of catalytic technologies – heterogeneous, homogeneous, and enzymatic – in organic synthesis. It is also necessary to look at the whole picture; not only the reaction steps but also the downstream processing. This necessitates the integration of a separation step for removal of the product and recovery and recycling of the catalyst into the overall scheme.

Elegant examples of multistep processes involving a series of catalytic steps are the Rhodia process for the manufacture of vanillin and the Lonza process for nicotinamide, discussed in Chapter 1. In these processes the product of each reaction is isolated, and perhaps purified, before going on to the next step. However, the ultimate in integration is to combine several catalytic steps into a one-pot, multistep catalytic cascade process [1–5]. This is truly emulating Nature, where metabolic pathways conducted in living cells involve an elegant orchestration of a series of biocatalytic steps into an exquisite multi-catalyst cascade, without the need for separation of intermediates. Such 'telescoping' of multistep syntheses into a one-pot catalytic cascade has several advantages (see Table 9.1).

They involve fewer unit operations, less solvent and reactor volume, shorter cycle times, higher volumetric and space–time yields and less waste (lower E

Table 9.1 Advantages and limitations of catalytic cascade processes.

Advantages	Limitations
Fewer unit operations	Catalysts incompatible
Drive equilibria in the desired direction	Rates very different
Less solvent/reactor volume	Difficult to find optimum pH, temperature, solvent, etc.
Shorter cycle times	
Higher volumetric and space–time yields	Catalyst recycle complicated
Less waste/lower E factor	Complicated reaction mixtures and work-up

factor). This translates to substantial economic and environmental benefits. Furthermore, coupling of reactions can be used to drive equilibria towards product, thus avoiding the need for excess reagents. On the other hand, there are several problems associated with the construction of catalytic cascades: catalysts are often incompatible with each other (e.g. an enzyme and a metal catalyst), rates are very different, and it is difficult to find optimum conditions of pH, temperature, solvent, etc. Catalyst recovery and recycle is complicated and downstream processing is difficult. Nature solves this problem by compartmentalization of the various biocatalysts. Hence, compartmentalization via immobilization is conceivably a way of solving these problems in cascade processes. It is also worth noting that biocatalytic processes generally proceed under roughly the same conditions – in water at around ambient temperature and pressure – which facilitates the cascading process.

9.2
Dynamic Kinetic Resolutions by Enzymes Coupled with Metal Catalysts

An excellent example of the successful combination of chemo- and biocatalysis in a two-step cascade process is provided by the dynamic kinetic resolutions (DKR) of chiral alcohols and amines. We first suggested [6], in 1993, that (de)-hydrogenation catalysts should be capable of catalyzing the racemization of chiral alcohols and amines via a dehydrogenation/hydrogenation mechanism as shown in Fig. 9.1.

Subsequently the groups of Williams [7] and Backvall [8] showed, in 1996 and 1997, respectively, that lipase-catalyzed transesterification of alcohols could be combined with transition metal-catalyzed racemization to produce an efficient dynamic kinetic resolution of chiral secondary alcohols (Fig. 9.2).

In the system described by Williams a rhodium catalyst was used, in the presence of an inorganic base and phenanthroline, and one equivalent of acetophenone, for the racemization, and a *Pseudomonas sp.* lipase for the acylation using vinyl acetate as the acyl donor (Fig. 9.3).

Backvall, in contrast, used the ruthenium complex **1**, which does not require the addition of an external base, in combination with *p*-chlorophenyl acetate as acyl donor (Fig. 9.4). The latter was chosen because it generates *p*-chlorophenol which does not react with the Ru catalyst. In contrast, vinyl acetate and analo-

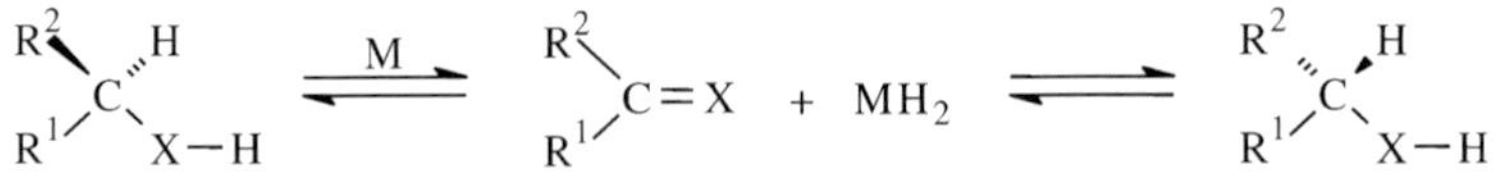

$$X = O, NR, etc$$

Fig. 9.1 Metal-catalyzed racemization of alcohols and amines.

(R)-selective enzyme
ROAc
fast
MLn
(S)-selective enzyme
ROAc
slow

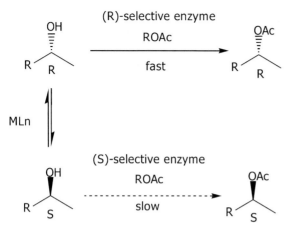

Fig. 9.2 Dynamic kinetic resolution of secondary alcohols.

Pseudomonas sp. lipase
Rh2(OAc)4 / o-phen
PhCOCH3 / KOH /c-C6H10
60% conv.
98% ee

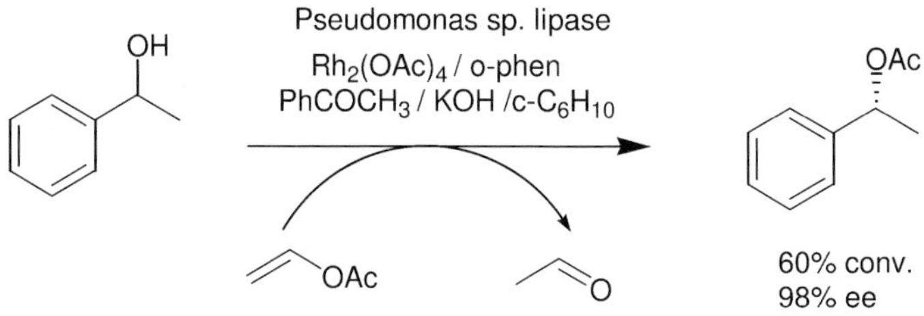

Fig. 9.3 Rh-catalyzed DKR of 1-phenylethanol.

+ ArOAc
(3 eq.)
Novozym 435
PhCOCH3 (1 eq.)
RuL (2 m%)
t-BuOH,70°C, 87h
+ ArOH
100% yield />99% ee
92% isolated yield
RuL =

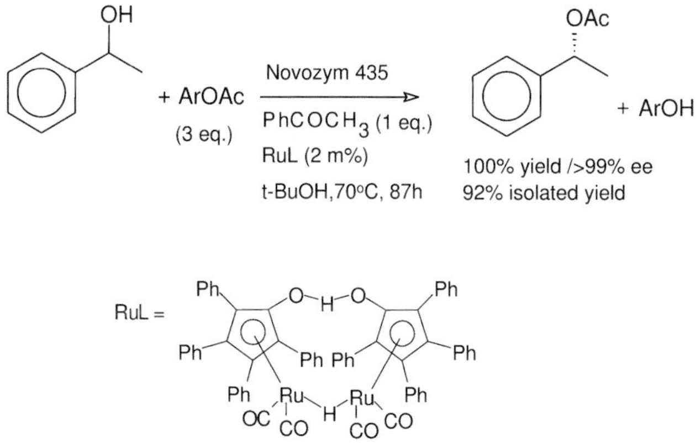

Fig. 9.4 Ru-catalyzed DKR of 1-phenylethanol.

gous acyl donors generate carbonyl compounds as the coproducts that compete for the catalyst. The need for up to one equivalent of additional ketone in these systems is to increase the overall reaction rate. This rate acceleration is a result of increasing the rate of the rate-limiting re-addition of the ruthenium hydride species to the ketone, owing to the higher concentration of ketone. Additional practical disadvantages of the ruthenium system are the need for 3 equivalents of acyl donor, which generate an extra 2 equivalents of *p*-chlorophenol on hydrolytic work-up.

Following on from this initial publication of Backvall, many groups have reported on a variety of ruthenium-based systems for the DKR of secondary alcohols [9–17] mainly with the goal of eliminating the need for added base and ketone and reducing the reaction time by increasing the rate of racemization. Some examples of ruthenium complexes (**1–8**) which have been used as the racemization catalysts in these systems are depicted in Fig. 9.5.

Some of these catalyze the smooth racemization of chiral secondary alcohols at room temperature. However, a major problem which needed to be solved in order to design an effective combination of ruthenium catalyst and lipase in a DKR of secondary alcohols was the incompatibility of many of the ruthenium catalysts and additives, such as inorganic bases, with the enzyme and the acyl donor. For example, the ruthenium catalyst may be susceptible to deactivation by the acetic acid generated from the acyl donor when it is vinyl acetate. Alternatively, any added base in the racemization system can catalyze a competing selective transesterification of the alcohol, resulting in a decrease in enantioselectivity. Consequently, considerable optimization of reaction protocols and conditions was necessary in order to achieve an effective DKR of secondary alcohols.

Several groups have reported [9] ruthenium-based systems that are compatible with the enzyme and acyl donor. The most active of these is the one based on catalyst **4** developed by Bäckvall [17, 18] which effects the DKR of secondary al-

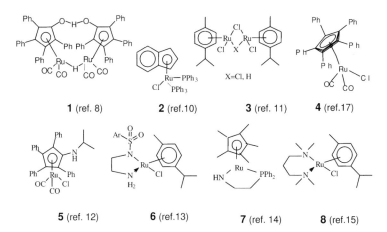

Fig. 9.5 Ruthenium complexes used as alcohol racemization catalysts.

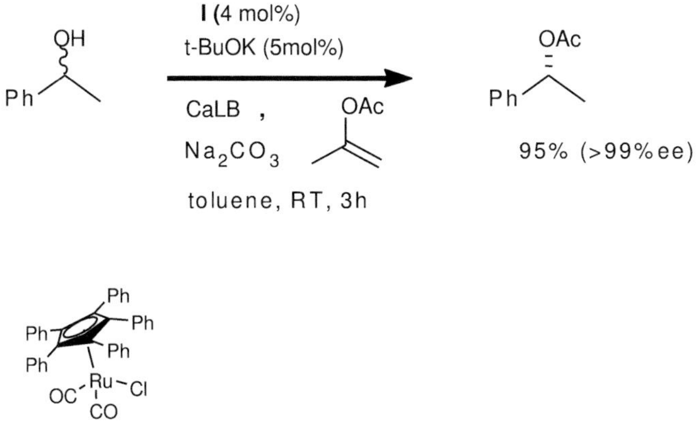

OH
Ph
I (4 mol%)
t-BuOK (5mol%)
CaLB ,
OAc
Na2CO3
toluene, RT, 3h
OAc
Ph
95% (>99%ee)
Ph
Ph
Ph
Ph
Ph
OC
Ru
Cl
CO

Fig. 9.6 Ru-catalyzed DKR at room temperature.

cohols in 3 h at room temperature, affording excellent enantioselectivities with a broad range of substrates (Fig. 9.6).

An important advance was made with the realization [18] that the function of the base, e.g. potassium *tert*-butoxide, was to convert the catalyst precursor to the active catalyst by displacing the chloride ion attached to ruthenium (Fig. 9.7). Consequently, only one equivalent of base with respect to the Ru catalyst is needed and superior selectivities are observed by performing a separate catalyst activation step, thereby largely avoiding competing base-catalyzed transesterification. The *tert*-butoxide ligand is then replaced by the alkoxide derived from the substrate which then undergoes racemization via a dehydrogenation/hydrogenation cycle as shown in Fig. 9.7.

t-BuOK
-KCl
OBut
OH
H
R
Me
η5→η3
β-elim
η3→η5
insertion

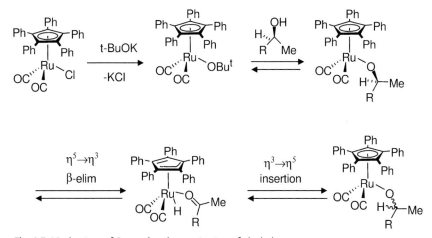

Fig. 9.7 Mechanism of Ru-catalyzed racemization of alcohols.

Notwithstanding these excellent results, there is still some room for further improvement. All of these systems require anaerobic conditions, owing to their air sensitivity, and consequently cannot be reused. Indeed all of these systems involve significant (from a cost point of view) amounts (2–5 mol%) of a homogeneous ruthenium catalyst, which is difficult to recycle. An interesting recent development in this context is the report, by Kim and coworkers [19], of an air-stable, recyclable catalyst that is applicable to DKR at room temperature (Fig. 9.8).

One place to look for good alcohol racemization catalysts is in the pool of catalysts that are used for hydrogen transfer reduction of ketones. One class of complexes that are excellent catalysts for the asymmetric transfer hydrogenation comprises the ruthenium complexes of monosulfonamides of chiral diamines developed by Noyori and coworkers [20, 21]. These catalysts have been used for the asymmetric transfer hydrogenation of ketones [20] and imines [21] (Fig. 9.9).

Hence, we reasoned that they should be good catalysts for alcohol (and amine) racemization but without the requirement for a chiral ligand. This indeed proved to be the case: ruthenium complex **6** containing the achiral diamine monosulfonamide ligand was able to catalyze the racemization of 1-phenylethanol [13]. Noyori and coworkers [22] had proposed that hydrogen transfer catalyzed by theses complexes involves so-called metal-ligand bifunctional catalysis, whereby hydrogens are transferred from both the ruthenium and the ligand (see Fig. 9.10).

By analogy, it seemed plausible that alcohol racemizations catalyzed by the same type of ruthenium complexes involve essentially the same mechanism (Fig. 9.11), in which the active catalyst is first generated by elimination of HCl by the added base. This 16e complex subsequently abstracts two hydrogens from the alcohol substrate to afford an 18e complex and a molecule of ketone. Reversal of these steps leads to the formation of racemic substrate.

This system was also shown to catalyze the DKR of 1-phenylethanol [13], interestingly in the presence of the stable free radical TEMPO as cocatalyst (Fig. 9.12). The exact role of the TEMPO in this system has yet to be elucidated and the reaction conditions need to be optimized, e.g. by preforming the active catalyst as in the Bäckvall system described above.

Similarly, the DKR of chiral amines can, in principle, be achieved by combining the known amine resolution by lipase-catalyzed acylation [23] with metal-cat-

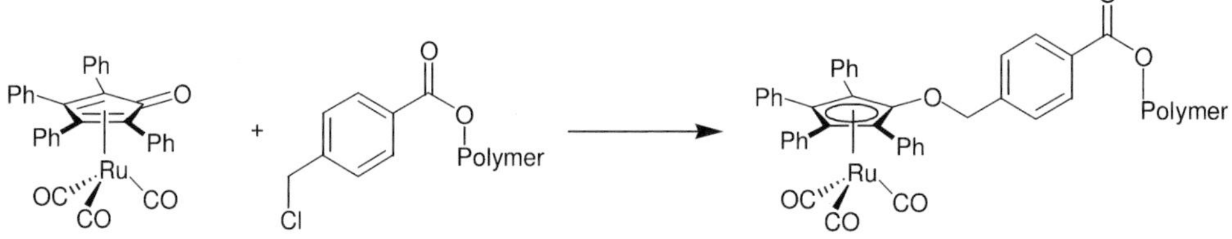

Fig. 9.8 Air stable, recyclable Ru catalyst for DKR of alcohols.

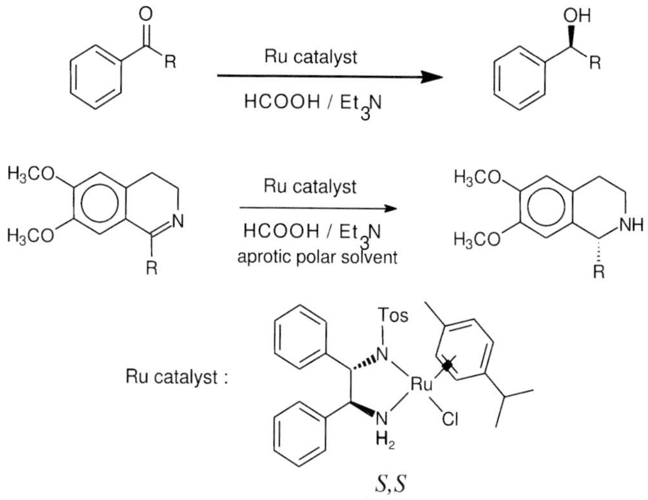

Fig. 9.9 Asymmetric transfer hydrogenation of ketones and imines.

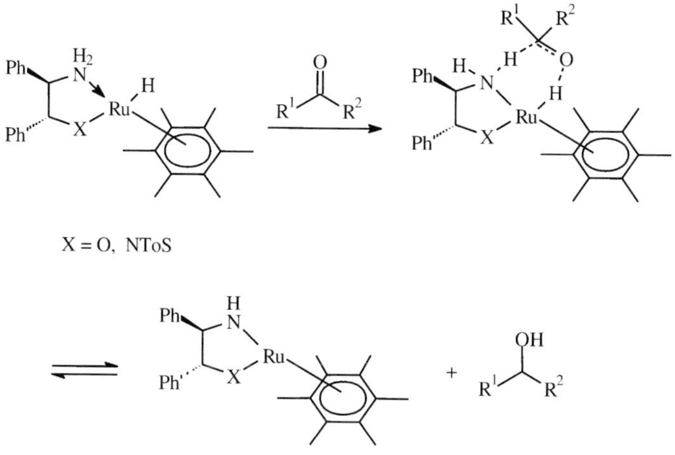

Fig. 9.10 Metal-ligand bifunctional catalysis.

alyzed racemization via a dehydrogenation/hydrogenation mechanism. However, although the DKR of alcohols is well established there are few examples of the DKR of amines. Reetz and Schimossek in 1996 reported the DKR of 1-phenethylamine (Fig. 9.13), using Pd-on-charcoal as the racemization catalyst [24]. However, the reaction was slow (8 days) and competing side reactions were responsible for a relatively low yield.

The racemization of amines is more difficult than that of alcohols owing to the more reactive nature of the intermediate (imine vs. ketone). Thus, imines readily undergo reaction with a second molecule of amine and/or hydrogenoly-

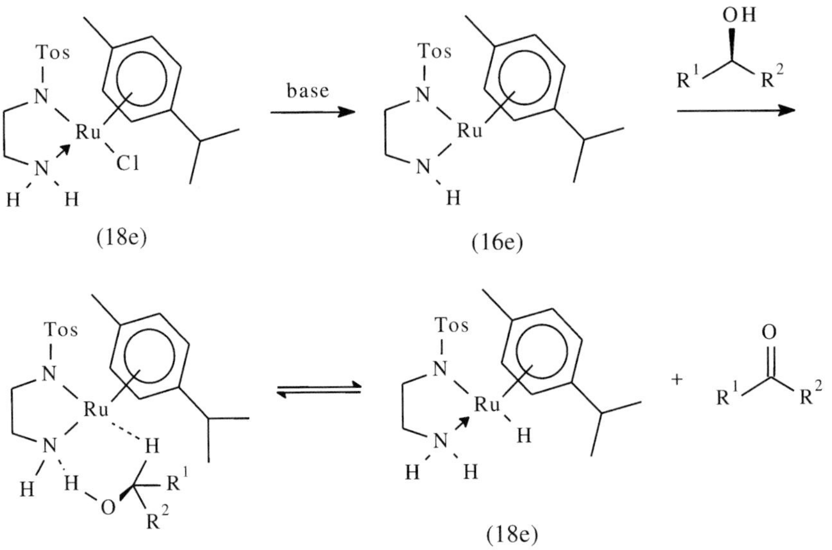

Fig. 9.11 Mechanism of racemization catalyzed by complex **6**.

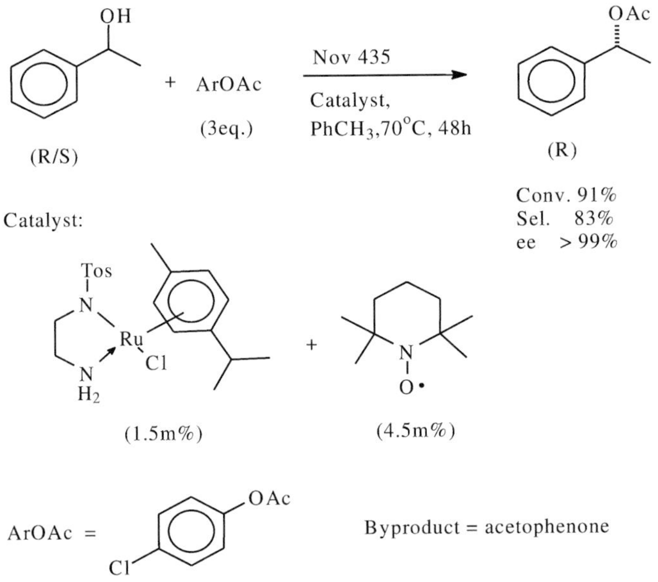

Fig. 9.12 DKR of 1-phenylethanol catalyzed by **6**/CaLB.

sis reactions as shown in Fig. 9.14. Furthermore, in the presence of water the imine can undergo hydrolysis to the corresponding ketone.

Recently, Jacobs and coworkers [25] showed that the use of alkaline earth supports, rather than charcoal, for the palladium catalyst led to substantial reduc-

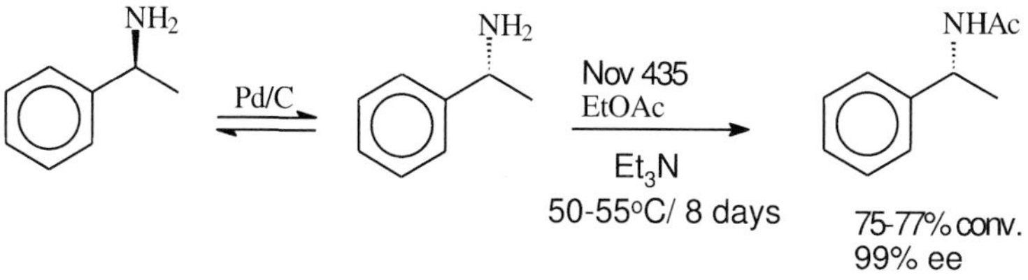

Fig. 9.13 DKR of 1-phenylethylamine catalyzed by a lipase/Pd-on-C.

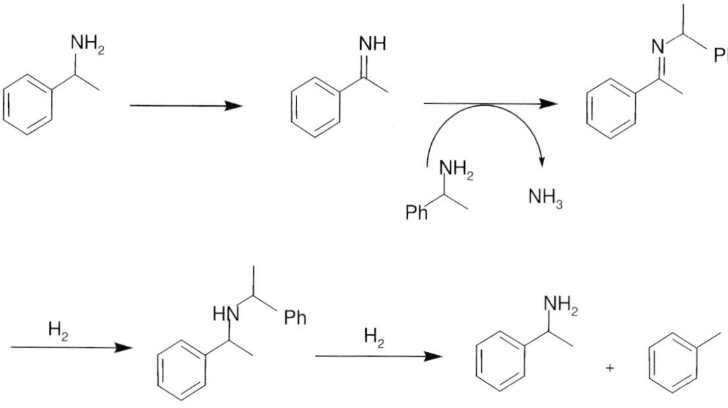

Fig. 9.14 Side reactions in metal-catalyzed racemization of amines.

tion in byproduct formation. It was suggested that the basic support suppresses the condensation of the amine with the imine (see Fig. 9.14). A 5% Pd-on-BaSO$_4$ catalyst was chosen for further study and shown to catalyze the DKR of amines with reaction times of 1–3 days at 70 °C. Good to excellent chemo- and enantioselectivities were obtained with a variety of chiral benzylic amines (see Fig. 9.15).

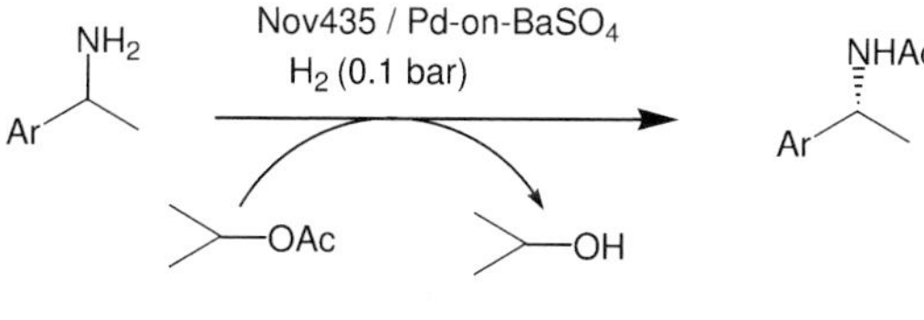

Ar	Time(h)	Conv.(%)	sel. amide (%)	ee (%)
C$_6$H$_5$	24	91	94	>99
4-MeC$_6$H$_4$	24	73	97	>99
4-MeOC$_6$H$_4$	48	90	98	>99
2-naphthyl	48	89	87	99
1-naphthyl	48	64	87	99

Fig. 9.15 DKR of primary amines catalyzed by Pd-on-BaSO$_4$/lipase.

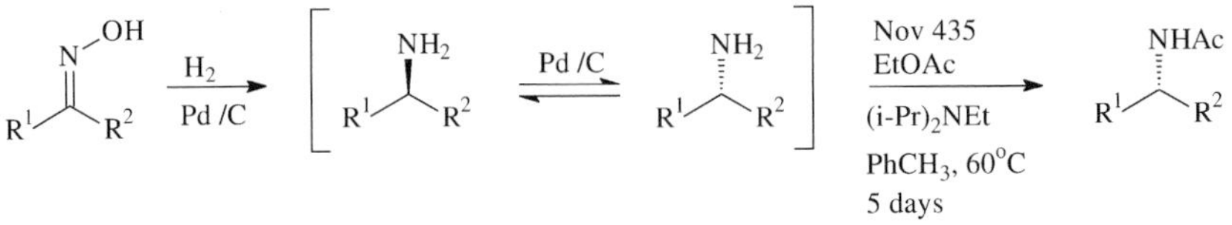

Fig. 9.16 Optically active amines from prochiral ketoximes.

Alternatively, Kim and coworkers showed [26] that prochiral ketoximes could be converted into optically active amines by hydrogenation over Pd-on-C in the presence of CaLB and ethyl acetate as both acyl donor and solvent, presumably via DKR of the amine formed *in situ* (see Fig. 9.16).

Paetzold and Backvall [27] have reported the DKR of a variety of primary amines using an analog of the ruthenium complex **1** as the racemization catalyst and isopropyl acetate as the acyl donor, in the presence of sodium carbonate at 90 °C (Fig. 9.17). Apparently, the function of the latter was to neutralize traces of acid, e.g. originating from the acyl donor, which would deactivate the ruthenium catalyst.

Another important difference between (dynamic) kinetic resolution of alcohols and amines is the ease with which the acylated product, an ester and an amide, respectively, is hydrolyzed. This is necessary in order to recover the substrate enantiomer which has undergone acylation. Ester hydrolysis is generally a facile process but amide hydrolysis, in contrast, is often not trivial. For example, in the BASF process [28] for amine resolution by lipase-catalyzed acylation the amide product is hydrolyzed using NaOH in aq. ethylene glycol at 150 °C (Fig. 9.18). In the case of phenethylamine this does not present a problem but it will obviously lead to problems with a variety of amines containing other functional groups.

We reasoned that, since an enzyme was used to acylate the amine under mild conditions, it should also be possible to employ an enzyme for the deacylation of the product. This led us to the concept of the Easy-on-Easy-off resolution of

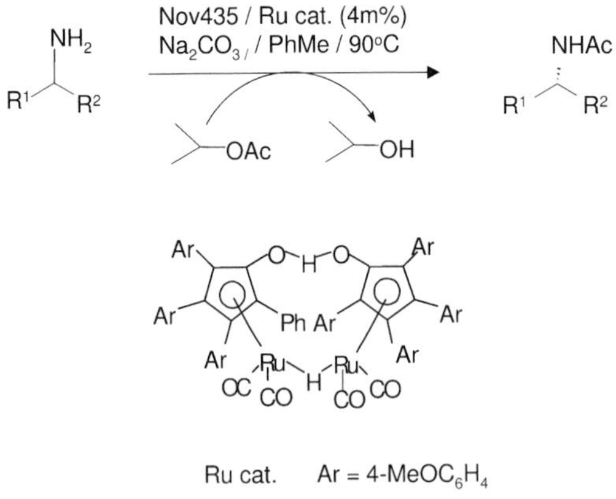

Fig. 9.17 DKR of amines catalyzed by Ru/CaLB.

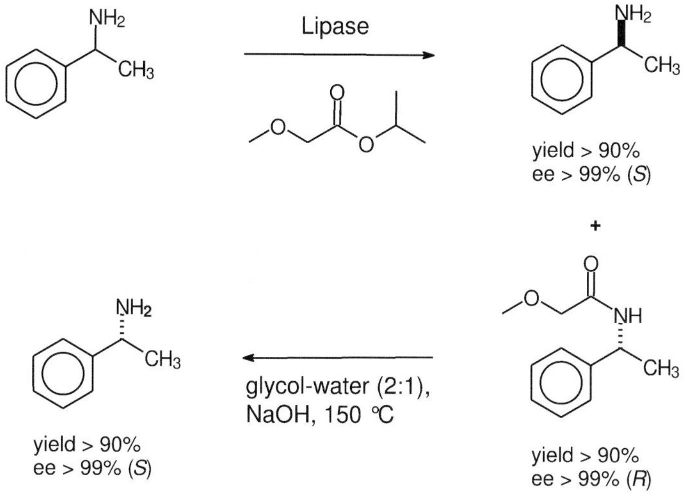

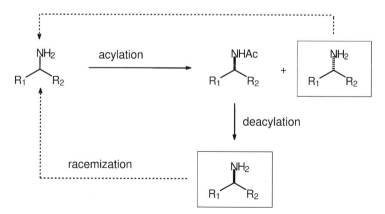

Fig. 9.18 Lipase-catalyzed resolution of amines: BASF process.

amines [29] which involves three steps: enzymatic acylation and deacylation and racemization of the unwanted enantiomer (Fig. 9.19). Combination of acylation and deacylation, in a DKR process as described above, leads to a two-step process in which only one enantiomer of the amine is formed.

For example, we showed that penicillin acylase from *Alcaligenes faecalis* could be used for acylation of amines in an aqueous medium at high pH. After extraction of the remaining amine isomer, the pH was reduced to 7 and the amide hydrolyzed to afford the other amine enantiomer (Fig. 9.20).

This one-pot process, performed in water, is clearly an attractive, green alternative to the above-mentioned BASF process. Unfortunately, high enantioselectivities are obtained only with amine substrates containing an aromatic moiety.

Fig. 9.19 The Easy-on-Easy-off resolution of amines.

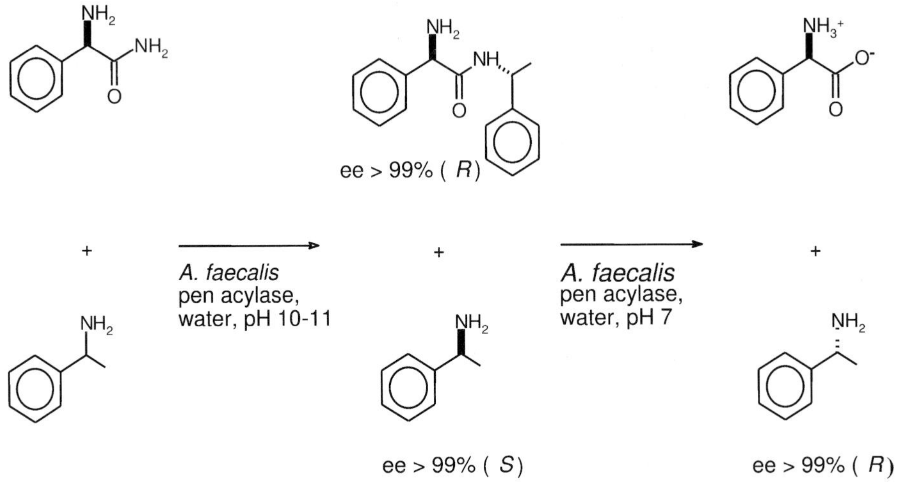

Fig. 9.20 Resolution of an amine with penicillin acylase.

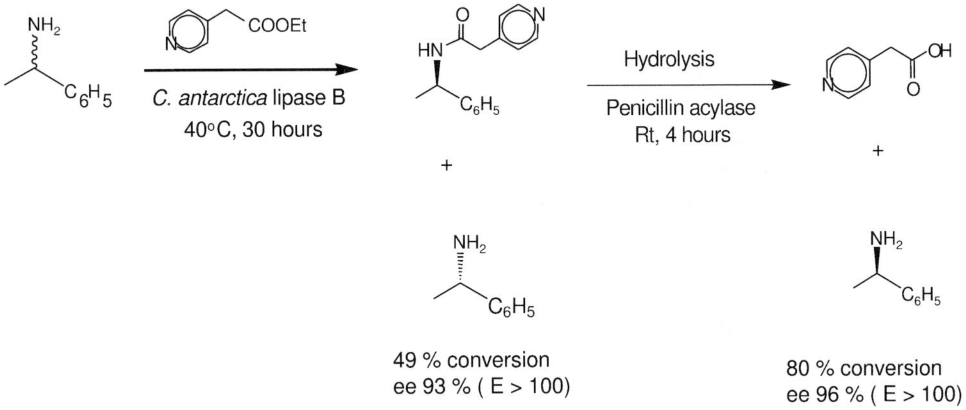

Fig. 9.21 Two-enzyme resolution of amines.

In order to broaden the scope we also examined [30] a combination of lipase-cat-alyzed acylation with penicillin acylase-catalyzed hydrolysis (deacylation). Good results (high enantioselectivity in the acylation and smooth deacylation) were obtained, with a broad range of both aliphatic amines and amines containing an aromatic moiety, using pyridylacetic acid ester as the acyl donor (Fig. 9.21).

We subsequently found that CaLB could be used for both steps (Fig. 9.22). That a lipase was able to effectively catalyze an amide hydrolysis was an unex-pected and pleasantly surprising result. Good results were obtained with CaLB immobilized as a CLEA (see later).

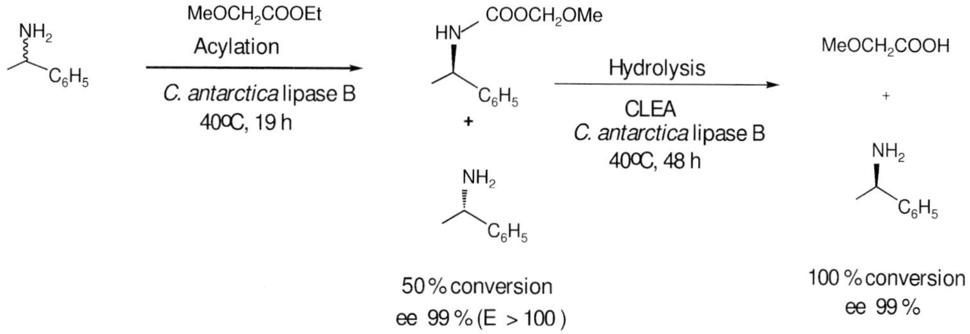

Fig. 9.22 CaLB for acylation and deacylation steps in amine resolution.

9.3
Combination of Asymmetric Hydrogenation with Enzymatic Hydrolysis

Asymmetric catalysis by chiral metal complexes is also an important technology for the industrial synthesis of optically active compounds [6, 31]. Most commercial applications involve asymmetric hydrogenation [31]. The classic example is the Monsanto process for the production of L-Dopa, developed in the 1960s [32]. Knowles was awarded the Nobel Prize in Chemistry for the development of this benchmark process in catalytic asymmetric synthesis. A closer inspection of the process (Fig. 9.23) reveals that, subsequent to the key catalytic asymmetric hydrogenation step, three hydrolysis steps (ester, ether, and amide hydrolysis), are required to generate the L-Dopa product. One can conclude, therefore, that there is still room for improvement e.g. by replacing classical chemical hydrolysis steps by cleaner, enzymatic hydrolyses under mild conditions, resulting in a reduction in the number of steps and the consumption of chemicals and energy. Furthermore, the homogeneous catalyst used in this and analogous

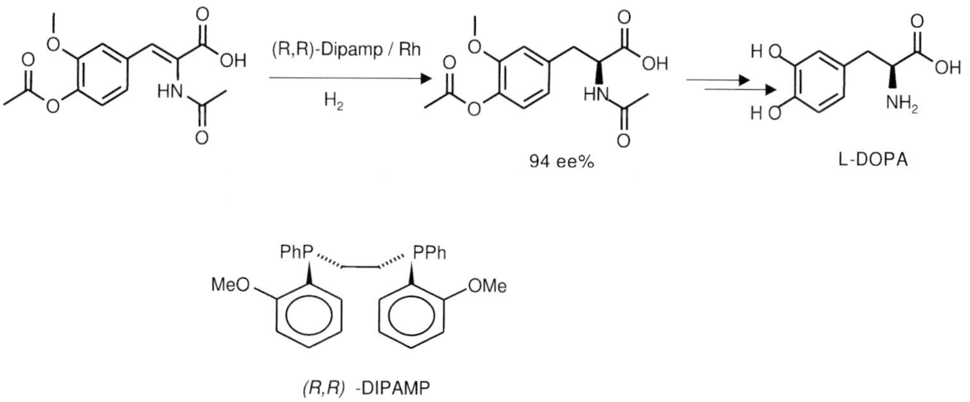

Fig. 9.23 Monsanto L-Dopa process.

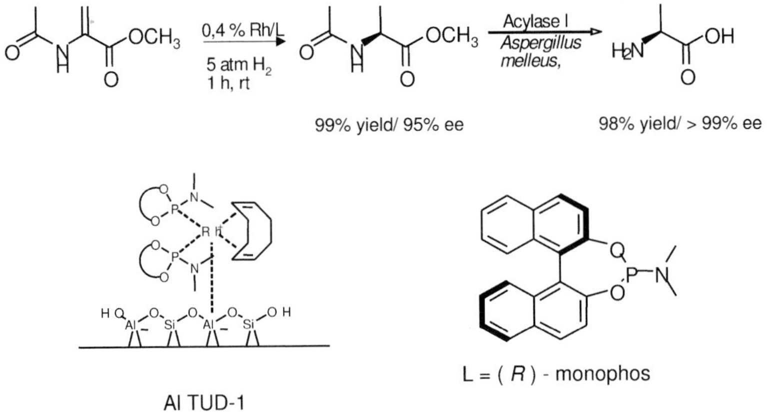
OCH3
0,4 % Rh/L
5 atm H2
1 h, rt
OCH3
Acylase I
Aspergillus melleus,
H2N
OH
99% yield/ 95% ee
98% yield/ > 99% ee
Al TUD-1
L = (R) - monophos

Fig. 9.24 Chemoenzymatic, one-pot synthesis of an amino acid.

asymmetric hydrogenations comprises two expensive components, a noble metal and a chiral diphosphine ligand. Hence, recovery and recycling of the catalyst is an important issue, as discussed in Chapter 7.

Consequently, much attention is currently being focused on telescoping such processes and integrating catalyst recovery and recycling into the overall process. For example, we have recently combined an analogous asymmetric hydrogenation of an *N*-acyl dehydroamino acid (Fig. 9.24), using a supported chiral Rh catalyst, with enzymatic hydrolysis of the product, affording a one-pot cascade process in water as the only solvent [33]. An additional benefit is that the enantiomeric purity of the product of the asymmetric hydrogenation is upgraded in the subsequent enzymatic amide hydrolysis step (which is highly selective for the desired enantiomer). The ester moiety in the intermediate also undergoes enzymatic hydrolysis under the reaction conditions.

9.4 Catalyst Recovery and Recycling

As noted above, recovery and recycling of chemo- and biocatalysts is important from both an economic and an environmental viewpoint. Moreover, compartmentalization (immobilization) of the different catalysts is a *conditio sine qua non* for the successful development of catalytic cascade processes. As discussed in Chapter 7, various approaches can be used to achieve the immobilization of a homogeneous catalyst, whereby the most well-known is heterogenization as a solid catalyst as in the above example.

An approach to immobilization which has recently become popular is microencapsulation in polymers, such as polystyrene and polyurea, developed by the groups of Kobayashi [34] and Ley [35], respectively. For example, microencapsulation of palladium salts or palladium nanoparticles in polyurea microcapsules affords recyclable catalysts for carbonylations, Heck and Suzuki couplings and (transfer) hydrogenations [36].

Another approach to facilitating catalyst separation while maintaining the benefits of homogeneous catalysis involves the use of thermoregulated biphasic catalysis, as already mentioned in Chapter 7. In this approach the catalyst dissolves in a particular solvent at one temperature and is insoluble at another. For example, a ligand or an enzyme can be attached to an ethylene oxide/propylene oxide block copolymer (see Chapter 7) to afford catalysts that are soluble in water at room temperature but precipitate on warming to 40 °C. The driving force for this inverted temperature dependence on solubility is dehydration of the ligand on heating. An added advantage is that runaway conditions are never achieved since the catalyst precipitates and the reaction stops on raising the temperature. An example of the application of this technique to biotransformations involves the covalent attachment of penicillin acylase to poly-N-isopropyl-acrylamide (PNIPAM) [37].

An interesting example of the use of a recyclable, thermoresponsive catalyst in a micellar-type system was recently reported by Ikegami et al. [38]. A PNIPAM-based copolymer containing pendant tetraalkylammonium cations and a polyoxometalate, $PW_{12}O_{40}^{3-}$, as the counter anion was used as a catalyst for the oxidation of alcohols with hydrogen peroxide in water (Fig. 9.25). At room temperature the substrate and the aqueous hydrogen peroxide, containing the catalyst, formed distinct separate phases. When the mixture was heated to 90 °C a

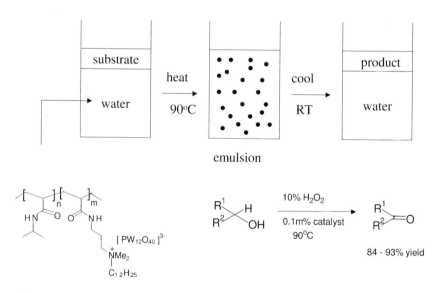

PNIPAAm-based thermoresponsive catalyst

Fig. 9.25 Oxidation of alcohols with hydrogen peroxide using a thermoresponsive catalyst in a micellar system.

stable emulsion was formed, in which the reaction took place with as little as 0.1 mol% catalyst. Subsequent cooling of the reaction mixture to room temperature resulted in precipitation of the catalyst, which could be removed by filtration and recycled.

Another interesting concept involves so-called organic aqueous tunable solvents (OATS) for performing (bio)catalytic processes [39]. The reaction takes place in a single liquid phase, comprising a mixture of water and a water miscible organic solvent, such as tetrahydrofuran or dioxane. When the reaction is complete, carbon dioxide is added to generate a biphasic system consisting of a gas-expanded liquid organic phase containing hydrophobic components and an aqueous phase containing the hydrophilic (bio)catalyst.

This is reminiscent of the 'miscibility switch', described in Chapter 7, whereby carbon dioxide and an ionic liquid are miscible at one pressure and become immiscible on reducing the pressure. This provided the possibility of performing an asymmetric hydrogenation in a monophasic ionic liquid/carbon dioxide mixture and, when the reaction is complete, reducing the pressure to afford a biphasic mixture in which the product is in the carbon dioxide phase and the rhodium catalyst in the ionic liquid phase, which can be recycled [40]. Another variation on this theme is a smart solvent that switches reversibly from a liquid with one set of properties to another with different properties. In the example presented, a mixture of an alcohol and an amine was allowed to react with carbon dioxide to generate an ionic liquid (Fig. 9.26) which reverts back to its nonionic form when purged with nitrogen or argon [41].

In Chapter 7 we have already discussed the use of fluorous biphasic systems to facilitate recovery of catalysts that have been derivatized with fluorous "ponytails". The relatively high costs of perfluoroalkane solvents coupled with their persistent properties pose serious limitations for their industrial application. Consequently, second generation methods have been directed towards the elimination of the need for perfluoro solvents by exploiting the temperature-dependent solubilities of fluorous catalysts in common organic solvents [42]. Thus, appropriately designed fluorous catalysts are soluble at elevated temperatures and essentially insoluble at lower temperatures, allowing for catalyst recovery by simple filtration.

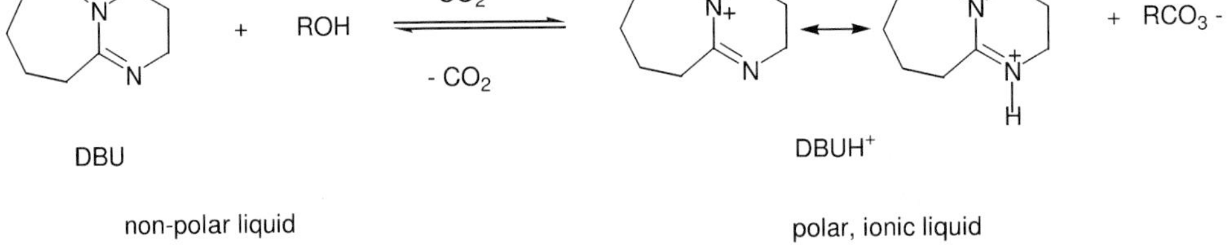

Fig. 9.26 Reversible non-polar to polar solvent switch.

R = $CH_2CH_2(CF_2CF_2)_n$
n = 6 or 8

Fig. 9.27 Catalyst-on-Teflon-tape for recycling of fluorous catalysts.

Alternatively, an insoluble fluorous support, such as fluorous silica [43], can be used to adsorb the fluorous catalyst. Recently, an eminently simple and effective method has been reported in which common commercial Teflon tape is used for this purpose [44]. This procedure was demonstrated with a rhodium-catalyzed hydrosilylation of a ketone (Fig. 9.27). A strip of Teflon tape was introduced into the reaction vessel and when the temperature was raised the rhodium complex, containing fluorous ponytails, dissolved. When the reaction was complete the temperature was reduced and the catalyst precipitated onto the Teflon tape which could be removed and recycled to the next batch.

9.5
Immobilization of Enzymes: Cross-linked Enzyme Aggregates (CLEAs)

We have recently developed an extremely effective method for immobilizing enzymes as so-called Cross-Linked Enzyme Aggregates (CLEAs) [45]. They exhibit high activity retention and stability and can be readily recovered and recycled without any loss of activity. Furthermore, the method is exquisitely simple – precipitation from aqueous buffer followed by cross-linking with, for example, glutaraldehyde – and is applicable to a broad range of enzymes. It does not require highly pure enzyme preparations and it actually constitutes a combination of purification and immobilization into one step. The methodology can also be applied to the co-immobilization of two or more enzymes to give 'combi CLEAs' which are more effective than mixtures of the individual CLEAs. These are ideally suited to conducting enzymatic cascade reactions in water, where an equilibrium can be shifted by removing the product in a consecutive biotransformation. For example, we have used a combi CLEA containing an S-selective nitrilase (from *Manihot esculenta*) and a non-selective nitrilase, in DIPE/Water (9:1) at pH 5.5, 1 h, for the one-pot conversion of benzaldehyde to S-mandelic acid (Fig. 9.28) in high yield and enantioselectivity [46].

A CLEA prepared from CaLB was recently shown to be an effective catalyst for the resolution of 1-phenylethanol and 1-tetralol in supercritical carbon dioxide in continuous operation [47]. Results were superior to those obtained with Nov 435 (CaLB immobilized on a macroporous acrylic resin) under the same

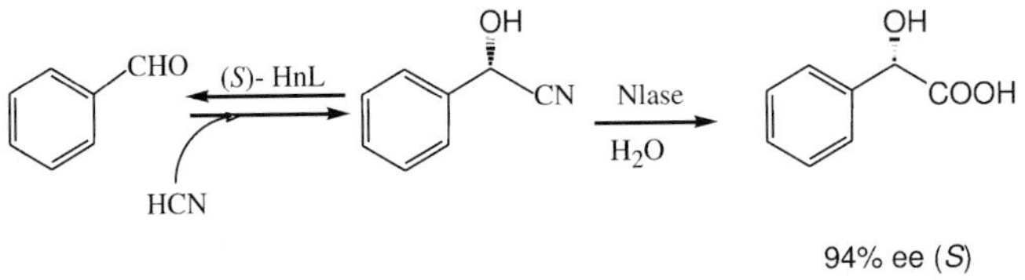

Fig. 9.28 One-pot conversion of benzaldehyde to *S*-mandelic acid with a combi CLEA.

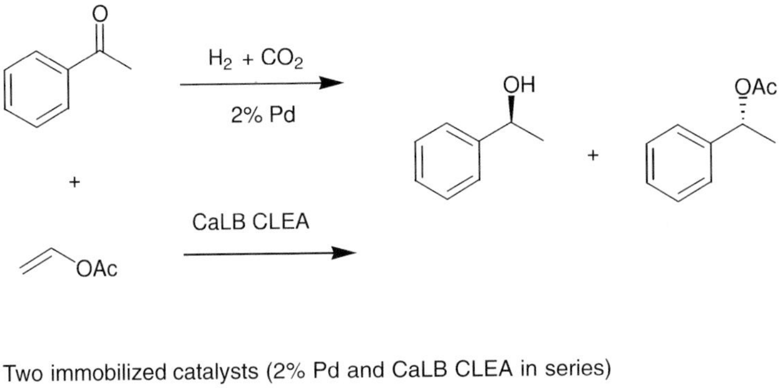

Two immobilized catalysts (2% Pd and CaLB CLEA in series)

Fig. 9.29 Continuous kinetic resolution catalyzed by a CaLB CLEA in scCO$_2$.

conditions. In addition, the palladium-catalyzed hydrogenation of acetophenone and the resolution of the resulting 1-phenylethanol were performed in series (Fig. 9.29), thereby achieving a reduction in energy consumption by obviating the need for de- and re-pressurization between the steps.

9.6
Conclusions and Prospects

The key to sustainability is catalysis and success will be dependent on an effective integration of catalysis in organic synthesis, including downstream processing. Traditional barriers between catalysis and mainstream organic synthesis are gradually disappearing, as are the traditional barriers separating the sub-disciplines of catalysis: heterogeneous, homogeneous and enzymatic. Hence, there is a trend towards the development of multistep syntheses involving a variety of catalytic steps. The ultimate in efficiency is to telescope these syntheses into catalytic cascade processes. A common problem encountered in the design of such processes is that of incompatibility of the different catalysts. A possible solution is to follow the example of Nature: compartmentalization as the key to compatibility. In practice, this means immobilization of the catalyst, which at the same time provides for its efficient recovery and recycling. As we have seen, various approaches can be envisaged for achieving this immobilization, including not only traditional immobilization as a filterable solid catalyst but also 'smart methods' such as attachment of catalysts to thermoresponsive polymers.

Finally, it is worth mentioning that a successful integration of catalytic reaction steps with product separation and catalyst recovery operations will also be dependent on innovative chemical reaction engineering. This will require the widespread application of sustainable engineering principles [48].In this context 'process intensification', which involves the design of novel reactors of increased volumetric productivity and selectivity with the aim of integrating different unit operations to reactor design, and miniaturization will play pivotal roles [49, 50].

References

1 A. Bruggink, R. Schoevaart, T. Kieboom, *OPRD*, **2003**, *7*, 622.

2 S.F. Meyer, W. Kroutil, K. Faber, *Chem. Soc. Rev.*, **2001**, *30*, 332.

3 F.F. Huerta, A.B.E. Minidis, J.-E. Bäckvall, *Chem. Soc. Rev.*, **2001**, *30*, 321.

4 L. Veum, U. Hanefeld, *Chem. Commun.*, **2006**, 825.

5 A. Ajamian, J.L. Gleason, *Angew. Chem. Int. Ed.*, **2004**, *43*, 3754.

6 R.A. Sheldon, in *Chirotechnology, the Industrial Synthesis of Optically Active Compounds*, Marcel Dekker, New York, 1993, pp. 90–91.

7 P.M. Dinh, J.A. Howard, A.R. Hudnott, J.M.J. Williams, W. Harris, *Tetrahedron Lett.*, **1996**, *37*, 7623.

8 A.L.E. Larsson, B.A. Persson, J.-E. Bäckvall, *Angew. Chem. Int. Ed.*, **1997**, *36*, 1211.

9 For reviews see O. Pamies, J.-E. Bäckvall, *Chem. Rev.*, **2003**, *10*, 3247; O. Pamies, J.-E. Bäckvall, *Trends Biotechnol.*, **2004**, *22*, 130; M.-J. Kim, Y. Ahn, J. Park, *Curr. Opin. Biotechnol.*, **2002**, *13*, 578 and [4].

10 J.H. Koh, H.M. Jung, J. Park, *Tetrahedron Lett.*, **1998**, *39*, 5545.

11 J.H. Koh, H.M. Jung, M.J. Kim, J. Park, *Tetrahedron Lett.*, 1999, *40*, 6281.

12 J.H. Choi, Y.H. Kim, S.H. Nam, S.T. Shin, M.J. Kim, J. Park, *Angew. Chem. Int. Ed.*, **2002**, *41*, 2373.

13 A. Dijksman, J.M. Elzinga, Yu-Xin Li, I.W.C.E. Arends, R.A. Sheldon, *Tetrahedron: Asymm.* **2002**, *13*, 879.

14 M. Ito, A. Osaku, S. Kitahara, M. Hirakawa, T. Ikariya, *Tetrahedron Lett.*, **2003**, *44*, 7521.

15 T.H. Riermeier, P. Gross, A. Monsees, M. Hoff, H. Trauthwein, *Tetrahedron Lett.*, **2005**, *46*, 3403.

16 G. Verzijl, J.G. de Vries, Q.B. Broxterman, *Tetrahedron Asymm.* **2005**, *16*, 1603.

17 B. Martin-Matute, M. Edin, K. Bogar, J.-E. Bäckvall, *Angew. Chem. Int. Ed.*, **2004**, *43*, 6535.

18 B. Martin-Matute, M. Edin, K. Bogar, F.B. Kaynak, J.-E. Bäckvall, *J. Am. Chem. Soc.*, **2005**, *127*, 8817.

19 N. Kim, S.-B. Ko, M.S. Kwon, M.-J. Kim, J. Park, *Org. Lett.*, **2005**, *7*, 4523.

20 A. Fujii, S. Hashiguchi, N. Uematsu, T. Ikariya, R. Noyori, *J. Am. Chem. Soc.*, **1996**, *118*, 2521.

21 N. Uematsu, A. Fujii, S. Hashiguchi, T. Ikariya, R. Noyori, *J. Am. Chem. Soc.*, **1996**, *118*, 4916.

22 R. Noyori, M. Yamakawa, S. Hashiguchi, *J. Org. Chem.*, **2001**, *66*, 7931.

23 F. van Rantwijk, R.A. Sheldon, *Tetrahedron*, **2004**, *60*, 501.

24 M.T. Reetz, K. Schimossek, *Chimia*, **1996**, *50*, 668.

25 A. Parvelescu, D. De Vos, P. Jacobs, *Chem. Commun.*, **2005**, 5307.

26 Y.K. Choi, M. Kim, Y. Ahn, M.J. Kim, *Org. Lett.*, **2001**, *3*, 4099.

27 J. Paetzold, J.-E. Bäckvall, *J. Am. Chem. Soc.*, **2005**, *127*, 17620.

28 G. Hieber, K. Ditrich, *Chim. Oggi*, June **2001**, 16.

29 D.T. Guranda, A.I. Khimiuk, L.M. van Langen, F. van Rantwijk, R.A. Sheldon, V.K. Svedas, *Tetrahedron Asymm.*, **2004**, *15*, 2901; see also D.T. Guranda, L.M. van Langen, F. van Rantwijk, R.A. Shel-

don, V. K. Švedas, *Tetrahedron Asymm.* **2001**, *12*, 1645.

30 H. Ismail, F. van Rantwijk, R. A. Sheldon, manuscript in preparation.

31 H. U. Blaser, E. Schmidt, Eds., *Asymmetric Catalysis on Industrial Scale*, Wiley-VCH, Weinheim, 2004.

32 W. S. Knowles, *Acc. Chem. Res.*, **1983**, *16*, 106.

33 C. Simons, U. Hanefeld, I. W. C. E. Arends, Th. Maschmeyer, R. A. Sheldon, *Adv. Synth. Catal.*, **2006**, *348*, 471.

34 S. Kobayashi, R. Akiyama, *Chem. Commun.*, **2003**, 449.

35 C. Ramarao, S. V. Ley, S. C. Smith, I. M. Shirley, N. DeAlmeida, *Chem. Commun.*, **2002**, 1132; S. V. Ley, C. Ramarao, R. S. Gordon, A. B. Holmes, A. J. Morrison, I. F. McConvey, I. M. Shirley, S. C. Smith, M. D. Smith, *Chem. Commun.*, **2002**, 1134.

36 D. A. Pears, S. C. Smith, *Aldrichim. Acta*, **2005**, *38*, 24; D. Pears, *Chim. Oggi*, **2005**, *23*, 29.

37 A. E. Ivanov, E. Edink, A. Kumar, I. Y. Galaev, A. F. Arendsen, A. Bruggink, B. Mattiasson, *Biotechnol. Progr.* **2003**, *19*, 1167.

38 H. Hamamoto, Y. Suzuki, Y. M. A. Yamada, H. Tabata, H. Takahashi, S. Ikegami, *Angew. Chem Int. Ed.*, **2005**, *44*, 4536.

39 J. M. Broering, E. M. Hill, J. P. Hallett, C. L. Liotta, C. A. Eckert, A. S. Bommarius, *Angew. Chem. Int. Ed.* **2006**, *45*, 4670.

40 M. C. Kroon, J. van Spronsen, C. J. Peters, R. A. Sheldon, G.-J. Witkamp, *Green Chem.*, **2006**, *8*, 246.

41 P. G. Jessop, D. J. Heldebrant, X. Li, C. A. Eckert, C. L. Liotta, *Nature*, **2005**, *436*, 1102.

42 J. Otera, *Acc.Chem.Res.*, **2004**, *37*, 288; M. Wende, J. A. Gladysz, *J. Am. Chem. Soc.*, **2003**, *125*, 5861.

43 A. Biffis, M. Braga, M. Basato, *Adv. Synth. Catal.*, **2004**, *346*, 451.

44 L. V. Dinh, J. A. Gladysz, *Angew. Chem. Int. Ed.*, **2005**, *44*, 4095.

45 L. Cao, L. M. van Langen, R. A. Sheldon, *Curr. Opin. Biotechnol.* **2003**, *14*, 387; R. A. Sheldon, R. Schoevaart, L. M. van Langen, *Biocatal. Biotrans.*, **2005**, *23*, 141.

46 C. Mateo, A. Chmura, S. Rustler, F. van Rantwijk, A. Stolz, R. A. Sheldon, *Tetrahedron Asymm.* **2006**, *17*, 320.

47 H. R. Hobbs, B. Kondor, P. Stephenson, R. A. Sheldon, N. R. Thomas, M. Poliakoff, *Green Chem.*, **2006**, *8*, 816.

48 M. A. Abraham (Ed.) *Sustainability Science and Engineering, Vol. 1, Defining Principles*, Elsevier, Amsterdam, 2006.

49 C. Tunca, P. A. Ramachandran, M. P. Dudokovic, in [48], pp. 331–347.

50 M. Spear, *Chem. Ind. (London)*, **2006**, (9), 16; J. C. Charpentier, *Chem. Eng. Technol.*, **2005**, *28*, 255; R. C. Costello, *Chem. Eng.*, **2004**, *111*(4), 27.

10
Epilogue: Future Outlook

10.1
Green Chemistry: The Road to Sustainability

There is no doubt that green chemistry is here to stay. Chemical companies and, indeed, companies in general are placing increasing emphasis on sustainability and environmental goals in their corporate mission statements and annual reports. The European Technology Platform on Sustainable Chemistry (SusChem) has recently published an Implementation Action Plan entitled "Putting Sustainable Chemistry into Action" [1]. The report defines three key technology areas: (i) industrial biotechnology, (ii) materials technology, and (iii) reaction and process design. The goal is "improving the eco-efficiency of products and processes to optimize the use of resources and minimize waste and environmental impact". Similarly, the US chemical industry produced a strategic plan – Technology Vision 2020 – which defined a long-term technology roadmap for the future [2]. Important goals were to "improve efficiency in the use of raw materials, the reuse of recycled materials, and the generation and use of energy and to continue to play a leadership role in balancing environmental and economic considerations".

To readers of this book this will all sound rather familiar. It contains an underlying and unifying theme of green chemistry as the means for achieving these noble goals. Initially, many people confused green chemistry with what is generally known as environmental chemistry which is concerned with the effects of chemicals on the environment and remediation of waste and contaminated land and water. In contrast, green chemistry is concerned with redesigning chemical products and processes to avoid the generation and use of hazardous substances and the formation of waste, thus obviating the need for a lot of the environmental chemistry.

The twelve principles of green chemistry, as expounded by Anastas and Warner in 1998 [3], have played an important role in promoting its application. They inspired others to propose additional principles [4] and, more recently, Anastas and Zimmerman [5] proposed the twelve principles of green engineering which embody the same underlying features – conserve energy and resources and avoid waste and hazardous materials – as those of green chemistry, but from an engineering viewpoint. More recently, a mnemonic, PRODUC-

Green Chemistry and Catalysis. I. Arends, R. Sheldon, U. Hanefeld
Copyright © 2007 WILEY-VCH Verlag GmbH & Co. KGaA, Weinheim
ISBN: 978-3-527-30715-9

P – Prevent waste

R – Renewable materials

O – Omit derivatization steps

D – Degradable chemical products

U – Use safe synthetic methods

C – Catalytic reagents

T – Temperature, pressure ambient

I – In- process monitoring

V – Very few auxiliary substances

E – E-factor, maximize feed in product

L – Low toxicity of chemical products

Y – Yes, it is safe

Fig. 10.1 Condensed principles of green chemistry.

TIVELY, has been proposed which captures the spirit of the twelve principles of green chemistry and can be presented as a single slide (Fig. 10.1) [6].

In the USA the Presidential Green Chemistry Challenge Awards [7] were introduced to stimulate the application of the principles of green chemistry and many chemical and pharmaceutical companies have received awards for the development of greener processes and products, e.g. Pfizer for developing a greener process for sildenafil manufacture (see Chapter 7).

Graedel [8] has reduced the concept of green chemistry and sustainable development to four key areas: (i) sustainable use of chemical feedstocks, (ii) sustainable use of water, (iii) sustainable use of energy and (iv) environmental resilience. These reflect the central tenets of sustainability, that is, "using natural resources at rates that do not unacceptably draw down supplies over the long term and (ii) generating and dissipating residues at rates no higher than can be assimilated readily by the natural environment".

10.2
Catalysis and Green Chemistry

Hopefully, this book has made clear that there is a common technological theme which underlies these concepts of green chemistry and sustainability in chemical products and processes, and that is the application of catalysis – homogeneous, heterogeneous and enzymatic. The application of catalytic technologies leads to processes that are more efficient in their use of energy and raw materials and generate less waste. The importance of catalysis in this context was recently underlined by the award of the 2005 Nobel Prize in Chemistry to Grubbs, Schrock and Chauvin for the development of the olefin metathesis reaction, a classic example of clean catalytic chemistry. According to the Swedish Academy olefin metathesis is "a great step forward for green chemistry". Similarly, the 2001 Nobel Prize in Chemistry was awarded to Noyori and

Knowles and Sharpless, for catalytic asymmetric hydrogenation and asymmetric oxidation, respectively.

In the future we can expect a shift towards the use of renewable raw materials which will completely change the technological basis of the chemical industry. It will require the development of catalytic processes, both chemo- and biocatalytic, for the conversion of these raw materials to fuels and chemicals. New coproducts will be generated which in turn can serve as feedstocks for other chemicals. A pertinent example is biodiesel manufacture which generates large amounts of glycerol, thus creating a need for new (catalytic) processes for the conversion of glycerol to useful products (see Chapter 8).

The shift towards biomass as a renewable raw material will also result in the development of alternative, greener products, e.g. biodegradable polymers, polymers derived from biomass to replace poorly degradable synthetic polymers. A good example of this is the production of carboxystarch by oxidation of starch. The product is a biodegradable water super absorbent which could replace the synthetic polyacrylate-based super adsorbents which have poor biodegradability. The traditional process to produce carboxystarch involves TEMPO catalyzed oxidation of starch with sodium hypochlorite, which generates copious amounts of sodium chloride as a waste product and possibly chlorinated byproducts. There is a need, therefore, for a greener, catalytic process using hydrogen peroxide or dioxygen as the oxidant.

It has been shown [9] that the copper-dependent oxidase enzyme, laccase, in combination with TEMPO or derivatives thereof, is able to catalyze the aerobic oxidation of the primary alcohol moieties in starch (Fig. 10.2). There is currently considerable commercial interest in laccases for application in pulp bleaching (as a replacement for chlorine) in paper manufacture and remediation of phenol-containing waste streams [10].

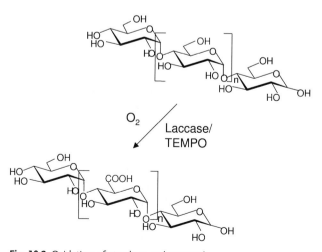

Fig. 10.2 Oxidation of starch to carboxystarch.

However, the relatively high enzyme costs form an obstacle to commercialization. Inefficient laccase use is a result of its instability towards the oxidizing reaction conditions. We have recently shown that the stability of the laccase under reaction conditions can be improved by immobilization as a cross-linked enzyme aggregate (see Chapter 9). It has also been shown that a water-soluble iron complex of a sulfonated phthalocyanine ligand is an extremely effective catalyst for starch oxidation with hydrogen peroxide in an aqueous medium [11].

It is also worth mentioning in this context that in the last few years a separate sub-division of catalysis has been gathering momentum: organocatalysis [12]. A wide variety of (enantioselective) reactions can be catalyzed and homogeneous organocatalysts can be heterogenized, on solid supports or in liquid/liquid biphasic systems, using similar methodologies to those used for metal complexes (see earlier chapters).

10.3
The Medium is the Message

As discussed at length in Chapter 7, the reaction medium is an important issue in green chemistry from two points of view: (i) the problem of solvent emissions and (ii) the problem of product separation and catalyst recovery and recycling. Many innovative approaches are being followed to solve these problems (see Chapters 7 and 9). Water and carbon dioxide have, in the grand scheme of things, obvious advantages and, in this context, it is worth mentioning the recent report on chemistry "on water" [13]. Several uni- and bimolecular reactions were greatly accelerated when performed in vigorously stirred aqueous suspensions. The efficiency is quite amazing but the origin of the rate enhancements remains unclear. Both heterogeneity and water appear to be required and the hydration layers of the particles may play a crucial role. We note, however, that the technique of so-called solid-to-solid transformations is quite often used for biocatalytic processes in water where both substrate and product are sparingly soluble in water [14]. Organocatalysis (see above) also lends itself, in some instances, to operation in aqueous media and recent reports include the use of praline-based organocatalysts for enantioselective aldol reactions [15] and Michael additions [16] in water. Even more simple is the recent report of asymmetric aldol reactions catalyzed by 10 mol% of the amino acid, tryptophan, in water as the sole solvent [17]. High enantioselectivities were obtained in the aldol reaction of a variety of cycloalkanones with aromatic aldehydes (see Fig. 10.3 for an example).

The use of ionic liquids as reaction media is undergoing an exponential growth with the emphasis shifting to the use of non-toxic, biodegradable ionic liquids, preferably from renewable raw materials [18]. There is even a IUPAC ionic liquids data base, "IL Thermo" [19].

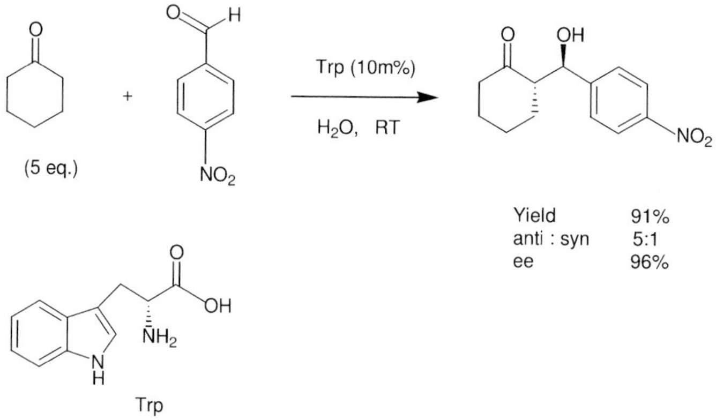

(5 eq.)

NO₂

Trp (10m%)

H₂O,   RT

O   OH

NO₂

Yield	91%
anti : syn	5:1
ee	96%

Trp

Fig. 10.3 Asymmetric aldol reaction catalyzed by tryptophan in water.

10.4
Metabolic Engineering and Cascade Catalysis

The need for novel catalytic processes is clear and, as discussed in Chapter 9, combining catalytic steps into cascade processes, thus obviating the need for isolation of intermediate products, results in a further optimization of both the economics and the environmental footprint of the process. *In vivo* this amounts to metabolic pathway engineering [20] of the host microorganism (see Chapter 8) and *in vitro* it constitutes a combination of chemo- and/or biocatalytic steps in series and is referred to as cascade catalysis (see Chapter 9). Metabolic engineering involves, by necessity, renewable raw materials and is a vital component of the future development of renewable feedstocks for fuels and chemicals.

10.5
Concluding Remarks

With sustainability as the driving force, the production and applications of chemicals are undergoing a paradigm change in the 21st century and green chemistry and catalysis are playing a pivotal role in this change. This revolutionary development manifests itself in the changing feedstocks for fuels and chemicals, from fossil resources to renewable feedstocks, and in the use of green catalytic processes for their conversion. In addition, there is a marked trend towards alternative, greener products that are less toxic and readily biodegradable. Ultimately this revolution will enable the production of materials of benefit for society while, at the same time, preserving the earth's precious resources and the quality of our environment for future generations.

References

1 www.suschem.org

2 www.chemicalvision2020.org

3 P. T. Anastas, J. C. Warner, *Green Chemistry: Theory and Practice*, Oxford University Press, Oxford, 1998, p. 30.

4 N. Winterton, *Green Chem.*, **2001**, *3*, G73.

5 P. T. Anastas, J. B. Zimmerman, in *Sustainability Science and Engineering: Defining Principles*, M. A. Abrahams (Ed.), Elsevier, Amsterdam, 2006, pp. 11–32.

6 S. L. Y. Tang, R. L. Smith, M. Poliakoff, *Green Chem.*, **2005**, *7*, 761.

7 www.epa.gov/greenchemistry

8 T. E. Graedel, in *Handbook of Green Chemistry and Technology*, J. Clark, D. Macquarrie (Eds.), Blackwell, Oxford, 2002, p. 56.

9 L. Viikari, M. L. Niku-Paavola, J. Buchert, P. Forssell, A. Teleman, K. Kruus, WO 992,324,0, 1999; J. M. Jetten, R. T. M. van den Dool, W. van Hartingsveldt, M. T. R. van Wandelen, WO *00/50621*, **2000**.

10 D. Rochefort, D. Leech, R. Bourbonnais, *Green Chem.* **2004**, *6*, 614.

11 S. L. Kachkarova-Sorokina, P. Gallezot, A. B. Sorokin, *Chem. Comm.* 2004, 2844; A. B. Sorokin, S. L. Kachkarova-Sorokina, C. Donze, C. Pinel, P. Gallezot, *Top. Catal.* 2004, *27* , 67.

12 For reviews see: A. Berkessel, H. Groerger, *Asymmetric Organocatalysis*, Wiley-VCH, Weinheim, 2005; J. Seayad, B. List, *Org. Biomol. Chem.*, **2005**, *3*, 719;

P. Dalko, L. Moisan, *Angew. Chem., Int. Ed.* **2004**, *43*, 5138; Y. Shi, *Acc. Chem. Res.* **2004**, *37*, 488; K. N. Houk, B. List, *Acc. Chem. Res.* **2004**, *37*, 487; P. Kocovsky, A. V. Malkov (Eds.),*Tetrahedron*, **2005**, *62(2/3)*, 260.

13 S. Narayan, J. Muldoon, M. G. Finn, V. V. Fokin, H. C. Kolb, K. B. Sharpless, *Angew. Chem. Int. Ed.*, **2005**, *44*, 3275; see also J. E. Klein, J. F. B. N. Engberts, *Nature*, **2005**, *435*, 746.

14 R. V. Ulijn, P. J. Halling, *Green Chem.*, **2004**, *6*, 488.

15 Y. Hayashi, T. Sumiya, J. Takahashi, H. Gotoh, T. Urishima, M. Shoji, *Angew. Chem. Int. Ed.* **2006**, *45*, 958.

15 N. Mase, K. Watanabe, H. Yoda, K. Takabe, F. Tanaka, C. F. J. Barbas, *J. Am. Chem. Soc.*, **2006**, *128*, 4966.

16 Z. Jiang, Z. Liang, X. Wu, Y. Lu, *Chem. Commun.*, **2006**, 2801.

17 G. Tao, L. He, W. Liu, L. Xu, W. Xiong, T. Wang, Y. Kou, *Green Chem.* 2006, *8*, 639; B. Ni, A. D. Headley, G. Li, *J. Org. Chem.* 2005, *70*, 10600.

18 http://ilthermo.boulder.nist.gov/ILThermo/mainmenu.uix

19 B. N. Kholodenko, H. V. Westerhoff (Eds.), *Metabolic Engineering in the Postgenomic Era*, Horizon Bioscience, Norfolk (UK), **2004**.

Subject Index

Green Chemistry and Catalysis. I. Arends, R. Sheldon, U. Hanefeld
Copyright © 2007 WILEY-VCH Verlag GmbH & Co. KGaA, Weinheim
ISBN: 978-3-527-30715-9

cirtonitrile 81
cis-methyl dihydrojasmonate 107
cis-pinane, autooxidation 376
cis-vitamin D, production 95
citral 19
– BASF synthesis 19
citronitrile, hydrocalcite-catalyzed condensation of benzylacetone with ethylcyanoacetate 79
Claisen-Schmidt condensations 78
– hydrotalcite-catalyzed aldol 78
– of substituted 2-hydroxyacetophenones with substituted benzaldehydes 78
Clariant, production of *o*-tolyl benzonitrile 25
clayzic 51
CLEA from CaLB 405
CLEAs *see* cross-linked enzyme aggregates
Club of Rome Report 329
clucose fermentation, chemicals from 333 ff.
CO insertion/R-migration 245
cobalt 137
Codexis 346
cofactor
– incompatibility 336
– pyridoxal phosphate (PLP) 242
– recycling 91 ff., 118, 120, 184
– regeneration 203, 208
combi CLEAs 405
– one-pot conversion of benzaldehyde to S-mandelic acid 406
combination of asymmetric hydrogenation with enzymatic hydrolysis 401 ff.
continuous kinetic resolution catalyzed by a CaLB CLEA in scCO$_2$ 406
copolymerization, alterning, of ethylene and CO in water 307
copper 142
– catalyst for alcohol oxidation 179 ff.
– catalyst for Baeyer-Villiger 211
Corynebacterium glutamicum 357
coumarin
– synthesis via zeolite-catalyzed intramolecular alkylations 62
– zeolite-catalysed synthesis 62
coupling reaction 24
cross metathesis (CM) 26
cross-linked enzyme aggregates (CLEAs) 405 ff.
Cu-phenanthroline 179
cyanoethylations, catalyzed by hydrotalcites 79

cyanohydrins 224 ff.
– (*R*)-mandelic acid 234
– nitrilase 234
3-cyanopyridine hydration 33
cyclialkylation, 4-phenyl-1-butene oxide 61
cyclic olefins, oxidative cleavage of 149
cyclizations 70
– zeolite-catalyzed 70
cycloalkanones, oxidation 188
cyclododecanol 163
cyclohexane monooxygenase 143, 144
cyclohexanediol, conversion 185
cyclohexanol 163
– zeolite-catalysed hydration 40
cyclohexanone 97, 163
– asymmetric Baeyer-Villiger oxidation 210
– from benzene 65
– hydrogenation 100
– monooxygenase 208
– two processes 66
cyclohexene, hydration to cyclohexanol ofer H-ZSM-5 65
cyclohexylbenzene 97
cysteine 283
cytochrome P450 144, 193
cytoxazone 231
– BAL 231

d

DAHP synthase 347, 352, 354
de novo fermentation 34
decarboxylase
– pyruvate decarboxylase (PDC) 230 ff.
– PDC from *Zymomonas mobilis* 232
– benzoylformate decarboxylase (BFD) 230 ff.
dehydration reactions, zeolites as catylysts 65
dehydrogenases 142
Deloxan® 72
deoxy sugar, synthesis 42
deprotection, peptide deformylase 285
D-erythrose-4-phosphate 347 ff.
desacetoxycephalosporinase 359
desymmetrization 268, 278
D-fructose 305
– transformation into 5-hydroxymethylfurfural 371
D-glucitol 305
β-D-glucose oxidation of 143
D-glyceraldehyde-3-phosphate 336
DHAP 239

q
quadrant rule 109
quinoxalinecarboxylic acid (2-), oxidation 168

r
racemases 280 ff.
– *Achromobacter obae* 281
– carbamoyl racemases 282
racemization 269, 274 f., 280
– catalyzed by ruthenium 394
– mechanism 396
– of alcohols and amines, metal-catalyzed 390
– of amines 395
– – side reactions in metal-catalyzed 397
– of chiral secondary alcohols 392
– of 1-phenylethanol 394
– via a dehydrogenation/hydrogenation mechanism 395
Rac-glycidol 240
Raney nickel 92
– chiral 101
raw materials, various, prices 330
rearrangement 67 ff.
– of epoxides 12
– of substituted styrene oxides 69
– α-pinene oxide 69
recycling 402 ff.
redox enzymes 141
redox molecular sieves 21
reductions 91 ff., *see also* hydrogenation
reductive alkylation
– of alcohols 98
– of amines 98
reductive amination 122
– of nitro derivatives 99
reductive elimination 106, 246, 248, 256
regulation, feedback mechanisms 334
Reichstein-Grussner process 364
renewable raw materials 34 ff., 329 ff.
– as catalysts 378 ff.
– carbohydrates 332 ff.
– green polymers from 379
Reppe carbonylation 253
resolution
– of an amine with penicillin acylase 400
– of 1-phenylethanol 405
– of 1-tetralol 405
resorcinol, reaction with ethyl acetoacetate 62
restricted transition state selectivity 58

reversible non-polar to polar solvent switch 404
Rh-MeDUPHOS 107
Rh-DOPAMP 104
Rh-DUPHOS 104
rhenium catalysts for epoxidation 150 f.
Rh-monophos 104
Rhodia 95
Rhodia process 10
– *p*-methoxyacetophenone 10
– vanillin manufacture 40
Rhodia vanillin process 40
Rhodococcus erythropolis 117
Rhodococcus rhodocrous 33
Rhodococcus ruber 184
– alcohol dehydrogenases 120
Rh-Xyliphos 108
riboflavin (vitamin B$_2$) 361 ff.
– biosynthesis 361
– biosynthesis in bacteria 363
– downstream processing 362
– fermentation 361
– from D-ribose 361
– Karrer-Tishler process 361, 362
– microbial production 361
Rieske
– cluster 146
– dioxygenase 143
ring closing metathesis (RCM) 26
– Ru-catalyzed 27
ring opening metathesis polymerisation (ROMP) 26, 259
risky reagents 38 ff.
Ritter reaction 66, 67
R-migration/CO insertion 245
Rosenmund reduction 17, 95
Ru/TEMPO catalyst 18
Ru-BINAP 103
Rubintrivir 122
Ru-catalysts for asymmetric epoxidation 199
Ru-catalyzed racemization of alcohols, mechanism 393
ruthenium catalyst
– for alcohol oxidation 172 ff.
– for epoxidation 151 f.
– for oxidative cleavage 158
rutheniumtetroxide 158

s
savinase – abacavir (ziagen) 283
Sabatier 9, 15
salt-free esterification of amino acids 13

Related Titles

Loupy, A. (ed.)

**Microwaves
in Organic Synthesis**

2006
Hardcover
ISBN 978-3-527-31452-2

Afonso, C. A. M., Crespo, J. P. G. (eds.)

Green Separation Processes
Fundamentals and Applications

2005
Hardcover
ISBN 978-3-527-30985-6

Berkessel, A., Gröger, H.

Asymmetric Organocatalysis
From Biomimetic Concepts to
Applications in Asymmetric Synthesis

2005
Hardcover
ISBN 978-3-527-30517-9

Tanaka, K.

**Solvent-free Organic
Synthesis**

2003
Hardcover
ISBN 978-3-527-30612-1

Wasserscheid, P., Welton, T. (eds.)

Ionic Liquids in Synthesis

2003
Hardcover
ISBN 978-3-527-30515-5